PSYCHOLOGY RESEARCH PROGRESS

HANDBOOK OF TECHNOLOGY IN PSYCHOLOGY, PSYCHIATRY AND NEUROLOGY

THEORY, RESEARCH, AND PRACTICE

Psychology Research Progress

Additional books in this series can be found on Nova's website
under the Series tab.

Additional e-books in this series can be found on Nova's website
under the e-book tab.

Psychiatry - Theory, Applications and Treatments

Additional books in this series can be found on Nova's website
under the Series tab.

Additional e-books in this series can be found on Nova's website
under the e-book tab.

PSYCHOLOGY RESEARCH PROGRESS

HANDBOOK OF TECHNOLOGY IN PSYCHOLOGY, PSYCHIATRY AND NEUROLOGY

THEORY, RESEARCH, AND PRACTICE

LUCIANO L'ABATE
AND
DAVID A. KAISER
EDITORS

Nova Science Publishers, Inc.
New York

For permission to use material from this book please contact us:
Telephone 631-231-7269; Fax 631-231-8175
Web Site: http://www.novapublishers.com

Additional color graphics may be available in the e-book version of this book.

LIBRARY OF CONGRESS CATALOGING-IN-PUBLICATION DATA

Handbook of technology in psychology, psychiatry and neurology : theory, research, and practice / editors, Luciano L'Abate, David A. Kaiser.
p. cm.
Includes bibliographical references and index.
ISBN 978-1-62100-004-4 (hardcover)
1. Neurosciences--Technological innovations. I. L'Abate, Luciano, 1928- II. Kaiser, David A.
RC337.H36 2011
616.89--dc23
2011028901

Published by Nova Science Publishers, Inc. † New York

To Nancy,
who helped us start the ball rolling

Contents

FOREWORD

Our world is changing and changing quickly. Who would have guessed a decade ago that approximately one billion more people would now have access to cell phones than clean water? One can be certain that the impact of technology will only become more pervasive in the decades to come and we cannot ignore how it will impact our mental health professions. The juggernaut of technological development has and will continue to dramatically alter how we as health professionals in the neurosciences pursue our efforts in research, training and treatment. The Handbook of Technology in Psychology, Psychiatry, and Neurology: Theory, Research and Practice provides a rare and needed analysis of the history of technology in our field, the current state-of-the-art as well as a vision for the future.

Drs. L'Abate and Kaiser have assembled a distinguished team of contributors who critically examine how recent technological developments can contribute to advances in a range of topics including educational technology, assessment and treatment. These advances include a wide range of new approaches to communication such as Wikis and Blogs and Telehealth Conferencing and extend to advances in clinical care such as Diffusion Tensor Imaging and Transcranial Magnetic and Deep-brain stimulation. One of the strengths of this broad cross-cutting presentation is that it provides the reader with the opportunity to consider how these diverse technologies might be used in new and different applications.

As discussed in the Handbook, technology is radically changing how we learn and how we teach. Because of the increasing speed of new technology development it is no longer possible to rely on training our new generation of professionals solely in the classroom; more than ever, students will need to learn how to learn and how to do it quickly on the job. Similarly, continuing education cannot be restricted to specialized courses taken place here and there but will become part of the professional's daily routine. The Handbook presents an array of new technical developments that will not only serve to keep the seasoned professional up to date but, perhaps more importantly, will provide new approaches to the education of our graduate students, residents and post-doctoral fellows.

The revolution in technology is also transforming clinical practice in ways that could not have been forecast just a few years ago. The key to improving the ability to correctly assess psychological and neurological disorders is the development of new ways to rapidly and accurately collect information on a patient's symptomology. For example, the use of ambulatory devices for patient self-report such as those described in the section on Advances in Assessment holds the promise to turn "snapshots" of information obtained from office visits into a continuous flow of information accessible in real time. Additional game-changing technologies are coming into their own in providing new dimensions to the treatment of a variety of disorders. For example, one can now literally manipulate a patient's physical and social environment with virtual reality, and as a result, create dramatic new opportunities to treat psychological disorders such as PTSD.

In summary, the Handbook is more than a welcome addition to new developments in Psychology, Psychiatry and Neurology; it is a call for a new approach to theory, research and practice that embraces a broad array of new technologies. In short, this is a book about innovation.

H. Elliott Albers, Ph.D. Director
Center for Behavioral Neuroscience
Regents Professor of Biology and Psychology
Georgia State University

PREFACE

> "…research results are always constrained by the methods (including the apparatus) used to produce them. The history of psychology…is rife with examples of paradigm shifts with the emergence of new technology (Rumbaugh and Washburn, 2003, p. 150).

Technology and America go hand in hand. We are a society that thrives on innovation and unrelenting change. Instead of wilting under the weight of this high-tech invasion, most mental health professionals welcome it and look towards technology for assistance. Going back to the days of the first psych laboratory at Clark University, American psychologists have always been early adopters of technology, and this handbook was conceived with this fact in mind. In the following chapters the authors present a wide array of technologies currently in service by most mental health researchers but very few practitioners. By reviewing these innovations under one cover and across many fronts –psychiatry, neurology, and psychology – an interested professional can grasp what technologies are available and which ones hold the most promise for his or her own clinical practice. The days of tea leaves and ink-blots interpretations are almost over even though many clinical training centers are still wed to such outdated, subjective, and expensive practices, including talk (L'Abate, 2012).

Attempts to reduce time pressure on psychologists by automating procedures is nothing new. More than 40 years ago Elwood (1969, 1973) devised spring-loaded drawers that delivered sections of the Weschler Adult Intelligence Scale at the push of a button. Mechanical interpretation of the Minnesota Multiphasic Personality Inventory also began in the 1960s, the same period of time when an automated playroom was built, one that allowed to monitor what each child did in two separate but connected playrooms (L'Abate, 2009). This approach was followed by written, reproducible enrichment programs for couples and families that eventually produced interactive, practice exercises and workbooks, many of them developed from the DSM-IV, forever linking evaluation with interventions (L'Abate and Weinstein, 1987; L'Abate and Young, 1987). With the advances in personal computers and Net connectivity, the floodgates opened on computerized service delivery of assessment, evaluation, and even intervention (e.g., L'Abate and Sweeney, 2011). Online assessment and evaluation of emotional, intellectual, and disordered functioning now evolves at such a pace that recent reviews are already out of date (Naglieri et al., 2004; Ritterband et al., 2003).

Online technology offers undreamed-of possibilities in health promotion, illness prevention, psychotherapy, and rehabilitation, as viewed by the 2011 Conference on

Evidence-Based Clinical Applications of Information Technology (website: http://www.interactivemediainstitute.com/CT16). With the rise of smart-phone and net connectivity, traditional one-to-one, face-to-face talk-based therapy is giving way to distance evaluations and treatments (L'Abate, 2008, 2009b; L'Abate and Sweeney, 2011), allowing mental health professionals access to anyone on the planet (in an industrialized culture). This means encountering head-on the full range of psychological, psychiatric, and neurological dysfunctionalities possible. Online services allow professionals to help participants through email and web-based interactions. It is possible to help troubled people without ever seeing them face to face or talking with them (Seligman, Steen, Park, and Peterson, 2005). Online interventions have expanded to include disease prevention and health promotion in addition to psychotherapy (Harwood and L'Abate, 2010; Kazantzis and L'Abate, 2007; L'Abate, 2007, 2011).

Electronic monitoring devices enable us to do more with more. Biofeedback, virtual reality therapy, and working memory training, among other innovations, are changing the way therapists go about promoting health and well-being. Populations who resisted talking or writing therapy in the past, due to language or cultural barriers, now benefit from tech-aided therapies. Many new interventions incorporate video gameplay or DVD-viewing as part of the treatment, increasing compliance and motivation among children in particular (Harwood and L'Abate, 2010; L'Abate, 2007).

There hve always been too few mental health professionals compared to the number of Americans who need assistance. If we want to make a dent in this problem, we need methods of administering effective treatments efficiently, to more individuals b units of professional time. In reviewing the present status of mental health professions and the popular use of least expensive providers, Robiner (2006) noted that:

> "As concerns about costs and quality are balanced, the roles of relatively more intensively trained professionals (i.e., psychologists) are likely to change. Increasingly, their roles may focus on their knowledge and skills in supervision, management, research, and program development, and less on the direct provision of clinical services. (p. 615)

The time of highly-trained professionals is highly valuable and to be distributed carefully. The use of portable devices such as a laptop computers and handheld monitors saves time – time better spent on advancing a team-effort or managing a group of specialists involved in many patients'care. On the basis of differentiating among professional, semi-professional, technical, and clerical skills, L'Abate (2002, p. 230) suggested a hierarchy of mental health personnel that was updated in more recent publications (L'Abate, 2008, 2012).

New technology solves many problems, and by the same token produces new problems. Communication technology such as videophones and social networks such as Facebook, Myspace, and Twitter change how individuals view themselves and others. The relationship between patient and therapist has also morphed in response to technology, with the ascendancy of directive, self-change, and interventions such as virtual reality therapy enabling the rise of master's-level psychotherapists in most areas (Norcross, Hedges, and Prochaska, 2002; L'Abate, in press).

With all of this sea changes and an explosion of clinical tools, one fact remains unclear: the role of reimbursement. Online self-help is not yet reimbursed by insurance companies, but what happens when treatments are automated but under the guidance of a specialist –will reimbursement remain the same per patient? Or will payments reflect the degree of involvement of the specialist, regardless of the effectiveness of such automated therapeutics. Are insurers justified in paying less if we are doing more but working less? Clearly, these issues deserve guidance and involvement of mental health professionals now and in the future.

This handbook reviews advances in the evaluation and treatment of neurological, psychological, and psychiatric disorders. These chapters are by no means exhaustive and many other technologies could be included in this review (e.g., for computer-driven stress-reduction programs, see Berger, 2004). This work follows a recent discussion on advances in clinical psychology, homework assignments (Kazantzis and L'Abate, 2007) and low-cost approaches to promote self- physical- and mental- health (Harwood and L'Abate, 2010; L'Abate, 2007, 2009a, 2012). Most of the instruments and processes discussed in this handbook remain outside of the mainstream of clinical and psychotherapy practice. When, where, and how to start training in this technology remains an open question left at this time in the hands of training centers and individual mental health professionals who may choose to remain and practice in the past century rather than in the present one .

PLAN FOR THIS HANDBOOK

Chapter 1 in Section I by the first editor of this volume serves as an introduction by to the historical applications of technology in the varies neurosciences, followed by Chapter 2' by Amy Przeworski and Michelle Newman about the strengths and limitations of technology in psychotherapy. In Chapter 3 T. Mark Harwood and Daniel Platt review the many ecological, environmental, and professsional issues facing the introduction of completely novel technological advances in mental health.

In Section II abut advances in educational technology, Anne M. Cleary and Edward DeLosh in Chapter 4 review personal response systems in the classroom. In Chapter 5, Robert Atkinson, Lijia Lin, and Stacey Joseph review animation and multimedia tutorials, while in Chapter 6, Bob Atkinson, Kent Sabo, and Quincy Conley review wikis and blogs in the participatory web. In Chapter 7, Henry Mahncke reviews memory training programs that increasingly assumes greater brain plasticity than it had been thought in the past. In Chapter 8, Lotfi B. Merabet and his coauthors include the development of navigation skills in the blind. In Chapter 9, Olga Solomon reviews much needed technology with autistic children, where repetition is necessary to make an imprint in their memories. In Chapter 10 primatologist David Washburn, Michael J. Beran, Theodore A. Evans, Megan L. Hoffman, and Timothy M. Flemming review the technology administered to non-human participants.

Section III is devoted to technological advances in assessment including Chapter 11 by Bob Atkinson, Andre Denham, and John M. Qick review ambulatory hand-held devices for self-report, that have reached astronomical highs in common usage. We are grateful to Bob Atkinson for helping us at the last minute to fill gaps by past contributors who left us suddenly on the lurch and enlisting some of his colleagues to complete those chapters. In Chapter 12 Yi-Tsi Seih and James W. Pennebaker introduce the revolutionary Language Index Word Count. In Chapter 13, Jason Moser and Tim P. Moran review a completely novel neurocognitive technology based on information-processing and Event Related Poterntials. In

Chapter 14 Hans-Peter Müller, Alexander Unrath, Axel Riecker, and Jan Kasubek review the use of diffusion tensor imaging with analytic methods for group comparison in neurology.

Section IV is devoted to advances in treatment including Chapter 15 on virtual reality with children and adults by Albert (Skip) Rizzo and his co-workers. In Chapter 16, Dafne C. Andrade, IgorC. Borges, Rodrigo F. Cury, and Filipe Fregni review the latest research on transcranial magnetic and deep-brain stimulation. In Chapter 17, David A. Kaiser and Andrea Meckley review research on various types of neurofeedback. In Chapter 18 Carlos de la Cuevas reviews the role of telehealth conferencing in the neurosciences. In Chapter 19, Luciano L'Abate reviews the background and research about programmed interactive practice exercises. In Chapter 20, José Gallego and Paul Emmelkamp review the latest research on psychological treatment through the internet.

In Section V, Savio Wong and Antoine Bechara in Chapter 21 conclude this volume with some predictions about the future of technology in psychology, psychiatry, and neurology in terms of their implications for training and treatment.

Readership

This Handbook has been produced primarily for research-oriented professionals in the neurosciences, including neurologists, psychiatrists, and psychologists in academic settings and clinical graduate training centers. Secondarily, it could be of interest to graduate students, residents, and post-doctoral fellows specializing the neurosciences, and in the three main mental health disciplines listed in the title of this volume. Thirdly, clinical training programs in the neurosciences may need to abandon traditional and time-consuming evaluation based on expert interpretations of projective techniques and learn to master technologies that are designed to reduce abstractions and generalities and increase the two characteristics of a professional science: specificity and concreteness.

References

Berger, T. (2004). Computer-based technological applications in psychotherapy training. *Journal of Clinical Psychology, 60*, 301–315.

Elwood, D. L. (1969). Automation in psychological testing. *American Psychologist, 24,* 287–289.

Elwood, D. L. (1973). Reliability of automated intelligence testing using a three- month test–-retest interval. *International Review of Applied Psychology, 22,* 157–162.

Harwood, T. M., and L'Abate, L. (2010). *Self-help in mental health: A critical evaluation.* New York: Springer Science.

Kazantzis, N., and L'Abate, L. (Eds.). (2007). *Handbook of homework in psychotherapy: Theory, research, and prevention.* New York: Springer Science.

L'Abate, L. (2002). *Beyond psychotherapy: Programmed writing and structured computer-assisted interventions.* Westport, CT: Ablex.

L'Abate, L. (Ed.). (2007). *Low-cost approaches to promote physical and mental health.* New York: Springer-Science.

L'Abate, L. (Ed.). (2008). *Toward a science of clinical psychology: Laboratory evaluations and interventions*. New York: Nova Science Publishers.

L'Abate, L. (2009a). *The Praeger handbook of play across the life cycle: Fun from infancy to old age*. Newport, CT: Praeger.

L'Abate, L. (2009b). *Toward a science of clinical psychology: Laboratory evaluations and interventions*. New York: Nova Science Publishers.

L'Abate, L. (2011). *Sourcebook of interactive practice exercises in mental health*. New York: Springer-Science.

L'Abate, L. (2012). *Clinical psychology and psychotherapy as a science: An iconoclastic prespective*. New York: Springer-Science.

L'Abate, L. (in press). Recommended online computer-assisted treatments. In G. P. Koocher, J. C. Norcross, & B. A. Greene (Eds.), *Psychologists' Desk Reference* (3rd Edition). New York: Oxford University Press.

L'Abate L., and Sweeney, L. G. (Eds.). (2011). *Research on writing approaches in mental health*. Bingley, UK: Emerald Group Publishing Limited.

L'Abate, L., and Weinstein, S. E. (1987). *Structured enrichment programs for couples and families*. New York: Brunner/Mazel.

L'Abate, L., and Young, L. (1987). *Casebook of structured enrichment programs for couples and families*. New York: Brunner/Mazel.

Naglieri, J. A., Drasgow, F., Schmit, M., Handler, L., Prifitera, A., Margolis, A., and Velasquez,, R. (2004). Psychological testing on the Internet: New problems, old issues. *American Psychologist, 59,* 150–162.

Norcross, J. C., Hedges, M., and Prochaska, J. O. (2002). The face of 2010: A Delphi poll on the future of psychotherapy. *Professional Psychology: Research and Practice, 33,* 316–322.

Ritterband, L. M., Gonder-Frederick, L. A., Cox, D. J., Clifton, A. D., West, R. W., and Borowitz, S. M. (2003). Internet interventions: In review, in use, and into the future. *Professional Psychology: Research and Practice, 34,* 527–534.

Robiner, W. N. (2006). The mental health professions: Workforce supply and demand issues and challenges. *Clinical Psychology Review, 26,* 600–625.

Rumbaugh, D. M., and Washburn, D. A. (2003). *Intelligence of ape and other rational human beings*. New Haven, CT: Yale University Press.

Seligman, M. E. P., Steen, T. A., Park, N., and Peterson, C. (2005). Positive psychology progress: Empirical validation of approaches. *American Psychologist, 60,* 410–421.

SECTION I. INTRODUCTION

In: Handbook of Technology in Psychology ...
Editor: Luciano L'Abate and David A. Kasier
ISBN: 978-1-62100-004-4

Chapter 1

TECHNOLOGICAL PRACTICES IN THE NEUROSCIENCES

Luciano L'Abate
Georgia State University, Atlanta, Georgia, US

The first part of this chapter may seem composed by personal impressions. It is drawn from many sources included in this volume and other sources, too many to cite individually, as well as in a chapter on information processing (De Giacomo, Mich, Santamaria, Sweeney, and De Giacomo, 2012). The second part concerning evaluation practices in the neurosciences includes quoting abstracts taken literally from the literature. I warn readers that in doing this, reporting these abstracts almost literally, may have violated the code of ethics of the American Psychological Association. I did not want to change the spirit and the letter of those abstracts, but I did change from first pronouns ("we", "us", "our'), referring to the authors directly ("they", "their") and from present to past tenses In spite of these changes, including quotes at the beginning and end of each abstract, should conform to the letter and the spirit of that code.

A BRIEF HISTORICAL BACKGROUND

Technology is a fuzzy set that has been evolving for centuries and continues to evolve at an astonishing rate. It can be argued that technology is situationally defined and that the situation informs and controls technology. Given that the history of technology must extend for centuries before the formal beginning of psychology, much of that history is not directly relevant to psychology, and thus, we will focus on the history of technology since 1879. Even in more recent times, technology has changed rapidly, and its use in psychology, psychiatry, and neurology has become increasingly important, as shown by all the chapters in this volume. Technology is related to automation, simulation, and cybernetics, all of which are dependent to some degree on technology, but they also may influence the development of technology. Long before the development of computers, simulations were preformed, and certainly, automation was a critical component of the industrial revolution. For example,

Norbert Weiner (1948) wrote *Cybernetics or Control and Communication in the Animal and the Machine* well before the advent of commercial computers or other devices that could accomplish the various tasks that he proposed. Although a history of automation, simulation, and cybernetics in a brief chapter is impractical if not impossible, we will include some information about those domains, particularly as they are utilized in psychology and in the other neurosciences. Technology may increase the accuracy of various tasks and may reduce the time required to perform some of those tasks. However, technology also may create tasks that could not have been undertaken prior to some technological developments. As a simple example, statistical analyses that were impractical in the 1940s or 1950s are now performed in a matter of seconds. Although technology may facilitate the solving of many problems, it must be recognized that technology also may create new problems. Biographical research about individuals is facilitated by the Internet, but the Internet contributes to the loss of privacy along with allowing for the acquisition of information that once was thought to be unobtainable (Carr, 2011; Morozov, 2010; Shirky, 2010; Siegel, 2009). As already noted at the beginning of this chapters, non-verbal and distance writing developments in information processing up-to-date can be found in a chapter by De Giacomo, Mich, Santamaria, Sweeney, and De Giacomo (2012). L'Abate and Sweeney (2012), in their final and concluding chapter in the same volume, declared information processing to be the reigning paradigm over competing paradigms in this century.

The functions and implementations of technology evolved, and in recent times, a distinction has begun to be made between technology and high technology. The windmill is an example of a technology that has changed structurally and functionally over time. Initially, windmills were used to provide power for mills in which substances were ground and to pump water, but recently, the use of windmills has been expanded to include generation of electricity. The structural changes are themselves aspects of technological developments that have made the windmill much more efficient than it was several centuries ago (Hills, 1994).

This distinction between technology and high technology perhaps is even fuzzier than the specification of technology, but high technology (hi-tech) usually refers to computer, digital, or electronic implementations. Clearly, technology, in general, includes mechanical as well as electronic developments and has evolved from the simplest tools to highly sophisticated devices, but much of the recent technological developments involve electronic hardware and the software that controls highly advanced hardware, such as cellular telephones, digital cameras, and remote sensing devices among an almost unlimited variety of devices.

In this chapter an historical description of the introduction of technology into a few areas in psychology, psychiatry, and neurolgy will be presented. In subsequent sections, we will discuss the role of technology in laboratory research, data analysis, tests and measurement, education, and clinical psychology in various forms of assessment, evaluation, diagnosis, and treatment of psychological problems. Few domains of psychology, psychiatry, and neurology have escaped technology's long reach, and in many cases, the introduction of technology has changed the nature of the domain.

Laboratory Research

Technological innovation has been ubiquitous in laboratory research, if not all psychological research, even if an exact beginning event may be difficult or impossible to identify because the innovations have been introduced gradually over many years. Prior to B. F. Skinner's development of the operant conditioning chamber, most animal research involved the time-consuming and labor-intensive observation of individual animals or of small groups of animals in mazes or other devices and a human observer who recorded the behavior of the animal (See Chapter 10 this volume for further expansion of this point).

For present purposes, laboratory research can be divided into the three general areas of stimulus presentation, response recording, and data analysis, with the data analysis meriting a section of its own. The situation in which observations can be made varies from naturalistic settings to highly stylized settings that are far removed from everyday life. For example, in a somewhat naturalistic setting in a laboratory, the subject may be asked to recall a passage of material after having read a passage of text on a sheet of paper or on a computer screen. Or, one might be shown a list of unrelated words after which the task is to recall those words. In early research in memory, perception, problem-solving, or other types of laboratories, the stimuli to be processed were presented by using flash cards, memory drums, tachistiscopes, or other devices so that the order of the stimuli could be varied and the presentation time could be controlled. The conversion to computer presentations was slow at the beginning because of a variety of technical problems such as the persistence of images on early monitors. With enhanced monitors and much more sophisticated software, the quality of the presentation and the timing of the presentation were brought into acceptable ranges for psychological research.

Pencil, paper, and stopwatch were replaced gradually by various types of equipment during the mid-twentieth century and by computers in most cognitive psychology laboratories by the 1990s. Stimuli of all kinds were presented on computer monitors, and participants could type their responses on the computer keyboard although the early keyboards were unsatisfactory for recording response times. With enhanced keyboards and appropriate software, response times now are recorded reliably and accurately. More stringent control and more reliable response recordings are two major advantages of the conversion to computer-control research.

However, programs to control the presentation of the stimuli and the recording of responses were needed. Initially, programs were written for specific studies in each laboratory, but gradually more general programs were developed and commercially available programs and systems were developed. Psychology Software Tools, Inc. was one of the first companies to develop general systems theory for research purposes. Initially, a DOS system (Micro-Electronic Laboratory—MEL) controlled stimulus presentations and recorded responses. Subsequent enhancements (E-Prime) on a Windows platform included modules for eye tracking research and fMRI studies in addition to traditional studies of memory, perception, and other cognitive functions. Numerous other companies produced similar software and hardware.

As the Internet has expanded and has been enhanced, much research that would have been performed in laboratories with an experimenter and a single participant or a small group of participants has been converted to research on the Internet, which allows for the presentation of many types of stimuli varying from a single stimulus to a vignette on video to many research participants as well as for the recording of responses. Obvious questions about this type research concern the selection of participants and the validity of the observations (Golling and Johnson, 2010).

In non-human research, learning studies often involved the use of mazes and choice boxes with recording being done by an experimenter who observed individual animals. As video recording became available, the behavior of animals in a maze and in other settings could be recorded so that observers later could score the responses, allowing for multiple observers and validation of the observations. Subsequently, photoelectric cells were placed strategically in a maze, and the animal's responses were recorded electronically for later analysis. More recently, various sensing devices have replaced the photocell, and direct recording in the computer has replaced paper and pencil recordings.

With the development of the operant conditioning chamber, popularly known as a Skinner box, Skinner proposed that an organism could produce a large number of responses (bar presses for rats and pecks for pigeons) without the need for an experi-menter to observe the animal. However, a device that could record a large number of discrete responses would be necessary. Skinner started with a kymograph, a standard recording device that had been invented by Karl Ludwig (1816-1895). The kymograph was a drum device on which smoked paper was wrapped. A stylus removed the smoke, leaving a white trail, as the drum turned and as responses were elicited. Skinner modified the device so that an ink pen moved across a continuous roll of paper that was fed through the machine at a constant speed. When a response was made, the pen was offset slightly, and thus, a cumulative record of responding was generated. The cumulative record is an example of a techn logical advance that facilitated the recording of responses and also changed the types of data that were available for analysis. The rate of responding as well as the number of responses could be determined directly from the cumulative record, but more importantly, the pattern of responding was visually displayed on the record (Lattal, 2004).

This is an instance of technology changing the nature of the observations. Prior to Skinner's innovation, the number of responses that were made was relatively small. For example, the number of correct responses or even the responses made by an animal at each choice-point in a multiple T-Maze were recorded with the numbers being very small. With the use of an operant conditioning chamber and the development of the cumulative recorder and other similar devices, hundreds of responses could be recorded for future analysis or description. Computer programs largely have replaced the cumulative record because of their versatility and because of the need to record more than the mere occurrence of a single type of response (see Chapter 10 for further details on this topic).

DATA ANALYSIS

The third component of research in a neuroscience or psychology laboratory is the analysis of the data, which has been influenced substantially by technology. In the early days of statistical analysis, the analyses were performed with paper and pencil, which limited the types of analyses that could be performed and the number of observations that could be analyzed within a reasonable amount of time. A variety of computational aids, such as logarithm and frequency distribution, existed, but the data analyses, nonetheless, were limited. With computers and analysis programs, researchers could use additional types of analyses (e.g., factor analysis became more practical) and could analyze larger numbers of observations than at previous times. For example, with mechanical calculators, one could enter pairs of values (x and y), which produced the computation of Σx^2, Σy^2, and Σxy, saving time and reducing the likelihood of errors because each number had to be entered only once.

Thus, correlations and factor analyses could be performed more quickly and accurately than they were with paper and pencil. Commonly, psychology laboratories contained Marchant (initially produced in 1911), Friden (initially produced in 1934 with a square root function added in 1952), or Monroe (beginning in 1912) calculators that allowed for many statistical computations.

Each machine had advantages and disadvantages as well as devoted admirers who wanted one and only of the standard types of calculators although each allowed for the critical computations. The addition of electric motors to drive the mechanical calculators made them more useful and faster than the manual versions had been (Redin, n.d.).

By the mid-1960s, the first electronic calculators, without mechanical parts, were being produced. The integrated circuit calculators initially were desktop machines that were faster than the mechanical calculators had been and included numerous mathematical functions in addition to the basic arithmetic operations.

A few years later, the size of the calculators had been reduced so that a calculator could be held in one's hand or even placed in a shirt pocket. These calculators soon replaced slide rules and other calculating devices and were important for psychological research because they allowed for the computation of means, standard deviations, and even correlations almost instantaneously. Furthermore, summary values (e.g., sums, sums of squares) needed to compute an analysis of variance were easily obtained.

Even before electronic calculators were developed, computers based on vacuum tubes were being constructed. These could be programmed to perform long sequences of computation at relatively fast rates, and the first commercially available computers were developed in the 1950s with various candidates for being the first commercially available computer (e.g., Univac, IBM 650). These machines could store programs that performed numerous sequence operations involving mathematical functions and conditional operations that depended upon a variety of specified values. In modern terms, these 1950s computers were very primitive, but in comparison to calculators and accounting machines that had been used previously to perform mathematical computation, they were amazingly fast, accurate, and versatile. Many mental health professionals learned to write programs on the IBM 650 as early as 1955.

When transistors replaced vacuum tubes, the speed, accuracy, and versatility of the computers increased by tremendous amounts. The programs stored in the computers could be longer and more complex than those in the 1950s. Along with hardware development, programming languages also were being developed, which allowed for increasingly complex computations.

In turn, more complex statistical analyses could be programmed, and soon statistical packages (e.g., BMDP, 1961; SPSS, 1968; and SAS, 1976) that could perform a variety of statistical analyses were developed. With the statistical packages that were available, analyses that previously would have taken days could be performed in a matter of minutes if not seconds. Both complex analyses and analyses of huge quantities of data could be performed quickly, and the output from those programs often was extensive. Unfortunately, those programs would produce analyses regardless of the nature of the data and regardless of the research. The task for the researcher became one of interpreting the analyses rather than producing them, and often the researcher did not have the knowledge and training that were needed to interpret the complex analyses. So, the ready available of programs that performed rapid and complex analyses still required a knowledgeable interpreter of those analyses.

Psychological Tests: Scoring, Interpretation, and Diagnosis

Long before the development of high technology, Paul Meehl (1954) claimed that statistical prediction and diagnosis were superior to clinical prediction and diagnosis. With the current developments in software and the availability of high-speed computers, the ease of making statistical predictions and diagnoses has increased substantially. These predictions and diagnoses can be combined with clinical judgments although many proponents of the statistical approach advise against the melding of methods. The statistical or actuarial approach includes various computer-based interpretations of psychological tests such as the Minnesota Multiphasic Personality Inventory (MMPI) and the Rorschach, among many others.

Perhaps the least controversial use of technology in psychological tests and measurement involves machine scoring of answer sheets of various kinds. IBM developed the IBM 805 Test Scoring Machine, which according to the company was more accurate and faster than manual scoring by a person. The IBM 805 remained in production from 1938 until 1963. More recently, machine-readable response forms, such as Scantron sheets, have become ubiquitous. Respondents mark their responses on a sheet that can be read by a special scanner, which produces a variety of information. The responses to each item (e.g., multiple choice question or true-false question) can be read, after which tabulations of the responses can be produced. These data then can be analyzed depending upon the nature of the questions and the responses. Scantron Corporation provides a dedicated single function system for reading the forms and for analyzing the responses. The software is embedded in the scantron device, but the examiner can choose to some extent the output from the system.

More recently, questions or stimuli of various kinds have been presented on a computer screen, and the responses of the participants have been recorded. For objective tests (personality, intelligence, achievement, attitude, et cetera), the automation of scoring, whether from subject-marked sheets or from responses on a computer, was one of many developments in the use of technology, which was very similar to the automation of other routine activities in domains other than psychology.

A more complex process involved the development of systems that interpreted those scores after the responses had been scored using a machine. Fowler (1985) referred to these interpretations as computer-based test interpretation (CBTI). According to Fowler (1985), who presented a history of computer-assisted psychological assessment, the earliest interpretation systems involved the MMPI. However, interpretation systems have been developed for many other tests, including projective tests such as the Rorschach.

In those early systems, reports were generated by selecting statements from a library of paragraphs and statements and assembling them into a narrative report about the examinee. The reports were designed to simulate a report that might have been written by a psychological expert, but little research has involved the comparison of CBTI reports and clinician-prepared reports. However, both psychologists and their clients apparently accepted the CBTI reports without reluctance.

Not everyone has been enthusiastic about CBTI. In particular, Matarazzo (1986) has criticized the commercially available software that many people who are untrained can purchase. Thus, an untrained individual may use the software to administer, to score, and to interpret psychological tests without any understanding of the meaning of the interpretation or the limitations of the system. This criticism is very similar to the criticism mentioned earlier in the data analysis section. In the data analysis setting as well as in the CBTI setting, the computer can produce reports that nonetheless require knowledge about the processes that produced those reports in order to produce a valid interpretation or diagnosis.

TECHNOLOGY AND OTHER DOMAINS

From the earliest days of clinical psychology, psychotherapy and other inter-ventions involved one therapist or counselor interacting with one patient or client, but group psychotherapy was established very soon thereafter, followed by couple and family therapy (Freedheim, 1992). Neither involved technology beyond the possible recording of therapy sessions. Along with computer-assisted psychological testing and interpretation, the question of whether therapy or counseling might be performed by a computer arose. Weizenbaum (1966) published a computer system to process natural language, and when a person interacted with ELIZA, his computer program, by using a computer terminal, the person might have thought that the interaction involved another person. In fact, many people have claimed that ELIZA was a computer simulation of a therapist, which was not Weizenbaum's objective. Subsequent attempts to utilize computers in clinical or counseling activities have been somewhat successful and likely will be even more successful as computers and programs become more powerful (see Chapter 2 of this volume).

One of the early introductions of technology, at least in one form, in clinical psychology was the development of a completely automated playroom by L'Abate (1973, 2008), who more recently developed a variety of workbooks designed to aid people in solving a variety of problems (Chapter 19 this volume). The conversion of workbooks from printed forms to computer-administered activities will be a relatively easy transition. Pennebaker (Chapter 12 this volume) also uses computers to collect and to analyze people's writings about emotional situations, which can be generalized to the analysis of any written material. The importance of distance writing rather than face-to-face talk in health promotion, sickness prevention, psychotherapy, and rehabilitation has been emphasized by L'Abate since 1986 and more recently by L'Abate and Sweeney (2011). Since face-to-face talk is a non replicable or at least difficult medium to duplicate and to replicate, L'Abate (1999, 2012) has been arguing that distance writing is the scientific, replicable cornerstone of psychological interventions when structured around a specific topic, a symptom, or a syndrome (see Chapter 19 in this volume)

In education, the change from blackboards and mimeographed forms to computer-assisted displays of various kinds has been occurring recently. However, technology in education has a relatively long history with flight simulators being an early form of technology in education. Clearly, the scoring of examinations now is highly dependent on machines, and some of the current ideas about teaching and learning can be traced to Skinner's development of teaching machines, which in turn were based on his research with non-human organisms. Computer controlled presentations in the classroom and the development of distance learning both are exceedingly dependent on technology, although correspondence courses of the 1940s and 1950s might be taken to be antecedents of the current distance learning activities (Sweeney, 2012).

Many other domains, such as biofeedback, clearly depend upon having technologies that allow for sensing bodily changes and providing rapid feedback. EEG recording and current fMRI recording could not exist without substantial technology, and at least one theoretical development, parallel distributed processing or connectionism, could occur only after technology had reached a high level of sophistication (see Chapters 16, 17, and 18 in this volume).

Major technologies in psychology, neuropsychology, psychiatry, and neurology refer to how evaluation of individual functioning is made in these disciplines, let alone couples and families (Hogan, 2005; Maruish, 1999; (Sperry, in press). However, there is no way we can review all the recent advances in evaluative practices in these disciplines and we shall limit ourselves to representative contributions.

INTELLIGENCE TESTING

"Much has changed in intelligence testing technology and application since the time of Binet's (1905) breakthrough (Kamphaus, Reynolds, and Vogel, 2009). Prior to Wech-sler's innovation of measuring verbal and "performance" abilities on a common test, intelligence tests of the first half of the 20th century typically offered one composite score and focused on assessment of the general intelligence construct. Edgar Doll (1953) identified the problem of over-application and limitations of intelligence testing for the therapeutic programming for individuals with developmental disabilities in the 1930s and provided the first measure of adaptive behavior, the Vineland Social Maturity Scales (Doll, 1935), to act as adjunct information more allied with day-to-day living skills than provi-ded by formal assessment of intellectual functions. During the latter half of the 20th century, intelligence tests began to offer an increasing array of composite or "part" scores intended to produce a more comprehensive evaluation of individual cognitive strengths and challenges. Consequently, interpretation focused more on patterns of abilities within individuals (ipsative test score interpretation) than just deviance from normative standards".

According to Kamphaus et al., (2009), "...at the outset of the 21st century the pendulum is returning to mid-swing with the concept of general intelligence gathering renewed favor in test interpretation, due in part to problems with new test overfactoring that has produced an ever-increasing array of composite scores of dubious clinical or scientific value.These trends have significant implications for the cognitive assessment of individuals with developmental disabilities. Although some concepts and principals such as general intelligence are largely unchanged, the tests themselves and interpretive practices and their uses have changed dramatically. This continuity and change is the focus of this chapter. A case example was used to demonstrate the principles of intellectual assessment for children with develop-mental disabilities".

"The Wechsler intelligence test has four factors representing four components of intellectual function. In China, there are marked cultural, educational, and economic disparities between rural and urban dwellers, which could lead to cultural bias.

Guo, Aveyard, and Dai (2009) applied the four-factor structure to responses to the Chinese Intelligence Scale for Young Children (CISYC) of 820 rural and 664 urban children aged 3 to 7 years. Measurement invariance testing using confirmatory factor analysis showed that the same four factors nested under a higher-order factor held for both rural and urban children. The general intelligence factor mean and variance were invariant and the means and range of scores for rural and urban children were similar. The results show no evidence of cultural bias and that the four factors can be interpreted similarly to those in the Wechsler intelligence test".

Naglieri's (2009) explained "...how Alan Kaufman has been a catalyst for change that has profoundly influenced the field of assessment.Kaufman helped initiate a change in a field of intelligence, which has been remarkably resistant to evolution, despite being described as one of the most important contributions psychology has made to society. His efforts helped start of a new era in intelligence testing and the work of his students has changed the face of assessment today. In order to segment the evolutionary nature of this revolution, Naglieri organized his chapter into two main sections; first Alan Kaufman's influence on the field and second his influence on Naglieri himself".

Personality Testing

"Face validity in personality testing is a controversial kind of test validity (Sartori, 2010). Personality tests are divided into two big categories: projective techniques and psychometric instruments. They differ also for face validity, which influences the judgments that people make about the tests themselves. Sartori reported about the scientific debate on face validity, and the results of a study carried out on naive participants in order to let them compare projective techniques and psychometric instruments on the mere basis of their surface. An ad hoc questionnaire was administered. It asked parti-cipants to compare projective techniques and psychometric instruments by using 13 adjectives. The sample, accidental, is composed of 238 participants, 45 males and 193 females. The data were analyzed through techniques of Correspondence Analysis. Personality tests are principally judged through two dimensions: the aesthetic and the efficaciousness. The first dimension characterized in particular projective tech-niques; the second describes psychometric instruments. Although participants acknowledged that psychometric instruments are credible and scientific, there is a clear preference for projective techniques, principally by females, people younger than 22 and participants with lower education. Personality tests have an appearance that is judged by those who look at them. The aesthetic seems to prevail on the efficacy perception, but it would be suitable to carry on the same research with a sample stratified in respect of the personal details measured by the questionnaire".

"Content validity can add little to the criterion validity of selection tests that show strong positive manifolds (O'Neill, Goffin, and Tett, 2009). However, the use of some non-cognitive tests, personality tests in particular, may not benefit from content validation in hiring situations. These writers argued that content validation has been strongly supported in so far as being associated with considerably larger criterion validity coefficients in personality testing. In these authors' views, content validation is the most direct way of identifying job-relevant personality traits and, accordingly, personality scales likely to demonstrate criterion validity. Goals in this commentary were twofold".

First, O'Neil et al. (2009) "..reviewed evidence bearing on the extent to which personality trait validities vary meaningfully across jobs in both magnitude and direction and whether experts can identify job-relevant traits a priori using content-oriented methods. This review leads to skepticism regarding the value of content validation applied to personality tests. Second, these authors linked their review to existing scientific theory that describes how and why content validation can identify the job relatedness of personality traits and suggest that the value of content validation as applied to personality testing is extensive".

"To discuss the nature of personality-performance relations in a representative study O'Neill, Goffin, and Gellatly (2010) assessed whether the predictive validity of personality scores is stronger when respondent test-taking motivation (TTM) is higher rather than lower. Results from a field sample comprising 269 employees provided evidence for this moderation effect for one trait, Steadfastness. However, for Conscien-tiousness, valid criterion prediction was only obtained at low levels of TTM. Thus, it assessed how TTM relates to the criterion validity of personality testing differently depending on the personality trait assessed. Overall, these and additional findings regarding the nomological net of TTM suggested that criterion validity is a unique construct that may have significant implications when personality assessment is used in personnel selection". This view is supported by the research of L'Abate, Cusinato, Maino, Colesso, and Scilletta, 2010).

Neuropsychological Testing

"Neuropsychological assessment aids in the diagnosis of Alzheimer's disease (AD) by objectively establishing cognitive impairment from standardized tests. Chapman, Mapstone, Poon, Gardner, et al. (2010) presented new criteria for diagnosis that use weighted combined scores from multiple tests. Our method employs two multivariate analyses: principal components analysis (PCA) and discriminant analysis. PCA (N = 216 participants) created more interpretable cognitive dimensions by resolving 49 test measures in our neuropsychological battery to 13 component scores for each participant. The component scores were used to build discriminant functions that classified each participant as either an early-stage AD (N = 55) or normal elderly (N = 78). Their discriminant function performed with high accuracy, sensitivity, and specificity (nearly all >90%) in the development, a cross-validation, and a new-subjects validation. When contrasted to two different traditional empirical methods for diagnosis (using cut scores and defining AD as falling below 5% on two or more test domains), our results suggested that the multivariate method was superior in classification (approximately 20% more accurate)".

Wachtel, Reti, Dhoche, Slomine, and Sanz (2011) "…presented a case report of a E., a 21 year-old autistic male with major depressive disorder with catatonic features complicated by extreme self-injury and suicidality whose acute electroconvulsive therapy (ECT) course was previously reported. Attempts at reducing M-ECT frequency below twice weekly resulted in prompt return of suicidal comments, posturing and dangerous self-harm despite concomitant usage of lorazepam, lithium, duloxetine and riluzole. On the three occasions of testing administration, E.'s performance was stable, and intellectual functioning fell within the baseline deficient range in all evaluations. 2003 testing using the Wechsler Intelligence Scale for Children had yielded a full-scale IQ=63. His index scores also remained notably stable over time, with no evidence for a decline in skill in any area. One patient in this series was unable to procure maintenance ECT services and did relapse; the other two patients remained in remission having received 286 and 156 ECT. While formal cognitive testing was not undertaken with these patients, their caregivers reported either stability or improvement in overall cognitive and adaptive daily functioning and no adverse effects.

"The basis of psychoanalysis and psychoanalytic psychotherapy is that of the therapeutic relationship (Pierce and Takahashi, 2010). One of the premises of psycho-analytically informed psychotherapies is that this relationship is the foundation that patients build on in order to deepen their understanding of themselves, which in turn develops insight and improves understanding of their functioning. Children or adole-scents who are the object of parental and community concern are most likely feeling troubled and highly anxious about most aspects of their lives. Pierce and Takahashi, in their chapter also reviewed currentthe thinking about learning disabilities, attention-deficit/hyperactivity disorder (ADHD), and other neuropsychiatric disorders as they present themselves in psychoanalysis and psychodynamic psychotherapy. The purpose of this discussion is to remind us of how we can begin to think about emotional conditions and subsequent behavioral functioning.

Psychiatric Testing

The interview has been the major form of evaluation in psychiatry. Unfortunately, its reliability and validity varies from questionable to poor (Edelbroock, 1985; Fisher, Epstein, and Harris, 1967; Malgady, Rogler, and Tryon, 1992; Robins, 1985; Stanghellini, 2004), even though, some claim that the face-to-face, talk-based interview may be basic to the creation of rapport and the therapeutic alliance (Safran and Muran, 1995). This claim, however, is no longer tenable given the widespread initiation of millions of quasi-semi- or actual therapeutic alliances made online every day.

Assuring research participants' capacity to provide informed consent has become increasingly important in health and mental health research, and each study faces unique capacity-assessment challenges, possibly requiring its own screening tool. Zayas, Cabassa, and Perez (2005) described the development and preliminary testing of a capacity-to-consent tool constructed for a study of psychiatric diagnosing in a community clinic. A 10-item screening device based on four legal standards for demonstrating capacity (understanding, appreciation, reasoning, and voluntarism) was created and tested with 68 adult patients entering the study. Only five participants (7%) failed the screen, 61 (93%) passed. No participants who passed at entry were later found in psychiatric evaluations to lack capacity. Apparently, this instrument was effective in identifying persons who could not demonstrate consent capacity, thereby protecting prospective participants.

Herbild, Bech, Gyrd-Hansen, Christensen, et al. (2011) attempted to identify the effects of local recommendations of pharmacogenetic testing in psychiatry with respect to treatment costs. Based on Danish patient registers, individual treatment costs within a 365-day period at three psychiatric hospitals recommending and using pharmacogenetic testing is compared retrospectively with treatment costs at other Danish psychiatric hospitals using alternate treatment strategies. Primary outcome of interest is total direct costs analyzed by multilevel modeling. Secondary outcome measures are healthcare consumption within specific sectors analyzed by Tobit-regressions. Costs among patients treated at hospitals recommending and using pharmacogenetic testing were not found to be statistically significantly different from costs among patients treated at sites using alternate treatment strategies. In spite of recommendations to test all patients the uptake of the test was, however, low (26-31 %). Treatment practice using routine therapeutic drug monitoring (in Århus) showed a trend towards lower costs.

Based on this natural experiment , Herbild, Bech, Gyrd-Hansen, Christensen, et al. (2011) were not able to document statistically significant differences in total costs between treatment sites that had guidelines recommending pharmacogenetic testing, relative to sites without such guidelines, over a period of one year. However, guidelines of pharmacogenetic testing and possibly also therapeutic drug monitoring seem to lead to reductions in costs for primary care services. In the case of the former, reductions do, however, seem to be outweighed by increases in costs for psychiatric and non-psychiatric inpatient stays. In conclusion, no statistically significant differences in total direct costs across sites with different treatment strategies were found.

Psychiatric evaluations are used to identify patients' needs, develop treatment plans, and monitor progress and outcomes. Sayer, Carlson, and Schnurr (2011) focused on assessment of functioning and disability as part of the psychiatric evaluation of individuals with possible posttraumatic stress disorder (PTSD). Although the importance of assessing functional status among individuals with serious mental illness, including schizophrenia, has received considerable attention in the clinical and scientific literature, the value of incorporating such assessments into standard clinical evaluations of individuals with possible PTSD has received limited attention. These investigators discussed the reasons clinicians should assess functioning and monitor treatment outcomes in patients with PTSD; they presented a conceptual framework to guide these evaluations and described the domains that should be considered, reviewing different methods of assessment.

NEUROLOGICAL TESTING

In its early beginning, neurological testing focused on visual acuity and simple movements: 10 upper extremity and 8 lower extremity neurological functions in normals and ambulatory multiple sclerotics are measureable quantities (Tourtellotte, 1965). Only hip flexor strength showed learning or fatigue effects. No significant differences in function were produced by a number of drugs, but changes in lower extremity functions were noted in subnormally functioning individuals with ACTH, steroid, and vitamin B[sub]12[/sub] administration.

Early diagnosis of PD [Parkinson's disease] is attainable by use of [visuo-manual coordination testing procedures (Hocherman and Giladi, 1996). these tests involve tracing and tracking of 2 dimensional paths with an unseen hand / the model paths are displayed on a computer monitor together with a cursor that provides feedback about movements of the hand / recorded data is subjected to off line analysis that generates several measures of performance as listed / a. total time of trial execution (TT) / b. vectorial error (VE) / c. the cumulative movement time with a VE which is greater than 50% / d. mean hand velocity in each trial / e. the number of events in which tracking is interrupted / the vast majority of the patients [were] readily distinguishable from the control Participants in the performance of both their symptomatic and nonsymptomatic hands 26 normal control participants and 21 PD patients [mean age 66.4 yrs] were studied

Most measures of stigma are illness specific and do not allow for comparisons across conditions. As part of a study of health-related quality of life for people with neurological disorders, Rao, Choi, Victorson, Bode, Peterman, Heinemann, and Cella (2009) developed an instrument to measure the stigma for people with chronic illness These investigators based item content on literature review, responses from focus groups, and cognitive interviews. They then administered the items to people with neurological disorders for psychometric testing. Five hundred eleven participants completed items of the stigma scale.

Exploratory factor analysis produced two factors that were highly correlated (r = 0.81). Confirmatory factor analysis produced high standardized loadings on an overall stigma factor (0.68–0.94), with poorer loadings on the two sub-domains (-0.12 to 0.53). These results demonstrated a sufficiently unidimensional scale that corresponded with the bifactor model. Item response theory modeling suggested good model fit, and differential item functioning analyses indicated that the 24-item scale showed potential for measure-ment equivalence across conditions. These efforts produced a stigma scale that had promising psychometric properties. Further study can provide additional information about its benefit in measuring the impact of stigma across conditions.

Conclusions

Intrusions of technology into almost every aspect of the world from game playing to life and death decision-making are all pervasive. Computers have beaten excellent chess players, and computer controlled systems now facilitate surgery. With the continued development of artificial intelligence and with increasingly powerful computers, the utilization of computers in all areas of psychology is likely to increase. Whether computer systems and other technology can replace psychological test examiners, clinical psychologists, industrial/organizational psychologists, experimental psychologists, and teachers of psychology cannot be determined at this time. In a recent article, Christian (2011) has argued that mind always will be superior to machine, but judges in an annual contest often have difficulty in distinguishing answers to questions generated by computers from answers generated by people.

Clearly, technology is useful in many ways, as described in this introduction, but as is usually the case, most entities have negative attributes as well as positive ones. Data analyses can be performed more accurately and quicker with computers than by humans alone. However, the interpretation of the output from the data analysis program remains with a human. So, statistical analyses that were impossible or impractical 20 years ago to the loss of skills (e.g., computational skills) that have been used for centuries. Rather than adding a few numbers or dividing two small numbers in one's head, a person now may reach for a calculator or even use a spreadsheet on a computer. Furthermore, invasions of privacy, new forms of bullying, and new ways of stalking come with the advancing technology.

Somewhat as Matarazzo claimed about automated or statistical interpretations of psychological tests, at least for a few years, humans will make decisions about how technology should be used, but in the future, technology may be used to make decisions about how technology should be used. It can be hoped that the technological system will not have as many biases as humans have.

REFERENCES

Carr, N. (2011). The shallows: What the Internet is doing to our brains. New York: Norton.

Chapman, R. M., Mapstone, M., Poon, A. P., Gardner, M. N., McCrary, J. W., DeGrush, E., nd ... Guillily, M. D. (2010). Diagnosis of Alzheimer's disease using neuropsychological testing improved by multivariate analyses. *Journal of Clinical and Experimental Neuropsychology*, 32, 793-808.

Christian, B. (2011). Mind vs. machine. *The Atlantic*, 307, 58-69.

De Giacomo, P., Mich, L., Santamaria, C., Sweeney, L. G., and De Giacomo, A. (2012). Information processing. In L. L'Abate (Ed.), *Paradigms in theory construction* (pp. 000-000). New York: Springer-Science.

Edelbrock, C. (1985). Age differences in the reliability of the psychiatric interview of the child. *Child Development*, 56, 265–275.

Fisher, J., Epstein, L., and Harris, M. (1967). Validity of the psychiatric interview: Predicting the effectiveness of the first Peace Corps volunteers in Ghana. *Archives of General Psychiatry*, 17, 744–750.

Fowler, R. D. (1985). Landmarks in computer-assisted psychological assessment. *Journal of Consulting and Clinical Psychology*, 53, 748-759.

Freedheim, D. (1992). History of psychotherapy: A century of change. Washington, D.C.: American Psychological Association.

Gosling, S. D., and Johnson, J. A. (2010). Advanced methods for conducting online behavioral research. Washington, DC: American Psychological Association.

Guo, B., Aveyard, P., and Dai, X. (2009). The Chinese Intelligence Scale for Young Children: Testing factor structure and measurement invariance using the framework of the Wechsler intelligence tests. *Educational and Psychological Measurement*, 69, 459-474.

Herbild, L., Bech, M., Gyrd-Hansen, D., Christensen, M., Werge, T., and Nielsen, K. (2011). Do guidelines recommending pharmacogenetic testing of psychiatric patients affect treatment costs and the use of healthcare services?. *Scandinavian Journal of Public Health*, 39, 147-155.

Hills, R. L. (1994). *Power from wind: A history of windmill technology*. New York: Cambridge University Press.

Hocherman, S. S., and Giladi, N. N. (1996). Early diagnosis of Parkinson's disease in new neurological patients by testing of visuo-manual coordination. In C. Ohye, M. Kimura, J. S. McKenzie, C. Ohye, M. Kimura, J. S. McKenzie (Eds.) , *The basal ganglia* 5 (pp. 427-431). New York, NY US: Plenum Press.

Hogan, T. P. (2005). 50 widely used psychological tests. In G. P. Koocher, J. C. Norcross, and S. S. Hill, III (Eds.), *Psychologists' desk reference* (2nd ed., pp. 101–104). New York: Oxford University Press.

Kamphaus, R. W., Reynolds, C. R., and Vogel, K. (2009). Intelligence testing. In J. L. Matson, F. Andrasik, M. L. Matson, J. L. Matson, F. Andrasik, M. L. Matson (Eds.),

Assessing childhood psychopathology and developmental disabilities (pp. 91-115). New York, NY US: Springer Science + Business Media.

L'Abate, L. (1973). The laboratory method in clinical child psychology: Three applications. *Journal of Clinical Child Psychology*, 2, 8-10.

L'Abate, L. (1986). Systematic family therapy. New York: Brunner/MazelL'Abate, L. 1999). Taking the bull by the horns: Beyond talk in psychological interventions. *The Family Journal: Therapy and Counseling for Couples and Families*, 7, 206-220.

L'Abate, L. (Ed.). (2008). *Toward a science of clinical psychology: Laboratory evaluations and interventions.* New York: Nova Science Publishers.

L'Abate, L. (2012). *Clinical psychology and psychotherapy as a science: An iconoclastic perspective.* New York: Springer-Science.

L'Abate, L., Cusinato, M., Maino, E., Colesso, W., and Scilletta, C. (2010). *Relational competence theory: Research and mental health applications.* New York: Springer-Science.

L'Abate, L., and Sweeney, L. G. (Eds.). (2011). *Research on writing approaches in mental health.* Bingley, UK: Emerald Group Publishing Limited.

Lattal, K. A. (2004). Steps and pips in the history of the cumulative recorder. *Journal of the Experimental Analysis of Behavio*r, 82, 329-355.

Malgady, R., Rogler, L., and Tryon, W. (1992). Issues of validity in the Diagnostic Interview Schedule. *Journal of Psychiatric Research*, 26, 59–67.

Maruish, M. E. (Ed.). (1999). *The use of psychological testing for treatment and outcomes assessment* (2nd ed.). Mahwah, NJ: Erlbaum.

Matarazzo, J. D. (1986). Computerized clinical psychological test interpretations: Unvalidated plus all mean and no sigma. *American Psychologist*, 41, 14-24.

Meehl, P. E. (1954). Clinical vs. statistical prediction: A theoretical analysis and a review of the evidence. Minneapolis, MN: University of Minnesota Press.

Morozov, E. (2010). *The net delusion: The dark side of internet freedom.* ???: Public Affairs.

Naglieri, J. A. (2009). Intelligent intelligence testing: The influence of Alan S. Kaufman. In J. C. Kaufman, J. C. Kaufman (Eds.) , *Intelligent testing: Integrating psycho- logical theory and clinical practice* (pp. 73-96). New York: Cambridge University Press.

O'Neill, T. A., Goffin, R. D., and Gellatly, I. R. (2010). Test-taking motivation and personality test validity. *Journal of Personnel Psychology*, 9, 117-125.

O'Neill, T. A., Goffin, R. D., and Tett, R. P. (2009). Content validation is fundamental for optimizing the criterion validity of personality tests. *Industrial and Organizational Psychology: Perspectives on Science and Practice*, 2, 509-513.

Pierce, M., and Takahashi, A. (2010). Neuropsychological testing and psychoanalysis in adolescent and young adult patients. In E. R. Arzubi, E. Mambrino, E. R. Arzubi, E. Mambrino (Eds.), *A guide to neuropsychological testing for health care professionals* (pp. 103-134). New York, NY US: Springer Publishing Co.

Rao, D., Choi, S. W., Victorson, D., Bode, R., Peterman, A., Heinemann, A., and Cella, D. (2009). Measuring stigma across neurological conditions: The development of the Stigma Scale for Chronic Illness (PARTICIPANTSCI). Quality of Life Research: *An International Journal of Quality of Life Aspects of Treatment, Care and Rehabilitation*, 18, 585-595.

Redin, J. (N.D.). A brief history of mechanical calculators: Part III, Getting ready for the 20th entury. http://www.xnumber.com/xnumber/mechanical3.htm. (Retrieved April 28, 2011)

Robins, L. (1985). Epidemiology: Reflections on testing the validity of psychiatric interviews. *Archives of General Psychiatry*, 42, 918–924.

Safran, J. D., and Muran, J. C. (Eds.). (1995). The therapeutic alliance. New York: Wiley.Sartori, R. (2010). Face validity in personality tests: Psychometric instruments and projective techniques in comparison. *Quality and Quantity: International Journal of Methodology*, 44, 749-759.

Sayer, N. A., Carlson, K. F., and Schnurr, P. P. (2011). Assessment of functioning and disability. In D. M. Benedek, G. H. Wynn, D. M. Benedek, G. H. Wynn (Eds.) , *Clinical manual for management of PTSD* (pp. 255-287). Arlington, VA US: American Psychiatric Publishing, Inc.

Shirky, C. (2010). *Cognitive surplus: Objectivity and generosity in a connected age*. ???: Penguin Press.

Siegel, L. (2009). *Against the MACHINE: How the web is reshaping culture and commerce—And why it matters.* New York: Spiegel and Grau.

Sperry, L. (2011). *Assessment of couples and families: Contemporary and cutting-edge strategies* (Second Edition). New York: Rutledge.

Stanghellini, G. (2004). The puzzle of the psychiatric interview. *Journal of Phenomenological Psychology*, 35, 173–195.

Sweeney, L. G. (2012). Education. In L. L'Abate (Ed.), *Paradigms in theory construction* (pp. 00-000). New York: Springer-Science.

Tourtellotte, W. W. (1965). Quantitative clinical neurological testing: I. A study of a battery of tests designed to evaluate in part the neurological function of patients with multiple sclerosis and its use in a therapeutic trial. *Annals of the New York Academy of Sciences*, 122, 480-505.

Wachtel, L. E., Reti, I. M., Dhoche, D. M., Slomine, B. S., and Sanz, J. (2011). Stability of neuropsychological testing during two years of maintenance electroconvulsive therapy in an autistic man. *Progress in Neuro-Psychopharmacology and Biological Psychiatry*, 35, 301-302.

Weiner, Norbert (1948). *Cybernetics or Control and Communication in the Animal and the Machine*. Cambridge, MA: MIT Press.

Weizenbaum, J. (1966). ELIZA—A computer program for the study of natural language communication between man and machine. *Communication of the ACM*, 9(1). disorders (4th ed.). Washington, DC: Author.

In: Handbook of Technology in Psychology ...
Editor: Luciano L'Abate and David A. Kasier
ISBN: 978-1-62100-004-4

Chapter 2

TECHNOLOGY IN PSYCHOTHERAPY: STRENGTHS AND LIMITATIONS

Amy Przeworski and Michelle G. Newman
Pennsylvania State University, PA, US

Technological advances have made computers so small, light, and convenient that that they have become indispensible parts of our lives. Seventy million American households have at least one computer in their household, and almost 62 million American homes have Internet access, numbers that are likely increasing (Cheeseman Day, Janus, and Davis, 2005). With such a widespread use of computers, it is no surprise that technology has been integrated into most aspects of human existence, from basic tasks such as shopping to more impactful aspects of life such as medical and psychological treatment.

Computer-assisted therapy has been applied for many psychological disorders using various formats, including Internet treatment, virtual reality treatment, treatments on CD rom, treatments on hand-held portable computers, videoconferencing, and more standard computer-assisted or computer guided packages that are implemented on desktop computers. Countless studies have demonstrated the efficacy of technology-based interventions for such disorders as anxiety, mood, substance use, and eating disorders (for reviews see Newman, Szkodny, Llera, and Przeworski, 2011a, 2011b, 2011c).

Computer-administered therapy addresses many of the barriers to treatment that often prevent potential clients from seeking therapy. However, despite the many advantages of computer-assisted therapy, some people remain skeptical of the use of technology in treatment. This chapter will provide a review of the many formats of technology-based therapies, the advantages of the use of technology in therapy and the disadvantages identified by skeptics.

BENEFITS OF TECHNOLOGY-BASED TREATMENTS

Technology-based therapies provide a means to overcome many of the barriers of traditional face-to-face therapies including feeling embarrassed or stigmatized, the expense of therapy, logistic issues, such as traveling to an appointment, finding child care, scheduling an appointment, and access to trained treatment providers in one's area. For example, frequency of treatment sessions can vary from a minimum of once per week, to several times per week and total number of required sessions to achieve a successful outcome often ranges from 10.8-46.4 (Turner, Beidel, Spaulding, and Brown, 1995). For some individuals, especially those living in rural areas, travel to a clinic or therapy office may require more time than they actually spend in therapy sessions. Public transportation in rural areas is often lacking and mental health facilities in rural areas often serve large geographic areas, requiring long-distance travel for appointments (Murray and Keller, 1991),

This leads many individuals in rural areas to receive mental health services from medical professionals rather than mental health providers (Fortney, Thill, Zhang, Duan, and Rost, 2001; Wells, Manning, Duan, Newhouse, and Ware, 1986). Primary care physicians are less likely to diagnose and offer treatment for psychological disorders than are mental health professionals (Harman, Rollman, Hanusa, Lenze, and Shear, 2002). When treatments are offered by medical professionals, they typically consist of medications, not therapy. Further, because primary care physicians do not have specialized training, they may have less knowledge regarding specific medications and their side effects and therefore may not provide the same level of care as a psychiatrist. Thus, rural populations are less likely to have access to optimal mental health care.

Even when clients do have access to therapists, many providers have not been trained in the evidence-based interventions that are most efficacious for specific disorders. For example, cognitive-behavioral interventions are the most evidence-based interventions for many anxiety and mood disorders (Barlow, Gorman, Shear, and Woods, 2000; Dobson, 1989; Hofmann and Spiegal, 1999; Newman, 2000); however, dissemination of these treatment techniques typically involves distributing therapy manuals and brief one or two day workshops. This does not provide the training necessary to implement these techniques effectively or to tailor the use of these techniques to specific disorders or populations. The use of technology-based treatments increases clients' access to therapists trained in the implementation of therapy techniques for specific disorders, even when therapists live remotely. Greater access to evidence-based treatments may increase the effectiveness of therapy and decrease the number of therapy hours that are necessary for clients' to achieve symptom relief. Technology also provides a new avenue for the training and supervision of therapists in specific techniques, thereby increasing the dissemination of the most efficacious interventions.

An additional barrier to face-to-face treatment is the cost of such services. Estimates of the cost for treatment for anxiety disorders vary from $1,260-$4,370 per client in the U.S. (Turner et al., 1995), severely limiting access to treatment for individuals without health insurance or who are of low socioeconomic status. Technology-based services have been estimated to provide a savings of $540-$630 per client when compared to face-to-face interventions (Newman, Consoli, and Taylor, 1999; Newman, Kenardy, Herman, and Taylor, 1997b). The decreased cost for such therapy may facilitate the dissemination of therapy to individuals experiencing barriers to treatment, thereby reaching an entirely new population of individuals suffering from psychological disorders.

Finally, some individuals do not seek therapy services due to the stigma involved in going to a therapist's office. The use of technology-based treatments reduces stigma as clients may complete many of these therapies within the comfort and privacy of their own home. When technology-based therapy relies on interactions with other clients, as in a chat room, clients may use an alias or private log-in name in order to increase their privacy. In comparison, face-to-face group therapy does not permit such anonymity.

CRITIQUES OF TECHNOLOGY-BASED THERAPY

Despite their promise, technology-based interventions have been received with skepticism and criticism by some psychologists. Psychotherapists have raised concerns that the availability of computer programs may diminish the ability of face-to-face clinicians' to practice their profession; however, many individuals who have participated in Internet therapy reported that they would not initially have sought face-to-face counseling and 65% of these Internet therapy clients later went on to use face-to-face counseling, suggesting that computer therapies actually provide a gateway to make use of face-to-face services (Metanoia, 2001). Other critics of technology-based therapy have raised concerns that it eliminates important predictors of therapeutic success such as therapist-client bond, facial cues, and body language. However, some technology-based services, such as videoconferencing, provide real-time video of clients and therapists and research has demonstrated that even those technologies that do not rely on video maintain many of the interpersonal factors that are important aspects of the therapeutic relationship, including trust and comfort in self-disclosure, empathy, and a therapeutic bond (Cook and Doyle, 2002; Knaevelsrud and Maercker, 2006). Users of technology-based therapies apply face-to-face interaction norms of politeness, notions of self and other, and gender stereotyping when interacting with computers (Nass, Fogg, and Moon, 1996; Nass, Steuer, Henriksen, and Dryer, 1994) and the breadth and depth of relationships formed in chat rooms are similar to that of face-to-face relationships (Parks and Roberts, 1998; Walther and Burgoon, 1992). Further, similar client reports of the therapeutic alliance have been found when using technology-based therapies as when participating in face-to-face therapy (Cook and Doyle, 2002; Schmidt, 2003). Clients have also been found to be more open in technology-based services when describing their alcohol intake (Erdman, Klein, and Greist, 1985), substance misuse (Supple, Aquilino, and Wright, 1999), suicidal ideation (Greist et al., 1973), and sexual experiences (Lapham, Kring, and Skipper, 1991; Millstein and Irwin, 1983; Romer et al., 1997). Such disinhibition permits increased honesty and depth of discovery in technology-based psychotherapy. Other critics have voiced concerns that clients may not want to receive therapy services using technology-based therapy or may be dissatisfied if they do participate in it. Some studies have shown that individuals had negative attitudes about seeking online help relative to face-to-face services (Chang and Chang, 2004; Rochlen, Beretvas, and Zack, 2004). However, when asked what the preferred method of delivery of self-help is, the majority of individuals would prefer to receive such services using a computer (Graham, Franses, Kenwright, and Marks, 2000, 2001). Further, those who did receive therapy using technology-based services have reported high satisfaction and credibility and similar rates of attrition to face-to-face therapy (Buglione, DeVito, and Mulloy, 1990; Carlbring, Ekselius, and Andersson, 2003; Carlbring, Westling, Ljungstrand, and Andersson, 2001; Carr, Ghosh, and Marks, 1988; Dolezal-Wood, Belar, and Snibbe, 1998; Escoffery, McCormick, and Bateman, 2004; Ghosh and Marks, 1987; Ghosh, Marks, and Carr, 1988; Newman, Consoli, and Taylor, 1997a; Proudfoot et al., 2003; Proudfoot et al., 2004; Richards and Alvarenga,

2002; Wright and Wright, 1997). Other critics have stated that technology-based treatments require clients to be technologically savvy and/or require substantial training in the use of the technology in order to benefit. However, even inexperienced Internet users benefitted as much as experienced users when participating in Internet treatment (Lange et al., 2000). Still others have identified ethical and legal concerns regarding technology-based treatments. For example, despite the increasing reliance on Internet-based mental health services, to date there are no agreed-upon guidelines for providing such services; nor is there a governing body that monitors the quality of these services or the competence of the professionals providing treatment. Web-based interventions present unique ethical and professional issues (Hsiung, 2002). For example, in the United States, if therapists practice in one state but provide therapy via videoconference, email, or another Internet medium, they may be providing services across state lines but may only be licensed to provide services within their home states. Additionally, it may be difficult for clients to verify the credentials of therapists when therapy is conducted via technology-based means. As such, there is an ongoing dialogue amongst U.S. professionals who provide Internet-based mental health services and regulating agencies such as the National Institutes of Mental Health (NIMH) and National Board for Certified Counselors (NBCC) regarding the ethical practice of providing online mental health services. Other agencies have been developed for the sole purpose of promoting health and mental health resources online and providing suggestions for the online provision of these services. These include the International Society for Mental Health Online (ISMHO) and Health on the Net Foundation (HON). However, to date, no agreed upon solutions to these ethical and professional concerns have been found. An additional challenge of technology-based therapy in the United States is the lack of insurance reimbursement for such services. It has been argued that such services should be covered by insurance (Williams, Remmes, and Thompson, 1996); however, to date, most insurance companies do not cover such services (Thompson and Fox, 2001). The decreased cost of technology-based services may lead clients to be able to pay out of pocket for these services; however, this may be difficult for individuals of lower socio-economic status. Below we review several of the most common uses of technology-based therapy as well as the strengths and limitations inherent in each. Further, several particularly novel uses of the technology in therapy are highlighted.

Video Teleconferencing

The use of video teleconferencing permits clients and therapy providers to have visually interactive electronic meetings from distant locations through the use of digital video cameras, web cameras, computer monitors, and the Internet. Live full-motion video images are sent via the Internet allowing users to have conversations in real time while being geographically distant.

Strengths

With more individuals having web-cameras on their personal computers, this form of technology-based therapy is becoming more accessible. The use of video teleconferencing has been found to produce equivalent therapeutic alliance to that of face-to-face therapy (Ghosh, McLaren, and Watson, 1997) and to be highly accepted by clients even when they are acutely or chronically psychotic or agitated (Jerome et al., 2000; Jerome and Zaylor, 2000). This form of computer-assisted therapy has been suggested as a potential low-cost solution to the limited mental health services in rural areas (Jerome and Zaylor, 2000). Further, this technology permits individuals to receive treatment from experts in the field and to gain access to therapists with highly specialized training in treatment techniques for particular disorders. This may be especially beneficial for disorders wherein dissemination of treatment techniques is lacking, such as obsessive-compulsive disorder. It may also permit therapists to conduct family meetings even when a family is located at distant locations. Videoconferencing may also provide a way for therapists to conduct therapy sessions that would otherwise require a home visit, such as exposure and response prevention to something specific in the home, such as the stove or a particular contaminant, therapy with an individual who is homebound due to agoraphobia, etc. Therapy conducted via video teleconferencing has been described in published case studies and multiple baseline studies of individuals with obsessive compulsive disorder (Himle et al., 2006), panic disorder (Cowain, 2001), social anxiety disorder (Pelletier, 2003), posttraumatic stress disorder (Deitsch, Frueh, and Santos, 2000), problem gambling (Oakes, Battersby, Pols, and Cromarty, 2008), bulimia and related disorders (Bakke, Mitchell, Wonderlich, and Erickson, 2001; Simpson et al., 2006). To date, however, few randomized controlled trials have been conducted examining video teleconferencing. Video teleconferencing also has been used effectively in the provision of treatment of various child and adult disorders including anxiety, mood, externalizing, and eating disorders, as well as gender identity difficulties (Bakke et al., 2001; Cowain, 2001; Deitsch et al., 2000; Himle et al., 2006; Manchanda and McLaren, 1998; Miller, Kraus, Kaak, Sprang, and Burton, 2002; Paul, 1997; Pelletier, 2003; Rendon, 1998; Simpson, Doze, Urness, Hailey, and Jacobs, 2001a) and has been demonstrated to be equally efficacious as face-to-face cognitive behavioral therapy (CBT) in randomized clinical trials for PTSD (Frueh et al., 2007; Germain, Marchand, Bouchard, Drouin, and Guay, 2009), panic disorder (Bouchard et al., 2004), and adjustment and interpersonal problems (Day and Schneider, 2002). To date, one study found videoconferencing CBT to be superior to face-to face therapy for childhood depression in terms of rate of improvement (Nelson, Barnard, and Cain, 2006); however, face-to face therapy was found to be superior to videoconferencing for bulimia in adults (Mitchell et al., 2008). Further, studies demonstrated client satisfaction and that positive therapeutic alliances were achieved through the use of videoconferencing in the treatment of these disorders (Bouchard et al., 2004; Ghosh et al., 1997; Himle et al., 2006; Manchanda and McLaren, 1998; Simpson, Doze, Urness, Hailey, and Jacobs, 2001b). Videoconferencing has also been used in training therapists (Rees and Gillam, 2001) as well as in supervision of psychiatric residents(Gammon, Sorlie, Bergvik, and Hoifodt, 1998) with both indicating satisfaction with these services.

Limitations

Many psychologists have a negative outlook on videoconferencing and have reported that they believe that therapy conducted via teleconferencing would be less effective than face to face therapy (Wray and Rees, 2003). One study found that therapists rated a videotape of a therapist and client as having poorer therapeutic alliance when a session was conducted via videoconferencing than when the exact same session was conducted in a face-to-face format (Rees and Stone, 2005). However, the predominance of the data indicates that video teleconferencing produces therapeutic alliance equivalent to that of face-to-face therapy (Day and Schneider, 2002; Frueh et al., 2007; Ghosh et al., 1997; Mallen, Day, and Green, 2003) and is highly accepted by clients even when they are acutely or chronically psychotic or agitated (Jerome et al., 2000; Jerome and Zaylor, 2000). Thus, therapists' negative views of videoconferencing may be unfounded.

There are additional considerations that therapists must make when using videoconferencing in therapy. First, therapists and clients must have access to the technology necessary to conduct such therapy. In order to participate in this form of therapy, it is necessary for both parties to have a web-camera, computer with monitor, and high speed Internet access. If clients do not have this technology within their homes, they may need to travel to facilities that have such technology, such as medical centers or clinics that may not afford the same level of privacy as a personal computer.

Therapists and clients must also be familiar with the technology required to connect. This may be accomplished through email systems that permit real-time video interactions or through a third party who coordinates telemedicine for a mental health facility. Care must be taken to coordinate the timing for the beginning and the end of sessions such that sessions are not abruptly discontinued by third parties who coordinate telemedicine. Additionally, clients and therapists must be familiar enough with the technology to know how to solve problems that may crop up, such as slowed Internet connections, loss of Internet connections, or poor resolution or visibility.

Further, it is necessary for both providers and clients to have high speed Internet in order to make use of this computer-based form of therapy. Even with the use of high speed Internet, there are often delays in audio and video reception, leading to clients and therapists speaking over one another, delayed responses to questions, and even delayed facial reactions. These difficulties can all be adjusted to through slow and careful speech and patience in waiting for responses to comments. An additional difficulty can be that the image resolution can render it difficult for subtle facial expressions and gestures to be perceived (Kuulasmaa, Wahlberg, and Kuusimaki, 2004). This makes the use of effective verbal communication essential and may require therapists to ask direct questions about clients' affective state, rather than relying on nonverbal cues.

Steps must be taken to ensure clients' confidentiality, including secure Internet connections and private rooms in which clients have access to a computer or videocamera with videoconferencing technology. It is especially important to consider the privacy of clients in the rooms where they are participating in videoconferencing. Also, therapists will only be able to see a portion of the room in which clients are sitting. It is also essential that clients have access to computers that are located in quiet rooms in which they are alone and feel that they can speak freely and without interruption or distraction.

Another complication in the use of videoconferencing is that therapists and clients cannot hand things to one another, such as worksheets, completed homework assignments, consent forms, etc. This difficulty may be solved through faxing or emailing documents to be discussed prior to sessions.

Video teleconferencing may not be optimal for clients who may experience frequent safety concerns such as homicidality or suicidality. If clients indicated suicidality during sessions conducted by videoconference, it would be essential for therapists to know where clients were currently located (that is, home addresses of clients or of clinics where clients participated in the videoconference), and for therapists to have been familiar with local resources that were available to help clients, including mobile crisis units and hospitals. These should be identified at the start of therapy. In addition, therapists should contract with clients on the issues of suicidality and homicidality as well as the steps that will be taken should this issue crop up.

A final, and potentially large disadvantage to this form of technology-based therapy, is that in the United States most insurance companies will not currently reimburse for services provided via videoconferencing (Thompson and Fox, 2001). This may be an insurmountable obstacle for many clients who rely on health insurance to pay for their mental health services. It is possible that this may change in the future, as videoconferencing becomes more common and accessible to clients and therapists.

INTERNET THERAPY

Internet therapy occurs in various formats including by electronic mail, chatting, and via multimedia websites. Each form of Internet therapy has unique strengths and limitations.

Email contact has been used as a stand-alone therapy or as an adjunct to face-to-face or Internet therapy. When used as a stand-alone therapy therapists email questions to clients if they want to conduct assessments. Therapists might also email instructions on how to implement homework techniques. Clients submit homework assignments and descriptions of their use of techniques as well as questions and comments regarding their experiences. When used in an adjunctive fashion, email may be used as a means to track homework assignments, check in with clients between sessions regarding their symptoms or use of techniques, and as a means to answer clients' questions between sessions. Therapy using email is currently the most commonly used form of asynchronous computer therapy (Cook and Doyle, 2002; Heinlen, Welfel, Richmond, and O'Donnell, 2003).

This practice, although becoming increasingly more common for individual therapy, has also been used by one group of psychologists for marital therapy. Jedlicka and Jennings (2001) successfully used email therapy with several couples with marital difficulties, many of whom were contemplating divorce. In this therapy each member of the couple sent emails to their therapist, who responded with suggestions for ways of increasing communication with their spouse, handling anger appropriately, and cooperating on budgeting and other domestic issues. The authors reported that email therapy provided couples with a means of examining the process of the marital difficulties and providing both members of the couple with an avenue for expressing their feelings without couples spiraling into arguments. The authors provide case summaries of various couples for whom this technique was helpful.

In yet another innovative use of email in therapy Murdoch and Connor-Green (2000) used email to prompt clients for homework reports and to provide feedback on homework assignments between therapy sessions. The authors presented three case studies in which they used email in homework assignments to increase therapy compliance and improve therapeutic alliance. The authors reported that email transactions required no more than ten minutes per day per client and were most effective when aimed at encouraging clients to take responsibility for making gains in therapy and problem solving during difficult situations. The authors also provide guidelines for determining which clients may be the most appropriate for this technique (Murdoch and Connor-Greene, 2000).

Strengths

Email therapy, either in an adjunctive or stand alone format, has been used for many psychological difficulties including weight loss (Tate, Jackvony, and Wing, 2003; Tate, Wing, and Winett, 2001), bulimia nervosa and binge eating disorder (Robinson and Serfaty, 2008), posttraumatic stress disorder (Lange, van de Ven, Schrieken, and Emmelkamp, 2001), panic disorder (Carlbring et al., 2001; Klein, Richards, and Austin, 2006). Therapy involving email allows therapists to give clients feedback multiple times between sessions thereby catching problematic behaviors quickly. This can be especially helpful for clients who are practicing a technique for the first time. Therapy including email also allows clients to initiate contact with therapists at times when they need their therapists the most, rather than having to wait until their next scheduled appointment. Asynchronous communication, such as email, does not require scheduled appointments and may permit clients time to reflect on their experiences and comments (Mora, Nevid, and Chaplin, 2008).

Limitations

Although increased contact is advantageous in some respects, it can also have several negative effects including (a) therapists needing to respond to clients' emails in a timely fashion, (b) clients may become confused about whether crises and difficulties should be conveyed via email, and (c) excessive uses of email by clients (Yager, 2001). Further, it can be difficult for therapists to understand the emotion and strength of emotion that a client is experiencing through expression via text alone. Subtle communications may be lost in email therapy.

Further, because email communication is asynchronous, it is difficult for therapists to follow-up on specific comments made in an email in order to gain further understanding about clients' meaning or experience. At times emails are blocked by spam filters if the filters rely on flagging emails through the use of particular words contained in the subject or body of the email. This can result in therapists and clients missing emails that are sent in the ongoing communication.

Finally, therapists must attend to issues of confidentiality and privacy, especially if clients are using email addresses that are also used by other individuals within the household or work-related email addresses.

Chatting

In therapy using chatting, therapists either use instant messaging to communicate with clients in real-time or a chat room to communicate with multiple clients. Chatting is the most commonly used synchronous computer-based therapy (Castelnuovo, Gaggioli, Mantovani, and Riva, 2003).

Strengths

Therapy via chat rooms also has many benefits, including synchronous communication with one's therapist from the comfort of home and possibly the opportunity to interact with others with similar difficulties (if the chatroom is public). Chatting also mimics face-to-face conversation in that there is a spontaneous give and take in which feedback is immediately provided to clients and clients are able to receive empathy and support. This may also lead to increased disclosure (Suler, 2004).

Limitations

Potential drawbacks include availability of therapists for limited times, therefore clients must schedule appointments, and lack of privacy if conducting therapy in a public chatroom. Additionally, the use of chat does not permit time for reflection and removes the possibility of therapists' noting clients' emotional expressions through body language or facial cues. It also relies on speed of typing so there are longer delays between each communication than is the case with video.

Multimedia Websites

Multimedia websites can provide clients with information regarding psychopathology and techniques, interactive exercises to learn and practice the techniques, and even video or audio files to teach techniques. Therapy on multimedia websites is often broken down into modules that clients complete in order at their own pace. At times multimedia websites are combined with chat rooms or the ability to contact therapists via email if necessary.

Strengths

Multimedia websites permit researchers and therapists to gather information from clients such as how effectively clients are using techniques (e.g., thoughts during a cognitive restructuring exercise or the speed of taking 10 breaths in diaphragmatic breathing), severity of clients' symptoms at pre- and post-therapy and at frequent intervals throughout therapy, and clients' compliance with completing therapy sessions and practicing techniques. Date stamping can provide information regarding the timing of the completion of modules in order to allow a better understanding of the frequency of web-based sessions that works best for clients.

Multimedia websites have been used with various psychological difficulties including body image disturbance (Winzelberg et al., 2000), panic disorder (Klein and Richards, 2001; Richards and Alvarenga, 2002), phobias (Kenwright, Marks, Gega, and Mataix-Cols, 2004), post traumatic stress disorder (Litz, Engel, Bryant, and Papa, 2007), pediatric encopresis (Ritterband et al., 2003), and social phobia (Andersson et al., 2006). Other advantages of multimodal websites include provision of services to individuals with minimal access to mental health services, the convenience of conducting therapy at home, and access to therapy components 24 hours a day.

Limitations

Multimodal website-based therapies usually have minimal, if any, contact with therapists and clients must complete most activities on their own; therefore, this form of therapy is often more of a self-help therapy than the other forms of Internet treatment. This brings with it additional challenges, including difficulty with clients remaining motivated and continuing to work through the program. Further, some of the components involved in these systems, such as video or audio files may require high speed Internet to download or specific software to play.

Virtual Reality Therapy

In virtual reality therapy (VRT) participants interact with a computer-generated 3-dimensional virtual world. Various types of virtual reality technology have been used. The most basic type consists of head-mounted displays with display screens for each eye and a head-tracking device that provides head orientation to a computer, which creates visual images on the display consistent with the direction clients are looking in the virtual environment. These images are projected onto the screens in the headset that clients are wearing and change to match changes in clients' direction of viewing.

The headset may also have earphones, which play audio cues consistent with the virtual environment. Some VRT technology also includes gloves equipped with position sensors, which allow clients to interact with stimuli in the virtual environment. In more advanced systems, clients stand in a booth surrounded by screens on which the computer-generated images are projected.

Clients do not wear any type of headgear other than shutter glasses, which have sensors that provide information regarding clients' head orientation. Clients may walk naturally and freely through the booth instead of remaining in one place as is typical of a head mounted display. This type of system permits greater immersion in the virtual environment (Krijn et al., 2004a).

VRT technology is most frequently used in exposure therapy with clients with anxiety disorders. In exposure therapy, clients expose themselves to situations that evoke anxiety or distress in order to provide the opportunity to habituate to the experience and to decrease avoidance of feared stimuli (Rothbaum, Hodges, and Kooper, 1997).

Strengths

VRT provides therapists with the ability to conduct exposure exercises that would present logistical difficulties if conducted in vivo including exposure to flying, combat situations, heights, public speaking situations, bridges, and animal phobias. Further, it permits therapists to tailor the virtual environment and stimuli within the environment to closely match clients' feared stimuli and to fully control the intensity of stimuli in order to proceed through exposure to feared stimuli in a graded manner.

Virtual reality exposure (VRE) has been used for many disorders including PTSD, specific phobias, social phobia, panic disorder, and agoraphobia (see meta-analyses (see meta-analyses by Parsons and Rizzo, 2008; Powers and Emmelkamp, 2008) and has found to be equally efficacious to in-vivo exposure in specific phobia (Emmelkamp, Bruynzeel, Drost, and Van Der Mast, 2001; Emmelkamp et al., 2002; Rothbaum et al., 1995).

Limitations

It is essential for clients to feel immersed in the virtual environment in order for VRE to be effective. To date, little research has examined what client factors may influence their experiences of immersion in this environment; however, the quality of a system may be an important variable (Krijn, Emmelkamp, Olafsson, and Biemond, 2004b; Regenbrecht, Schubert, and Friedmann, 1998). One study showed an association between attrition and diminished immersion in the virtual environment (Krijn et al., 2004a). Further, most practitioners will not have access to a virtual reality system; therefore, this type of therapy can only be conducted by a select number of therapists.

Hand-Held Portable Devices, CD Rom and Desktop Computer

Hand-Held Portable Devices

Palmtop computers were some of the earliest uses of technology in therapy. Palmtops were small computers that fit in the palm of one's hand and looked like small laptops, with a flip open screen and keyboard. The computers were programmed with therapeutic software programs that often consisted of several modules that provided instructions for the use of techniques as well as ecological momentary assessment (EMA) modules (Stone and Shiffman, 1994) consisting of pre-programmed alarms that would ask clients to rate their severity of symptoms and help clients attempt to identify cues that exacerbated their symptoms.

The computer recorded clients' symptom levels as well as data regarding what modules the client used and when. This data could be downloaded to a desktop computer by a therapist. These computers have often used in an adjunctive form to face-to-face therapy. More recently, with the development of smaller and more compact computers, this technology has become more accessible to clients. Many individuals use a personal digital assistant (PDA) in their personal lives and more than half of individuals in the U.S. have used wireless devices (most of which have used cell phones (Eadie, 2001). Therapy software applications are currently being created for PDAs and cell phones.

In one novel use of portable devices in therapy, Morris and colleagues (2010) used cell phones equipped with visual displays for cognitive-behavioral therapy in individuals with high levels of stress. Participants were provided with mobile phones with a therapy application loaded on them. Participants were prompted at various points throughout the day to report on their mood using various scales. Once participants reported on their mood, they could use therapy modules on the phone that consisted of breathing, physical relaxation and cognitive reappraisal techniques. Five case studies illustrate the use of this mobile therapy to cope with stress and increase awareness of their moods.

Strengths

Therapy using handheld portable devices has most commonly been used in the treatment of anxiety disorders (e.g., Baer and Surman, 1985; Gruber, Moran, Roth, and Taylor, 2001; Newman, 1999; Newman et al., 1999; Newman, Kenardy, Herman, and Taylor, 1996; Newman et al., 1997b; Przeworski and Newman, 2004, 2006), but has also been used for weight loss (Agras, Taylor, Feldman, Losch, and Burnett, 1990), bulimia (Norton, Wonderlich, Myers, Mitchell, and Crosby, 2003), and drinking-related problems (Weitzel, Bernhardt, Usdan, Mays, and Glanz, 2007). The portability of hand held computer therapy is an obvious advantage. Clients may use the computer to guide them through the use of techniques in the environments in which symptoms emerge. The use of pre-programmed alarms on hand-held devices may also improve client compliance with homework assignments to practice therapy techniques. Further, EMA may provide the most accurate assessment of client symptom severity and improve client symptom monitoring in order to identify temporal patterns of symptoms as well as symptom cues. This type of treatment is becoming increasingly more accessible and cost-effective for clients due to the large number of individuals who use PDAs on a daily basis.

Limitations

Therapy using hand-held devices has been used primarily as an adjunct to face-to-face therapy; therefore, it is unclear how efficacious this therapy would be as a stand-alone therapy. Further, the small screen that is used on these devices means that therapies must rely primarily on concise text.

CD Rom and Desktop Computers

Desktop computers have been used for computer therapy in various ways, including therapy programs on CD rom and software installed on the desktop computer. They have been used for the treatment of anxiety, depression, bulimia, and weight loss (Bara Carril et al., 2004; Carr et al., 1988; Gega, Norman, and Marks, 2007; Kenwright, Liness, and Marks, 2001; Marks et al., 2003; Murray et al., 2007; Selmi, Klein, Greist, and Harris, 1982; Whitfield, Hinshelwood, Pashely, Campsie, and Williams, 2006; Yates, 1996) and were among some of the earliest uses of technology in psychotherapy. CDs and desktop computers have often been used as a method of guiding clients through cognitive-behavioral techniques such as relaxation, cognitive restructuring, and self-monitoring.

However, in one particularly novel use of computers in therapy, Ahmed and colleagues (Ahmed, 2002; Ahmed, Bayog, and Boisvert, 1997; Ahmed and Boisvert, 2006) have used computers as a means of visually representing the therapy dialogue for individuals with cognitive and attentional difficulties who may have difficulty attending to therapy fully. During this type of computer-assisted therapy, the therapist types the therapy dialogue verbatim or highlights the main themes and issues of the conversation. Clients may then review dialogues on computer screens if they lose focus to review the last statements that were made.

This form of therapy has been used with individuals with schizophrenia who experience cognitive and auditory processing deficits secondary to intrusive hallucinations and delusions (Berman et al., 1997; Blackwood et al., 1987; Braff, 1993; Catts et al., 1995; Corrigan and Storzbach, 1993; Harris, Ayers, and Leek, 1985; Morice and Delahunty, 1996; Perry and Braff, 1994; Strauss, 1993). These cognitive deficits may render it difficult for clients with schizophrenia to understand and participate in therapy. Visually presenting therapeutic interactions may provide clients with cognitive deficits an avenue to increase their abilities to attend to therapy.

When clients become delusional or disorganized, they are directed to a computer screen to review the last statement of the therapeutic dialogue (Ahmed and Boisvert, 2006). One study examined this technique in three inpatients with schizophrenia in a multiple baseline design and found a significant reduction in the frequency of delusions relative to traditional therapy (Ahmed et al., 1997). This form of computer-assisted therapy has also been used to assist clients in identifying realistic therapy goals and with adolescents with oppositional traits (Ahmed, 2002).

STRENGTHS

Because these programs do not rely on transmission via the Internet, the software may include many visual elements. Further, these programs are available to individuals who do not have high speed Internet connections but who have a computer with a CD drive or access to such a computer. Because they rely on less advanced technology, they may be more accessible to most individuals than other forms of technology-assisted therapy.

LIMITATIONS

One obvious limitation of this technology is that it is less portable than other forms of technology-assisted therapy and therefore may not permit a client to use the technology in situations in which symptoms arise. Another limitation is that many desktop computers have multiple users, for example if there is one desktop computer within a home, an entire family may use this computer. This can lead to difficulties in maintaining confidentiality if clients are recording confidential information on the computer.

Conclusion

Technology-assisted therapy is an efficacious and cost-effective alternative to face-to-face therapy with many benefits, including improving accessibility to rural individuals with limited access to such services, portability, and reducing stigma that may be associated with face-to-face services. The use of technology in therapy can also improve client self-monitoring, practice of therapy skills, and use of therapy skills in the specific situations in which symptoms arise and may therefore improve the efficacy of existing therapies. Despite some psychologists' concerns regarding technology-based therapy removing important aspects of therapy, such as the therapeutic alliance and therapist ability to see and interpret client body language and subtle cues, research has suggested otherwise (Cook and Doyle, 2002; Parks and Roberts, 1998; Schmidt, 2003; Walther and Burgoon, 1992). Newer, lighter, and more portable technologies are being developed every day and with this new technology will come even more novel and creative uses of technology in therapy.

References

Agras, W. S., Taylor, C. B., Feldman, D. E., Losch, M., and Burnett, K. F. (1990). Developing computer-assisted therapy for the treatment of obesity. . *Behavior Therapy, 21*(1), 99-109. doi: 10.1016/S0005-7894(05)80191-1.

Ahmed, M. (2002). Computer-facilitated dialogue with patients who have schizophrenia. *Psychiatric Services, 53*(1), 99-100. doi: 10.1176/appi.ps.53.1.99-a.

Ahmed, M., Bayog, F., and Boisvert, C. M. (1997). Computer-facilitated therapy for inpatients with schizophrenia. *Psychiatric Services, 48*(10), 1334-1335.

Ahmed, M., and Boisvert, C. M. (2006). Using computers as visual aids to enhance communication in therapy. *Computers in Human Behavior, 22*(5), 847-855. doi: 10.1016/j.chb.2004.03.008.

Andersson, G., Carlbring, P., Holmström, A., Sparthan, E., Furmark, T., Nilsson-Ihrfelt, E.,. Ekselius, L. (2006). Internet-based self-help with therapist feedback and in vivo group exposure for social phobia: A randomized controlled trial. [JOUR]. *Journal of Consulting and Clinical Psychology, 74*(4), 677-686. doi: 10.1037/0022-006X.74.4.677.

Baer, L., and Surman, O. S. (1985). Microcomputer-assisted relaxation. *Perceptual and Motor Skills, 61*(2), 499-502.

Bakke, B., Mitchell, J., Wonderlich, S., and Erickson, R. (2001). Administering cognitive-behavioral therapy for bulimia nervosa via telemedicine in rural settings. *International Journal of Eating Disorders, 30*(4), 454-457. doi: 10.1002/eat.1107.

Bara Carril, N., Williams, C. J., Pombo Carril, M. G., Reid, Y., Murray, K., Aubin, S., Schmidt, U. (2004). A preliminary investigation into the feasibility and efficacy of a CD-ROM-based cognitive-behavioral self-help intervention for bulimia nervosa. *International Journal of Eating Disorders, 35*(4), 538-548. doi: 10.1002/eat.10267.

Barlow, D. H., Gorman, J. M., Shear, M. K., and Woods, S. W. (2000). Cognitive-behavioral therapy, imipramine, or their combination for panic disorder: A randomized controlled trial. *Journal of the American Medical Association, 283*(19), 2529-2536. doi: 10.1001/jama.283.19.2529.

Berman, I., Viegner, B., Merson, A., Allan, E., Pappas, D., and Green, A. I. (1997). Differential relationships between positive and negative symptoms and neuropsychological deficits in schizophrenia. *Schizophrenia Research, 25*(1), 1-10. doi: 10.1016/s0920-9964(96)00098-9.

Blackwood, D. H., Whalley, L. J., Christie, J. E., Blackburn, I. M., St Clair, D. M., and McInnes, A. (1987). Changes in auditory P3 event-related potential in schizophrenia and depression. *British Journal of Psychiatry, 150*(2), 154-160. doi: 10.1192/bjp.150.2.154.

Bouchard, S. p., Paquin, B., Payeur, R., Allard, M., Rivard, V., Fournier, T., . . . Lapierre, J. (2004). Delivering Cognitive-Behavior Therapy for Panic Disorder with Agoraphobia in Videoconference. *Telemedicine Journal and e-Health. Special Issue: Telemedicine in Canada, 10*(1), 13-24. doi: 10.1089/153056204773644535.

Braff, D. L. (1993). Information processing and attention dysfunctions in schizophrenia. [Journal; Peer Reviewed Journal; Original Journal Article]. *Schizophrenia Bulletin, 19*(2), 233-259.

Buglione, S. A., DeVito, A. J., and Mulloy, J. M. (1990). Traditional group therapy and computer-administered treatment for test anxiety. *Anxiety Research, 3*(1), 33-39.

Carlbring, P., Ekselius, L., and Andersson, G. (2003). Treatment of panic disorder via the Internet: A randomized trial of CBT vs. applied relaxation. *Journal of Behavior Therapy and Experimental Psychiatry, 34*(2), 129-140. doi: 10.1016/S0005-7916(03)00026-0.

Carlbring, P., Westling, B. E., Ljungstrand, P., and Andersson, G. (2001). Treatment of panic disorder via the Internet: A randomized trial of a self-help program. *Behavior Therapy, 32*(4), 751-764. doi: 10.1016/S0005-7894(01)80019-8.

Carr, A. C., Ghosh, A., and Marks, I. M. (1988). Computer-supervised exposure treatment for phobias. *Canadian Journal of Psychiatry, 33*(2), 112-117.

Castelnuovo, G., Gaggioli, A., Mantovani, F., and Riva, G. (2003). From psychotherapy to e-therapy: The integration of traditional techniques and new communication tools in clinical settings. *CyberPsychology and Behavior, 6*(4), 375-382. doi: 10.1089/1094931 03322278754.

Catts, S. V., Shelley, A. M., Ward, P. B., Liebert, B., McConaghy, N., Andrews, S., and Michie, P. T. (1995). Brain potential evidence for an auditory sensory memory deficit in schizophrenia. *American Journal of Psychiatry, 152*(2), 213-219.

Chang, T., and Chang, R. (2004). Counseling and the Internet: Asian American and Asian international college students' attitudes toward seeking online professional psychological help. [Journal; Peer Reviewed Journal; Original Journal Article]. *Journal of College Counseling, 7*(2), 140-149.

Cheeseman Day, J., Janus, A., and Davis, J. (2005). Computer and internet use in the United States, 2003 [electronic resource]. In A. Janus, J. Davis and U. S. C. Bureau (Eds.), *Current population reports. Series P-23, Special studies; no. 208.* Washington, D.C.: U.S. Dept. of Commerce, Economics and Statistics Administration, U.S. Census Bureau.

Cook, J. E., and Doyle, C. (2002). Working alliance in online therapy as compared to face-to-face therapy: Preliminary results. *CyberPsychology and Behavior, 5*(2), 95-105. doi: 10.1089/109493102753770480.

Corrigan, P. W., and Storzbach, D. M. (1993). The ecological validity of cognitive rehabilitation for schizophrenia. [Journal; Peer Reviewed Journal]. *Journal of Cognitive Rehabilitation, 11*(3), 14-21.

Cowain, T. (2001). Cognitive-behavioural therapy via videoconferencing to a rural area. *Australian and New Zealand Journal of Psychiatry, 35*(1), 62-64. doi: 10.1046/j.1440-1614.2001.00853.x.

Day, S. X., and Schneider, P. L. (2002). Psychotherapy using distance technology: A comparison of face-to-face, video, and audio treatment. *Journal of Counseling Psychology, 49*(4), 499-503.

Deitsch, S. E., Frueh, B. C., and Santos, A. B. (2000). Telepsychiatry for post-traumatic stress disorder. *J Telemed Telecare, 6*(3), 184-186. doi: 10.1258/1357633001935194.

Dobson, K. S. (1989). A meta-analysis of the efficacy of cognitive therapy for depression. *Journal of Consulting and Clinical Psychology, 57*(3), 414-419. doi: 10.1037/0022-006X.57.3.414.

Dolezal-Wood, S., Belar, C. D., and Snibbe, J. (1998). A comparison of computer-assisted psychotherapy and cognitive-behavioral therapy in groups. *Journal of Clinical Psychology in Medical Settings, 5*(1), 103-115. doi: 10.1023/A:1026210020906.

Eadie, A. (2001). How wireless connects in North America *The Globe and Mail (Toronto), January 9*, T2.

Emmelkamp, P. M. G., Bruynzeel, M., Drost, L., and Van Der Mast, C. A. P. G. (2001). Virtual reality treatment in acrophobia: A comparison with exposure in vivo. *CyberPsychology and Behavior, 4*(3), 335-339. doi: 10.1089/109493101300210222.

Emmelkamp, P. M. G., Krijn, M., Hulsbosch, A. M., de Vries, S., Schuemie, M. J., and van der Mast, C. (2002). Virtual reality treatment versus exposure in vivo: A comparative evaluation in acrophobia. *Behaviour Research and Therapy, 40*(5), 509-516. doi: 10.1016/S0005-7967(01)00023-7.

Erdman, H. P., Klein, M. H., and Greist, J. H. (1985). Direct patient computer interviewing. *Journal of Consulting and Clinical Psychology, 53*(6), 760-773. doi: 10.1037/0022-006X.53.6.760.

Escoffery, C., McCormick, L., and Bateman, K. (2004). Development and process evaluation of a web-based smoking cessation program for college smokers: Innovative tool for education. [Peer Reviewed Journal; Empirical Study; Followup Study; Qualitative Study; Quantitative Study; Journal Article]. *Patient Education and Counseling, 53*(2), 217-225. doi: 10.1016/s0738-3991(03)00163-0.

Fortney, J., Thill, J. C., Zhang, M., Duan, N., and Rost, K. (2001). Provider choice and utility loss due to selective contracting in rural and urban areas. *Medical Care Research and Review, 58*(1), 60-75.

Frueh, B. C., Monnier, J., Yim, E., Grubaugh, A. L., Hamner, M. B., and Knapp, R. G. (2007). A randomized trial of telepsychiatry for post-traumatic stress disorder. *J Telemed Telecare, 13*(3), 142-147. doi: 10.1258/135763307780677604.

Gammon, D., Sorlie, T., Bergvik, S., and Hoifodt, T. S. (1998). Psychotherapy supervision conducted by videoconferencing: a qualitative study of users' experiences. *Journal of Telemedicine and Telecare, 4(Suppl. 1)*, 33-35. doi: 10.1258/1357633981931353.

Gega, L., Norman, I. J., and Marks, I. M. (2007). Computer-aided vs. tutor-delivered teaching of exposure therapy for phobia/panic: Randomized controlled trial with pre-registration nursing students. [Peer Reviewed]. *International Journal of Nursing Studies, 44*(3), 397-405. doi: 10.1016/j.ijnurstu.2006.02.009.

Germain, V., Marchand, A., Bouchard, S. p., Drouin, M.-S., and Guay, S. p. (2009). Effectiveness of cognitive behavioural therapy administered by videoconference for posttraumatic stress disorder. *Cognitive Behaviour Therapy, 38*(1), 42-53. doi: 10.1080/16506070802473494.

Ghosh, A., and Marks, I. M. (1987). Self-treatment of agoraphobia by exposure. *Behavior Therapy, 18*(1), 3-16. doi: 10.1016/S0005-7894(87)80047-3.

Ghosh, A., Marks, I. M., and Carr, A. C. (1988). Therapist contact and outcome of self-exposure treatment for phobias: A controlled study. *British Journal of Psychiatry, 152*, 234-238. doi: 10.1192/bjp.152.2.234.

Ghosh, G. J., McLaren, P. M., and Watson, J. P. (1997). Evaluating the alliance in videolink teletherapy. *J Telemed Telecare, 3(Suppl. 1)*, 33-35. doi: 10.1258/1357633971930283

Graham, C., Franses, A., Kenwright, M., and Marks, I. (2000). Psychotherapy by computer: A postal survey of responders to a teletext article. *Psychiatric Bulletin, 24*(9), 331-332. doi: 10.1192/pb.24.9.331.

Graham, C., Franses, A., Kenwright, M., and Marks, I. (2001). Problem severity in people using alternative therapies for anxiety difficulties. *Psychiatric Bulletin, 25*(1), 12-14. doi: 10.1192/pb.25.1.12.

Greist, J. H., Gustafson, D. H., Stauss, F. F., Rowse, G. L., Laughren, T. P., and Chiles, J. A. (1973). A computer interview for suicide-risk prediction. *The American Journal of Psychiatry, 130*(12), 1327-1332.

Gruber, K., Moran, P. J., Roth, W. T., and Taylor, C. B. (2001). Computer-assisted cognitive behavioral group therapy for social phobia. *Behavior Therapy, 32*(1), 155-165. doi: 10.1016/S0005-7894(01)80050-2.

Harman, J. S., Rollman, B. L., Hanusa, B. H., Lenze, E. J., and Shear, M. K. (2002). Physician office visits of adults for anxiety disorders in the United States, 1985-1998. *Journal of General Internal Medicine, 17*(3), 165-172. doi: 10.1046/j.1525-1497.2002.10409.x.

Harris, A., Ayers, T., and Leek, M. R. (1985). Auditory span of apprehension deficits in schizophrenia. [Journal; Peer Reviewed Journal]. *Journal of Nervous and Mental Disease, 173*(11), 650-657.

Heinlen, K. T., Welfel, E. R., Richmond, E. N., and O'Donnell, M. S. (2003). The nature, scope, and ethics of psychologists' e-therapy Web sites: What consumers find when surfing the Web. *Psychotherapy: Theory, Research, Practice, Training, 40*(1-2), 112-124. doi: 10.1037/0033-3204.40.1-2.112.

Himle, J. A., Fischer, D. J., Muroff, J. R., Van Etten, M. L., Lokers, L. M., Abelson, J. L., and Hanna, G. L. (2006). Videoconferencing-based cognitive-behavioral therapy for obsessive-compulsive disorder. *Behav Res Ther, 44*(12), 1821-1829. doi: 10.1016/j.brat.2005.12.010.

Hofmann, S. G., and Spiegal, D. A. (1999). Panic control treatment and its applications. *Journal of Psychotherapy Practice and Research, 8*(1), 3-11.

Hsiung, R. C. (2002). *e-Therapy: Case studies, guiding principles, and the clinical potential of the Internet*. New York, NY: Norton and Co.

Jedlicka, D., and Jennings, G. (2001). Marital therapy on the Internet. *Journal of Technology in Counseling, 2*(1), 15.

Jerome, L. W., DeLeon, P. H., James, L. C., Folen, R., Earles, J., and Gedney, J. J. (2000). The coming age of telecommunications in psychological research and practice. *American Psychologist, 55*(4), 407-421. doi: 10.1037//0003-066X.55.4.407.

Jerome, L. W., and Zaylor, C. (2000). Cyberspace: Creating a therapeutic environment for telehealth applications. *Professional Psychology: Research and Practice, 31*(5), 478-483. doi: 10.1037/0735-7028.31.5.478.

Kenwright, M., Liness, S., and Marks, I. (2001). Reducing demands on clinicians by offering computer-aided self-help for phobia/panic: Feasibility study. *British Journal of Psychiatry, 179*(5), 456-459. doi: 10.1192/bjp.179.5.456.

Kenwright, M., Marks, I. M., Gega, L., and Mataix-Cols, D. (2004). Computer-aided self-help for phobia/panic via Internet at home: A pilot study. *British Journal of Psychiatry, 184*(5), 448-449. doi: 10.1192/bjp.184.5.448.

Klein, B., and Richards, J. C. (2001). A brief Internet-based treatment for panic disorder. *Behavioural and Cognitive Psychotherapy, 29*(1), 113-117. doi: 10.1017/S135246 5801001138.

Klein, B., Richards, J. C., and Austin, D. W. (2006). Efficacy of Internet therapy for panic disorder. [JOUR]. *Journal of Behavior Therapy and Experimental Psychiatry, 37*(3), 213-238. doi: 10.1016/j.jbtep.2005.07.001.

Knaevelsrud, C., and Maercker, A. (2006). Does the quality of the working alliance predict treatment outcome in online psychotherapy for traumatized patients? *Journal of Medical Internet Research, 8*(4), e31. doi: 10.2196/jmir.8.4.e31.

Krijn, M., Emmelkamp, P. M. G., Biemond, R., de Wilde de Ligny, C., Schuemie, M. J., and van der Mast, C. A. P. G. (2004a). Treatment of acrophobia in virtual reality: The role of immersion and presence. *Behaviour Research and Therapy, 42*(2), 229-239. doi: 10.1016/S0005-7967(03)00139-6.

Krijn, M., Emmelkamp, P. M. G., Olafsson, R. P., and Biemond, R. (2004b). Virtual reality exposure therapy of anxiety disorders: A review. *Clinical Psychology Review, 24*(3), 259-281. doi: 10.1016/j.cpr.2004.04.001.

Kuulasmaa, A., Wahlberg, K. E., and Kuusimaki, M. L. (2004). Videoconferencing in family therapy: a review. *Journal of Telemedicine and Telecare, 10*(3), 125-129. doi: 10.1258/135763304323070742.

Lange, A., Schrieken, B., van de Ven, J. P., Bredeweg, B., Emmelkamp, P. M. G., van der Kolk, J., . . . Reuvers, A. (2000). "Interapy": The effects of a short protocolled treatment of posttraumatic stress and pathological grief through the Internet. *Behavioural and Cognitive Psychotherapy, 28*(2), 175-192.

Lange, A., van de Ven, J. P., Schrieken, B., and Emmelkamp, P. M. G. (2001). Interapy. Treatment of posttraumatic stress through the Internet: A controlled trial. *Journal of Behavior Therapy and Experimental Psychiatry, 32*(2), 73-90. doi: 10.1016/S0005-7916(01)00023-4.

Lapham, S. C., Kring, M. K., and Skipper, B. (1991). Prenatal behavioral risk screening by computer in a health maintenance organization-based prenatal care clinic. *American Journal of Obstetrics and Gynecology, 165*(3), 506-514.

Litz, B. T., Engel, C. C., Bryant, R. A., and Papa, A. (2007). A randomized, controlled proof-of-concept trial of an Internet-based, therapist-assisted self-management treatment for posttraumatic stress disorder. [Peer Reviewed]. *The American Journal of Psychiatry, 164*(11), 1676-1683. doi: 10.1176/appi.ajp.2007.06122057.

Mallen, M. J., Day, S. X., and Green, M. A. (2003). Online versus face-to-face conversation: An examination of relational and discourse variables. [Journal; Peer Reviewed Journal;

Original Journal Article]. *Psychotherapy: Theory, Research, Practice, Training, 40*(1-2), 155-163. doi: 10.1037/0033-3204.40.1-2.155.

Manchanda, M., and McLaren, P. (1998). Cognitive behaviour therapy via interactive video. *Journal of Telemedicine and Telecare, 4(Suppl. 1)*, 53-55. doi: 10.1258/13576 33981931452.

Marks, I. M., Mataix Cols, D., Kenwright, M., Cameron, R., Hirsch, S., and Gega, L. (2003). Pragmatic evaluation of computer-aided self-help for anxiety and depression. *British Journal of Psychiatry, 183*(1), 57-65. doi: 10.1192/bjp.183.1.57.

Metanoia. (2001, 2001). E-therapy history and survey. Retrieved June 1st, 2001, from http://www.metanoia.org/imhs/history.htm.

Miller, T. W., Kraus, R. F., Kaak, O., Sprang, R., and Burton, D. (2002). Telemedicine: A Child Psychiatry Case Report. *Telemedicine Journal and E-Health, 8*(1), 139-141. doi: 10.1089/15305620252933482.

Millstein, S. G., and Irwin, C. E., Jr. (1983). Acceptability of computer-acquired sexual histories in adolescent girls. *Journal of Pediatrics, 103*(5), 815-819. doi: 10.1016/S0022-3476(83)80493-4.

Mitchell, J. E., Crosby, R. D., Wonderlich, S. A., Crow, S., Lancaster, K., Simonich, H., Myers, T. C. (2008). A randomized trial comparing the efficacy of cognitive-behavioral therapy for bulimia nervosa delivered via telemedicine versus face-to-face. *Behaviour Research and Therapy, 46*(5), 581-592. doi: 10.1016/j.brat.2008.02.004.

Mora, L., Nevid, J., and Chaplin, W. (2008). Psychologist treatment recommendations for Internet-based therapeutic interventions. *Computers in Human Behavior, 24*(6), 3052-3062. doi: 10.1016/j.chb.2008.05.011.

Morice, R., and Delahunty, A. (1996). Frontal/executive impairments in schizophrenia. [Journal; Peer Reviewed Journal]. *Schizophrenia Bulletin, 22*(1), 125-137.

Morris, M. E., Kathawala, Q., Leen, T. K., Gorenstein, E. E., Guilak, F., Labhard, M., and Deleeuw, W. (2010). Mobile Therapy: Case Study Evaluations of a Cell Phone Application for Emotional Self-Awareness. *Journal of Medical Internet Research, 12*(2), e10.

Murdoch, J. W., and Connor-Greene, P. A. (2000). Enhancing therapeutic impact and therapeutic alliance through electronic mail homework assignments. *Journal of Psychotherapy Practice and Research, 9*(4), 232-237.

Murray, J. D., and Keller, P. A. (1991). Psychology and rural America: Current status and future directions. *American Psychologist, 46*(3), 220-231. doi: 10.1037/0003-066X.46.3.220.

Murray, K., Schmidt, U., Pombo-Carril, M.-G., Grover, M., Alenya, J., Treasure, J., and Williams, C. (2007). Does therapist guidance improve uptake, adherence and outcome from a CD-ROM based cognitive-behavioral intervention for the treatment of bulimia nervosa? [Article]. *Computers in Human Behavior, 23*(1), 850-859. doi: 10.1016/j.chb.2004.11.014.

Nass, C., Fogg, B. J., and Moon, Y. (1996). Can computers be teammates? *International Journal of Human-Computer Studies, 45*(6), 669-678. doi: 10.1006/ijhc.1996.0073.

Nass, C., Steuer, J., Henriksen, L., and Dryer, D. C. (1994). Machines, social attributions, and ethopoeia: Performance assessments of computers subsequent to "self-" or "other-" evaluations. [Journal; Peer Reviewed Journal]. *International Journal of Human-Computer Studies, 40*(3), 543-559. doi: 10.1006/ijhc.1994.1025.

Nelson, E.-L., Barnard, M., and Cain, S. (2006). Feasibility of telemedicine intervention for childhood depression. *Counselling and Psychotherapy Research, 6*(3), 191-195. doi: 10.1080/14733140600862303.

Newman, M. G. (1999). The clinical use of palmtop computers in the treatment of generalized anxiety disorder. *Cognitive and Behavioral Practice, 6*(3), 222-234. doi: 10.1016/S1077-7229(99)80080-7.

Newman, M. G. (2000). Recommendations for a cost-offset model of psychotherapy allocation using generalized anxiety disorder as an example. *Journal of Consulting and Clinical Psychology, 68*(4), 549-555. doi: 10.1037/0022-006X .68.4.549.

Newman, M. G., Consoli, A., and Taylor, C. B. (1997a). Computers in assessment and cognitive behavioral treatment of clinical disorders: Anxiety as a case in point. *Behavior Therapy, 28*(2), 211-235. doi: 10.1016/S0005-7894(97)80044-5.

Newman, M. G., Consoli, A. J., and Taylor, C. B. (1999). A palmtop computer program for the treatment of generalized anxiety disorder. *Behavior Modification, 23*(4, 597-619. doi: 10.1177/0145445599234005.

Newman, M. G., Kenardy, J., Herman, S., and Taylor, C. B. (1996). The use of hand-held computers as an adjunct to cognitive-behavior therapy. *Computers in Human Behavior, 12*(1), 135-143. doi: 10.1016/0747-5632(95)00024-0.

Newman, M. G., Kenardy, J., Herman, S., and Taylor, C. B. (1997b). Comparison of palmtop-computer-assisted brief cognitive-behavioral treatment to cognitive-behavioral treatment for panic disorder. *Journal of Consulting and Clinical Psychology, 65*(1), 178-183. doi: 10.1037//0022-006X.65.1.178.

Newman, M. G., Szkodny, L. E., Llera, S. J., and Przeworski, A. (2011a). A review of technology-assisted self-help and minimal contact therapies for anxiety and depression: Is human contact necessary for therapeutic efficacy? *Clinical Psychology Review, 31*(1), 89-103. doi: 10.1016/j.cpr.2010.09.008.

Newman, M. G., Szkodny, L. E., Llera, S. J., and Przeworski, A. (2011b). A review of technology-assisted self-help and minimal contact therapies for drug and alcohol abuse and smoking addiction: Is human contact necessary for therapeutic efficacy? *Clinical Psychology Review, 31*(1), 178-186. doi: 10.1016/j.cpr.2010.10.002.

Newman, M. G., Szkodny, L. E., Llera, S. J., and Przeworski, A. (2011c). A review of technology-assisted self-help and minimal contact therapies for for eating disorders and related problems: Is human contact necessary for therapeutic efficacy? *Manuscript under review*. doi: 10.1016/j.cpr.2010.10.002.

Norton, M., Wonderlich, S. A., Myers, T., Mitchell, J. E., and Crosby, R. D. (2003). The use of palmtop computers in the treatment of bulimia nervosa. [Article]. *European Eating Disorders Review, 11*(3), 231-242.

Oakes, J., Battersby, M. W., Pols, R. G., and Cromarty, P. (2008). Exposure therapy for problem gambling via Videoconferencing: a case report. *J Gambl Stud, 24*(1), 107-118. doi: 10.1007/s10899-007-9074-4.

Parks, M. R., and Roberts, L. D. (1998). "Making MOOsic": The development of personal relationships on line and a comparison to their off-line counterparts. *Journal of Social and Personal Relationships, 15*(4), 517-537. doi: 1177/0265407598154005.

Parsons, T. D., and Rizzo, A. A. (2008). Affective outcomes of virtual reality exposure therapy for anxiety and specific phobias: A meta-analysis. *Journal of Behavior Therapy and Experimental Psychiatry, 39*(3), 250-261. doi: 10.1016/j.jbtep.2007.07.007.

Paul, N. L. (1997). Telepsychiatry, the satellite system and family consultation. *Journal of Telemedicine and Telecare, 3*(suppl_1), 52-53. doi: 10.1258/1357633971930364.

Pelletier, M. H. (2003). Cognitive-behavioral therapy efficacy via videoconferencing for social (public speaking) anxiety disorder: Single case design. *Dissertation Abstracts International: Section B: The Sciences and Engineering, 63*(12-B).

Perry, W., and Braff, D. L. (1994). Information-processing deficits and thought disorder in schizophrenia. [Journal; Peer Reviewed Journal]. *American Journal of Psychiatry, 151*(3), 363-367.

Powers, M. B., and Emmelkamp, P. M. G. (2008). Virtual reality exposure therapy for anxiety disorders: A meta-analysis. *Journal of Anxiety Disorders, 22*(3), 561-569. doi: 10.1016/j.janxdis.2007.04.006.

Proudfoot, J., Goldberg, D., Mann, A., Everitt, B., Marks, I., and Gray, J. A. (2003). Computerized, interactive, multimedia cognitive-behavioural program for anxiety and depression in general practice. *Psychological Medicine, 33*(2), 217-227. doi: 10.1017/S0033291702007225.

Proudfoot, J., Ryden, C., Everitt, B., Shapiro, D. A., Goldberg, D., Mann, A., . . . Gray, J. A. (2004). Clinical efficacy of computerised cognitive-behavioural therapy for anxiety and depression in primary care: Randomised controlled trial. *British Journal of Psychiatry, 185*(1), 46-54. doi: 10.1192/bjp.185.1.46.

Przeworski, A., and Newman, M. G. (2004). Palmtop computer-assisted group therapy for social phobia. *Journal of Clinical Psychology, 60*(2), 179-188. doi: 10.1002/jclp.10246.

Przeworski, A., and Newman, M. G. (2006). Efficacy and utility of computer-assisted cognitive behavioural therapy for anxiety disorders. *Clinical Psychologist, 10*(2), 43-53. doi: 10.1080/13284200500378779.

Rees, C. S., and Gillam, D. (2001). Training in cognitive-behavioural therapy for mental health professionals: a pilot study of videoconferencing. *Journal of Telemedicine and Telecare, 7*(5), 300-303. doi: 10.1258/1357633011936561.

Rees, C. S., and Stone, S. (2005). Therapeutic Alliance in Face-to-Face Versus Videoconferenced Psychotherapy. *Professional Psychology: Research and Practice, 36*(6), 649-653. doi: 10.1037/0735-7028.36.6.649.

Regenbrecht, H. T., Schubert, T. W., and Friedmann, F. (1998). Measuring the sense of presence and its relations to fear of heights in virtual environments. *International Journal of Human-Computer Interaction, 10*(3), 233-249. doi: 10.1207/s15327590ijhc1003_2.

Rendon, M. (1998). Telepsychiatric treatment of a schoolchild. *J Telemed Telecare, 4*(3), 179-182. doi: 10.1258/1357633981932172.

Richards, J. C., and Alvarenga, M. E. (2002). Extension and replication of an Internet-based treatment program for panic disorder. *Cognitive Behaviour Therapy, 31*(1), 41-47. doi: :10.1080/16506070252823652.

Ritterband, L. M., Cox, D. J., Walker, L. S., Kovatchev, B., McKnight, L., Patel, K., Sutphen, J. (2003). An Internet intervention as adjunctive therapy for pediatric encopresis. *Journal of Consulting and Clinical Psychology, 71*(5), 910-917.

Robinson, P., and Serfaty, M. (2008). Getting better byte by byte: A pilot randomised controlled trial of email therapy for bulimia nervosa and binge eating disorder. *European Eating Disorders Review, 16*(2), 84-93. doi: 10.1002/erv.818.

Rochlen, A. B., Beretvas, S. N., and Zack, J. S. (2004). The online and face-to-face Counseling Attitudes Scales: A validation study. *Measurement and Evaluation in*

Counseling and Development. Special Issue: Use of Technology in Assessment, 37(2), 95-111.

Romer, D., Hornik, R., Stanton, B., Black, M., Li, X., Ricardo, I., and Feigelman, S. (1997). "Talking" computers: A reliable and private method to conduct interviews on sensitive topics with children. *Journal of Sex Research, 34*(1), 3-9. doi: 10.1080/00224499709551859.

Rothbaum, B. O., Hodges, L., and Kooper, R. (1997). Virtual reality exposure therapy. *Journal of Psychotherapy Practice and Research, 6*(3), 219-226.

Rothbaum, B. O., Hodges, L. F., Kooper, R., Opdyke, D., Williford, J. S., and North, M. (1995). Effectiveness of computer-generated (virtual reality) graded exposure in the treatment of acrophobia. *American Journal of Psychiatry, 152*(4), 626-628.

Schmidt, U. (2003). Getting technical. *European Eating Disorders Review, 11*(3), 147-154.

Selmi, P. M., Klein, M. H., Greist, J. H., and Harris, W. G. (1982). An investigation of computer-assisted cognitive-behavior therapy in the treatment of depression. *Behavior Research Methods and Instrumentation, 14*(2), 181-185.

Simpson, J., Doze, S., Urness, D., Hailey, D., and Jacobs, P. (2001a). Evaluation of a routine telepsychiatry service. *Journal of Telemedicine and Telecare, 7*(2), 90-98.

Simpson, J., Doze, S., Urness, D., Hailey, D., and Jacobs, P. (2001b). Telepsychiatry as a routine service - the perspective of the patient. *Journal of Telemedicine and Telecare, 7*(3), 155-160.

Simpson, S., Bell, L., Britton, P., Mitchell, D., Morrow, E., Johnston, A. L., and Brebner, J. (2006). Does Video Therapy Work? A Single Case Series of Bulimic Disorders. *European Eating Disorders Review, 14*(4), 226-241. doi: 10.1002/erv.686.

Stone, A., and Shiffman, S. (1994). Ecological momentary assessment (EMA) in behavioral medicine. *Annals of Behavioral Medicine, 16*, 199-202.

Strauss, M. E. (1993). Relations of Symptoms to Cognitive Deficits in Schizophrenia. *Schizophrenia Bulletin, 19*(2), 215-231.

Suler, J. (2004). The online disinhibition effect. *CyberPsychology and Behavior, 7*(3), 321-326. doi: 10.1089/1094931041291295.

Supple, A. J., Aquilino, W. S., and Wright, D. L. (1999). Collecting sensitive self-report data with laptop computers: Impact on the response tendencies of adolescents in a home interview. *Journal of Research on Adolescence, 9*(4), 467-488. doi: 10.1207/s15327795jra0904_5.

Tate, D. F., Jackvony, E. H., and Wing, R. R. (2003). Effects of Internet behavioral counseling on weight loss in adults at risk for type 2 diabetes: A randomized trial. *Journal of the American Medical Association, 289*(14), 1833-1836. doi: 10.1001/jama.289.14.1833.

Tate, D. F., Wing, R. R., and Winett, R. A. (2001). Using Internet-based technology to deliver a behavioral weight loss program. *Journal of the American Medical Association, 285*(9), 1172-1177. doi: 10.1001/jama.285.9.1172.

Thompson, T. G., and Fox, C. E. (2001). Telemedicine Report to Congress, Health Resources and Services Administration Retrieved December 5, 2010, from Office for the Advancement of Telehealth http://www.hrsa.gov/telehealth/pubs/report2001.htm.

Turner, S. M., Beidel, D., Spaulding, S., and Brown, J. (1995). The practice of behavior therapy: A national survey of cost and methods. *Behavior Therapist, 18*, 1-4.

Walther, J. B., and Burgoon, J. K. (1992). Relational communication in computer-mediated interaction. *Human Communication Research, 19*(1), 50-88. doi: 10.1111/j.1468-2958.1992.tb00295.x.

Weitzel, J. A., Bernhardt, J. M., Usdan, S., Mays, D., and Glanz, K. (2007). Using wireless handheld computers and tailored text messaging to reduce negative consequences of drinking alcohol. *Journal of Studies on Alcohol and Drugs, 68*(4), 534-537.

Wells, K. B., Manning, W. G., Duan, N., Newhouse, J. P., and Ware, J. E., Jr. (1986). Sociodemographic factors and the use of outpatient mental health services. *Med Care, 24*(1), 75-85. doi: 10.1097/00005650-198601000-00008.

Whitfield, G., Hinshelwood, R., Pashely, A., Campsie, L., and Williams, C. (2006). The Impact of a Novel Computerized CBT CD Rom (Overcoming Depression) Offered to Patients Referred to Clinical Psychology. *Behavioural and Cognitive Psychotherapy, 34*(01), 1-11. doi: doi:10.1017/S135246580500250X.

Williams, M. E., Remmes, W. D., and Thompson, B. G. (1996). Nine reasons why healthcare delivery using advanced communications technology should be reimbursed. *J Am Geriatr Soc, 44*(12), 1472-1475.

Winzelberg, A. J., Eppstein, D., Eldredge, K. L., Wilfley, D., Dasmahapatra, R., Dev, P., and Taylor, C. B. (2000). Effectiveness of an Internet-based program for reducing risk factors for eating disorders. *Journal of Consulting and Clinical Psychology, 68*(2), 346-350. doi: 10.1037/0022-006X.68.2.346.

Wray, B. T., and Rees, C. S. (2003). *Is there a role for videoconferencing in cognitive-behavioural therapy?* Paper presented at the 11th Australian Association for Cognitive and Behaviour Therapy State Conference, Perth, Western Australia, Australia.

Wright, J. H., and Wright, A. S. (1997). Computer-assisted psychotherapy. *Journal of Psychotherapy Practice and Research, 6*(4), 315-329.

Yager, J. (2001). E-mail as a therapeutic adjunct in the outpatient treatment of anorexia nervosa: Illustrative case material and discussion of the issues. *International Journal of Eating Disorders, 29*(2), 125-138. doi: 10.1002/1098-108X(200103)29:2<125::AID-EAT1002>3.0.CO;2-S.

Yates, F. (1996). Developing therapeutic computer programs with particular reference to a program to teach coping strategies to problem drinkers. [Article]. *Journal of Mental Health, 5*(1), 57-64. doi: 10.1080/09638239650037199.

In: Handbook of Technology in Psychology … ISBN: 978-1-62100-004-4
Editor: Luciano L'Abate and David A. Kasier

Chapter 3

Ecological, Environmental and Professional Issues

T. Mark Harwood[1] and Daniel Pratt
[1]Eureka, California, US

In 1597, Sir Francis Bacon stated, "Knowledge is power"—of course, this is no less true today than it was when he made his pronouncement centuries ago. By logical extension, we may conceptualize technology in much the same manner—technology has the power to educate/transmit knowledge and store knowledge or information with greater efficiency than ever before. The applications for technology are virtually limitless and perhaps only limited by our imagination. For example, technology is increasingly being used in myriad military applications with greater emphasis on combat training, medical/psychiatric diagnosis, and treatment than ever before. Additionally, the employment of technology in general medical and various mental health fields is increasing by leaps and bounds. Moreover, technology is finding great acceptance in education and professional training programs. The foregoing small sample of technological applications increases efficiency, saves time and money, and increases the likelihood that positive or desired outcomes will be achieved. In this chapter, we will focus on technology as it applies to the field of mental health.

One sophisticated application of technology specifically crafted for the mental health field embodies the desired ethical considerations and, most importantly, the psychologists' version of the Hippocratic Oath; to wit: *Do no harm*; therefore, it bears brief mention here. More specifically, InnerLife (Beutler, Williams, and Norcross, 2008; www.innerlife.com) is an empirically supported, comprehensive, web-based application of cutting-edge technology. It is the result of decades of on-going systematic research employing RCT ATI design and methodology. The program is in a constant state of improvement/refinement as more research support emerges and more patients are added to the ever-growing data-base. Greater precision in decision-making and treatment planning is the result of refinement.

InnerLife is primarily a patient-driven program and as such, it contains self-help elements (Harwood and L'Abate, 2010). Systematic Treatment Selection (STS; Beutler and Clarkin, 1990; Beutler, Clarkin, and Bongar, 2000; Beutler et al., 2003) provides the theoretical underpinnings of Innerlife and has the same capability; however, STS is designed for use by the treating or referring clinician rather than the patient. The InnerLife program walks the patient through a comprehensive self-report assessment, informs the patient (or clinician) if treatment is indicated, identifies patient problem areas, aids in diagnosis, tracks treatment progress, displays predicted course of treatment on a variety of dimensions based upon patients with similar diagnostic, demographic, and problem profiles, provides a variety of self-help resources that are specifically formulated for the patient's presenting problem(s), identifies empirically supported treatment manuals that are appropriate for the particular patient, and identifies treatments to avoid or treatments that are contraindicated for the patient. Projected change trajectories are provided to the patient and/or therapist, and each can track the patient's progress through the program. Research suggests that interventions like this that are tailored to the individual seeking help can improve therapeutic outcomes (Beutler and Harwood, 2000; Beutler et al., 2000; Harwood and Beutler, 2008; Harwood and L'Abate, 2010; Harwood and Williams, 2003; Pulier et al., 2007). Also, Percivic, Lambert, and Kordy (2004) reviewed information technology programs and explained how they can enhance psychotherapy outcomes in everyday practice by providing clinicians with immediate feedback as to the patient treatment response through continuous monitoring.

The foregoing merely represents a sample of InnerLife's technological power and clinical application—the parameters of this chapter limit us to this brief description—the interested reader is directed to Harwood, Beutler, and Groth-Marnat (2011) for a detailed description of the technology involved, patient and clinician applications, and the strong empirical underpinnings of InnerLife. Other related, albeit older and different methodologically, computer-monitoring systems exist for clinical quality assurance. See Beutler (2001) for a comparison of these systems,

As technology develops, so grows its incorporation in the mental health disciplines. A goal of this chapter is to review relevant literature and raise important issues in order to discuss the implications of technology's inclusion within the practice and research of psychology, psychiatry, and neurology. This critical review attempts to concisely examine technology's current contribution and potential future impact on the patients, practitioners, researchers, and students within these disciplines. It is important that we approach emerging and complex technologies with a grain of salt, and evaluate the cost and benefit of every application carefully based on the particular context of intended use and the degree of impairment or distress experienced by the patient. Ultimately, we want to ensure a fit between technology and each particular environment, patient, and presenting problem, so that it facilitates research, training, and clinical application.

Ecological and Ethical Considerations

How will technology be received by the mental health profession? What impact can we anticipate from the use of technology in mental health treatment, training, and our interface with medical professionals? Will technology ever become refined enough to provide an acceptable substitute for or replacement for face-to-face talk therapy in all treatment aspects?

With respect to "distance therapy", the field has plenty of "technology eschewers" who fear or reject the contribution that technology provides for the mental health disciplines. Then there are the "technological doubters" who recognize that technology already has important application in a variety of areas (e.g., phobias, training) but have difficulty accepting the potential of technology promised by "technological dogmatists" or "visionaries". For most psychotherapists who have remained involved in face-to-face treatment, the idea that technology can ever re-create the necessary feelings of intimacy, trust, and basic human connection that develop when patient and therapist interact and occupy the same relative space is difficult to accept.

For these technological eschewers or doubters, the sophistication of technology is appreciated but deemed inadequate in the absence of an environment where the subtleties of human interaction are likely to be missed or misinterpreted. For example, can technology ever capture the information that seasoned therapists "sense" when conducting psychosocial treatment in proximity that fosters the ephemeral connection between patient and therapist—a connection that helps produce change through mechanisms such as corrective emotional experiences (CEEs) and allows one to attend to process in a fine-grained manner? How well can distance-based technology convey important but often subtle patient cues such as a blush, quavering voice, pupil dilation, perspiration, body movements, and eye gaze with therapist or significant other that help inform clinical judgment and empirically-supported decision making? Are the prognostic and decision-making qualities of patient coping style, resistance or reactance, functional impairment, and subjective distress altered by technological application of distance-psychotherapy and specific interventions?

We believe these are important questions, worthy of empirical investigation. We hope that therapists/researchers will maintain a healthy skepticism that characterizes scientific inquiry and reserve judgment on each application of technology until a large body of literature has been developed and methodologically rigorous meta-analyses have been performed. Using the foregoing as a point of departure, there is little doubt that the technological application of psychotherapy is appropriate in a variety of venues by shear necessity. As an example, psychotherapists are in low supply in many prisons. Additionally, the potential for danger that a psychotherapist faces from a health and physical injury perspective would support the use of technology as a primary, or adjunct to, treatment in many prisons. Another area where technology-driven distance-therapy may be especially useful is rural locations where the population may not be great enough to support a practice or where many professionals may not wish to live. Finally, some elderly individuals are frail and unable to travel to an office or they are able to travel but with great difficulty and inconsistency. In each of the foregoing, technology may be used at least on a part-time, supplemental basis because a careful examination of the cost-benefit ratio supports the addition of technological-based intervention over no or very little intervention.

The application of technology to psychotherapy is a growing practice. Experts in the field predict "computerized therapies, use of virtual reality, self-help resources and self-help techniques will substantially increase in the next 10 years" (Norcross, Hedges, and Prochaska, 2002; as cited in Newman, 2004, p. 144). A special issue of the *Journal of Clinical Psychology* (*JClP*) was dedicated to this topic (Newman, 2004). The technological applications covered by the issue included a review of computer-aided self-help treatments (Gega, Marks, and Mataix-Cols, 2004), a computer-administered treatment for problem drinkers (Squires and Hester, 2004), an internet self-help therapy for treating tinnitus (Andersson and Kaldo, 2004), using palmtop computers as an adjunct to group psychotherapy for generalized anxiety disorder and social phobia (Przeworski, 2004), a virtual reality (VR) treatment case study for severe burn pain (Hoffman et al., 2004), a respiratory biofeedback technique for treating panic disorder (Meuret, Wilhelm and Roth, 2004), and a review of using computers in psychotherapy (Tate and Zabinsky, 2004). This issue of the Journal of Clinical Psychology offers a step in the direction of empirical support for the use of these emerging technologies.

In an earlier volume of *JClP*, Beutler and Harwood (2004) published an article examining the application of virtual reality technology for use by fledgling therapists (i.e., trainees). I (the senior author) am reminded of a particularly disturbing incident involving a trainee who was seeing his first "real" client in his Master's program practicum—with no apparent indication of what was to transpire, the client left his first session with this trainee and promptly committed suicide by firearm in the university parking lot. The emotional fall-out of this incident resulted in a great deal of subjective distress for the fledgling student counselor who ultimately dropped out of the program. Virtual patients may be useful in the early stages of training to help foster the development of important, essential skills (i.e., an acceptable level of clinical competence). Moreover, trainees would be more likely to attain increased levels of confidence in their ability to apply empirically-supported change strategies and interventions. There are also applications for more advanced skills training. Of course, there is no way to determine if VR training would have prevented the tragic incident mentioned above; however, a more seasoned trainee might have picked up on something that the patient stated, he may have called in his supervisor for direction, or he may have noticed something about how the client presented that may have warranted a query of suicidality—a topic that fledgling therapists often shy away from.

Harwood and L'Abate (2010) also present some general ramifications that stem from the movement toward an increased use of technology in mental health interventions. They see potential for enhancing cost-effectiveness in treatment, increasing access to mental health care, and being able to offer "step-by-step progression from least to most expensive approaches" (p. 42). They also recognize potential challenges, like insurance coverage and incorporation in professional training, but see these challenges as surmountable, as discussed in different ways by most chapters within this volume.

Ethical/Professional Issues

It is questionable whether or not ethical and legal implications are being considered in a systematic, appropriately rigorous, and comprehensive manner before the actual application of emerging technologies. Barnett and Scheetz (2003) expressed that technological "advances have moved forward much more rapidly than people's understanding of the technologies ethical implication" (p. 87). Perhaps this is an inevitable result of profit motive and career advancement. Further, "Rather than eschew technological advances altogether, ...psychotherapists must as a profession and as individual practitioners work to advance the understanding of the ethical and legal issues associated with these technological advances and create new standards and mechanisms for addressing them" (p. 87). In the mean time, the APA (2002) states that, ''in those emerging areas in which generally recognized standards for preparatory training do not yet exist, psychologists nevertheless take reasonable steps to ensure the competence of their work and to protect clients, students, research participants, and others from harm.'' Today, even technologies many have grown comfortable using regularly, like the telephone and Internet, present unique ethical dilemmas when applied to the clinical setting.

TELETHERAPY AND TELEMEDICINE

Teletherapy and telemedicine are the application of therapy and medical services from a distal location and primarily via telephone. Nickelson (1998) defined telehealth as "the use of telecommunications and information technology to provide access to health assessment, intervention, consultation, supervision, education, and information across distance (p. 527)" (as cited in Barnett and Scheetz, 2003, p. 88). This includes more longstanding technologies like the telephone. In fact, VandenBos and Williams (2000) found that out of the clinicians who reported using telehealth as a means of delivering clinical services, 98% reported using the telephone compared to 2% using the internet or satellite technology. This difference in usage rates may have become more balanced in the intervening years since VandenBos and Williams conducted their survey; however, there is likely to be differential rates of use between various forms of technology due to refinement or availability. That is, internet and satellite technology/equipment is more expensive than telephones and "skyping" (essentially, voice-calls over the internet with other features including videoconferencing capability) is free but not yet well-perfected and fails to provide a smooth, fine-grained communication experience; however, improvements continue and we cannot predict how well this technology will ultimately replicate the therapist's office experience.

Working from a distance brings up unique ethical challenges. Koocher and Keith-Spiegel (2008) offer an account of several ethical dilemmas that arise in the use of teletherapy and emergent technologies in psychology and psychiatry (pp. 142-146). They ask informed questions and offer some general suggestions for those engaging with new technologies in practice or research. They discuss the fact that the American Psychiatric Association has decided not to directly address teletherapy in its ethics code (APA Ethics Committee, 1997), and that "no clear professional consensus or detailed ethical guidelines currently exist" (p. 143). Each practitioner or researcher enters the domain of distance-work at his or her own risk. The lack of clear guidelines does not prohibit the use of teletherapy or similar services, but it does put the professional at risk for encountering unforeseen ethical and/or legal dilemmas. Moreover, the use of teletherapy or distance therapy may put patients at greater risk than traditional face-to-face psychotherapy. Some of the ethical dilemmas to consider before offering therapeutic services primarily or solely via telephone include: 1) lacking nonverbal cues as data available for use in thorough assessment and diagnosis (e.g., body posture, body movement, eye contact, grooming), 2) means of responding in emergency situations such as those that occur when presented with a suicidal patient during a distance-session, 3) understanding clinician liability and risk management in working with patients at risk for harming themselves or others, 4) licensure outside the jurisdiction of where the patient resides, 5) ensuring confidentiality, and 6) having appropriate malpractice insurance for offering this type of service (Barnett and Scheetz, 2003, p. 88).

Using e-mail as means of communication or ongoing counseling services also poses ethical concerns such as: 1) the high likelihood that the written word will be misinterpreted at least some of the time, 2) the inability to guarantee confidentiality or difficulty responding effectively to emergencies, 3) absence of visual and verbal cues as data for thorough assessment and diagnosis, and 4) the patient misunderstanding the nature or proven effectiveness of the services to be rendered (Barnett and Scheetz, 2003, p. 90). An elaboration on the last point:

There also is no clear understanding of the effectiveness or appropriate use of e-mail as a therapeutic medium (Maheu and Gordon, 2000). Thus, little empirical support exists to guide clinicians in deciding for which individuals and for which difficulties e-mail-based interventions may be appropriate. E-mail's utility and effectiveness are unknown. Although accurately advertising these services as not being psychotherapy is a helpful step, the likelihood of alternative assumptions by consumers is great..." (Barnett and Scheetz, 2003, p. 90).

Liability is also important to consider in our litigious society. In terms of services rendered with Internet as a medium, "Several courts have ruled that a professional relationship does exist when an individual pays a fee to a professional for advice given" (e.g., Hunt v. Disciplinary Board, 1980). It is important to review the most recent legal precedents set where you practice and consult with legal professionals before starting to or continuing to practice professionally over the Internet.

Although interactive televideo (IATV), including "skyping", offers visual and verbal cues for more thorough assessment and diagnosis, it still poses ethical obstacles related to: 1) mandatory reporting guidelines (i.e., do the laws of the patient's or therapist's location apply?), 2) emergency response procedures, 3) licensure jurisdiction, and 4) unintended interruptions at the client's location of internet or satellite service and therein the potential breach of confidentiality (Barnett and Scheetz, 2003, p. 88). Licensing clarifications and further elaborated ethical guidelines will help the mental health field safely apply this growing medium of communication; however, for now, there is still concern for the "total lack of consumer-oriented regulation" (Koocher and Keith-Spiegel, 2008, p. 146; Koocher and Morray, 2000).

For teletherapy, a suggested guide to consider is the four Cs: contracting, competence, confidentiality, and control (Koocher, 2007). Koocher and Keith-Spiegel (2008) offer questions that should be considered regarding these four Cs:

- "What contracts or agreements for providing distance services will we make with our clients?"
- "What competencies will we need to offer services remotely?"
- "What new factors will constrain confidentiality protections?"
- "Who will control the practice of teletherapy (i.e., the regulation of practice and data access)"(p. 143)

Contrastingly, working at distance from patients may improve a clinician's assessment of the presenting problem. It "may increase objectivity and avoid personal whims and wills that may affect our judgment" (L'Abate, 2008a, 2008b; as cited in Harwood and L'Abate, 2010, p. 14). Harwood and L'Abate (2010) critically review what the research has to say about different methods for working at a distance from participants. This book is a helpful guide to begin critically appraising when to use self-help or distance interventions, and how to ensure quality of these interventions according to the needs of the patient. The authors consider the limitations of these approaches, but emphasize the need for professionals to learn how to use emerging technologies to enhance the reach of mental health. They provide many examples of successful technological applications, one example being a telephone-administered psychotherapy that significantly reduced depressive symptoms and attritions rates (Mohr, Vella, Hart, Heckman, and Simon, 2008). It will be important for clinicians to learn how to conceptualize client problems thoroughly enough to decide whether potentially beneficial and money saving interventions like distance writing, online support groups, and online therapy would be more appropriate in addressing their needs (Harwood and L'Abate, 2010). This can save a client time and money, and free up time for clinicians to intervene with more severely disturbed patients.

An important area for the application of technology in mental health and medicine is in the service of underserved populations. As mentioned previously, those who reside in rural areas often represent an underserved population where health services could be offered from a distance via tele-technologies. Nelson and Bui (2010) review the telepsychology research in rural areas and apply it to a case study in an effort to "approximate evidence-based child psychotherapy from face-to-face practice using the video-conferencing technology" (p. 490). More research is required in this area, but this article is an example of the continued research that is needed to impact underserved populations while ensuring quality service. Technology can allow the valuable services of psychology, psychiatry, and neurology to impact a wider range of populations than the urban or affluent.

Emerging Technologies

VR is likely to grow into a prominent domain of technological application in the future mental health field. It may not be long before we see VR technology available for treating multiple psychological disorders. Exposure treatment for certain anxiety disorders is an area that lends itself to the investigation of using VR as a medium of exposure. For example, VR for treating combat related PTSD is an example where technology can provide a means, with the possibility of being more engaging and effective than imaginal exposure, to reprocess a traumatic event in a safe environment that would be unreasonable to recreate otherwise (Rizzo, Rothbaum, and Graap, 2007). More research is needed in this area to ensure the benefits outweigh the costs when applying VR to mental health interventions. It is important to empirically validate and replicate studies on VR as a treatment modality. Also, as the previous sections discuss, it is important to ensure the ecological fit and appropriate ethical application of VR on a case-by-case basis.

VR can also offer unique training opportunities. As mentioned earlier, Beutler and Harwood (2004) explored the benefits of using virtual reality (VR) to train psychotherapists. They challenge that our current training models raise practical and ethical concerns. As cited in Beutler and Harwood (2004), a 1994 research review showed nonsignificant or negative correlations between supervision and patient improvement (Beutler, Machado, and Neufeldt, 1994). VR training can increase the "ecological validity" of treating difficult and dangerous patients (Beutler and Harwood, 2004, p. 318). It can also help therapists learn the ability to maintain the "needed flexibility and effectiveness" in anxiety provoking situations like threats of suicide, threats of violence, chemical intoxication, and patient requests for personal information from the therapist (p. 319).

Specifically, contemporary technology offers a means for instituting training procedures that (a) are standardized, (b) maintain a realistic and reliable portrayal of the training stimuli, (c) allow immediate feedback on skill development, (d) can provide corrective feedback for therapist arousal of defense or anxiety, (e) are safe for both patient and therapist, and (f) are cost efficient to deliver....The principle difficulties include (a) the initial cost of developing a complete VR system, which may be significant given the uncertain benefits that might be realized; and (b) virtual technologies have not yet mastered a way to model complex social interchange. (pp. 320-321)

Psychologists-in-training and trainers are ethically mandated to ensure the beneficence, competence, and non-maleficence of services rendered (American Psychological Association, 2002). As such, there is potential to offer more effective, and therefore more ethical, therapy to the clients of psychotherapists-in-training with increasing advancements in VR technology.

Digital immersive virtual environment technology (IVET) is one example where technology is being used to study human behavior that can inform our discussions around VR applications. Computer programs (i.e., agents) or human users (i.e., avatars) control the virtual humans in the IVEs. Bailenson, Blascovich, Beall, and Loomis (2003) used this technology to study the social influence virtual humans (both agents and avatars) exert on real human subjects in the IVE by measuring how the subject manages interpersonal space (i.e., proxemics) in response to the virtual human. The results imply that humans manage proxemics similarly in an IVE with a computer controlled virtual human as they would in real life. One limitation of this finding is that proxemics is considered to be a low-level response system; meaning that it is more automatic, reflexive, and unconscious (Bailenson, et al., 2003; Blascovich and Beall, 2010). High-level responses are more consciously controlled responses like those in meaningful conversations (Bailenson, et al., 2003, p. 820). The high-level response systems are more relevant to the complex and meaningful interactions that would need to occur in a VR training or therapeutic environment. Blascovich and Beall (2010) describe a four-factor model of social influence in IVEs (cf., Blascovich et al., 2002). One conclusion of the article is that a host of nonverbal characteristics must appear realistic to the participant for proper social influence processes to occur. What we might take from this preliminary research is that the complexity of nonverbal cues needed to communicate nonverbal emotional states in computer agents is one area where VR technology will need to continue improving in order for VR training of therapists to reach the potential benefits Beutler and Harwood (2004) discuss and foresee.

It is also possible that technology may not improve the training of mental health professionals, even with difficult clients. Even if technology progresses to the point where trainees can be convinced the computerized patient is somewhat "real", is human behavior predictable enough for it to be translated into a computer program? Could interaction in a virtual environment produce in trainees a false confidence in their competence to work in a real clinical environment? Would VR training offer an increase in the quality of training that warrants its high cost to create and evaluate in outcome research?

Psychotherapeutic devices are another example of emerging technologies in mental health research and practice. Examples include growth in psychiatry's attention to a much safer and less intense application of electroconvulsive therapy (ECT) for the treatment of severe refractory depression (Dukakis and Tye, 2006), and collaboration with neurosurgeons to investigate the benefits of deep brain stimulation (DBS) in certain severe psychopathologies (Carpenter, 2006; Fitzgerald, 2006; Glannon, 2006; as cited in Koocher and Keith-Spiegel, 2008, p. 141).

Ethically, Koocher and Keith-Spiegel (2008) suggest that therapists, and even psychiatrists with limited internal medicine training, should always recognize the boundaries of their competence when using new technologies, avoid the use of unsafe or unproven devices, and collaborate with qualified physicians when treating conditions with possible organic roots (p. 141). Again, we encourage readers to ask hard questions and reserve judgments until appropriate levels of empirical evidence and meta-analyses converge to support context-specific applications of emerging technologies.

Neuropsychological Applications

Even something as exciting as the advances in neuroimaging that will continue to enhance our understanding of biological bases of behavior needs to be interpreted carefully, especially when there is an attempt to integrate behavioral and neuroimaging findings (Fins, 2008).

Although technological advances in neuroimaging are occurring at a rapid rate, neuroimaging is limited to providing a picture of the intracranial realm and detecting some changes or pathology in brain morphology. Unfortunately, visible lesions or even gross abnormal morphology does not necessarily translate to detectible/measureable behavioral or cognitive changes. The reverse is also true—that is, gross behavioral or cognitive changes may occur sans detectible lesions or morphology.

As such, technology-based assessment is often helpful for the neuropsychologist because the primary objective is to determine functional status (strengths and weaknesses both ipsative and normative) on a variety of cognitive and behavioral dimensions. A number of computerized measures are available for neuropsychologists including the Multi-digit Memory Test (Niccolls and Bolter, 1991), the Wisconsin Card Sorting Test (WCST; Heaton, Chelune, Talley, Kay, Curtiss, 1993), and the MicroCog (Powell et al., 2004). Increasingly, neuropsychological assessment is becoming technology-driven.

The benefits of technology as applied to neuropsychological assessment include: 1) accurate measurement of reaction times, 2) reduced administrator error, 3) quick comprehensive scoring/interpretation of complicated test data, 4) the presentation of stimuli that could not be presented practically in any other format, 5) easy establishment of baseline functioning (as in military applications) that allow a fine-grained and comprehensive comparison of baseline functioning and functioning post-deployment, 6) easy and accurate assessment of change over time on myriad dimensions of cognitive and behavioral functioning, and 7) often participants enjoy a reduction in test administration times.

There is even growing evidence that computer games may be used to improve cognitive functioning, delay rate of decline, or reverse some lost abilities among the elderly (Basak, Boot, Voss, and Kramer, 2008). The application of this type of technology to younger patients has also been investigated (Dahlin, Nyberg, Bäckman, and Neely, 2008) with promising results.

This chapter presents only a brief and general overview of some of the technological applications available to the neuropsychologist. The interested reader is directed to Chapters 13, 14, and 16 in this volume for more detailed information on specific technological applications to neuropsychology.

Conclusion

There is still much work to be done to understand the impact of technology on the fields of psychology, psychiatry, and neurology. Overall, the benefits may outweigh the costs, but the cost-benefit analyses must occur on a case-by-case basis for each new technology. Potential benefits include: 1) cost savings to patients, practitioners, insurance companies, and society, 2) improved functional status for a larger population of individuals, 3) efficiency of test administration with respect to scoring, interpretation, and time, 4) presentation of unique stimuli, and 5) reductions in error rates for both administration and scoring.

It is difficult to calculate the potential savings in time at this relatively nascent stage in the application of technology as it relates to mental health. Likewise, the potential benefits to patients in terms of an individual's functional status and productivity is unknown but likely to be significant, especially when benefits are considered on a societal/national level. Given the positive trajectory that technology currently enjoys in both innovation and refinement, the future is promising for our patients and it portends increased applications for technology in mental health assessment and treatment.

References

American Psychological Association. (2002). Ethical principles of psychologists and code of conduct. *American Psychologist, 57,* 1060–1073.

American Psychological Association Ethics Committee. (1997). *Statement on services by telephone, teleconferencing, and Internet.* Retrieved January 18, 2011. From http://www.apa.org/ethics/education/telephone-statement.aspx.

Andersson, G., and Kaldo, V. (2004). Internet-based cognitive behavioral therapy for tinnitus. *Journal of Clinical Psychology*, *60*(2), 171-178. doi:10.1002/jclp.10243.

Bacon, F. (1597). *Religious meditations of heresies.*

Bailenson, J., Blascovich, J., Beall, A., and Loomis, J. (2003). Interpersonal distance in immersive virtual environments. *Personality and Social Psychology Bulletin, 29*, 819-833. doi:10.1177/0146167203029007002.

Barkham, M., Margison, F., Leach, C., Lucock, M., Mellor-Clark, J., Evans, C., Benson, L., Connell, J., Audin, K., and McGrach, G. (2001). Service profiling and outcome benchmarking using the CORE–OM: Toward practice-based evidence in the psychological therapies. *Journal of Consulting and Clinical Psychology*, *69*, 184–196.

Barnett, J., and Scheetz, K. (2003). Technological advances and telehealth: Ethics, law, and the practice of psychotherapy. *Psychotherapy: Theory, Research, Practice, Training*, *40*, 86-93. doi:10.1037/0033-3204.40.1-2.86.

Basak, C., Boot, W. R., Voss, M. W., and Kramer, A. F. (2008). Can training in a real-time strategy video game attenuate cognitive decline in older adults? *Psychology and Aging, 23*, 765-777.

Beutler, L. E. (2001). Comparisons among quality assurance systems: From outcomes assessment to clinical utility. *Journal of Consulting and Clinical Psychology, 69*(2), 197.

Beutler, L. E. and Clarkin, J. F. (1990) *Systematic Treatment Selection: Toward targeted therapeutic interventions.* New York, NY: Brunner/Mazel.

Beutler, L. E., Clarkin, J. F., and Bongar, B. (2000). *Guidelines for the systematic treatment of the depressed patient.* New York, NY: Oxford University Press.

Beutler, L. E., and Harwood, T. M. (2004). Virtual reality in psychotherapy training. *Journal of Clinical Psychology, 60*(3), 317-330.

Beutler, L. E., Machado, P. P. P., and Neufeldt, S. A. (1994). Therapist variables. In A. E. Bergin and S. L. Garfield (Eds.), *Handbook of psychotherapy and behavior change* (4th ed., pp. 229–269). New York, NY: Wiley.

Beutler, L. E., Moleiro, C., Malik, M., Harwood, T. M., Romanelli, R., Gallagher-Thompson, D. and Thompson, L. (2003). A comparison of the Dodo, EST, and ATI factors among co-morbid stimulant dependent, depressed patients. *Clinical Psychology and Psychotherapy, 10*, 69-85.

Beutler, L. E., Williams, O. B., and Norcross, J. N. (2008). *Innerlife.com.* A copyrighted software package for treatment planning. At www.innerlife.com.

Blascovich, J., and Beall, A. (2010). Digital Immersive Virtual Environments and instructional computing. *Educational Psychology Review, 22*(1), 57-69. doi:10.1007/s10648-010-9120-0.

Blascovich, J., Loomis, J., Beall, A., Swinth, K., Hoyt, C., and Bailenson, J. (2002). Immersive virtual environment technology as a research tool for social psychology. *Psychological Inquiry, 13,* 103–125.

Carpenter, L. L. (2006). Neurostimulation in resistant depression. *Journal of Psychoparmacology, 20,* 35-40.

Dahlin, E., Nyberg, L, Bäckman, L., and Neely, A. S. (2008). Plasticity of executive functioning in young and older adults: Immediate training gains, transfer, and long-term maintenance. *Psychology and Aging, 23*, 720-730.

Dukakis, K., and Tye, L. (2006). *Shock: The healing power of electroconvulsive therapy.* New York, NY: Avery/Penguin.

Fins, J. (2008). Neuroethics and neuroimaging: Moving toward transparency. *American Journal of Bioethics, 8*(9), 46-52. doi:10.1080/15265160802334490.

Fitzgerald, P. B. (2006). A review of developments in brain stimulation and the treatment of psychiatric disorders. *Current Psychiatry Reviews, 2,* 199-205.

Gega, L., Marks, I., and Mataix-Cols, D. (2004). Computer-aided CBT self-help for anxiety and depressive disorders: Experience of a London clinic and future directions. *Journal of Clinical Psychology, 60*(2), 147-157. doi:10.1002/jclp.10241.

Glannon, W. (2006). Neuroethics. *Bioethics, 20,* 37-52.

Harwood, T. M., Beutler, L. E., and Groth-Marnat, G. (2011). *Integrative assessment of adult personality* (3rd ed.). New York, NY: Guilford Press.

Harwood, T. M., and L'Abate, L. (2010). *Self-help in mental health: A critical review.* New York, NY: Springer.

Heaton, R. K., Chelune, G. J., Talley, J. L., Kay, G. G., and Curtiss, G. (1993). *Wisconsin Card Sorting Test Manual–Revised and Expanded (WCST)*. Odessa, FL: Psychological Assessment Resources.

Hoffman, H. G., Patterson, D. R., Magula, J., Carrougher, G. J., Zeltzer, K., Dagadakis, S., and Sharar, S. R. (2004). Water-friendly virtual reality pain control during wound care. *Journal of Clinical Psychology*, *60*(2), 189-195. doi:10.1002/jclp.10244.

Hunt V. Disciplinary Board, 381 S. 2d 52 (Ala. 1980).

Koocher, G. P. (2007). Twenty-first century ethical challenges for psychology. *American Psychologist, 62,* 375-384.

Koocher, G. P., and Keith-Spiegel, P. (2008). *Ethics in psychology and the mental health professions: Standards and cases.* (3rd ed.). New York: Oxford University Press.

Koocher, G. P. and Morray, E. (2000). Regulation of telepsychology: A survey of state attorneys general. *Professional Psychology, 31,* 503-508.

L'Abate, L. (2008a). Proposal for including distance writing in couple therapy. *Journal of Couple and Relationship Therapy, 7,* 337-362.

L'Abate, L. (2008b). Working at a distance from participants: Writing and nonverbal media. In L. L'Abate (Ed.), *Toward a science of clinical psychology: Laboratory evaluations and interventions* (pp. 355-383). New York, NY: Nova Science.

Maheu, M. M., and Gordon, B. L. (2000). Counseling and therapy on the Internet. *Professional Psychology: Research and Practice*, *31*, 484–489.

Meuret, A. E., Wilhelm, F. H., and Roth, W. T. (2004). Respiratory feedback for treating panic disorder. *Journal of Clinical Psychology*, 60(2), 197-207. doi:10.1002/jclp.10245.

Mohr, D. C., Vella, L., Hart, S., Heckman, T., and Simon, G. (2008). The effect of telephone-administered psychotherapy on symptoms of depression and attrition: A meta-analysis. *Clinical Psychology: Science and Practice, 15,* 243-253.

Nelson, E., and Bui, T. (2010). Rural telepsychology services for children and adolescents. *Journal of Clinical Psychology*, *66*(5), 490-501.

Newman, M. (2004). Technology in psychotherapy: An introduction. *Journal of Clinical Psychology*, *6*, 141-145. doi:10.1002/jclp.10240.

Niccolls, R., and Bolter, J. (1991). *Multi-Digit Memory Test (computer version).* Los Angeles, CA: Western Psychological Services.

Nickelson, D. W. (1998). Telehealth and the evolving health care system: Strategic opportunities for professional psychology. *Professional Psychology: Research and Practice, 29,* 527-535.

Norcross, J. C., Hedges, M., and Prochaska, J. O. (2002). The face of 2010: A Delphi poll on the future of psychotherapy. *Professional Psychology: Research and Practice*, *33*, 316-322.

Percevic, R., Lambert, M., and Kordy, H. (2004). Computer-supported monitoring of patient treatment response. *Journal of Clinical Psychology*, *60*(3), 285-299.

Powell, D., Kaplan, E., Whitla, D., Weintraub, S., Catlin, R., and Funkenstein, H. (2004). *MicroCog™: Assessment of Cognitive Functioning Windows® Edition* (MicroCog™ for Windows®). San Antonio, TX : Pearson.

Przeworski, A., and Newman, M. G. (2004). Palmtop computer-assisted group therapy for social phobia. *Journal of Clinical Psychology*, *60*(2), 179-188. doi:10.1002/jclp.10246.

Rizzo, A., Rothbaum, B. O., *and* Graap, K. (2007). Virtual reality applications for the treatment of combat-related PTSD. In C. R. Figley, and W. P. Nash (Eds.), *Combat stress injury: Theory, research, and management* (pp. 183-204). New York, NY: Routledge.

Squires, D. D., and Hester, R. K. (2004). Using technical innovations in clinical practice: The Drinker's Check-Up software program. *Journal of Clinical Psychology, 60*(2), 159-169. doi:10.1002/jclp.10242.

Tate, D. F., and Zabinski, M. F. (2004). Computer and Internet applications for psychological treatment: Update for clinicians. *Journal of Clinical Psychology, 60*(2), 209-220. doi:10.1002/jclp.10247.

Vandenbos, G. R., and Williams, S. (2000). The Internet versus the telephone: What is telehealth, anyway? *Professional Psychology: Research and Practice, 31*, 490–492.

Section II.
Advances in Educational Technology

In: Handbook of Technology in Psychology ...
Editor: Luciano L'Abate and David A. Kasier
ISBN: 978-1-62100-004-4

Chapter 4

WIRELESS RESPONSE SYSTEMS: 'CLICKER' USES AND BENEFITS IN AND OUT OF THE CLASSROOM

Anne M. Cleary and Edward L. DeLosh
WaveAtlas, Online Case Study, US

Wireless response systems (WRSs) allow audiences or groups to respond to questions or prompts via wireless remotes, or "clickers". In this chapter we discuss the existing research on the benefits of clickers in classroom and learning settings, and we suggest potential uses of clicker systems outside of the classroom, including corporate and clinical settings. We also discuss how clicker uses may change with the advent of new technologies for allowing remote responding from any geographic location, and for incorporating clicker capabilities into smartphones.

HISTORICAL BACKGROUND

Through a receiver unit connected to a computer, clicker responses are collected rapidly and can be presented on the screen immediately after all of responses are collected. Responses can even be displayed in real time; for example, if the question requires a yes-no answer, the two bars on the screen representing the number of yes versus no responses can continue to move up and down as people in the audience continue clicking in with their responses. Different types of software can be used with the clickers. For example, many systems include their own software for displaying questions, and many are designed to interact with Powerpoint so that questions or prompts can be displayed within Powerpoint and the responses collected with the Powerpoint slide on the screen. System software can also be used to collect and store all of the individual responses from each unique remote that was used in a given session, allowing the administrator to review individuals' responses at a later time, if desired.

Clickers are widely used in university settings (e.g., Caldwell, 2007; Carnevale, 2005; Cleary, 2008; Muir and Cleary, 2011; Reay, Pengfei and Bao, 2008) and K-12 classrooms (e.g., Nocente and Belostotski, 2009; Ozel, Yetkiner and Capraro, 2008) across the country. The use of clickers is even beginning to spread beyond classroom settings. For example, at least one company, H-ITT, explicitly markets them for boardroom use in corporate settings. At the authors' institution, clickers are used in faculty meetings to anonymously poll faculty, gather opinions, and take formal votes on issues.

There are many different ways that clickers can be used. They can be used to take a poll, to take a vote, to take attendance, or to quiz the audience on their knowledge (e.g., Beatty, 2004). In quizzing, an advantage of clicker systems is that they allow immediate feedback, both for the instructor or speaker and for the class or audience, which in turn allows the instructor or speaker to identify problem areas and continue further discussion on those areas. In this way, clicker systems facilitate dynamic interactions between the speaker and the audience. Other uses include replicating behavioral research studies (Cleary, 2008; Langley, Cleary and Kostic, 2007), involving the audience in live demonstrations of statistical analyses (Muir and Cleary, 2011), and even involving the audience in a novel behavioral research study that has never before been conducted (Muir and Cleary, 2011).

Objectives: Why Introduce Clickers?

There are many reasons why the introduction of clickers to classroom, audience, or group settings can be beneficial. Prior research on learning and memory suggests that the use of clickers in classroom or educational settings has the potential to enhance learning and retention. One pertinent phenomenon from research on learning and memory is the testing effect (e.g., Carrier and Pashler, 1992; Roediger and Karpicke, 2006a, 2006b), which is the finding that being tested on material improves long-term retention of the material relative to not being tested or restudying the material. Clickers provide a means to regularly incorporate quizzes or questions into lectures or presentations. Instructors, may for example, use clickers to ask questions periodically throughout their lectures. Alternatively, they may use clickers to administer quizzes at the end of a unit or class session. As such, clickers provide a means to capitalize on the testing effect to enhance long-term retention of the material being presented.

Another pertinent area from research on learning and memory has to do with self-generation and elaboration. Memory is generally better for material that was self-generated and elaborated on than for material that was passively observed (e.g., Hyde and Jenkins, 1973; Slamecka and Graf, 1978). Clickers, by encouraging the audience or group to actively think, generate information, and produce a response, encourage elaborative processing and self-generation (Lanz, 2010). Similar to this idea is the notion from the education literature (e.g., McKeatchie, 1999; Lanz, 2010) that active participation in the presentation (i.e., active learning) benefits retention. This may be because of elaboration and self-generation, although there may also be a benefit of clickers on attention as well (Lanz, 2010), as it is harder to daydream and let one's attention wander when being asked to actively provide a response.

Another reason why clickers may be beneficial is that they help to ensure participation in the discussion. Part of the reason for this is that clickers can provide anonymity, and by providing anonymity, people who might not otherwise participate in a discussion are more likely to respond (Lanz, 2010). This benefits those individuals, but also benefits the discussion leader and the rest of the group, who might not otherwise have been provided with the opinions of those who would not have responded had it not been an anonymous process. The anonymity provided by clicker systems may particularly benefit people who are especially shy, people who fear speaking in public, or people who are especially sensitive to criticism. It may also particularly benefit people in situations where the discussion is controversial, or where the opinions or answers provided might be considered controversial or embarrassing to admit publicly.

An additional reason to introduce clickers is that they provide immediate feedback. This allows the audience or group to get a sense of what the rest of the group is thinking. Some applications by the first author (c.f., Cleary, 2008; Muir and Cleary, 2011) were replications of existing phenomena and illustrations of statistical analyses that allowed the group to immediately see, even statistically, how the group performed or responded. For example, Muir and Cleary (2011) describe a clicker demonstration in which females are first asked to click in with their height, then males, and the two distributions are presented on the screen to be compared for discussion. In another example, Muir and Cleary illustrated correlation analysis using clickers. Students were asked to click in with their responses to the questions, "The invasion of Iraq was a mistake" and "The country made the right choice by returning George W. Bush to the presidency." They were told to rate the strength by which they either agree or disagree with each statement using a scale of 1-9 (where 1 means "strongly disagree" and 9 means "strongly agree"). What was shown to the entire group was the correlation between responses to the first question and responses to the second question. These examples illustrate how clickers can allow the group leader or instructor, as well as the rest of the group, to get a sense of what the group is thinking regarding an issue. Furthermore, the data collected from the group can then be subjected to statistical analysis, the results of which can be presented on the screen for viewing and discussion. This, in turn, introduces a dynamic that can affect the group interactions themselves.

Finally, as indicated above, clicker systems provide an instant archive of responses. The responses are saved to the computer, and this can be done in either an anonymous fashion, or in such a way that the identities of responders are preserved within the archive. Instructors and discussion leaders can take advantage of this feature in multiple ways. By referencing archival files of clicker response data, there is a record that can be used to guide future discussions or the design of future activities. With a just a show of hands or a verbal discussion, the instructor or discussion leader would have to rely on memory when later considering how the discussion went. Instructors can also take advantage of the archival feature of clicker systems to formally assess student learning. Because the system records each individual's responses to all queries within a session, and because the system can be set up to link the identity of an individual with the responses from a particular remote, it is possible to administer in-class quizzes using clickers, then subsequently grade those quizzes. This quick and efficient method of testing allows instructors to administer quizzes frequently. Frequent quizzing, in turn, encourages students to pay attention and keep up with course material, and also takes advantage of the testing effect, as described above.

CLINICAL POPULATIONS MOST EXPECTED TO BENEFIT

Because the use and investigation of clicker systems has primarily occurred within classroom and teaching settings, the most obvious clinical populations to benefit from them at this point in time are people with learning disabilities. Those with attention-deficit disorder

(ADD) are an example. People with ADD have difficulty concentrating and maintaining the focus of their attention. This may make it difficult for people with ADD to learn from a lecture or presentation, or to benefit from an ongoing group discussion. Because the use of clickers are thought to create a level of engagement beyond that in non-clicker discussions or presentations (Lanz, 2010), this could potentially help those with ADD to maintain the focus of their attention. More specifically, because clickers continually demand responses, it is more difficult for a person's mind to wander; the requirement of a response with the clicker can serve as a trigger or cue to refocus attention again on the presentation or discussion. This continual triggering or refocusing of attention may help those with ADD glean more from a presentation or discussion than would otherwise be the case.

Because clickers have also been shown to be useful as a tool for collecting data from large group samples (Bunz, 2005; Cleary, 2008; Cleary et al., 2007; Muir and Cleary, 2011), they may also be beneficial for conducting research on clinical populations. For example, researchers can use clickers as a tool for rapidly collecting survey responses (Bunz, 2005), or even for administering tests, such as cognitive tests, and examining performance (Cleary et al., 2007; Muir and Cleary, 2011). They present an efficient means of collecting data from large groups of people at a time. A particular benefit is that the data are stored in a text file that can be opened using Excel, and little to no reorganizing or recoding of the data is necessary. Thus, researchers who study clinical populations, such as those with learning disabilities or those with cognitive or memory problems, may be able to use clickers as a methodological tool for conducting research studies on these populations.

Although the primary use and investigation of clickers has been in areas of education and research, as the use of clicker systems continues to spread beyond the classroom, they may confer benefits to clinical populations outside of academic settings as well. As mentioned, some companies are marketing clickers for boardroom use in corporate settings. One possibility regarding the clinical use of clickers pertains to support groups. It is possible that, because of the anonymity that clickers allow, they could be useful for facilitating discussions among clinical populations in support group settings. Furthering this possibility is that some companies are now marketing mobile versions of clickers. These mobile versions are designed to allow people to participate in the clicker activities and discussions without actually being present in the room. This means that people can be at home or away on travel yet still participate in clicker activities and discussions; they can send their responses to be included in the graphical displays of responses, and they can participate in discussion of those responses via texting. This allows a new level of anonymity that may be particularly useful for clinical populations in support group settings. For example, people with addictions may not feel comfortable attending support group meetings for fear of seeing someone that they know at the meeting, or of having others find out about their addiction problems. If people can participate in support groups anonymously by "clicking in" to the meetings, that may encourage participation by people who would not otherwise participate in a support group.

Recent Applications: Successes and Failures

The majority of research on the use of clicker systems has taken place in educational settings—primarily in the classroom. Overall, this research suggests that the use of clickers can indeed enhance learning and retention in educational settings. For example, as mentioned above, research on the testing effect in psychology suggests that quizzing students periodically on the material should enhance learning and retention. Additionally, there may be some benefit to quizzing with clickers in particular, because of the deeper level of engagement that may be elicited by clicker use (Lanz, 2010). In general, the research supports these ideas.

Mayer et al. (2008) performed a study in which they compared a situation in which students were quizzed with 2-4 questions per lecture with clickers; this condition was compared to one in which students were given the same quiz questions but without clickers, and with a condition in which students were not quizzed at all. Students quizzed with clickers performed better on their course exams than students in the other two groups, which in turn, did not differ from one another. This empirically supports the idea that quizzing with clickers does confer an advantage to learning and retention over traditional methods.

Along similar lines, Reay, Li and Bao (2008) compared a situation in which physics students answered a set of questions in which concepts were presented in different contexts using clickers with a situation in which students did not use clickers. Students using clickers performed better on exam questions as well as on later “concept inventories”. Furthermore, the typically-shown gap between male and female performance was diminished in the clicker group. Similarly, Crossgrove and Curran (2008) found that students in an introductory biology course who used clickers during lectures performed better on exams than students in the same course taught without clickers. This advantage was even greater among non-biology majors than among biology majors. These studies add to a growing body of literature suggesting that quizzing with clickers actually does confer a benefit on learning and retention, even when compared to quizzing with other non-clicker methods.

As described above, clicker systems are especially effective for giving quizzes on a frequent basis because they allow quizzes to be adminstered quickly and efficiently. One of the co-authors has, in fact, successfully moved to a model of quizzing college students in almost all (25 of 28) class periods during the semester. Contrary to what one might assume, students come to favor this approach over traditional methods of testing, in which they might be given just two or three semester exams (Sensenig and DeLosh, 2011). On course surveys, students report that with frequent testing, they learn and remember the material better, they keep up with the material better, and they have a better insight into which concepts need more study. After having experienced frequent testing, students also recommend that courses adopt this approach over traditional approaches. Thus, frequent testing may not only confer a benefit to learning and retention, it is also very well received, at least by college students.

Other research on successes with clickers has suggested that clickers are useful for collecting data for research purposes. Bunz (2005) compared the scantron method with the clicker method for obtaining survey responses. Overall, the results suggested that the clicker method was a viable alternative to using the scantron method, with the added benefit that participants reported liking the clicker method better (i.e., it was reported to be more fun). Another benefit to collecting survey data with clickers over scantrons is that the data are available electronically immediately following the test, without the need for hand coding and scoring or for scanning documents into a scanner for coding performance.

Langley et al. (2007) replicated known empirical phenomena typically shown through experimentation (i.e., the false memory effect and the levels-of-processing effect) using clickers. Not only did they demonstrate that clickers could be used as an alternative to traditional data collection methods (i.e., individual testing on computers), but the effects sizes found were comparable to those reported in the literature using the traditional methods. This suggests that clickers are a viable alternative for data collection in experimental approaches. Thus, taken together with Banz' (2005) study, these results suggest that wireless response systems may be used successfully as a research tool for use in both survey and experimental studies for the purposes of data collection from large groups of people at once.

One potential drawback to carrying out research studies using clickers is that if everyone is in the same room at the same time, responding to the same items on the screen, the researcher can notcounterbalance across items, presentation order, and other factors, among those in the room. Langley et al. (2007) describe some ways in which this drawback can be handled, however, such as running different groups of participants such that the conditions or stimuli for the second group have been changed from those of the first group for counterbalancing purposes. A similar idea could be applied in survey research, where the order of the questions might be different for a subsequent group of people compared to the first.

Some studies have taken the idea that clickers can be used for research data collection a step further, showing that clickers can be capitalized on to actively engage students in the research process, or the processes by which researchers collect data. Cleary (2008) describes using clickers in the classroom to replicate known behavioral research findings as a class. Students respond with clickers to stimuli on the screen and the data are routed to a text file, to be opened up in Excel or a statistical software package for analysis. The statistical analyses on the data are performed, resulting in statistically significant findings that replicate prior studies, and the resulting graphs are displayed in front of the class for discussion. Graphs from prior classes' in-class replications may also be shown for comparison.

In one example, Cleary (2008) replicated the false memory effect (Roediger and McDermott, 1995) in class. Students were read a study list of words taken from Roediger and McDermott's word lists and told to remember them. They were then given a recognition test on the screen and asked to respond A for "old" (to indicate that the item had come from the study list) and B for "new" (to indicate that the item was new and had not been presented on the study list). The false memory effect was shown by the fact that students tended to "false alarm" to test items that had not been studied, but that were strongly related to the list of studied items. For example, although the word "anger" had not been presented, students were highly likely to call it "old" because it related strongly to a set of words that *was* presented. Statistical analyses confirmed significant false memory effects on the recognition test.

Muir and Cleary (2011) go on to describe many other ways in which students can be actively drawn into the processes by which behavioral scientists carry out research. For example, they describe collecting responses for the purposes of demonstrating how statistical analyses are carried out. If students were to rate, using their clickers, the extent to which they agree or disagree with the statement, "The invasion of Iraq was a mistake" followed by "The country made the right choice by returning George W. Bush to the presidency," a subsequent scatterplot can be shown on the screen before the class, illustrating the negative correlation. Moreover, the actual statistical analysis can be performed before the class. That is, the actual Pearson's r value can be computed and the resulting level of significance determined using statistical software. This actively draws students into the process of applying statistical analyses to data.

As Muir and Cleary (2011) point out, this principle can apply to many types of statistical analyses as well, including experimental situations that might be analyzed with t-test. For example, following from the types of activities described in Cleary (2008) and Muir and Cleary, if students are asked to respond "old" or "new" to recognition test items on a screen after being presented with a study list, if two within-subjects conditions are being compared (e.g., pictures vs. words at study), a paired samples t-test could be the analysis being illustrated using clickers. Muir and Cleary also describe using clickers in similar ways to illustrate distributions, mean, median and mode, as well as the concept of variability and standard deviation.

Finally, Muir and Cleary (2011) also describe using clickers to go a step beyond merely replicating known findings in the classroom, but also for testing novel hypotheses as a class for the purposes of actively involving students in novel hypothesis testing and the scientific method. Muir and Cleary describe an activity in which students were asked to submit ideas for a research study that could be carried out in class using clickers. The instructor and the TA jointly chose a winning idea, and that experiment was carried out in class. The results of the experiment were available during the same class period, presented in graph format on the screen before the class. The statistical analyses were also performed before the class. In short, given that clicker systems can be used for actual data collection purposes, producing effect sizes comparable to those produced using standard methods (Langley et al., 2007), they can also be used to actively draw students into the research process in the classroom.

There have also been some failures, or difficulties, reported with clickers. Most of these have been technical difficulties. For example, Hatch, Jensen and Moore (2005) describe their experiences with the use of clickers in a college anatomy class. They report many logistical problems with using clickers in a large lecture hall. Among the problems that they encountered were discovering that they needed to purchase a second receiver unit to accommodate the extremely large number of students in the lecture hall (150 students). They also reported a steep learning curve with regard to learning how to use the software. Additional problems involved setting up and taking down the system, which was a portable system that the instructors brought to class with them. However, on this issue, they report that having the clicker system permanently installed in the classroom would make its use easier. Currently, on many college campuses (including the authors' campus) these systems are now permanently installed in classrooms.

Hatch et al. (2005) also describe problems with students purchasing their clickers on time. As is typical on college campuses, students were required to purchase their individual clickers through the campus bookstore and to pay the registration fee for that particular course. The majority of students had not purchased their clickers by the first class period, thus that day's activities were carried out with less than 20% of the class participating. By the second week, two thirds had purchased their clickers. This highlights another potential issue with the student-owned clicker approach: Students may forget to bring their clickers to class.

With the advent of mobile clicker systems, which allow people to remotely participate in clicker activities, it may become even more important for people to purchase and carry with them their own clickers. However, it is possible that the future may see a turning toward mobile phone applications that function analogously to clickers. This may cut down on the likelihood that people will forget to carry them with them, and may make the use of clicker systems more prevalent in academic and non-academic settings alike.

Finally, one issue noted by the first author with regard to clicker activities in her own classes is that carrying out the activities does detract from class time. Thus, there may be a trade-off between using clickers and covering what needs to be covered. The key is to effectively integrate the clicker activities into a larger framework.

FUTURE APPLICATIONS

As mentioned, some clicker companies now market clickers for use in non-academic settings, such as in corporate boardroom settings. Thus, future applications of clicker systems may include more use in company boardroom discussions, more mobile integration for virtual interactive meetings in which very large groups of people are involved, and perhaps even (as suggested above), more application to clinical settings such as support group settings or educational seminars. More research needed in these areas in particular.

TECHNICAL INFORMATION

Many specific models of clicker systems exist today and are manufactured by different companies. There are two general ways in which the system can be implemented. It can be a portable system in which the instructor or discussion leader carries the receiver unit and/or the clickers in a laptop carry-bag. Alternatively, an organization (such as a University) can more permanently install the receiver unit to a computer system in a particular room and attendees can be expected to purchase their own clicker units.

Older clicker systems tended to only allow single key responses (e.g., A, B, C, D or E), but newer systems allow for text responses to be given. As mentioned above, some newer systems also allow mobile access, whereby a person can "click in" to the receiver unit without being physically present in the room. Some newer systems also have the capability to allow for the collection of the response time to press a key, as well as the response itself.

CONCLUSION

Clickers represent a promising means of actively involving everyone in a group discussion or presentation. They have had particular utility in academic settings, both college-level and K-12, and research with them suggests that their use in classroom learning sessions general benefits learning and retention of the material. Because of their benefit on learning and retention, clickers may be especially beneficial for people with learning disabilities or who have difficulty with concentration, such as those with attention deficit disorder. Future research should examine the use of clickers with these populations.

Clickers have additional benefits as well, such as with the protection of anonymity of responses. This can increase participation, especially in discussions that may be heated or controversial. Furthermore, clicker systems allow for an archived record of the responses and the overall response data, which can be useful in many settings. With the advent of mobile clicker systems, whereby people can respond remotely, without being physically present in the room holding the receiver unit, clickers may become more prevalent, both within and outside of academic settings.

REFERENCES

Carrier, M. and Pashler, H. (1992). The influence of retrieval on retention. *Memory and Cognition*, *20*, 632-642.

Beatty, I. (2004). Transforming student learning with classroom communication systems. Educause Research Bulletin, 2004, 1-13. Retrieved March 24, 2007, from http://www.educause.edu/ir/library/pdf/ERB0403.pdf

Bunz, U. (2005). Using scantron versus an audience response system for survey research: Does methodology matter when measuring computer-mediated communication competence? *Computers in Human Behavior*, *21*, 343-359.

Caldwell, J. E. (2007). Clickers in the large classroom: Current research and best-practice tips. *Life Science Education*, *6*, 9-20.

Carnevale, D. (2005). Run a class like a game show: 'Clickers' keep students involved. *The Chronicle of Higher Education*, *51*, B3.

Cleary, A. M. (2008). Using wireless response systems to replicate behavioral research findings in the classroom. *Teaching of Psychology*, *35*, 42-44.

Crossgrove, K. and Curran, K. L. (2008). Using clickers in nonmajors- and majors-level biology courses: Student opinion, learning, and long-term retention of course material. *Life Science Education*, *7*, 146-154.

Hatch, J., Jensen, M., and Moore, R. (2005). Manna from heaven or "clickers" from hell: Experiences with an electronic response system. *Journal of College Science Teaching*, *34*, 36-39.

Hyde, T. S., and Jenkins, J. J. (1973). Recall of words as a function of semantic, graphic, and syntactic orienting tasks. *Journal of Verbal Learning and Verbal Behavior*, *12*, 471-480.

Langley, M. M., Cleary, A.M., Kostic, B. (2007). On the use of wireless response systems in experimental psychology: Implications for the behavioral researcher. *Behavior Research Methods*, *39*, 816-823.

Lanz, M. E. (2010). The use of 'clickers' in the classroom: Teaching innovation or merely an amusing novelty? *Computers in Human Behavior*, *26*, 556-561.

Mayer, R. E., Stull, A., DeLeeuw, K., Almeroth, K., Bimber, B., Chun, D., Bulger, M., Campbell, J., Knight, A. and Zhang, H. (2008). Clickers in college classrooms: Fostering learning with questioning methods in large lecture classes. *Contemporary Educational Psychology*, *34*, 51-57.

McKeachie, W. J. (1999). *Teaching Tips* (10th Ed.). Boston: Houghton Mifflin.

Muir, G. M., and Cleary, A. M. (2011). Enhancing Student Engagement and Learning Using "Clicker"-based Interactive Classroom Demonstrations. In Dunn, D.S, Wilson, J. C., Freeman, J., and Stowell, J. R. (Eds.) *Best Practices for Technology-Enhanced Teaching and Learning*. Oxford University Press.

Nocente, N. and Belostotski, G. (2009). Elementary and Junior High School Use of Clickers. In T. Bastiaens et al. (Eds.), *Proceedings of World Conference on E-Learning in*

Corporate, Government, Healthcare, and Higher Education 2009 (pp. 999-1008). Chesapeake, VA: AACE.

Reay, N.W., Pengfei, L., Bao, L. (2008). Testing a new voting machine question methodology. *American Journal of Physics*, *76*, 171.

Roediger, H.L. and Karpicke, J.D. (2006a). Test-enhanced learning: Taking memory tests improves long-term retention. *Psychological Science*, *17*, 249-255.

Roediger, H. L. and Karpicke, J. D. (2006b). The power of testing memory: Basic research and implications for educational practice. *Perspectives on Psychological Science*, *1*, 181-210.

Roediger, H. L. III., and McDermott, K. B. (1995). Creating false memories: Remembering words not presented in lists. *Journal of Experimental Psychology: Learning, Memory, and Cognition*, *21*, 803-814.

Sensenig, A. E., and DeLosh, E. L. (2011). *Student's Impressions of Frequent Testing as a Learning Tool*. Manuscript submitted for publication.

Slamecka, N. J., and Graf, P. (1978). The generation effect: Delineation of a phenomenon. *Journal of Experimental Psychology: Human Learning and Memory*, *4*, 592-604.

Ozel, S., Yetkiner, Z.E., and Capraro, R. M. (2008). Technology in K-12 mathematics classrooms. *School Science and Mathematics, 108*, 80-85.

In: Handbook of Technology in Psychology …
Editor: Luciano L'Abate and David A. Kasier
ISBN: 978-1-62100-004-4

Chapter 5

ANIMATIONS AND MULTIMEDIA TUTORIALS

Robert K. Atkinson*[1], *Lijia Lin and Stacey Schink Joseph
[1]Arizona State University, Tempe, Arizona, US

In an era of swiftly evolving digital technologies, computer use has become an integral part of our everyday life and computer-based programs designed to instruct, communicate, and gather information are more available than ever before. Typically such programs are designed to include animation and multimedia. These types of treatments provide an environment with words and pictures and have the potential to promote interest and deep learning (Mayer, 2005). Teachers, researchers, and other practitioners have utilized multimedia programs and tutorials to teach and communicate a wide variety of subject matter. The effectiveness of multimedia communication and instruction depends largely upon the use of appropriate design guidelines for instruction, multimedia and user interface. A considerable amount of research over the past 15 to 20 years has focused on developing and testing guidelines in a variety of disciplines and content areas to determine how best to build words and pictures into effective learning environments. Understanding the current research and best practices will help practitioners who use multimedia tutorials to choose appropriate products as well as create them. In this chapter, we discuss the current research on animation and multimedia, provide a series of research-based design principles and highlight some potential factors that moderate learning and interaction in multimedia environments. In addition, we discuss the limitations of the current research and future directions. Finally, we present tools and applications that can be used to create animations and multimedia tutorials.

MULTIMEDIA RESEARCH

Definitions

For the purposes of this chapter, multimedia is defined as a combination of words and pictures (Mayer, 2005) used to convey a message. Words are verbal information that can either be represented as text (visual input) or as narration (auditory input). Pictures are visual representations of graphical information that can, for example, take the form of line graphs, illustrations, diagrams, photos, animations and videos. In practice, animations are typically incorporated into multimedia environments. An animation is the presentation of "a series of frames, so that each frame appears as an alternation of the previous one" (Betrancourt and Tversky, 2000, p.313). In essence, animations are moving static pictures that are shown dynamically to demonstrate temporal movements, processes and procedures.

A multimedia tutorial is an instructional program that provides a learner information about a topic or skill and guides him/her through initial acquisition of the information or skill. While tutorials are not typically used to provide extended practice and assessment, a tutorial should include both presentation and guidance (Alessi and Trollip, 2001). Tutorials can be delivered to learners in digital format to virtually any device that can be connected to the internet. This includes not only desktop computers and laptops, but smartphones and other handheld devices such as an iPod and iPad.

Theoretical Framework

Why use multimedia tutorials? The purpose of using multimedia to teach and communicate is that it can be designed to accommodate how the human mind works (Mayer, 2001) and ultimately optimize how people learn. For example, words and pictures can be presented in ways that facilitate how people use their cognitive processes to perceive information and process it with their sensory channels, working memory and prior knowledge. Cognitive theories of multimedia learning have been used as research frameworks to explore working memory capacity and determine how best to combine words and pictures into effective learning environments. Highlights from two cognitive learning theories, cognitive theory of multimedia learning and cognitive load theory, are described below.

The cognitive theory of multimedia learning (CTML; Mayer, 2005a) describes the cognitive mechanisms of multimedia learning using cognitive learning theory assumptions. CTML presupposes at least three basic assumptions about learning and human memory: a) there are two information processing channels, one for visual information and one for auditory information (Paivio, 1979, 1986), b) memory capacity is limited, as each channel can receive and process limited amounts of information at one time, and c) for learning to occur, a learner must actively process information by selecting, organizing and connecting it with prior knowledge. The supposition is that deep learning can be facilitated by effectively presenting instructional information within these constraints. For example, given the limited capacity of each sensory channel, people may learn better in certain circumstances when instruction is provided via both sensory channels as opposed to one, such as when presenting images and auditory narration rather than text alone. Presenting text-only information may cause capacity limitations in the visual channel and hinder learning. More recently, the role of affect (*cognitive-affective theory of learning with media*, cf. Moreno and Mayer, 2007) has been included in CTML as there is research to support that motivation influences learning (Boekaerts, 2007; Husman and Hilpert, 2007). With this addition, learners' motivation, affective states and meta-cognition are incorporated into the learning model to theoretically explain the interrelationships between cognitive learning theory and affect, as well as to make accommodations for other types of media (i.e., virtual reality and animated agent-based) that may provide learners with instructional information besides words and pictures.

Similar to CTML, cognitive load theory (CLT; Paas, Renkl, and Sweller, 2003; Sweller, 1994; Sweller, van Merriënboer, and Paas, 1998) is based upon a multi-component working memory model (Baddeley, 1992) and assumes humans have limited working memory capacity. It also assumes there are two partially independent subsystems for temporarily storing and processing different types of sensory input: the visuospatial sketchpad and the phonological loop. The visuospatial sketchpad is a storage system for visual and spatial information, while the phonological loop stores auditory information as well as representations of verbal information. Cognitive load theory assumes that these subsystems are controlled by long-term memory schema and that "organized information in long-term memory directs the manner in which information is processed in working memory" (Sweller, 2005a, p. 25). In theory, for learning to occur, information in working memory must be connected to prior knowledge schema and converted to long-term memory storage (Sweller, 2005a). This aspect of the theory is based upon schema theory (Chi, Glaser, and Rees, 1982) and the idea that the ultimate goal of learning is schema construction in long-term memory and subsequent schema automation to minimize working memory load.

Well-designed multimedia instruction can facilitate this process as well as the construction of a learner's missing long-term memory schema. However, since working memory has a limited capacity and information is processed in separate temporary storage systems, a learner may experience cognitive load to an extent that interferes with learning. Cognitive load is considered to be the mental load imposed on working memory that is required to process information. There are three categories of cognitive load—intrinsic load, extraneous load and germane load. Intrinsic cognitive load is caused by the inherent difficulty of a learning task and the amount of load is determined by the number of interacting elements (element interactivity). More interacting elements may require more working memory processing and lead to a higher level of intrinsic load (cf. Sweller et al., 1998). However, the level of intrinsic load can remain stable or unchanged for a designated task or instruction when learners' prior knowledge is unchanged (Schnotz and Kurschner, 2007). Extraneous cognitive load is considered to hinder the learning process as it is the mental effort that is used to process irrelevant information. It can be caused by the ineffective design of instruction; for instance, in a complex animation, the thematically important information may not be salient enough to be easily perceived by a learner. This may require the learner to intensively search the animation, an activity irrelevant and extraneous to learning, and use additional working memory capacity to identify useful information. Germane cognitive load is the mental effort necessary to promote the development of cognitive schema (Sweller et al., 1998). For instance, prompting learners to explain what they have learned to themselves is a domain-general technique designed foster deep learning and germane cognitive load (Chi, 2000; Roy and Chi, 2005). Since these sources of cognitive load are assumed to be additive, the challenge for researchers and multimedia designers is how best to create learning environments to manage cognitive load regardless of the source so a learner's working memory and cognitive capacity does not become overloaded.

A NOTE ON ANIMATION

As noted earlier, animation is typically considered a "picture" component of a multimedia environment. However, we will report specifically on animations to shed light on the current controversy in the research literature regarding its learning benefits. Research conducted to characterize the learning benefits of animation has produced inconclusive results–instruction that utilizes animation is not necessarily more effective than instruction that utilizes static pictures. We provide a brief history and report relevant research findings.

One of the affordances of digital technology is the ease with which instructional animations can be created and distributed. The goal of early animation research conducted in the 1990s was to demonstrate the effectiveness of animations. A substantial number of these studies compared animations to static graphics. For instance, as early as 1988, Baek and Layne investigated high school students' knowledge acquisition on a mathematical rule (rate times time equals distance) by using a computer-assisted learning environment in which animations or static graphics were presented. The results of the study revealed that students learning with animations outperformed their peers who learned from static graphics. Rieber (1990) conducted a similar study. Elementary school students studied a computer-based lesson which used either static or animated graphics to describe concepts of Newton's law of motion. The findings of this study were the same as Baek and Layne's study—participants in the animated graphics condition developed a better conceptual understanding than those in the static graphics condition. Other empirical research at that time also provided evidence to support the use of animations for instructional purposes (Park and Gittleman, 1992; Rieber, 1991a, 1991b; Thompson and Riding, 1990).

However, after closely examining the animations and statics graphics in the published literature, Tversky, Morrison and Betrancourt (2002) found that the two formats differed to great extent. For example, animations delivered more information than static graphics and contained opportunity for learner interactivity while static graphics did not. Therefore, the animation effect was confounded and learning effects may be attributed to the chance that the animation delivered more information or it caused learner interactivity. Based on their detailed analysis, Tversky et al. suggested that the design and development of animations should be consistent with the content and structure of how they will be perceived internally and should facilitate learners' perception.

Subsequently, many researchers (e.g., Ayres, Marcus, Chan, and Qian, 2009; Hegarty, Kriz, and Cate, 2003; Mayer, Hegarty, Mayer and Campbell, 2005; Wong et al, 2009) made attempts to control the animation condition(s) and the static graphics condition(s) to make them as equivalent as possible with respect to the amount of information delivered. However, it is of note that animations involve moving frames that cause visual-perceptual changes, whereas static images do not. Therefore, it appears, in practice, to be very challenging to control the amount of information communicated and create fundamentally equivalent animations and static pictures.

Trends in recent empirical research reveal a pattern that animations may be effective at facilitating the acquisition of procedural knowledge (Arguel and Jamet, 2009; Ayres et al., 2009; Michas and Berry, 2000; Wong et al, 2009). In two experiments (Ayres et al., 2009), students learned to tie knots and complete puzzle rings from either animations or static representations. The results showed that animations enhanced students' acquisition of procedural knowledge. van Gog, Paas, Marcus, Ayres and Sweller (2009) provided an explanation of this effect from a neuroscience perspective. When a learner views an animation demonstrating a procedure, his/her *mirror neuron system,* a hypothesized system in a human's brain that plays a major role in understanding and internalizing actions, is activated. Thus, animations that demonstrate procedures are potentially as effective as observing a live human's actions. However, it is unknown whether an animation delivering other types knowledge can activate the mirror neuron system. Empirical results from studies testing the use of animations to present knowledge and skills other than procedures tend to be diverse and inconclusive. Some studies revealed a positive impact (Catrambone and Seay, 2002; Lin and Atkinson, 2011) while other studies did not (Kim, Yoon, Whang, Tversky and Morrison, 2007; Kühl, Scheiter, Gerjets, and Gemballa, 2011). It is evident that additional research is necessary to clarify the conditions for which animations are most effective.

MULTIMEDIA DESIGN PRINCIPLES AND GUIDELINES

Theoretically-Based Principles

For several decades, educational researchers have used the theoretical frameworks mentioned to investigate the instructional conditions under which students learn best with multimedia. This research has resulted in a series of basic multimedia design principles that are intended to guide the design and development of multimedia environments and tutorials. These principles include, but are not limited to, split-attention, modality, coherence, redundancy, visual cueing, and social-agency. We will briefly describe these principles and review the guidelines specific to animation.

The *split-attention principle* (Ayres and Sweller, 2005) describes a condition in which different sources of essential instructional information, such as graphics and descriptive on-screen text, are presented apart from one another. As a result, learners, especially novices, must conduct repeated visual searches to find and mentally integrate relevant information from separate sources. This activity is a potential cause of cognitive load, as it requires the learner to use additional working memory capacity to integrate individual pieces of information during a learning task. Empirical evidence supports the recommendation to integrate information sources and/or present different sources of instructional information in close proximity to one another (Chandler and Sweller, 1996; Cierniak, Scheiter, and Gerjets, 2009; Sweller and Chandler, 1994; for a review, see Ginns, 2006). This is especially true for complex learning materials with interactivity. An integrated design may reduce cognitive load by reducing the amount of information, or number of elements interacting in working memory that are required to process the learning task. Consequently learners may have more cognitive resources for germane processing.

Another approach designed to reduce potential cognitive load is to provide multimedia instructional information to both sensory channels instead of only to one channel. This can be accomplished by presenting graphic information visually while providing audio narration instead of on-screen text. For example, a learner can view animations presented on the computer screen while listening to narrated instructional explanations via headphones. This practice demonstrates the *modality principle*, which is derived from empirical research (Kühl, Scheiter, Gerjets, and Gemballa, 2011; Mousavi, Low, and Sweller, 1995; Tindall-Ford,

Chandler, and Sweller, 1997). For instance, Mousavi, Low and Sweller (1995) created a learning task where learners were given a geometry diagram with either visual or auditory instruction. They obtained strong evidence of the modality effect from a series of six experiments where learners performed best on the learning task when presented with the diagram and the audio narration of the instructions. It is of note that certain types of audio input do not have a positive impact on learning performance. For example, background music and sounds have potential to be processed by learners as extraneous elements (seductive details that are perhaps interesting but not essential for understanding the learning activity) and interfere with learning (Moreno and Mayer, 2000).

Not only have results from research studies shown that multimedia instruction is most effective when background music and sounds are excluded, they also indicate that interesting texts, visualizations and videos can interfere with learning. By including only the information essential to the learning task, the learner is receiving "coherent" instructional messages (*coherence principle*, Mayer, 2005b). The coherence principle is somewhat counter-intuitive, as in many situations instructional designers and multimedia developers would prefer to utilize some design elements to make the learning material appear more interesting. However, from a theoretical and conceptual perspective, it is reasonable to conclude that providing irrelevant information via visual and audio formats may overload the two sensory channels that are responsible for processing information. In fact, the results from empirical research provide strong evidence to support the coherence principle. Mayer, Heiser and Lonn (2001) added interesting narrations or video clips to an animation that showed how lightning forms. The added interesting multimedia elements contained information unrelated to the learning material. The results showed that learners who viewed animation with the extraneous elements performed significantly lower on the transfer test compared to their counterparts who viewed multimedia instruction without such elements.

As indicated by the coherence principle, more is not necessarily better. That is, adding non-essential information does not necessarily lead to more effective instruction and better learning outcomes. This also holds true when onscreen text and an auditory narration of the onscreen text are combined. Even though the verbal information is essential for learning, combining similar instruction in both visual and auditory format is considered redundant. This concept is reflected in the *redundancy principle*. The redundancy principle suggests that the presentation of redundant information in visual and auditory formats may increase working memory load, interfere with information transfer to long-term memory (Sweller, 2005) and hinder learning (Mayer et al., 2001). However, a low degree of redundancy may be benign, as the additional information may guide learners' attention (Mayer and Johnson, 2008). These results are potentially conflicting and indicate the need to consider moderators (or individual differences among learners) that may impact the extent to which the redundancy principle is a useful guideline. As we discuss in the next section, learners' prior knowledge may impact the applicability of multimedia principles.

Learners' attention is integral to the learning process, because attending to and selecting relevant stimuli are part of an active process necessary for learning. In a multimedia environment that delivers instruction via complex animations with audio narration, it is likely that a learner will have to perform a visual search to find important information within an animation and link it to verbal information that he/she hears from narrations. As indicated previously, we should minimize extraneous cognitive load by reducing the visual search activity. *Visual cueing* (or *signaling*, see Mayer, 2005b) is one of the techniques for adding non-content visual devices to cue thematically important information and make it salient to learners. Research from the 1980s and 1990s shows the benefits of cueing in texts (Loman and Mayer, 1983; Lorch, 1989; Lorch and Lorch, 1996; Lorch, Lorch, and Inman, 1993). Recent research has investigated the visual cueing effect in animations and multimedia. Specifically, researchers have studied various types of cueing devices, such as arrows (Lin and Atkinson, 2011; Boucheix and Lowe, 2010), color coding (Jamet, Gavota, and Quaireau, 2008), spotlight highlighting (de Koning, Tabbers, Rikers, and Paas, 2007, 2010), zooming (Amadieu, Mariné, and Laimay, 2011) and flashing (Atkinson, Lin, and Harrison, 2009; Jeung, Chandler, and Sweller, 1997). These studies found that the cueing techniques promote effective and efficient learning. A review of empirical studies (de Koning, Tabbers, Rikers and Paas, 2009; Mayer and Moreno, 2003; Wouters, Paas and van Merriënboer, 2008) suggests that visual cueing is effective in directing learners' attention and potentially reducing extraneous cognitive load.

The *social-agency principle* (Atkinson, Mayer and Merrill, 2005; Mayer, Sobko and Mautone, 2003; Moreno, Mayer, Spires, and Lester, 2001) helps to explain the effects of human-computer interaction on learning in multimedia environments. The hypothesis supporting this principle asserts that learning can be enhanced when learners perceive their experience with the computer as social communication, involving reciprocal interaction (Moreno et al., 2001). For example, the presence of verbal and visual social cues in computer-based environments (i.e., spoken narration coupled with the narrator's image) can foster the development of a human-computer partnership by encouraging learners to experience their interaction with the computer in a similar way that they experience human-to-human interaction (Atkinson et al., 2005). In research conducted on multimedia learning, interactivity and learner control are features often used to foster social agency (Bodemer, Ploetzner, Feuerlein, and Spada, 2004; Corbalan, Kester, and van Merriënboer, 2011; Mayer and Chandler, 2001; Schwan, and Riempp, 2004). Researchers and practitioners also create and embed animated pedagogical agents in multimedia environments for the same purpose. An animated pedagogical agent is a lifelike character that "interacts" with a learner by providing verbal (e.g., feedback) and/or non-verbal social cues (e.g., gestures). The empirical research evidence for this principle includes (a) visually presenting an agent promotes learning (Atkinson, 2002; Dunsworth and Atkinson, 2007); (b) an agent with a human voice promotes better learning than an agent with a computer synthesized voice (Atkinson et al., 2005); (c) an agent providing elaborate feedback promotes better learning than an agent only providing simple feedback (see Figure 1 for an example of agent; Atkinson, Lin, Christopherson, Joseph, and Harrison, 2011; Moreno, 2004; Moreno and Mayer, 2005). However, it is of note that an animated pedagogical agent must be designed to encourage and motivate learners to make sense of the contents and materials. Otherwise, it may be a extraneous information source that distracts learners' attention.

Best Practice Guidelines

Cognitive learning theory-based practices are useful for understanding how and when to present words and pictures to create effective multimedia. While these practices can be used to create multimedia elements in tutorials, attention must also be paid to the practices used to design instruction and develop the user interface. Instructional design practices are used to create effective instruction (e.g., learning content, instructional strategies, evaluation plan). Similarly, user interface design practices are utilized to enhance users' experience by attending to how users interact with the tutorial on a computer or digital device. In the following sections, suggested guidelines for instructional design and user interface design are offered.

Instructional Design Guidelines

Alessi and Trollip (2001) outline guidelines for the sequence of events and instructional elements that should be included in a tutorial. The authors recommend that a tutorial include four components: 1) tutorial introduction, 2) learning content, 3) short practice, and 4) closing. Each component is described below.

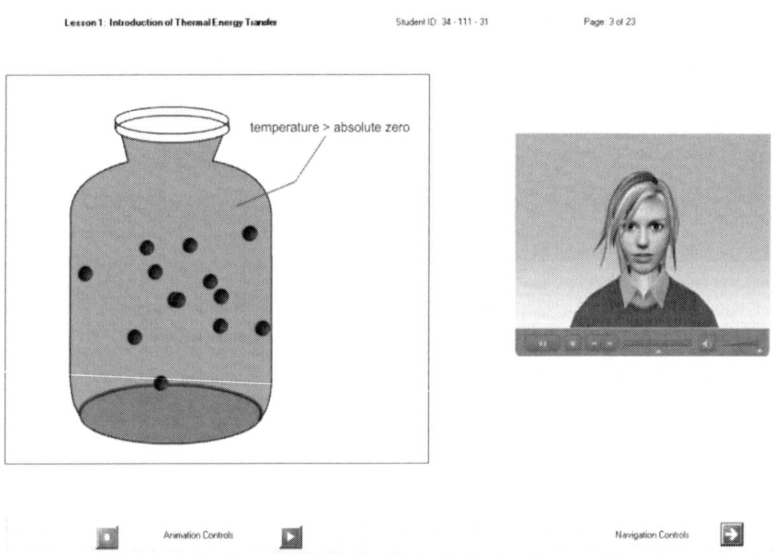

Figure 1. An animated agent providing instructional explanations.

First, a tutorial introduction should include a title page and the tutorial's learning objectives. Learning objectives are presented at the beginning of the tutorial to inform participants of exactly what they will learn or be able to do when they finish the tutorial. Generally, objectives are focused on behavioral outcomes and include information about performance conditions and criteria. The following are examples of objectives: a) "By the end of this tutorial, you will be able to accurately complete an electronic billing form;" b) "The following tutorial is designed to teach you how to connect your computer modem on an office intranet;" and c) "In the following lesson, you will learn how to create a short digital tutorial with four instructional elements." Objectives also serve to activate prior knowledge, which has been shown to facilitate the acquisition of new knowledge and skills (Duchastel and Merril, 1973).

Second, the tutorial should present the learning content. This section contains the key information that participants will use to perform the behaviors outlined in the objectives. It is common practice for instructional designers to map the instructional content to the objectives as a means to ensure adequate and relevant content is provided for each objective.

Third, the tutorial should provide practice, which gives participants an opportunity to interact with the learning content by answering questions about the content and receiving corrective feedback. Practice activities reinforce learning and provide an opportunity for the tutorial developer and the participant to assess learning progress.

Last, the tutorial should deliver a clear closing. This final component provides a summary of the learning content. It is intended to give participants an opportunity to review their progress towards the tutorial's objectives and enhance their learning retention. It also signals the end of the tutorial.

User Interface Design Guidelines

The overall effectiveness of a tutorial is contingent upon a well-designed user interface. A tutorial interface is characterized as the points where humans interact with the tutorial on a digital device, for example, a screen on a computer or mobile device and a mouse or touch pad. Typical interface design goals include improving user productivity and increasing satisfaction by reducing visual search and mental workload as well as reducing memory and motor work (Galiz, 2007, p. 131). In the next two sections, general guidelines to accomplish these goals via navigation control and screen design are outlined.

Navigation. Users advance through a tutorial using mouse clicks or screen taps. Clear, logical, and intuitive navigation is essential to the successful use of a tutorial. In fact, navigation is so fundamental to tutorial design that it is necessary to address it prior to designing almost any other feature (Tullis & Albert, 2008). To develop a logical navigation plan for the tutorial, create a site map that outlines the screen sequence and includes each of the four tutorial components. It is critical to make the site map available to users. When designing navigation elements for each page, consider the recommendations provided by Galiz (2007, p. 139):

- Provide the most important or most frequently used navigation controls consistently at the top left of a page.
- Present information in a flow that reflects how tutorial users typically read (i.e., from top-to-bottom in left-to-right fashion).
- Create a logical tab order that guides users to subsequent sequentially displayed information, which eliminate users' need to backtrack through information with the tab key.
- Display command buttons at the end of a tabbed sequence (e.g., when designing an input form, display the "submit" button at the bottom of the form).

Screen design. Although there are a host of design guidelines that can be used to make screen design decisions (see Lidwell, Holden, & Butler, 2003, for example), this section highlights four general graphic design guidelines that can be used to create visually appealing, user-friendly, text and graphics. These guidelines are contrast, alignment, repetition, and proximity, which are described in more detail below.

- Contrast refers to creating visual difference or distinctions among text or graphic elements. It can be used to direct attention, organize information, or create visual appeal. For instance, to draw attention to the title or header of a screen, use a bold color font and a different (contrasting) but complementary background color. In the body of a screen, a tutorial designer can ease the effort required by participants to find information and create distinct headings and subheadings by using different weighted fonts (e.g., bold vs. non-bold) and font sizes.

- Alignment is the organization of text or graphic element such that each edge lines up with other elements along a column, row or common center (Lidewell et al., 2003). Tutorial designers should vertically and horizontally align elements to reduce visual search requirements and to reduce the complexity of the screen (Galitz, 2007). For example, since participants typically read screen text from left-to-right, consider left aligning text elements.
- When repetition is used, text and graphic features are consistently reused to create a unified cohesive look and feel throughout the tutorial. Examples include using the same font style and font weight for each subheading on each screen, repeating the same style graphics on each page, and consistently utilizing the same area of the screen for navigation buttons.
- Finally, proximity refers to grouping related text and graphic elements together on a page. As indicated previously, keeping related pieces of information close together may reduce visual search and mental workload while enhancing participants' ability to mentally integrate and process information. However, it is common practice in print design to refer readers from the body of the text to a graphic element located in the margin of the page or on another page. This design is likely to cause readers to split their attention among discontinuous information sources. Consider instead locating text that describes a graphic element directly next to the graphic and using white space or lines to create "margins" around related elements.

CAVEAT: MODERATORS IN MULTIMEDIA LEARNING

A moderator is a variable that "affects the direction and/or strength of the relation between an independent or predictor variable and a dependent or criterion variable" (Baron and Kenny, 1986, p.1174). In multimedia learning research, individual difference variables and instructional content (the moderators) have been shown to impact the applicability of multimedia design principles. Prominent moderators include the complexity of learning tasks and individual learner characteristics, such as prior knowledge, spatial ability, and age related cognitive decline. Each variable requires an explanation of conditions that influence learning and cognition in both research and practice.

Prior knowledge is an influential moderator that has been investigated by many researchers (ChanLin, 1998, 2001; Kriz and Hegarty, 2007; Lee, Plass, and Homer, 2006). It has the potential to influence the effectiveness of the multimedia learning principles across different learner groups, such as experts and novices. Theoretically, the difference between novice and expert learners lies in their schema structure. For example, experts who already have sufficient prior knowledge or skills have high-level, well-organized schemas. Therefore, less mental effort is required in working memory to process information compared to novices. High-knowledge learners or experts may experience a paradoxical learning effect referred to as the *expertise reversal effect* (Kalyuga, 2007; Kalyuga, Ayres, Chandler, & Sweller, 2003).

For example, instruction that includes detailed domain specific information may promote learning for novices but hinder learning for experts. Novices, who have limited or absent schemas (experience, knowledge, or skills), may need detailed instructional explanations to learn a new skill. However, learners with subject matter expertise may process the details as extraneous information that increases cognitive load and hinders learning. Empirical evidence for this principle has been found in a number of studies across several domains (Kalyuga, Chandler, & Sweller, 1998, 2000; Kalyuga & Sweller, 2004). Of specific relevance to designing multimedia tutorials, Kalyuga (2008) conducted a study to investigate the instructional effectiveness of static and animated algebraic diagrams for learners with varied levels of knowledge. He found that while novice learners benefited most from the static diagrams, knowledgeable learners benefited most from the animated diagrams. The author explained this result by theorizing that processing requirements for the animated diagrams may have exceeded the novice's working memory capacity. Ultimately, the animated diagrams overloaded their working memory and hindered learning. Likewise, knowledgeable learners may have processed the static diagrams as extraneous details and experienced a similar working memory capacity bottleneck that interfered with learning.

Spatial ability is another learner characteristic that has the potential to moderate learning in multimedia environments. Spatial ability involves the ability to mentally rotate objects and to mentally imagine the manipulation processes. Research reveals that learners' spatial abilities play a role in learning from animations, as well as navigating in the environment. Mayer and Sims (1994) found that the instructional benefits of presenting an animation and narration simultaneously were profound for learners with high spatial ability but not for those with low spatial ability. Mayer, Mautone and Prothero (2002) found similar results—pictorial scaffolding was more beneficial for learners with high spatial ability than those with low spatial ability. Waller (2000) tested a structural equation model in order to shed light on the relationships between spatial ability, ability to acquire spatial knowledge from a virtual environment (VE), proficiency with the navigation interface of a VE, and other factors. The researcher found that spatial ability was a significant predictor for ability to acquire spatial knowledge from a VE and was significantly correlated to proficiency with the navigation interface of a VE.

A person's age, or natural age-related cognitive decline, is also a factor that influences how people learn in multimedia environments. As people advance into their elder years, their working memory capacity (Salthouse and Babclcok, 1991) and the ability to inhibit distracting information (Allen, Madden, Groth, and Crozier, 1992) tends to decrease. Consequently, older learners may experience cognitive overload and decreased performance in solving complex tasks or learning in a complex learning environment without instructional aids. Results of a few studies provide tentative evidence for this potential moderating variable (Van Gerven, Paas, Van Merriënboer, Hendriks, and Schmidt, 2003; Van Gerven, Paas, Van Merriënboer, and Schmidt, 2002, 2006). For instance, Van Gerven, Paas, Van Merriënboer and Schmidt (2002) used either worked examples or conventional problems to teach elderly and young adults to solve water-jug problems. They found elderly participants benefited more than the young adults from studying worked examples—the elderly sample had a substantially lower level of perceived cognitive load and were quicker than their younger counterparts at learning from the worked examples.

In addition to learner characteristics (prior knowledge, spatial ability, and cognitive aging) the *complexity of the instructional materials or tasks* moderate learning in multimedia.

From a cognitive load perspective, intrinsic cognitive load is determined by the inherent complexity of a task. If, for example, the visualizations are simple with merely a few elements, the intrinsic load is low. Learners may have sufficient working memory capacity to process and integrate such information, even if it causes split-attention or increases visual search requirements. As a result, techniques intended to reduce extraneous cognitive load may become ineffective. For instance, Jeung et al. (1997) found that flashing indicators (one type of visual cueing) promoted student learning for geometry concepts when instructional information required a high degree of visual search, but not for material that required less visual search. In two meta-analyses, Ginns (2005, 2006) revealed that the modality effect and the contiguity effect[1] had a larger impact on results when instructional materials were high in element interactivity than when they were low in element interactivity.

Limitations in Multimedia Research

We would like to highlight some limitations in the current educational research literature regarding learning with animations and multimedia. We are presently challenged by research that has been limited to a few content domains. In addition, there are few studies that systematically study the impact of mediators. Whereas moderating variables (i.e., prior knowledge) influence the strength of a relationship between independent or predictor variable (i.e., using a multimedia tutorial) and a dependent or criterion variable (i.e., learning the tutorial content), mediators explain the relationship between these two variables (i.e., learners' perception about the relevance or value of the tutorial's content).

It is notable that most studies in the literature investigate multimedia learning in a limited content domain range, such as humans' cardiovascular system, mechanical systems (pulley system, brake system and flushing system), lightning formation, geometry word problems, and weather maps. Only a small number of studies used other content domains, such as geography (Lin and Atkinson, 2011), chemistry (Yang, Andre and Greenbowe, 2003), and social sciences (Westelinck, Valcke, De Craene, and Kirschner, 2005). Therefore, the generalizability of those multimedia design principles is unknown. The question of whether these principles are domain-specific still has not been answered. Future research should investigate the external validity of the multimedia principles.

The second limitation is that the research-based multimedia design principles have only been validated in a few populations, such as college students (e.g., Mayer et al., 2001), high school students (Atkinson et al., 2005) and elderly people (e.g., Van Gerven et al., 2002) and results are usually based on data collected from homogenous samples. There is both a benefit and a drawback in studying an intervention effect in a relatively homogeneous sample or population. The benefit is that researchers may increase their likelihood of detecting a significant intervention effect from a sample of people with similar background and characteristics, as homogeneity can help reduce the error and noise in the analysis. A large number of the current empirical studies revealed large effect sizes for the multimedia intervention, a positive indication that multimedia is indeed effective. However, there is a potential obvious drawback. That is, the conclusions about the effectiveness of multimedia

[1]The contiguity effect occurs when instructional materials are integrated and split-attention is avoided.

lack generalizability. Therefore, we recommend researchers overcome this issue in future investigations.

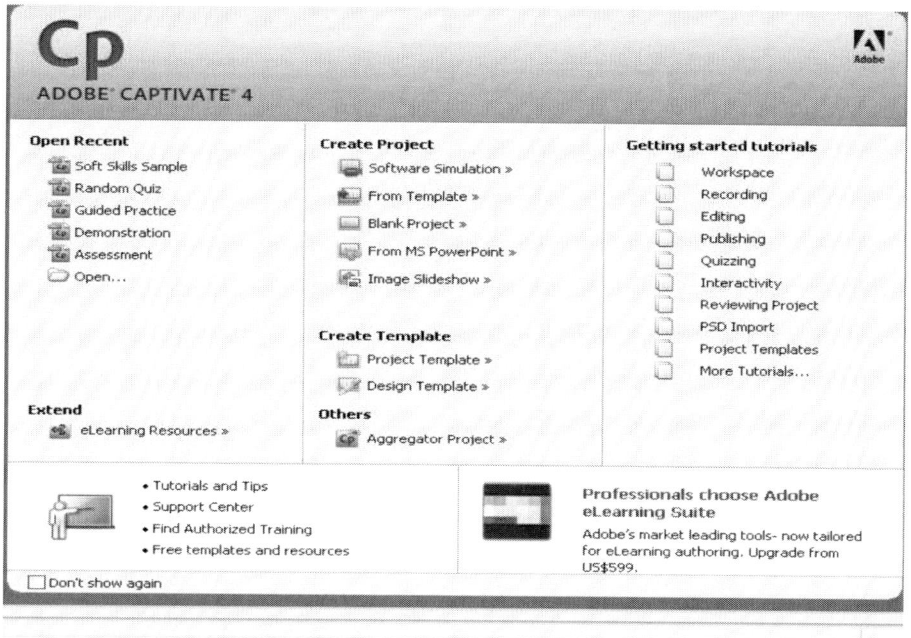

Figure 2. Start screen of Adobe Captivate.

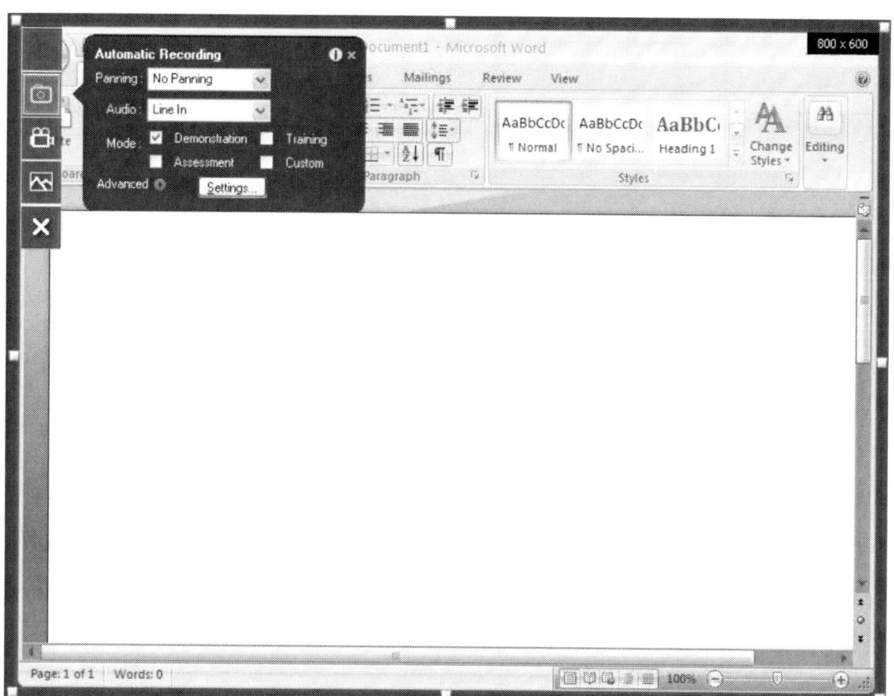

Figure 3. Screen capture recording phase in Adobe Captivate.

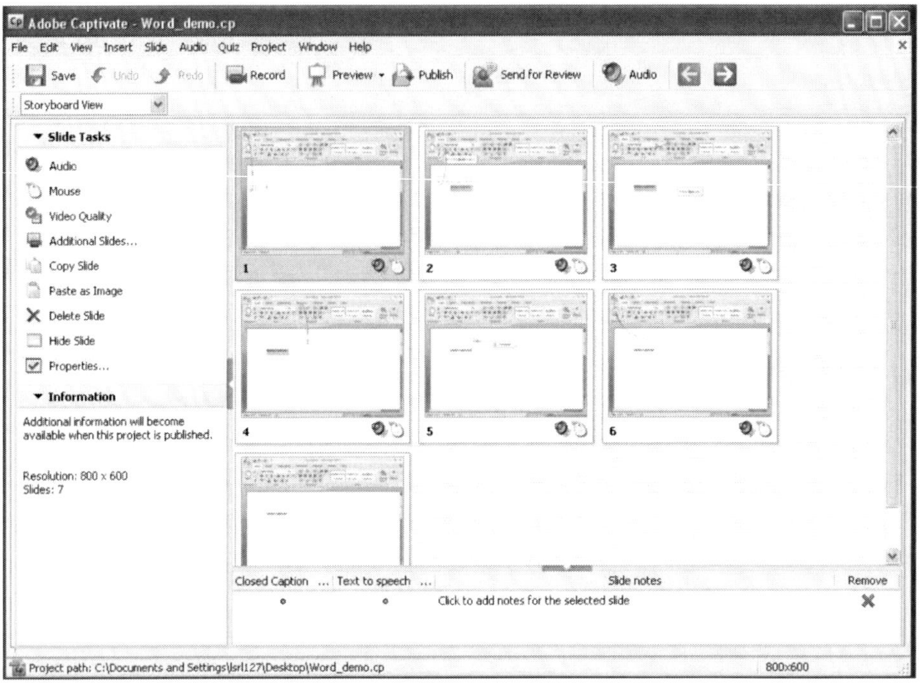

Figure 4. End state of recording in Adobe Captivate.

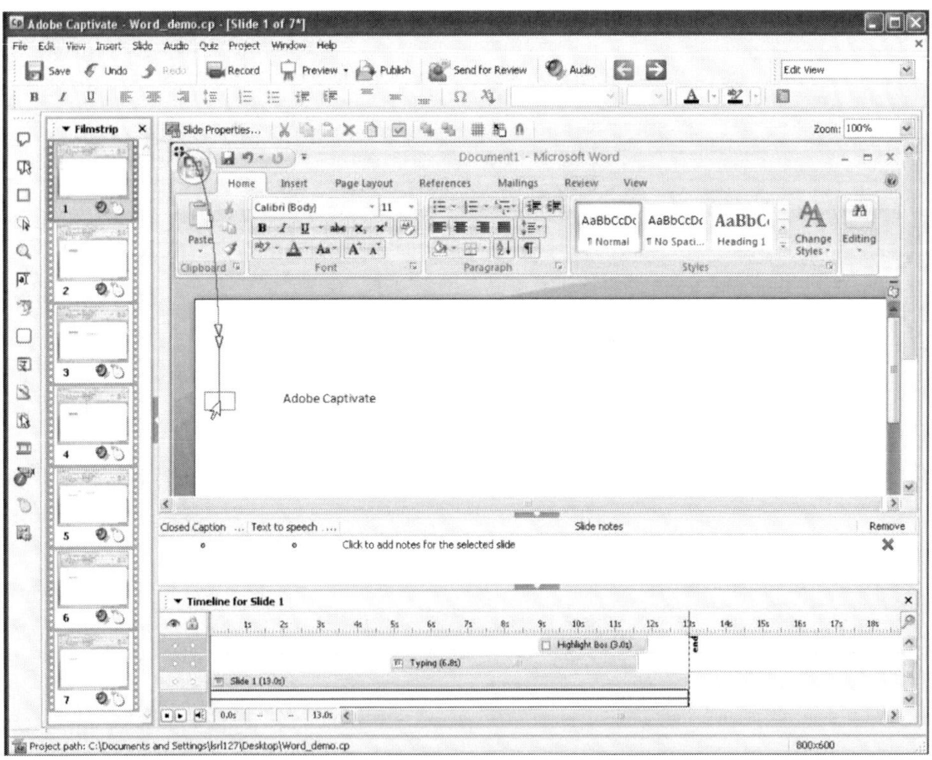

Figure 5. Editing a recording in Adobe Captivate.

The third limitation is that researchers have not fully investigated the role of mediators such as cognitive load, motivation, meta-cognition, and affect in multimedia learning. Moreno and Mayer (2007) extended the cognitive theory of multimedia learning by incorporating affective, cognitive, and meta-cognitive factors—cognitive-affective theory of leaning with media. However, the inter-relationships among these factors are still unknown. Some empirical studies were conducted to address these research questions. For instance, Moreno (2009) investigated the meta-cognitive prompts in learning from animated classroom exemplars. The results revealed meta-cognitive prompts had a positive impact on learning. However, the potential moderating or mediating effect of meta-cognition on learning and motivation is unknown, as the study did not address these variables. Similarly, the study conducted by Moreno (2007) only investigated the impact of some multimedia design techniques on learning (i.e., cognitive load and intrinsic motivation) without researching the inter-relationships between these factors. Therefore, we encourage future research to directly address the potential moderating and/or mediating effects to shed light on the ambiguous inter-relationships among these factors in multimedia environments.

Animations and Multimedia Applications

Currently, Adobe Flash and Captivate are among the most influential commercial software packages used to produce multimedia tutorials. Adobe Captivate allows users to create multimedia presentations by inserting images, audio recordings, and/or computer screen captures. Specifically, users can choose "Software Simulation" option after launching Adobe Captivate, if they want to create a multimedia tutorial or presentation (see Figure 2). Then, they can choose their preferred movie type and movie size, as well as recording options (such as sound and recording mode), according to their own goals and needs. If the goal is to present a series of procedures or operations, the Demonstration Mode may be a good choice. If the goal is to prompt learners to perform some actions as indicated, the Training Mode is a good choice. Finally, if the goal is to assess skill acquisition, the Assessment Mode is preferred (see Figure 3). After this set-up, users can click the Record button to start recording. Every computer action will be recorded as a snapshot during the process of recording (see Figure 4).

After completing the recording, users can edit it by adding text labels, audio or animations (see Figure 5). By clicking the Publish button, users can export the recorded presentation or tutorial in various file format based on their needs. On the other hand, Adobe Captivate allows users to create a presentation or tutorial by importing existing PowerPoint slides (see Figure 2). Animation effects in the PowerPoint slides can also be transferred to Captivate. After importing, users can also edit the presentations within Captivate.

In Adobe Flash, users can create changes in motion, color and shape in just a few mouse clicks. For instance, users can set a rectangle in the first frame and change it to a circle in the 100th frame. Then, by choosing the Tween Mode to Shape, users can successfully create simple animation showing a process that a rectangle gradually changes to a circle. Videos can also be imported into Flash to add more realism to the content. Interactivity can be incorporated by using Components in Flash combined with Action Script programming.

There are also many online applications that have great potential for creating multimedia tutorials. Some (e.g., Jing) provide user-friendly interface for animated screen captures. Users can record a series of their computer operations, highlight the recording by arrows, explain the recording using textbox and incorporate their narrations for further explanations. And those movies of screen captures can be easily shared online with interactive features. There are also some other applications (e.g., SitePal) that allow users, who do not have computer programming skills or 3D modeling skills, to create an animated virtual agent with voice-over narrations, facial expressions and gaze movements, and embed it into their website, flash programs, emails or PowerPoint. As a result, the interaction between the learners and the computer has the potential to be enhanced, as presenting an animated virtual agent can motivate learners and engage them in a learning activity.

REFERENCES

Alessi, S.M., and Trollip, S.R. (2001). *Multimedia for learning: Methods and development.* Needham, MA: Allyn and Bacon.

Amadieu, F., Mariné, C., and Laimay, C. (2011). The attention-guiding effect and cognitive load in the comprehension of animations. *Computers in Human Behavior, 27*, 36-40.

Arguel, A., and Jamet, E. (2009). Using video and static pictures to improve learning of procedural contents. *Computers in Human Behavior, 25*, 354-359.

Atkinson, R. K. (2002). Optimizing learning from examples using animated pedagogical agents. *Journal of Educational Psychology, 94*, 416-427.

Atkinson, R. K., Lin, L. and Harrison, C. (2009). Comparing the efficacy of different signaling techniques. *Proceedings of World Conference on Educational Multimedia, Hypermedia and Telecommunications 2009* (pp. 954-962). Chesapeake, VA: AACE.

Atkinson, R. K., Lin, L., Christopherson, R., Joseph, S. and Harrison, C. (2011). *Animated agents and learning: Does the type of verbal feedback they provide matter?* Manuscript under revision.

Atkinson, R. K., Mayer, R. E., and Merrill, M. M. (2005). Fostering social agency in multimedia learning: Examining the impact of an animated agent's voice. *Contemporary Educational Psychology, 30*, 117-139.

Ayres, P., Marcus, N., Chan, C., and Qian, N. (2009). Learning hand manipulative tasks: When instructional animations are superior to equivalent static representations. *Computers in Human Behavior, 25*, 348-353.

Ayres, P., and Sweller, J. (2005). The split-attention principle in multimedia learning. In R. E. Mayer (Ed.), *The Cambridge handbook of multimedia learning.* (pp. 135-146). New York, NY, US: Cambridge University Press.

Baddeley, A. (1992). Working memory. *Science, 255*, 556-559.

Baek, Y. K. and Layne, B. H. (1988). Color, graphics, and animation in a computer-assisted learning tutorial lesson. *Journal of Computer-Based Instruction, 15*, 131–135.

Baron, R. M., and Kenny. D. (1986). The moderator-mediator variable distinction in social psychological research: Conceptual, strategic, and statistical considerations. *Journal of Personality and Social Psychology, 51*(6), 1173-1182.

Bétrancourt, M., and Tversky, B. (2000). Effect of computer animation on users' performance: A review. *Le Travail Humain, 63*(4), 311-329.

Bodemer, D., Ploetzner, R., Feuerlein, I., and Spada, H. (2004). The active integration of information during learning with dynamic and interactive visualisations. *Learning and Instruction, 14*(3), 325-341.

Boekaerts, M. (2007). What have we learned about the link between motivation and learning/performance? *Zeitschrift für Pädagogische Psychologie, 21*, 263-269.

Boucheix, J., and Lowe, R. K. (2010). An eye tracking comparison of external pointing cues and internal continuous cues in learning with complex animations. *Learning and Instruction, 20*(2), 123-135.

Catrambone, R., and Seay, A. F. (2002). Using animation to help students learn computer algorithms. *Human Factors, 44*(3), 495-511.

Chandler, P., and Sweller, J. (1996). Cognitive load while learning to use a computer program. *Applied Cognitive Psychology, 10*(2), 151-170.

ChanLin, L. (1998). Animation to teach students of different knowledge levels. *Journal of Instructional Psychology, 25*(3), 166-175.

ChanLin, L. (2001). Formats and prior knowledge on learning in a computer-based lesson. *Journal of Computer Assisted Learning, 17*(4), 409-419.

Chi, M. T. H. (2000). Self-explaining expository texts: The dual processes of generating inference and repairing mental models. In R. Glaser (Ed.), (pp. 161-238). Mahwah, NJ, US: Lawrence Erlbaum Associates Publishers.

Chi, M., Glaser, R., and Rees, E. (1982). Expertise in problem solving. In Sternberg, R. (Ed.), *Advances in the Psychology of Human Intelligence*, Erlbaum, Hillsdale, NJ.

Cierniak, G., Scheiter, K., and Gerjets, P. (2009). Explaining the split-attention effect: Is the reduction of extraneous cognitive load accompanied by an increase in germane cognitive load? *Computers in Human Behavior, 25*(2), 315-324.

Corbalan, G., Kester, L., and van Merriënboer, J. J. G. (2011). Learner-controlled selection of tasks with different surface and structural features: Effects on transfer and efficiency. *Computers in Human Behavior, 27*, 76-81.

de Koning, B. B., Tabbers, H. K., Rikers, R. M. J. P., and Paas, F. (2007). Attention cueing as a means to enhance learning from an animation. *Applied Cognitive Psychology.Special Issue: A Cognitive Load Approach to the Learning Effectiveness of Instructional Animation, 21*, 731-746.

de Koning, B. B., Tabbers, H., Rikers, R. M. J. P., and Paas, F. (2009). Towards a framework for attention cueing in instructional animations: Guidelines for research and design. *Educational Psychology Review, 21*, 113-140.

de Koning, B. B., Tabbers, H. K., Rikers, R. M. J. P., and Paas, F. (2010). Attention guidance in learning from a complex animation: Seeing is understanding? *Learning and Instruction, 20*, 111-122.

Duchastel, P. C., and Merril, P. F. (1973). The Effects of Behavioral Objectives on Learning: A Review of Empirical Studies. *Review of Educational Research, 43*, 53-69.

Dunsworth, Q., and Atkinson, R. K. (2007). Fostering multimedia learning of science: Exploring the role of an animated agent's image. *Computers and Education, 49*, 677-690.

Galitz, W. O. (2007). *The essential guide to user interface design: An introduction to GUI design principles and techniques.* Indianapolis, IN:Wiley Publishing.

Ginns, P. (2005). Meta-analysis of the modality effect. *Learning and Instruction, 15*, 313-331.

Ginns, P. (2006). Integrating information: A meta-analysis of the spatial contiguity and temporal contiguity effects. *Learning and Instruction, 16*, 511-525.

Hegarty, M., Kriz, S., and Cate, C. (2003). The roles of mental animations and external animations in understanding mechanical systems. *Cognition and Instruction, 21*, 325-360.

Husman, J. and Hilpert, J. (2007). The intersection of students' perceptions of instrumentality, self-efficacy, and goal orientations in an online mathematics course. *Zeitschrift für Pädagogische Psychologie, 21*, 229-239.

Jamet, E., Gavota, M., and Quaireau, C. (2008). Attention guiding in multimedia learning. *Learning and Instruction, 18*, 135-145.

Jeung, H., Chandler, P., and Sweller, J. (1997). The role of visual indicators in dual sensory mode instruction. *Educational Psychology, 17*, 329-343.

Kalyuga, S. (2007). Expertise reversal effect and its implications for learner-tailored instruction. *Educational Psychology Review, 19*, 509-539.

Kalyuga, S. (2008). Relative effectiveness of animated and static diagrams: An effect of learner prior knowledge. *Computers in Human Behavior, 24*, 852–861

Kalyuga, S., Ayres, P., Chandler, P., and Sweller, J. (2003). The expertise reversal effect. *Educational Psychologist, 38*, 23-31.

Kalyuga, S., Chandler, P., & Sweller, J. (1998). Levels of expertise and instructional design. *Human Factors, 40*(1), 1-17.

Kalyuga, S., Chandler, P., & Sweller, J. (2000). Incorporating learner experience into the design of multimedia instruction. *Journal of Educational Psychology, 92*(1), 126-136.

Kalyuga, S., & Sweller, J. (2004). Measuring knowledge to optimize cognitive load factors during instruction. *Journal of Educational Psychology, 96*(3), 558-568.

Kim, S., Yoon, M., Whang, S. M, Tversky, B., and Morrison, J. B. (2007). The effect of animation on comprehension and interest. *Journal of Computer Assisted Learning, 23*(3), 260-270.

Kühl, T., Scheiter, K., Gerjets, P., and Gemballa, S. (2011). Can differences in learning strategies explain the benefits of learning from static and dynamic visualizations? *Computers and Education, 56*(1), 176-187.

Lee, H., Plass, J. L., and Homer, B. D. (2006). Optimizing cognitive load for learning from computer-based science simulations. *Journal of Educational Psychology, 98*(4), 902-913.

Lidwell, W., Holden, K., & Butler, J. (2003). *Universal principles of design.* Beverly, MA: Rockport Publishers Inc.

Lin, L. and Atkinson, R. K. (2011). Using Animations and Visual Cueing to Support Learning of Scientific Concepts and Processes. *Computers and Education, 56*(3), 650-658.

Loman, N. L., and Mayer, R. E. (1983). Signaling techniques that increase the understandability of expository prose. *Journal of Educational Psychology, 75*, 402-412.

Lorch, R. F. (1989). Text signaling devices and their effects on reading and memory processes. *Educational Psychology Review, 1*, 209–234.

Lorch, R. F., and Lorch, E. P. (1996). Effects of organizational signals on free recall of expository text. *Journal of Educational Psychology, 88*, 38-48.

Lorch, R. F., Lorch, E. P., and Inman, W. E. (1993). Effects of signaling topic structure on text recall. *Journal of Educational Psychology, 85*, 281-290.

Mayer, R. E. (2001). *Multimedia learning*. New York: Cambridge University Press.

Mayer, R. E. (2005a). Cognitive theory of multimedia learning. In R. E. Mayer (Ed.), *The Cambridge handbook of multimedia learning*. (pp. 31-48). New York, NY, US: Cambridge University Press.

Mayer, R. E. (2005b). Principles for reducing extraneous processing in multimedia learning: Coherence, signaling, redundancy, spatial contiguity, and temporal contiguity principles. In R. E. Mayer (Ed.), *The Cambridge handbook of multimedia learning*. (pp. 183-200). New York, NY, US: Cambridge University Press.

Mayer, R. E., and Chandler, P. (2001). When learning is just a click away: Does simple user interaction foster deeper understanding of multimedia messages? *Journal of Educational Psychology, 93*(2), 390-397.

Mayer, R. E., Hegarty, M., Mayer, S., and Campbell, J. (2005). When static media promote active learning: Annotated illustrations versus narrated animations in multimedia instruction. *Journal of Experimental Psychology: Applied, 11*(4), 256-265.

Mayer, R. E., Heiser, J., and Lonn, S. (2001). Cognitive constraints on multimedia learning: When presenting more material results in less understanding. *Journal of Educational Psychology, 93*(1), 187-198.

Mayer, R. E., and Johnson, C. I. (2008). Revising the redundancy principle in multimedia learning. *Journal of Educational Psychology, 100*(2), 380-386.

Mayer, R. E., Mautone, P., and Prothero, W. (2002). Pictorial aids for learning by doing in a multimedia geology simulation game. *Journal of Educational Psychology, 94*(1), 171-185.

Mayer, R. E., and Moreno, R. (2003). Nine ways to reduce cognitive load in multimedia learning. *Educational Psychologist, 38*(1), 43-52.

Mayer, R. E., and Sims, V. K. (1994). For whom is a picture worth a thousand words? Extensions of a dual-coding theory of multimedia learning. *Journal of Educational Psychology, 86*(3), 389-401.

Mayer, R. E., Sobko, K., and Mautone, P. D. (2003). Social cues in multimedia learning: Role of speaker's voice. *Journal of Educational Psychology*, *95*(2), 419-419.

Michas, I. C., and Berry, D. C. (2000). Learning a procedural task: Effectiveness of multimedia presentations. *Applied Cognitive Psychology, 14*(6), 555-575.

Moreno, R. (2004). Decreasing cognitive load for novice students: Effects of explanatory versus corrective feedback in discovery-based multimedia. *Instructional Science, 32*(1-2), 99-113.

Moreno, R. (2007). Optimising learning from animations by minimising cognitive load: Cognitive and affective consequences of signalling and segmentation methods. *Applied Cognitive Psychology, 21*(6), 765-781.

Moreno, R. (2009). Learning from animated classroom exemplars: the case for guiding student teachers' observations with metacognitive prompts. *Educational Research and Evaluation, 15,* 487-501.

Moreno, R., and Mayer, R. E. (2000). A coherence effect in multimedia learning: The case for minimizing irrelevant sounds in the design of multimedia instructional messages. *Journal of Educational Psychology, 92*(1), 117-125.

Moreno, R., and Mayer, R. E. (2005). Role of guidance, reflection, and interactivity in an agent-based multimedia game. *Journal of Educational Psychology, 97*(1), 117-128.

Moreno, R., and Mayer, R. (2007). Interactive multimodal learning environments. *Educational Psychology Review, 19*(3), 309-326.

Moreno, R., Mayer, R. E., Spires, H. A., and Lester, J. C. (2001). The case for social agency in computer-based teaching: Do students learn more deeply when they interact with animated pedagogical agents? *Cognition and Instruction, 19*(2), 177-213.

Mousavi, S. Y., Low, R., and Sweller, J. (1995). Reducing cognitive load by mixing auditory and visual presentation modes. *Journal of Educational Psychology, 87*(2), 319-334.

Paas, F., Renkl, A., and Sweller, J. (2003). Cognitive load theory and instructional design: Recent developments. *Educational Psychologist*, 38(1), 1-4.

Paivio, A. (1979). *Imagery and verbal processes*. Hillsdale, NJ: Erlbaum.

Paivio, A. (1986). *Mental representations*. New York: Oxford University Press.

Park, O.-C. and Gittelman, S. S. (1992). Selective use of animation and feedback in computer-based instruction. *Educational Technology, Research, and Development, 40*, 27–38.

Rieber, L. P. (1990). Using computer animated graphics with science instruction with children. *Journal of Educational Psychology, 82*(1), 135-140.

Rieber, L. P. (1991a). Animation, incidental learning, and continuing motivation. *Journal of Educational Psychology, 83*(3), 318-328.

Rieber, L. P. (1991b). Effects of visual grouping strategies of computer-animated presentations on selective attention in science. *Educational Technology Research and Development, 39*, 5-15.

Roy, M., and Chi, M. T. H. (2005). The self-explanation principle in multimedia learning. In R. E. Mayer (Ed.), *The Cambridge handbook of multimedia learning* (pp. 271-286). New York, NY, US: Cambridge University Press.

Schnotz, W., and Kurschner, C. (2007). A reconsideration of cognitive load theory. *Educational Psychology Review, 19*(4), 469-508.

Schwan, S., and Riempp, R. (2004). The cognitive benefits of interactive videos: Learning to tie nautical knots. *Learning and Instruction, 14*(3), 293-305.

Sweller, J. (1994). Cognitive load theory, learning difficulty, and instructional design. *Learning and Instruction, 4*(4), 295-312.

Sweller, J. (2005). The redundancy principle in multimedia learning. In R. E. Mayer (Ed.), *The Cambridge handbook of multimedia learning.* (pp. 159-167). New York, NY, US: Cambridge University Press.

Sweller, J. (2005a). Implications for cognitive load in multimedia learning. In R. E. Mayer (Ed.), *The cambridge handbook of multimedia learning* (pp. 19-30). New York, NY: Cambridge University Press.

Sweller, J., and Chandler, P. (1994). Why some material is difficult to learn. *Cognition and Instruction, 12*(3), 185-233.

Sweller, J., van Merrienboer, J. J. G., and Paas, F. (1998). Cognitive architecture and instructional design. *Educational Psychology Review, 10*(3), 251-296.

Tindall-Ford, S., Chandler, P., and Sweller, J. (1997). When two sensory modes are better than one. *Journal of Experimental Psychology: Applied, 3*(4), 257-287.

Thompson, S. V. and Riding, R. J. (1990). The effect of animated diagrams on the understanding of a mathematical demonstration in 11- to 14-year-old pupils. *British Journal of Educational Psychology, 60*, 93–98.

Tullis, T., & Albert, B. (2008). Measuring the user experience: Collecting, analyzing and presenting usability metrics. Burlington, MA: Morgan Kaufman van Gog, T., Paas, F., Marcus, N., Ayres, P., and Sweller, J. (2009). The mirror neuron system and observational learning: Implications for the effectiveness of dynamic visualizations. *Educational Psychology Review, 21*(1), 21-30.

Waller, D. (2000). Individual differences in spatial learning from computer-simulated environments. *Journal of Experimental Psychology: Applied, 6*, 307-321.

Westelinck, K. D., Valcke, M., De Craene, B., and Kirschner, P. (2005). Multimedia learning in social sciences: Limitations of external graphical representations. *Computers in Human Behavior, 21*(4), 555-573.

Wong, A., Marcus, N., Ayres, P., Smith, L., Cooper, G. A., Paas, F., et al. (2009). Instructional animations can be superior to statics when learning human motor skills. *Computers in Human Behavior, 25*(2), 339-347.

Wouters, P., Paas, F., and van Merriënboer, J. J. G. (2009). Observational learning from animated models: Effects of modality and reflection on transfer. *Contemporary Educational Psychology, 34*(1), 1-8.

Yang, E., Andre, T., and Greenbowe, T. J. (2003). Spatial ability and the impact of visualization/animation on learning electrochemistry. *International Journal of Science Education, 25*(3), 329.

In: Handbook of Technology in Psychology …
Editor: Luciano L'Abate and David A. Kasier
ISBN: 978-1-62100-004-4

Chapter 6

THE PARTICIPATORY WEB

Robert K. Atkinson*[1], *Kent Sabo and Quincy Conley
[1]Arizona State University, Tempe, Arizona, US

The purpose of this chapter is to introduce the Participatory Web to mental health professionals. We begin by providing a definition and classification for the term by reviewing its historical roots, technical specificities, and how Web 1.0 gave rise to Web 2.0, social media and ultimately the Participatory Web.

Although Facebook and Twitter seem pervasive in popular culture, the same is not true in the fields of psychology, psychiatry, and neurology. Anecdotal evidence suggests these tools are being used in clinical contexts, but very little empirical research exists. We describe some of the pioneering research in the field that is establishing the case for Participatory Web tools in evidence-based practice.

We then discuss the untapped potential that the participatory web holds for mental health. Due to the lack of research, there is a huge opportunity for researchers and practitioners in the mental health field to determine the most effective uses of these technologies as they did with Web 1.0 technologies.

Tools enabled by Participatory Web technology are ubiquitous. Data shows that more and more adults are using the technology; social networks are not just for teenagers anymore. In the second half of the chapter, we describe how to sign up and use some of the most popular Participatory Web tools including wikis, blogs, and social networking sites. The first step in leveraging these tools in clinical practice is to become familiar with how they work. It is time to participate.

HISTORICAL BACKGROUND

Social media is currently one of the hottest buzzwords in popular culture. The term is typically used when describing asynchronous web-based technologies used for communicating. To better explain what social media is and how it came to be, it may be helpful to first understand the history.

The term social media is believed to stem from the second generation of Web design and development called Web 2.0. The term Web 2.0 was coined after experts recognized that the

Web was changing the way people used technology. Darcy DiNucci (1999) was the first to use the term Web 2.0 in her response to this change: "The Web will be understood not as screenfulls of text and graphics but as a transport mechanism, the ether through which interactivity happens. It will… appear on your computer screen… on your TV set… your car dashboard… your cell phone… hand-held game machines … maybe even your microwave oven (p. 32)."

In 2005, technologist and innovator Tim O'Reilly (2005) further defined the term Web 2.0 as "the Web as a platform" (p. 1). He also describes the distinguishing features between the Web 2.0 from Web 1.0 as shown in Tables 1. and 2. O'Reilly explains that the difference between Web 1.0 and Web 2.0 is at the technological and ideological foundations. Web 1.0 is a series of separate computers that allowed users to surf Web pages in isolation; whereas, Web 2.0 technology is much more powerful and social in nature.

Table 1. Characteristics of Web 1.0 and Web 2.0

Web 1.0 Characteristics	Web 2.0 Characteristics
Web pages were static. Web pages weren't interactive. Applications were proprietary.	User experience is dynamic. Web content is dynamic. Rich user participation.

Table 2. Examples of Web 1.0 and Web 2.0 (O'Reilly, 2005, p. 1)

Web 1.0 Examples	Web 2.0 Examples
Personal Websites	Blogging
Ofoto	Flickr
Britannica Online	Wikipedia
MP3.com	Napster
Domain Name Speculation	Search Engine Optimization (SEO)
Content Management Systems	Wikis

Table 3. Examples of the Participatory Web Tools (Beer and Burrows, 2010)

Tool	Examples
Social Network	Facebook
Blog	Blogger
Micro-blog	Twitter
Wiki	Wikipedia
Streaming Video	YouTube

Table 4. Characteristics of the Participatory Web (Jenkins, 2006)

Characteristics	
Innovative	Networked
Convergent	Global
Everyday	Generational
Appropriative	Unequal

The Participatory Web

There is no single accepted term that fully captures the concept of Web 2.0 or social media. However, one phrase that is currently being used because of the increasing use of wikis, blogs, and social networking technologies is the Participatory Web. Grabe and Grabe (2008) call it the Participatory Web because users participate on the Web by sharing information, generating content, and collaborating. It is a relatively new idea, but still relates to the aforementioned theory that the Web is a participation platform. We prefer this term because it best embodies the "opportunities to interact and contribute" for anyone with access to participate through the Web (Grabe and Grabe, 2008).

Even if the concept of the Participatory Web is new to you, you have more than likely encountered some of the most common tools that fall under its umbrella. Chances are that you have a social networking site profile that you use to communicate with friends, family and colleagues.

Or perhaps you have experience using other Participatory Web tools such as wikis while looking up the details of a historical event that you could not remember such as Wikipedia. Another common tool is the blog, where you'll find people interacting through posts and leaving opinions using comments. We will talk about the tools in more detail later in the chapter, but in Table 3 you'll find a list of the most common Participatory Web tools with the most popular examples (Beer and Burrows, 2010).

Characteristics of the Participatory Web

According to Henry Jenkins (2006), the characteristics of the contemporary Participatory Web is important to understand so that we can better identify opportunities for facilitating the development of new skills. These characteristics of Participatory Web tools are provided in Table 4. Grabe and Grabe (2008) suggest that we consider these characteristics over just focusing on the examples because the characteristics will better provide us with the empirical support for how to use the Web to our benefit.

Who is Participating?

Although there is a long way to go before everyone is participating on the Web, recent research shows that an increasing amount of adults are using Participatory Web tools. Every 10 minutes, at least 10 hours of content is uploaded through tools like blogs, wikis, and video sharing platforms.

Surprisingly, it is not just limited to teenagers anymore; significant contributions are coming from other generations as well. Research shows that usage trends among adults in the United States have increased from 2007 to 2009 in various Participatory Web tools and activities.

Therefore, it is fair to say that Participatory Web tools are more than just a fad and should be considered by anyone interested in leveraging the potential of the Web (Anderson, Reitsma, and Liackman, 2009; Kaplan and Haenlein, 2009; Ostrow, 2009).

Opportunity for Scholarship

For the moment, scholarship pertaining to Participatory Web tools relevant to the fields of psychology, psychiatry, and neurology is sparse. However, emerging research on the Participatory Web is beginning to investigate how we can leverage the technology in many fields, addressing a range of topics while contributing to the larger body of knowledge. To date, the bulk of research on how to use Participatory Web tools is focused on making social connections, marketing, and how it impacts privacy (Boyd and Ellison, 2008).

White (2010) describes how blog postings with narratives and images can inform those that encounter that content. This suggests that blogging could be used as a new medium for learning. However, White said further research should be conducted on to provide a framework for judgments about cultural norms and values. In an article about constructing a universal rating and reviewing system for health professionals and services, Hardey (2010) discusses the way in which people participate online using established forms of interaction. Hardey emphasizes the importance of allowing people to be able to rate and review without censorship. This could be important for implementing a rating and reviewing system for health care providers using the same Participatory Web rating tools that allow for books, movies, and hotels to be reviewed online. In more general terms, this article could be used as a springboard to investigate the impact the Participatory Web has on health professionals and the possible consequences it has for those who provide services (Hardey, 2010).

Another potential research avenue hinges on how organizations can capitalize on the user created data from the Participatory Web. This concept is known as "knowledgeable capitalism" (Thrift, 2005, p. 21). The premise is that information generated using Participatory Web tools is transactional data that could be mined. The data could then be used to predict behavior, identify patterns, make recommendations and discriminate between potential customers. However, no significant evidence has been found yet about the transactional data sources generated by Participatory Web applications. These sources of data are now providing a variety of entities with valuable informational archives upon which they can draw. Nevertheless, health sciences have yet to investigate the predictive value of data mining (Savage and Burrows, 2007). Lastly, O'Reilly (2005) provides a list of potential areas for exploration. He suggests that Participatory Web tools can impact society in the following ways:

- Fosters learning
- Enhances collaboration
- Improves knowledge sharing
- Promotes ease of resource location
- Greatly improves communication skills
- Provides personal and professional enrichment and focus
- Excellent tool for establishing individual and organizational expertise
- Facilitates opportunities to build personal and professional relationships

It is our hope that those encouraged by this chapter will pursue opportunities that support these notions with empirical data. Marketing researchers have already determined that the Participatory Web is impacting people's decision-making processes. The tools are enabling people to make decisions on where they shop, what to buy, and how much to pay for products.

Clinical Populations most Expected to Benefit

There are few studies that investigate the clinical use of Participatory Web technologies in psychology, psychiatry or neurology. This makes it difficult to provide a definitive list of clinical populations expected to benefit from the use of Participatory Web technologies in a mental health context. Three such studies are mentioned below and are described in more detail later in this chapter. Konovalov, Scotch, Post and Brandt (2010) analyzed the content of blogs written by active-duty military to find health-related information. Ko and Kuo (2009) surveyed bloggers about their self-disclosure and their perceptions of social capital. Finally, a study on a social network site developed for people with self-reported depression reported mixed results (Takahashi et al., 2009).

There exists a much larger body of work that studied the clinical benefits of Internet-based interventions (Web 1.0). The interventions within these studies addressed issues related to depression, anxiety disorders in adolescents and adults, bulimia nervosa and alcohol abuse. A meta-analysis was conducted to examine the results of studies focused on the treatment and prevention of depression and anxiety disorders using cognitive behavior therapy delivered over the Internet (Spek, Cuijpers, Kyklicek, Keyzer and Pop, 2006). Researchers found a moderate mean effect size for the studies included in the meta-analysis. Studies that investigated prevention interventions had a small effect size while treatment studies showed a large effect size Research shows that adolescents (aged 14-21) participating in an online depression intervention program demonstrated significant declines in depressed mood with moderate to large effect sizes (Van Voorhees et al., 2009). Use of cognitive behavior therapy to treat bulimia nervosa showed significant decreases in psychopathological levels and bulimic behavior when compared to a wait list control (Fernanded-Aranda et al., 2009). The Internet is a popular resource for patients to congregate and communicate; therefore one would assume that the social support provided by online groups would help to address various mental health issues.

However, a meta-analysis of research on Internet support groups, found a lack of methodologically sound research conducted in this area (Griffiths, Calear and Banfield, 2009). Questionable research designs were more likely to present positive results bringing into question the true effectiveness of Internet support groups.

In a review of the research on Internet-based alcohol interventions, Bewick et al. (2008) found that the interventions were well received but evidence supporting electronic screening and brief interventions was inconsistent. Another meta-analysis of Internet-based alcohol intervention programs showed that patients benefit from the program and that these online programs are useful in groups that are unlikely to seek traditional treatment like women, young people and at-risk users (White et al., 2010).

We suggest that mental health professional use the established research on Web 1.0 Internet-based interventions as a guide for researching and developing uses of Participatory Web tools in clinical mental-health contexts.

Recent Applications

Blogs

Blogs are a specific type of website or are frequently integrated into a website. Typically, one or more individuals use them to post comments about a particular subject, with the posts consisting of text, graphics, links, or video. Both health care providers and their patients contribute to blogs. Although there is ample anecdotal evidence of their use as well as opinion on their benefits and pitfalls, empirical studies on the use of blogs are few and far between. The empirical studies that do investigate blogs in clinical settings, studied the content of blogs written by military personal, how blogs may enhance well being through self-disclosure and surveyed medical bloggers and blog content. Health care practitioners that choose to write blogs or read those of their patients need to be cognizant of the associated ethical and professional implications.

Konovalov, Scotch, Post and Brandt (2010) were interested in studying the content of blogs written by active-duty soldiers because blogs can contain personal health-related information. The mental health-related information may be similar to information shared during a psychological consultation. Using the tools of biomedical informatics, information retrieval techniques and natural language processing, the researchers were able to aggregate the feelings and emotions of deployed military bloggers. The researchers identified both physical and emotional descriptions of combat. They found that descriptions of physical experiences were often followed by descriptions of the emotional reaction within two sentences (Konovalov et al., 2010).

Analyzing text contained within blogs is challenging. Blogs are often poorly structured, contain field-related terms, and spelling variations. The tools and processes used by the researchers to parse the contents of blogs could help therapists easily determine textual material that is relevant to mental health issues stemming from combat exposure. The tools could be used to train mental health providers working with the military or could be used by mental health providers in focus groups as discussion material (Konovalov et al., 2010).

Ko and Kuo (2009) investigated how bloggers' level of self-disclosure influenced their social capital and how their social capital influenced their perception of subjective well being. The researchers analyzed the responses to a survey from 596 participants with personal journal-type blogging experience. The items in the questionnaire were based on the constructs of self-disclosure, social integration, social bonding, social bridging and subjective well being. The results suggested that a bloggers' level of self-disclosure significantly affects their perception of social integration, bonding social capital and bridge social capital. The bloggers' perceptions promoted their subjective well being. Most bloggers responded that they share their moods or feelings in their blogs. Research suggests that those that share their innermost thoughts can gain social support and improve their social integration. Bloggers self-disclosure could also help to build intimate relationships (Ko and Kuo, 2009).

Lagu, Kaufman, Asch and Armstrong (2008) sought to report the approximate number and content of blogs that described patient interactions, identify violations of privacy or confidentiality and determine if blogs followed professional expectations. The researchers inspected the blogs of nurses and doctors (including residents and fellows). They analyzed 271 blogs using qualitative analysis techniques and found that over half of the blogs included identifiable information (image, sub-specialty, location and name). "Individual patients were described in 114 (42.1%) blogs. Patients were portrayed positively in 43 blogs (15.9%) and negatively in 48 blogs (17.7%). Of blogs that described interactions with individual patients, 45 (16.6%) included sufficient information for patients to identify their doctors or themselves" (Lagu et al., 2008, p. 1642). The researchers suggest that doctors and nurses who want to share their narratives through the use of a blog should practice caution. Care should be taken to avoid the disclosing of confidential information or the publishing of content that reflects on the author and the medical profession poorly (Lagu et al., 2008).

Kovic, Lulic and Brumini (2008) examined medical blogs and their bloggers. Medical bloggers responded to a thirty-seven item survey designed to gather information about their demographics, habits, characteristics and motivation. Medical bloggers had a high level of education and most worked in the medical industry. Most were experienced bloggers, having blogged for more than two years. The majority published their real name (75%) in their blog. More than half of the bloggers had published a scientific paper and 44% had written a book or book chapter.

Medical bloggers were motivated to blog by the desire to share their knowledge and skill and influence their reader's thinking. More than half of the respondents' blogged to express themselves creatively and 50% blogged to document personal experiences. Of the bloggers surveyed, 66% received attention from the news media regarding their blogs. In addition it was determined that medical bloggers were avid consumers of medical news from both online and traditional sources (Kovic, Lulic and Brumini, 2008).

Blog Concerns

Concerns about health professionals blogging include confidentiality, boundary issues, establishing treatment relationships, public forum implications and anonymity (Klumpp, 2010). Some recommend approaching blogging as one would any other publication in traditional media (Dainton, 2009) while others recommend avoiding blogging completely until the medical bodies provide clear directives for bloggers in the health care professions (Baerlocher and Detsky, 2008). Tunick and Mednick (2008) suggest the psychology profession should develop guidelines regarding blogging and pursue research to investigate the use of blogs in clinical settings.

"Advances in communication and technology have posed serious challenges to some of the fundamental ethical principles upon which psychologists operate, including the protection of patient privacy and confidentiality and the ideals of beneficence and nonmaleficence." (Tunick and Mednick, 2008, p. 585). Bloggers who breach confidentiality are subject to discipline from the licensing board, governmental and federal agencies. A civil lawsuit could also be filed. Bloggers should be concerned with the level of self-disclosure in their blog. Bloggers need to determine an appropriate level of self-disclosure for any patient that may come across the blog. Bloggers should refrain from posting any content that could be perceived as medical advice and therefore establish a treatment relationship. Mental health professionals should have a clear purpose for blogging and be careful to provide professional content (Klumpp, 2010).

Blogs can have benefits for patients and their families. Blogs can function as a form of cathartic narration, which facilitates adaptive coping and efficient communication with friends and family, which provides emotional support. However, patient and family blogs also raise ethical and professional concerns including privacy of the patient and family, the privacy of other patients, health-care providers' reactions or reputations and therapeutic boundaries (Tunick and Mednick, 2008).

There are potential benefits from psychologists reading their patients' blogs. For example there would be increased information sharing and alliance building as well as insight into patient and family coping. Other mental health concerns may be found in a blog that would not otherwise come to the psychologist's attention. The psychologist could then act on the information appropriately. If a psychologist chooses to read a patient's blog it should be clear to the psychologist that reading could provide a clinical benefit. The psychologist should also address the decision with the family/patient. It is important that the psychologist receives permission to the read the blog and goes to great lengths to explain their intentions. (Tunick and Mednick, 2008).

Wikis

Healthcare professionals use wikis to collaboratively create, synthesize, share and disseminate knowledge. The usefulness and quality of wiki content depends entirely on users. This requires users to develop collaboration behaviors (Archambault et al., 2010). Anecdotally, wikis are used in healthcare practices; however, most studies have focused on the educational uses of wikis (Kardong-Edgren, et al., 2009).

A group of pathologists created a pathology informatics curriculum for their residents using content from Wikipedia and organizing it in their own wiki. They found the comprehensiveness, quality, currency and the utility of Wikipedia articles to be high and useful for beginning and advanced learners. The wiki allowed health professionals to keep the information current and accurate (Kim, Gudewicz, Dighe and Gilbertson, 2010).

Reports suggest that wikis are being used both by healthcare professionals and organizations to share information (Kardong-Edgren, et al., 2009). Professionals suggest that wikis are useful tools for research teams and can help encourage collaboration, the sharing of documents and study implementation experiences Although there is anecdotal evidence of the use of wikis in healthcare settings "there is a paucity of literature on the use of wikis in the mental health sector, suggesting wiki use is not widespread" (Bastida, McGrath and Maude, 2010, p. 144).

Other Participatory Web Technologies

As terms like Health 2.0 and Medicine 2.0 enter popular usage, the technologies are still being examined and have yet to find a place in clinical psychiatric practice (Yellowlees and Nafiz, 2010). In fact researchers found thirty-six unique definitions of health 2.0, eight definitions of Health 2.0/Medicine 2.0 as equivalent terms and only two definitions of medicine 2.0. Researchers identified seven main concepts within the forty-six definitions. The seven concepts were: patients or consumers, Web 2.0 or technology, health professionals, social networking, changing health care, collaboration and health content. Only nine of the definitions were found in peer-reviewed journals and the rest were found online in grey literature (Van De Belt, Engelen, Berben and Schoonhoven, 2010). The definitions for Web

2.0 and the Participatory Web are up for debate, so it is likely that definitions for Health 2.0 and Medicine 2.0 will remain in a similar state in the near future.

Eysenbach (2008) outlined five major themes of Participatory Web technologies in healthcare: social networking, participation, apomediation, collaboration and openness. Social networking involves links among a group of individuals, which creates a Web of interrelationships. Participation involves both health consumers and providers. Participatory Web technologies rely on participants to create and manage the content while the technology simply acts as a platform for the users' actions. Apomediation is a new term that describes how Participatory Web technologies act as a "guide on the side" for users seeking information and services. Apomediation is in contrast with intermediation is which a person or tools stands "in between" a user and the information or service e.g., a travel agent. Collaboration refers to Participatory Web technologies ability to bring individuals or groups together to produce something despite geography or time. Openness refers to the open availability and ownership of personal health data as well as research data (Eysenbach, 2008).

The Internet has solidified itself as an important health communication channel. Interventions delivered via traditional Internet technology were shown to be as effective as face-to-face therapy for depression and anxiety (Yellowlees and Nafiz, 2010). However, with the growing popularity of Participatory Web tools, health professionals need to acquire the requisite skills to leverage the power of these tools. Health professionals need these skills to effectively promote health by acting as a resource and creating a platform for reaching a larger audience (Hanson, Thackeray, Barnes Neiger and McIntyre, 2008). Experienced clinical practitioners (doctors, nurses, etc.) had positive attitudes toward a Participatory Web portal, which they helped design, specific to their field and location. Further education of clinicians is needed in the use and benefits of Participatory Web technologies (Nordqvist, Hanberger, Timpka and Nordfelt, 2009). Patients also have positive attitudes toward the use of technology in mental health. In a survey of 352 regular, active-duty U.S Soldiers, 84% were willing to engage in technology delivered mental health care. Thirty-three percent of soldiers who were not willing to engage in traditional face-to-face therapy were willing to engage in technology-delivered therapy. This result suggests that technology-delivered therapy could reach more than a third of soldiers who would not otherwise seek traditional mental health treatment (Wilson, Onorati, Mishkind, Reger and Gahm, 2008).

As with wikis and blogs, research is sorely needed on the various aspects of patient and healthcare professional use of healthcare related Participatory Web technologies like social networks and podcasts (Randeree, 2009). Two such studies examined the content of medical podcasts and the potential beneficial mental health effects of social networks.

Wilson, Petticrew and Booth (2009) examined the content of medical podcasts produced by top medical journals and found there were few. Journals that do produce podcasts typically provide a brief summary of the articles in the journal and may have an interview with the author of the featured article. Although some of the podcasts were of mediocre quality, podcasts produced by journals were a valuable resource for those looking to keep up with the latest research without reading through all the articles. Journals can use podcasts as a means to entice listeners to read the journal articles. This is a promising way to increase readership as podcasts tend to be typically free and made readily available from the journal websites (Wilson, Petticrew and Booth, 2009).

Online communities focused on depression include a large number of untreated and undiagnosed members. Providing online peer support is an important strategy for patients without physical access due to lack of either human and/or financial sources. With the advent of social networking sites, specific social networks now focus on various health-related purposes. However there is very little research on the effectiveness of health-related social networks (Takahashi et al., 2009).

Takahashi et al. (2009) conducted two surveys on a Japanese social network created for people with self-reported depression. The survey revealed that more than 90% of respondents could be diagnosed with mood disorders or had depressive tendencies. Potential benefits of this depression social network included peer support. Some users found strength in the fact that others in the social network who were more depressed, were persevering. Users were also empowered by providing support to other users. The variety of communication options provided by the social network (passive, active or interactive) allows a user to access peer support in a way they are most comfortable. Takahashi et al. (2009) also found that some users may experience a "downward depressive spiral" due to participation in the social network. This could be due to greater dependency in people with depressive tendencies or the mood state while using the social network.

FUTURE APPLICATIONS

Participatory Web technologies have the potential for many uses in mental health practices from analyzing patient's posts, internal health organization communication, and the delivery of interventions. Video and podcasting could be used as vehicles for engaging and effective communication. Blogs and forums could connect healthcare professionals from within and across disciplines. Wikis are knowledge resources that could be reviewed and updated by a group of professionals (Kubben, 2010).

Wikis could be used to enhance the efficiency of communication among health-care providers with complex patients. Wikis enable asynchronous communication among multiple health-care providers working with one patient. Asynchronous communication is efficient and is not interruptive. An example could be a medical team working with a military service member with both physical combat injuries and posttraumatic stress disorder. Each member of the health-care team could update a patient's wiki page with the latest information. This information would be immediately available to all the other members of the multi-disciplinary healthcare team making it easier for all members to make the appropriate decisions for a complex problem (Naik and Singh, 2010). Wikis could be used in collaborative recovery plans and with therapeutic groups. Therapeutic groups share experiences and learning about mental health. A secure wiki could enable an interactive and collaborative recovery plan. This method would require less confrontation and would help avoid scheduling conflicts. Within a wiki, each member of a therapeutic group could be as involved as they wanted. They could just read the wiki or contribute heavily. Mental health care providers could monitor the frequency of use and content of the recovery plan wiki to evaluate any changes in mental state. The group's use of the wiki would result in a shared knowledge and experience base (Bastida, McGrath and Maude, 2010). Participatory Web technologies allow greater access to personal health information. They do not rely on one person or organization to create value but on a community of individuals. These technologies allow patients to find and support each other. Just as face-to-face support groups can provide health benefits, so to could virtual support groups enabled by Participatory Web technologies. The technologies allow patients to share knowledge and experience of their health care. Patients can provide feedback on the quality, location and services available for various health issues. Patients could collaboratively create and manage health information. Research shows that collectively created content can be accurate (Deshpande and Jadad, 2006). Participatory Web technologies allow healthcare providers to meet the needs of underserved health groups. By creating intact communities around uncommon health issues, the technologies enable healthcare providers and companies to access and assess these groups. Companies otherwise do not tend to address rare health issues due to a smaller dispersed population (Deshpande and Jadad, 2006).

The Internet allows psychiatrist and patient relationships to become collaborative. In fact, patients' use of the Internet is so pervasive that it has become part of the psychiatrist-patient relationship (Yellowlees and Nafiz, 2010). Social networking sites further affect the psychiatrist-patient relationship. Social networks allow many more types of interactions than in the past. They include: psychiatrist-patient, patient-patient, psychiatrist-psychiatrist, psychiatrist-patient-caregiver-almost anyone. In the future the psychiatrist-patient relationship components could involve electronic data collection and analysis. Data could be sourced from personal medical records, blog posts and social network communication. Data analysis will be aided with decision support systems, communication systems and pattern-recognition systems. Despite all the potential benefits, psychiatrists should be careful when using social networking sites for therapy, as they have not been designed for this purpose. Security and confidentiality are significant issues within social networks. Participatory Web technologies could easily enable a non-specialist healthcare provider to collaborate asynchronously with a psychiatrist who could provide an expert consultation and send it back to the healthcare provider. As participatory collaborative technologies change the psychiatrist-patient relationship, psychiatrists should be aware of boundary violations caused by the fluidity of the relationship in these environments (Yellowlees and Nafiz, 2010).

Technical Information

Participatory Web Technology

Before the evolution of Web 2.0 and the Participatory Web, there was little need for different websites to communicate with each other. In the latest iteration of the Web however, there arose a need for the design of tools to enable sharing of information between tools. As a result, the hardware and software of Participatory Web have become more sophisticated in the way that they communicate with other technologies. Now, the communication compatibility of hardware and software has been moved to the forefront for developers of Participatory Web tools (Kamel Boulos and Wheelert, 2007).

The most technical aspect of participatory technology is the software running behind the scenes. Because it was essential that the software be able to interact with multiple tools, innovative programming languages have emerged. The new universal programming languages allow multiple tools to share data with other tools. The most notable programming languages are, but not limited to, AJAX, XML and RSS. When data from a tool is available in one of these formats, a second tool can use the data to integrate a portion of the second tool's functionality by linking the two tools together. When this software design philosophy is implemented, the intention is to make it easier to find and organize data. This is a major premise behind the Participatory Web movement (Kamel Boulos and Wheelert, 2007; O'Reilly, 2005). The hardware consideration from a user's perspective is that the personal computing device, or point of access to the Web, must be compatible with their lifestyle. Users usually determine this compatibility by assessing the connection speed, wired or wireless, and by the form factor. These qualifiers are usually based on user preference (Kamel Boulos and Wheelert, 2007).

Social Networks

Social networks are one of the most popular Participatory Web tools. A social network as defined by Boyd and Ellison (2008) is an application that allows individuals to (p. 210):

1. Construct a public or semi-public profile within a bounded system
2. Articulate a list of other users with whom they share a connection
3. View and traverse their list of connections and those made by others with the system

One of the main characteristics of this category of tool is that it can be easily integrated into users' daily lives. Various technological affordances such as tagging, liking, or following allow users to associate with a wide range of interests and people. Most social networks share similar features; however the cultures within the social networks are varied. For instance, Facebook, the largest social networking site, is known for connecting with friends from the past in a casual yet personal manner.

However, Linked In has similar features but is used primarily for professional purposes. Social networks can both help users maintain real-world social networks, and help strangers connect (Boyd and Ellison, 2008). Some social networks aim to capture the largest audience while others cater to specific group characteristics including language, race, or sexual orientation.

So far, the bulk of the research done related to the Participatory Web had been done on social networks. Currently, researchers from different fields are pursuing opportunities to investigate social networks in order to "better understand the practices, implications, culture, and meaning of these sites, as well as users' engagement with them" (Boyd and Ellison, 2008, pp. 211).

Facebook

Facebook is the most popular social networking website with over 500 million users. Half of those users log in every day (Hepburn, 2011). Facebook is also second only to Google as the most visited website on the Internet (Alexa, 2011). With so many active users, it is likely that a percentage of them would like information about or benefit from your expertise, research or practice.

The fundamental purpose of Facebook as a social network is to connect people, groups and organizations. As such, you have three options to connect with users on Facebook; through a profile, a page or a group.

Profiles

A profile is the primary element in Facebook (see Figure 1). It is a personal profile that users share with friends, family and colleagues. If you want to create a profile for your practice, group or other organization, a profile would not be appropriate. Facebook has features for non-personal profiles that we will discuss later. Profiles contain personal information that you choose to share. If you do not currently have a Facebook profile, you can sign up for one on Facebook's landing page by providing your first and last name, email, password, gender and birth date. From there, you can begin editing your profile. In your profile you can include location and contact information, your education and work history, your activities and interests and a profile picture.

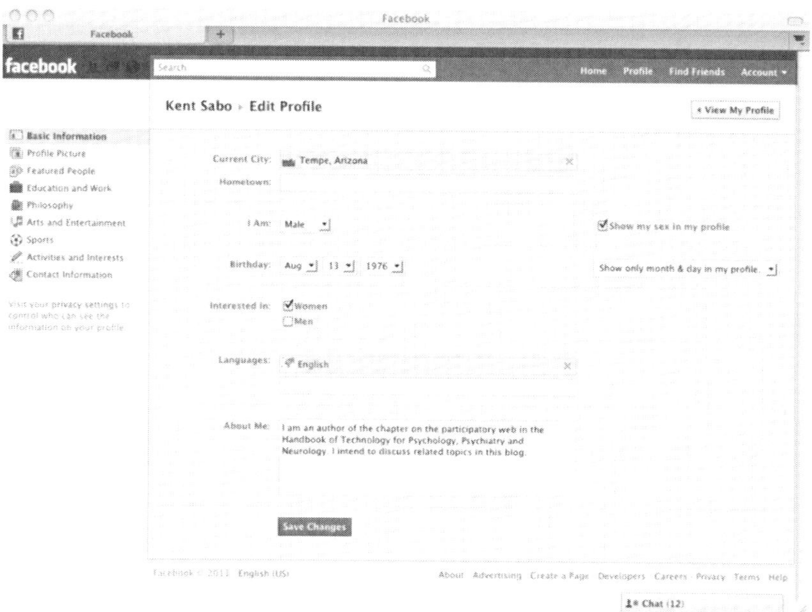

Figure 1. Editing a Facebook profile.

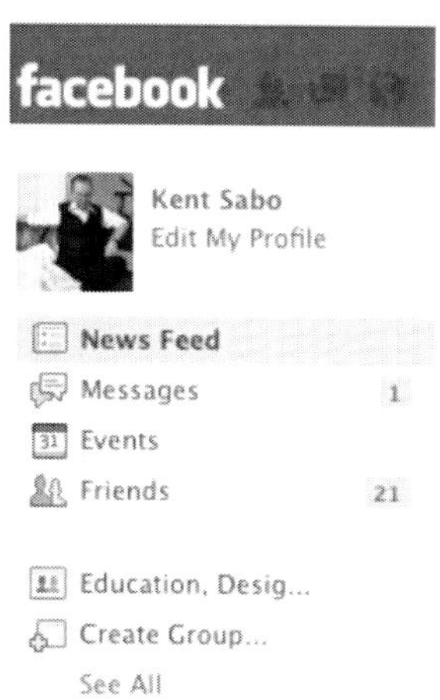

Figure 2. Facebook features.

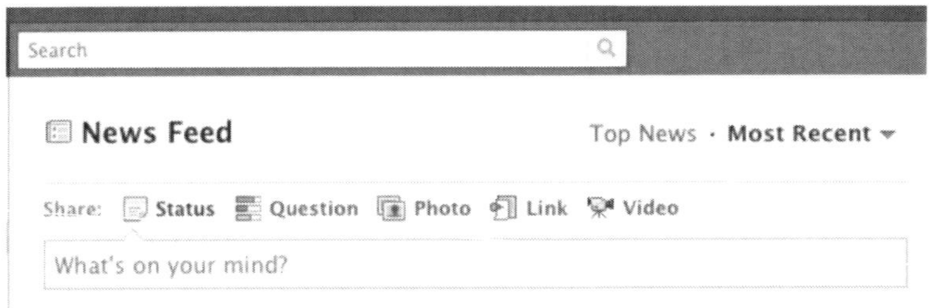

Figure 3. Facebook news feed.

After you've edited your profile information you will want to seek out friends, family and colleagues that also have Facebook profiles. Initially, you can use the search bar at the top of the page to search for people you know to have a Facebook profile. When you find those people, click the Add as Friend button at the top of their profile. This will send a request to the profile owner for approval. When they approve the request, they will show up in your Friends column on the left side of your profile (see Figure 2).

As you gain more friends, click on Find Friends in the upper right corner of the page to see a grid of people who are friends of your friends. This feature allows you to filter the grid by information you have included in your profile including: hometown, current city, high school, mutual friend, university and employer. You too will inevitably receive friend requests and friend suggestions. You can confirm, ignore or deny friend requests by clicking on Friends or the friend request icon in the upper left of the page.

News Feed

As you begin to acquire friends, you will see their status updates in your news feed which takes center stage on your Facebook home page (see Figure 3). Facebook users can share their status through text, photos, links or videos by using the text box beneath the news feed header that asks, “What's on your mind?” Top-level options for viewing your news feed include the links Top News and Most Recent. Top News includes only the most interesting updates from friends as determined by Facebook. Most Recent chronologically lists all updates from your friends. Click the drop-down arrow next to Most Recent to filter and select options for your news feed.

For each status update, you will see links to Like the post or comment on the post. For updates that include external links, you will see the option to Share. Clicking Like will add your name to the list of others who also like the post. Clicking the comment link or writing in the comment text box will add your comment to a thread of comments below the original post. Clicking the Share link will post the original link to your profile Wall and will be available to view in your friend's news feeds.

The Wall takes the same position on your profile as the news feed does on your Facebook home page. Your Wall features your status updates, shared photos and links as wells as posts written directly on your wall by your friends. To write on your friend's wall, view their profile and begin typing in the text box that reads “write something...” Be aware that walls are not private and others are able to read what you post. For messages you do not want to make public, use the private message feature by clicking on Messages in the upper left of your Facebook home page.

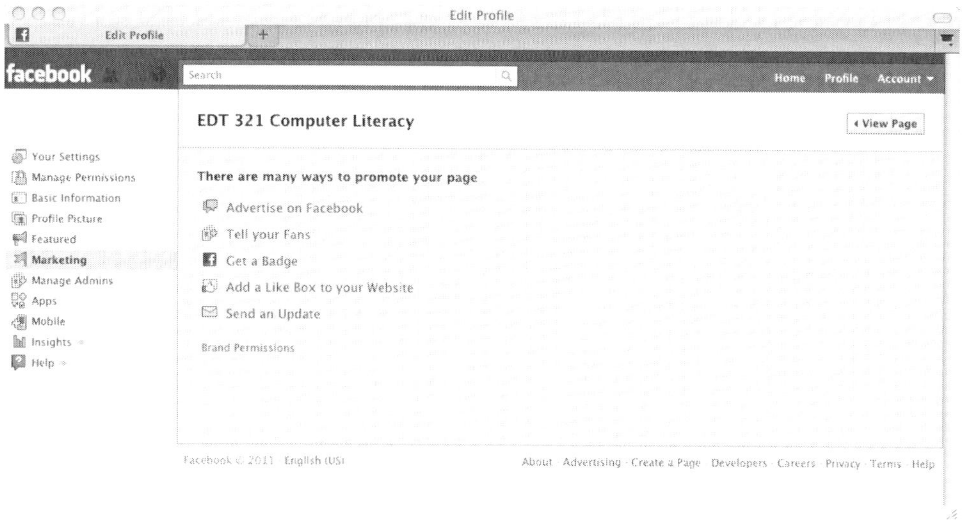

Figure 4. Facebook marketing options.

Pages

After you create a personal profile, you can create a Facebook Page by visiting facebook.com/pages, clicking Create Page and selecting the type of page you want to create. Among the types of pages that can be created are local business, organizations, products and causes. Pages work much in the same way as profiles although pages have features uniquely suited for organizations. Similar features include the Wall, info page and photos. Unique features include marketing options, analytics and management of multiple administrators for the page. While logged in, you can choose to use Facebook through your personal profile or through your Page. While using your personal profile, all your posts and activity will be identified with your personal profile information. If you would like to post as your organization's Page, click Account in the upper right and select Use Facebook as Page.

As with your profile, you should edit the page to include your organizations information by clicking the Edit Page button. After editing basic information and uploading pictures, click on Marketing. From here you have the option to advertise on Facebook, upload a contact list to notify your contacts about your page and to add a Like box to your organization's external website (see Figure 4). While still in the editing view, click Insights to view the analytics for your page. You will see the number of Likes, post views, comments and Wall posts graphically displayed and filtered by week or month. These analytics allow you to quantify the impact of your organization's Facebook page.

Groups

Many of the features of Facebook are geared towards communicating with a large number of people. Facebook groups are pages for a smaller group of people. To create a group, click the Create Group link from the left column on your home page. Groups can be secret, closed or open. If a group is designated as secret then only the members of the group can see it.

Closed groups can be viewed on Facebook, but only the members of the group can read the posts. An open group is available for anyone to view. Group pages have a similar format, as do profiles and pages and shares similar features. Features unique to groups include the ability to chat with members of the group (if they are logged on), create a group document and communicate outside of Facebook through the group's Facebook email address. To add members to your groups, click on Add Friends to Group and select the friends you would like to invite to the group. If you are interested in joining a group, click on Ask to Join a Group on the group's page.

Facebook is a rich and multi-faceted platform for communication with an unprecedented reach. We only touched on the basics and high-level features in this chapter to introduce those unfamiliar with the service. As more mental health practitioners are familiarized with Facebook, we are sure to see unique and innovative uses just as we saw with Web 1.0 technologies.

Blogs

One of the most talked about tools of the Participatory Web is blogs. Even though people were long creating webpages, there has been a ground swell among researchers and technologists around the potential for blogs. We feel that buzz stems from the realization that blogs could have a bigger impact beyond just being a personal diary or daily opinion column. To better understand the impact blogs could have, let’s take a moment to define what they are.

The term blog comes from combination of the words “Web” and “log”. A blog is a webpage used to share, reflect, and engage in commentary on news and events. Blogs have been called online journals where the entries are listed chronologically, this aspect doesn’t really account for the attention they are getting (O’Reilly, 2005).

Perhaps the buzz can be attributed to the ability to contribute and participate on the Web without the need for programming skills, as was necessary with Web 1.0 technology. With advancements in the technology supporting blog hosting sites, users with basic computer skills can be up and running on the Web in no time. Using built-in templates and What You See Is What You Get (WYSIWYG) editors, users can design professional looking blog sites with little experience.

Blogger

Blogger is a free web-based service, owned by Google, which allows users to easily publish content to the Web (see Figure 5). Blogger is one of the most popular blog services with 39% of bloggers surveyed by Technorati using Blogger to host their blog (Sobel, 2010). Creating a blog with blogger is a simple process. If you have a current Google account, you can sign into Blogger with your Google account email and password from the Blogger landing page.

Otherwise, you will need to create a new Google account by clicking the Get Started link from the landing page. When you sign up for a new Google account you will be asked for an email address, a name you would like displayed when you make posts, your gender and your birthday. In addition you will create a password and indicate that you accept the Blogger terms of service.

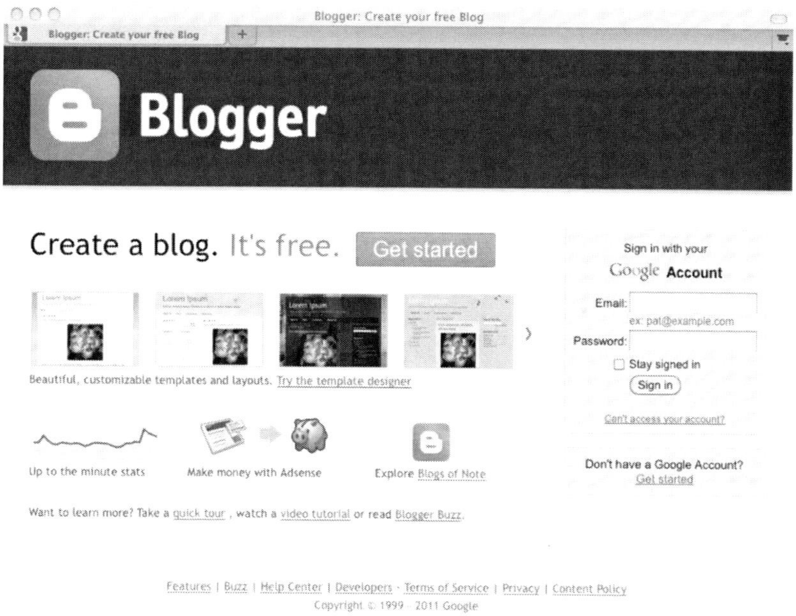

Figure 5. Blogger home page.

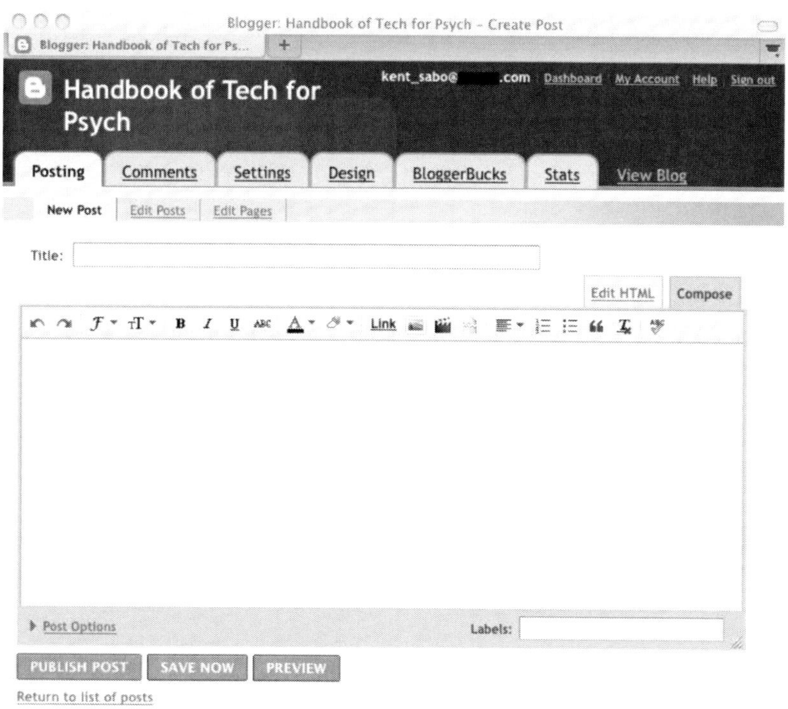

Figure 6. Blogger new post form.

After providing your sign in information, you will be taken to the Dashboard where you can create your blog by clicking Create Your Blog Now. Depending on the purpose for your blog, you will choose an appropriate name. Will your blog describe your experiences as a psychiatrist, or as a place for future and current patients to find information about your practice?

Before you decide on the name of your blog spend some time thinking about the blog's purpose and your goals for the blog. Some blog names that you create may already be taken since Blogger is such a popular service. Once you select the blog's name, you will then select the template for your blog. You can change the template at any time in the future so you can choose the first template to catch your eye and move on to publishing your first post.

New Posts

Immediately after you click Start Blogging you will see the Posting form. The posting form includes a text box for you to type the title of your blog post and a text box for the body of the message (see Figure 6).

The text box for the body of the blog post includes a WYSIWYG text editor with functions similar to those of popular word processing programs. These functions allow you to edit font, font size, text color, and to insert links, images and video. If you are familiar with HTML, you can edit the HTML code by clicking on the Edit HTML tab above the text box. In the lower right of the New Post form, you'll see a text box for Labels. Labels are key words that describe the topic of your post. Readers can search and browse your blog posts using the labels you provide for each post.

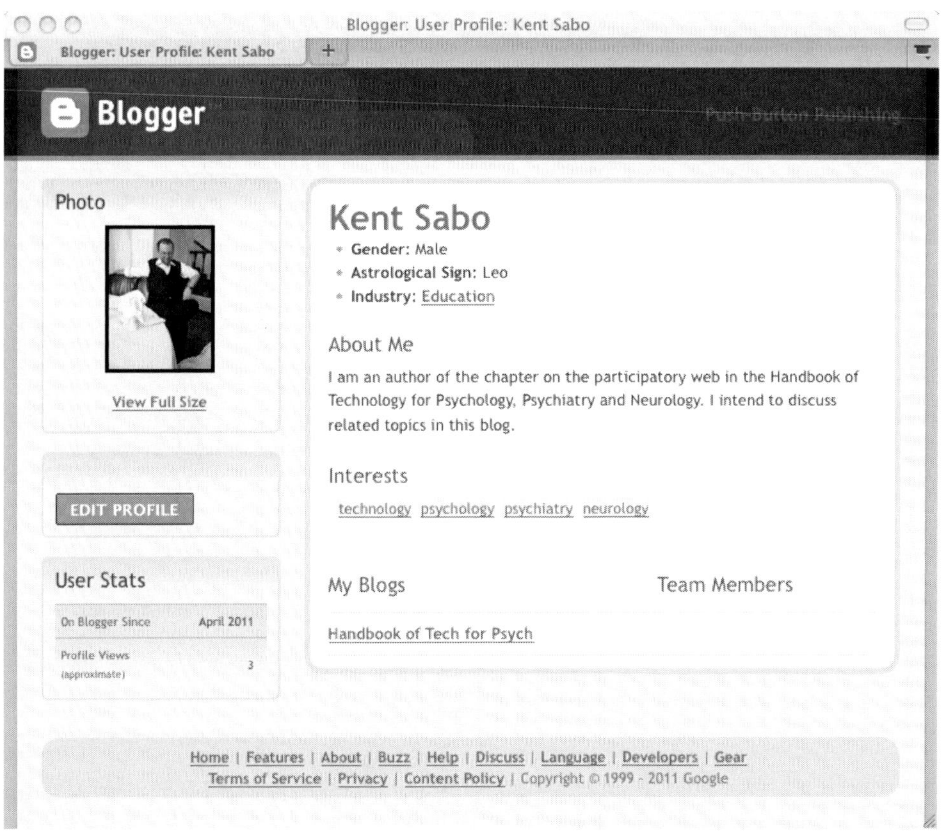

Figure 7. Blogger profile.

At the bottom of the New Post form are three buttons: Publish post, Save Now and Preview. Click Save Now to save the current state of your post. Your content will be saved even if your browser or operating systems crashes and you will be able to start where you left off when you log back into Blogger. Click the Preview button if you wish to see exactly how your post will look when it is published to your blog. Click Publish post if your post is ready to publish to your blog. You will then have the option to view your post, edit your post or create a new post.

After you've published a few posts and informed your friends, family and colleagues of your blog, you'll want to check to see if any of your posts have comments from readers. To view your comments, click the Comments link from the Dashboard. It is good practice to respond to readers that have questions or to simply thank them for commenting. This activity may encourage them to continue reading your blog.

Profile

When you log into Blogger, your first stop will be the Dashboard. From the dashboard you will see all the blogs that you manage or to which you contribute (you can create as many blogs as you wish in Blogger).

You will also see links related to your profile. Your Blogger profile can contain information about you and your blog (see Figure 7). In addition to the description you provide, when you click Edit profile, you can upload an image or audio clip to your profile. You can also choose to provide demographic information and interests that will become links in your profile.

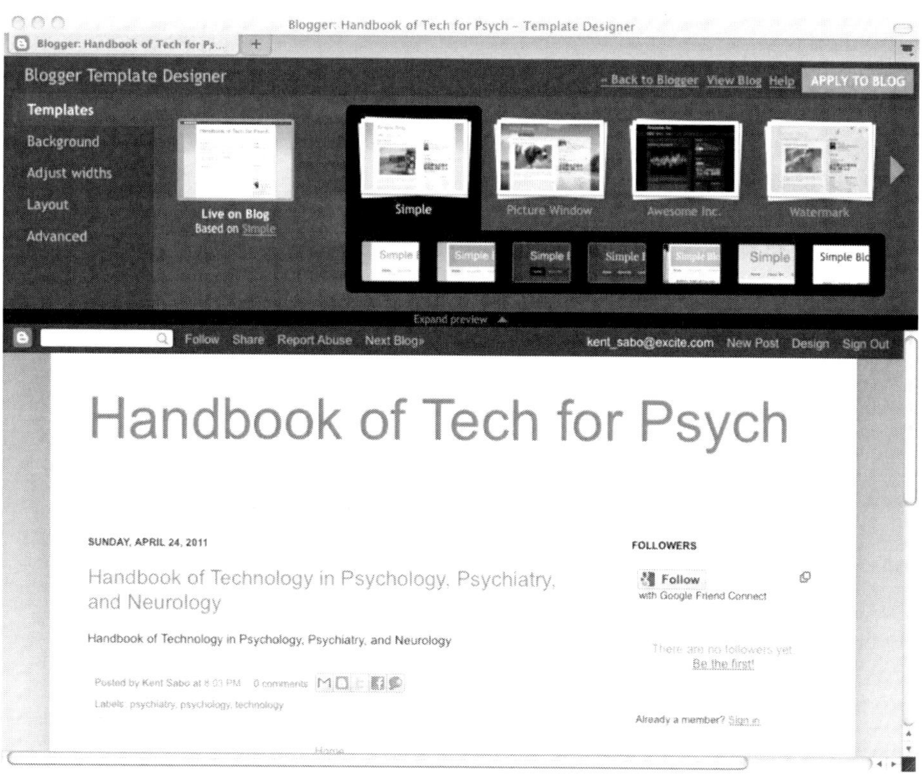

Figure 8. Blogger template designer.

Design

Clicking the Design link displays options for adding and arranging page elements, editing HTML and designing your blog template (see Figure 8). The Add and Arrange Page Elements page displays the generic layout of your blog including the title, the area for blog posts and default elements like Followers, Blog Archive and About Me. The Followers element lists the readers who choose to follow your blog. Blog Archive is a chronological list of all your blog posts.

About Me displays the about me text from your profile. Click on Edit to configure any one of the elements on the page. For the elements in the right column, you can click and drag the elements to sort them as you wish. Click Add a Gadget to display a host of gadgets that can make your blog more functional for you and your readers. Examples of basic gadgets include Follow by Email, Popular Posts and Blog's Stats. The Follow by Email gadget automatically sends your newly published posts to reader's email inboxes when they choose that option from your blog.

The Popular Posts feature displays a list of the most popular posts on your blog and Blog's Stats displays your blog's page views. Click on the Template Designer link to edit your blog's template, background image, element size, layout and other advanced features. In the Blogger template designer, any changes made to the aforementioned options are made in the preview pane. Click the Apply to Blog button in the upper right to apply the changes you selected in the template designer to your published blog.

Monetize

Click on the BloggerBucks link from the Dashboard and you have the option to add Google AdSense to your blog. AdSense is an advertising service that automatically analyzes the content of your blog and displays relevant advertisements. You can choose the color format and placement of advertisements in your blog. Depending on the types of advertisements that are served to your blog, you can earn money when they are viewed or when they are clicked.

Deciding to serve advertisements on your blog is a significant decision as they can affect your readers' experience. Although you can control the location of the advertisements, you cannot directly control the content. Your choice to advertise should align with your purpose for blogging and the goals for your blog.

Twitter

Twitter is a micro-blogging tool for sending short messages called tweets. Through tweets you can learn about what is happening in the world and follow people you find interesting. Twitter is a great way to share ideas, synthesize events, and network with people.

All you need to get started on Twitter is to create a free account by visiting Twitter.com. Once you have signed up, you can immediately follow and communicate with people. All you need is an Internet connection on your computer or a mobile phone that accepts text messages. After you register, users can create their first tweet by typing something in the "What are you doing" box (see Figure 9) and click update.

Figure 9. Twitter profile page.

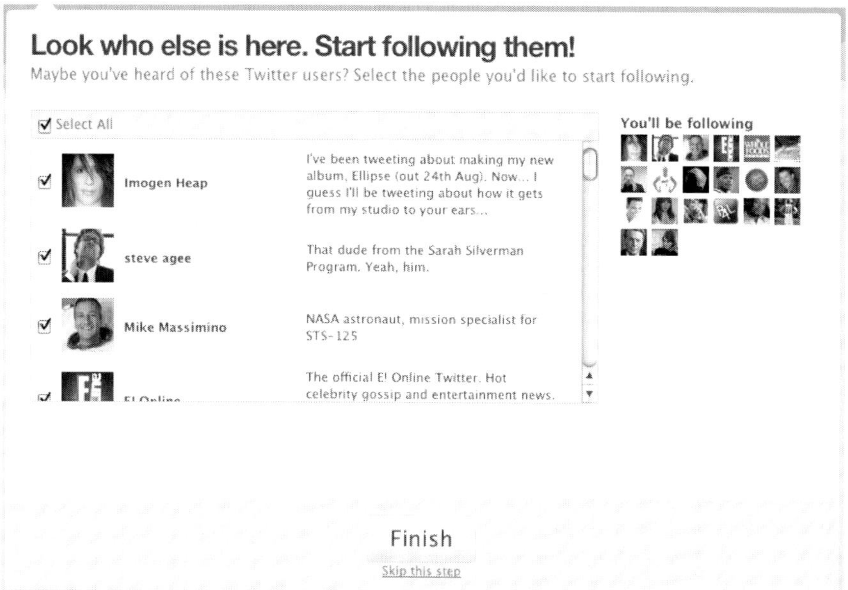

Figure 10. Twitter find followers page.

The goal is to get people to follow you. As shown in Figure 10, the best place to start is to start following people you know, maybe a friend, family member, or a colleague. As a courtesy, they usually follow you. You can find someone to follow in one of two ways. You can start by following the people who are following someone of common interest. To see other people who they are following, all you have to do is click “Followers”. You can also look someone up using the search feature in Twitter. The other way to follow someone is if you know a person’s user name or email address you can find him or her using the search feature in Twitter. Once you find them, click on their username and then click the "Follow" button to begin to follow them. It is important to learn some of the rules and guidelines of Twitter to utilize it effectively. The rules of Twitter are:

- Be polite; always be respectful
- Be appropriate; no offensive language, pictures or links
- Be concise, because you are limited to 140 characters per Tweet
- Be sure to thank the people who retweet (RT) something you wrote

Part of the fun in using Twitter is how easy and intuitive it is. The four basic features to master are:

- @username - Creates a link to that user in your post.
- RT - Retweet, to copy someone else's post in a new update. Give them credit by adding their @username.
- # - A hashtag helps to organize your tweets into categories for easier searching.
- DM - Direct message, send a tweeter a private message instead of an update that all your followers can read.

RSS READERS

One of the most unfamiliar and undervalued Participatory Web tool is the news aggregator. News aggregators, or Really Simple Syndication (RSS) readers, are used to view syndicated news and other Web content.

RSS readers can automatically aggregate regularly updated Web content such as blog entries, news headlines, audio, and video. With all the content available through social networks, wikis, blogs, podcasts, etc., you need a tool to help you quickly evaluate what information is important to you.

Benefits of RSS readers:

- No spam
- RSS is secure
- 100% delivery rate
- Unsubscribe with one click
- Subscriptions are anonymous

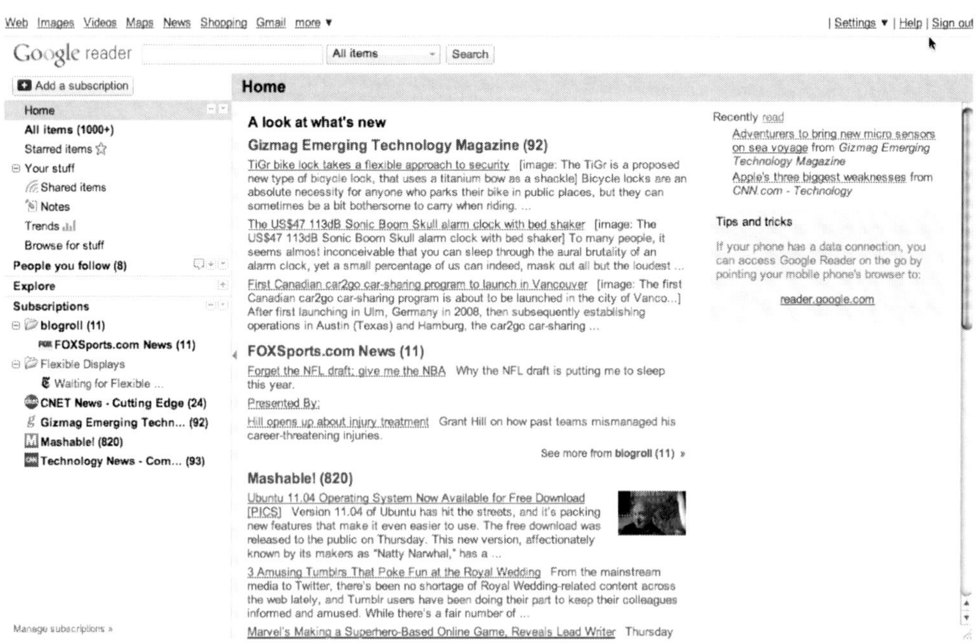

Figure 11. Google Reader.

Figure 12. RSS Icons.

RSS feeds are an excellent way to find all types of information. Via RSS feeds, you can receive news updates and announcements, keep up with your favorite websites and identify new trends. RSS readers can also serve as an organizational tool. For instance, scholars who use the Internet to conduct research can use RSS feeds to help them collect vital research materials from multiple sites and categorize them by pre-defined criteria. This way the RSS feeds are automating work that is usually done manually by organizing data in a repository that is readily available when needed. This feature frees up valuable time and resources for researchers. Nearly anything you can find on the Web can be delivered via an RSS feed.

Most RSS readers like Google Reader, FeedDemon, and NewsCrawler, are free. However, some offer additional services for a nominal fee. A feature to consider when selecting a RSS reader is how feeds are viewed. You might need to combine several feeds into a single view, hiding items that the viewer has already seen, and categorizing feeds and items with search terms.

RSS readers give you many ways to set up a subscription, but the most common way to add content is searching by key words or specific Web addresses. To set up a subscription using key words, the RSS reader scans for related content. Once the RSS reader finds matches, it will feed the information to your RSS reader. The other common way to set up a subscription for most RSS readers is to select specific webpages you want the reader to scan for you. All you need is the URL. A third option for subscribing is to look for an orange or gray icon labeled RSS or XML to indicate that the content is available via a feed (see Figure 12). Simply click the RSS icon and the RSS reader automatically starts monitoring it for updates.

Wikis

According to the creators Leuf and Cunningham (2001), wikis are a content management systems in the form of a collaborative website. The concept of wikis originated in 1995 when computer programmers Bo Leuf and Ward Cunningham created a database allowing fellow programmers to collect their knowledge. Subsequently, numerous wiki websites were created for similar purposes. Wikis were created for tasks such as "personal note taking to collaborating online, creating an internal knowledge base, assembling an online community, and managing a traditional website" (Matias, 2003, p. 1).

It is common for a group or organization to use wikis for developing collaborative communities related to a topic or purpose. One of the major benefits of wikis is that anyone with access to the Web can create or edit a wiki using a Web browser. Different from a blog, multiple individuals are able to create and edit wiki content at the same time. Wikis allow users to record document histories and revert to an older version of a document if necessary. Wikis assist users in simplifying and organizing topic pages (Matias, 2003). The benefits of using wikis include:

- Easy editing
- Simple markup language
- Document histories
- Easy linking
- Simple page creating
- Automated site organization
- Document management
- Enables collaboration

In addition to plain text, wiki pages can include hyperlinks, embedded images, audio, and video. Some wikis also have a discussion or comment option. The combination of these features can be used to provide feedback, identify a connection, or identify changes that have been made. Additions and changes can be tracked using history tools. Wikis are fully searchable and are ideal spaces for accessing useful and current resources. The best-known example of a wiki is the massively collaborative online encyclopedia Wikipedia (O'Reilly, 2005).

Innovative organizations are making their mark on the Web by utilizing the power of wikis and looking for ways to extend it even further. Wikipedia, being the most famous wiki is based on the idea that any user can add and edit any entry. This radical methodology is an experiment in trust, grounded on Eric Raymond's theory about open source software. He believes that the Web and all things on it belong to the collective and that the content will self-correct. This premise has yet to be proven; nevertheless, it is fun to think about in other applications (O'Reilly, 2005).

PBworks

PBworks has subscription options for business, education and personal purposes. The differences between subscriptions are the storage and performance levels of the account. PBworks also offers a free option, which is more than adequate for most people or organizations. If necessary, you can easily upgrade to a subscription level that better meets your needs. Another unique aspect of PBworks is that they refer to their wikis as workspaces. The workspaces are a collection of files and pages hosted by PBworks. When you sign up for the service, you can create your own working area to store and create information. PBworks even provides you with your own distinctive Web address with the PBWorks.com extension. For example, if you named your workspace "myfirstwiki" the location of your workspace would be: http://myfirstwiki.pbworks.com.

Even with the free subscription on PBworks, you can control who has access your workspace and all the data on it. When you sign up for an account, you'll receive an email with a link to set up the permissions and security settings. You can make your workspace only viewable by people you invite or approve so that your data is completely private.

Getting started with PBworks requires following a straightforward registration process similar to most Participatory Web tools. All that is required is a unique account name, which will serve as the account Web address, and to provide an email address, user name and password as shown in Figure 13.

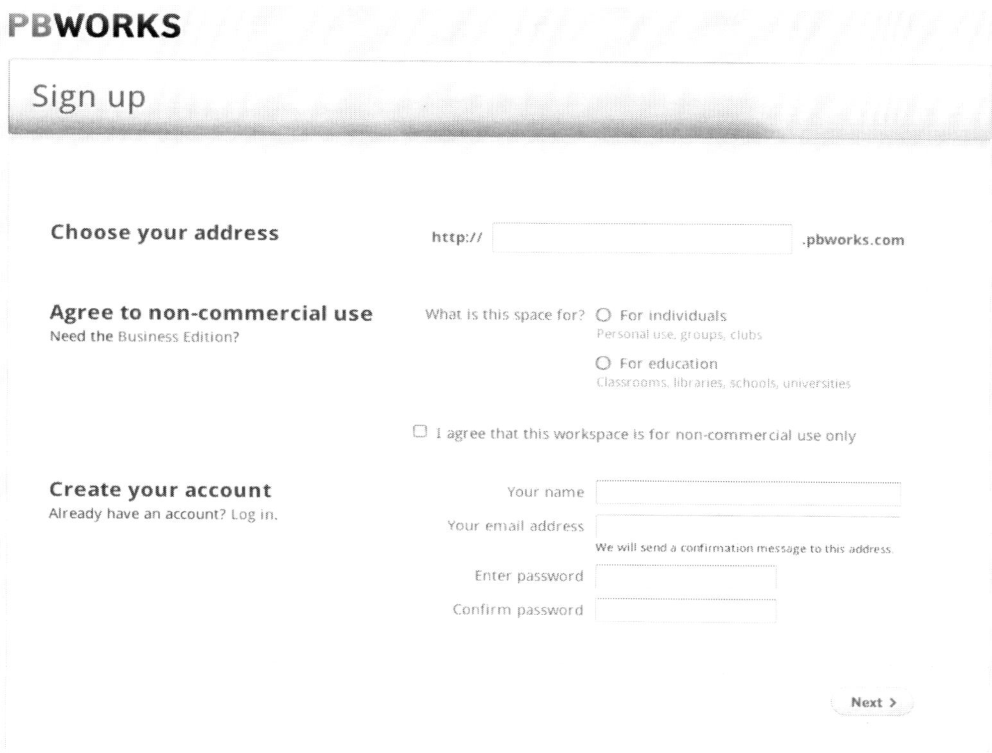

Figure 13. PBworks sign up page.

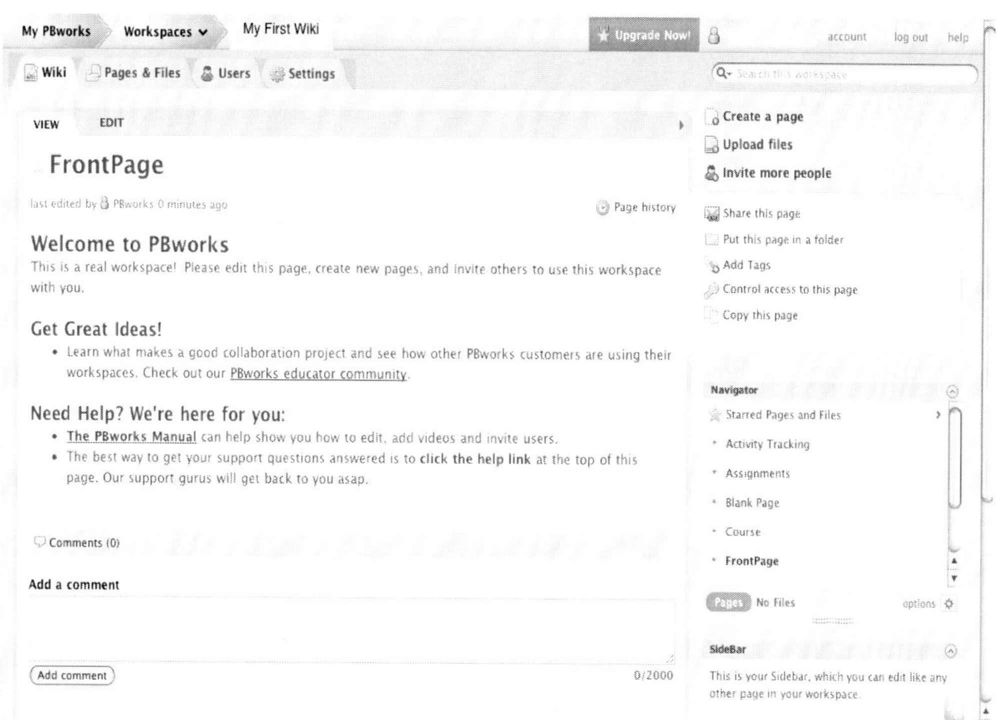

Figure 14. PBworks workspace page.

Once that part of the registration process is completed, you will receive an email with the link to access your very own wiki page. To finish signing up and continue to your workspace, click the link in the confirmation email sent to your email address and click the link provided. After clicking the link to activate your PBworks account, you will be taken to a workspace similar to the one shown in Figure 14.

You'll be able to log in to your workspace from any Web browser using just your email address and password. Click on the word "workspaces" in the top navigation bar to see all of your workspaces.

Conclusion

The Participatory Web is full of untapped potential for use in clinical settings, particularly in psychology, psychiatry and neurology. We indicated some of the key issues that arise for mental health practitioners as a consequence of the Participatory Web revolution. The most glaring of those issues is the need for empirical data to be gathered to determine the value added benefit of using these tools in mental health contexts. No doubt this will be a key issue for those interested in academic research and evidence-based clinical practice.

Current research has just scratched the surface by analyzing blog content for mental-health related content, investigating bloggers' perception of well being and the effects of social network participation on depression. Participatory Web tools are unparalleled platforms for communication. For example, designing and researching social-network based interventions for the treatment and prevention of depression and anxiety or addiction to smoking and alcohol could be an appropriate way to leverage the technology. By recognizing the potential in these tools, we hope mental health practitioners are inspired to pursue research that offers empirically based evidence supporting the use of Participatory Web tools in mental health practice.

Beyond direct clinical uses, mental health professional can use Participatory Web tools in their day-to-day professional dealings. We've provided a jumping-off point for parties interested in participating in the Web with our brief how-to section. They could use RSS feed readers to keep up with the latest news in the field, collaborate with colleagues across the globe using a wiki, and disseminate quality information on mental health through a blog. We look forward to more psychologists, psychiatrists and neurologists becoming involved in the evolution of the Participatory Web and observing its impact on the field of mental health and healthcare.

References

Alexa. (2011, April 25). Alexa Top 500 Global Sites. *Alexa*. Retrieved April 26, 2011, from http://www.alexa.com/topsites.

Anderson, J., Reitsma, R., and Liackman, B. (2009). Consumer behavior online: A 2009 deep dive. Retrieved from http://www.forrester.com/rb/Research/consumer_behavior_online_2009_deep_dive/q/id/54327/t/2.

Archambault, P. M., Légaré, F., Lavoie, A., Gagnon, M.-P., Lapointe, J., St-Jacques, S., and Pham-Dinh, M. (2010). Healthcare professionals' intentions to use wiki-based reminders to promote best practices in trauma care: a survey protocol. *Implementation Science, 5*, 45. doi:10.1186/1748-5908-5-45.

Baerlocher, M. O., and Detsky, A. S. (2008). Online medical blogging: don't do it! *Canadian Medical Association Journal, 179*(3), 292. doi:10.1503/cmaj.080757.

Bastida, R., McGrath, I., and Maude, P. (2010). Wiki use in mental health practice: Recognizing potential use of collaborative technology. *International Journal of Mental Health Nursing, 19*(2), 142-148. doi:10.1111/j.1447-0349.2009.00655.

Beer, D. and Burrows, R. (2010). Consumption, prosumption and Participatory Web cultures: An introduction. *Journal of Consumer* Culture, *10*(3), 3-12, doi: 10.1177/1469540509354009.

Bewick, B. M., Trusler, K., Barkham, M., Hill, A. J., Cahill, J., and Mulhern, B. (2008). The effectiveness of web-based interventions designed to decrease alcohol consumption–a systematic review. *Preventive Medicine, 47*(1), 17-26. doi:10.1016/j.ypmed.2008.01.005.

Boyd, D. M. and Ellison, N. B. (2008). Social network sites: definition, history, and scholarship. *Journal of Computer-Mediated Communication, 13*, 210–230. doi: 10.1111/j.1083-6101.2007.00393.x.

Campbell, C. (2005). The craft consumer: Culture, craft and consumption in a postmodern society, *Journal of Consumer Culture, 5*(1), 23–42.

Dainton, C. (2009). Physician-writers in the age of blogging. *Canadian Medical Association Journal, 181*(5), 348-349. doi:10.1503/cmaj.090305.

Deshpande, A., and Jadad, A. (2006). Web 2.0: Could it help move the health system into the 21st century? *The Journal of Men's Health and Gender, 3*(4), 332-336. doi:10.1016/j.jmhg.2006.09.004.

DiNucci, D. (1999). "Fragmented Future". *Print 53*(4), 32.

Eysenbach, G. (2008). Medicine 2.0: Social Networking, Collaboration, Participation, Apomediation, and Openness. *Journal of Medical Internet Research, 10*(3), 22. doi: 10.2196/jmir.1030.

Fernandez-Aranda, F., Nunez, A., Martinez, C., Krug, I., Cappozzo, M., Carrard, I., . . . Lam, T. (2009). Internet-Based Cognitive-Behavioral Therapy for Bulimia Nervosa: A Controlled Study. *CyberPsychology and Behavior, 12*(1), 37-41.

Grabe, M. and Grabe, C. (2008). Meaningful learning and the participatory web. [Web log post]. Retrieved from http://learningaloud.com/participatoryweb/.

Griffiths, K. M., Calear, A. L., and Banfield, M. (2009). Systematic Review on Internet Support Groups (ISGs) and Depression (1): Do ISGs Reduce Depressive Symptoms? *Journal of Medical Internet Research, 11*(3), 40. doi:10.2196/jmir.1270.

Hanson, C., Thackeray, R., Barnes, M., Neiger, B., and McIntyre, E. (2008). Integrating Web 2.0 in Health Education Preparation and Practice. *American Journal of Health Education, 39*(3), 157-166.

Hardey, M. (2010). Consuming professions: User-review websites and health services. *Journal of Consumer Culture,10*(1), 129-149. doi: 10.1177/1469540509355023.

Hepburn, A. (2011, January 18). Facebook Statistics, Stats and Facts For 2011 | Digital Buzz Blog. *Digital Buzz*. Retrieved April 26, 2011, from http://www.digitalbuzzblog.com/facebook-statistics-stats-facts-2011/.

Jenkins, H. (2006). Eight traits of the new media landscape. [Web log post]. Retrieved from http://henryjenkins.org/2006/11/eight_traits_of_the_new_media.html.

Kamel Boulos, M. N. and Wheeler, S. (2007). The emerging Web 2.0 social software: an enabling suite of sociable technologies in health and health care education. *Health Information and Libraries Journal, 24*(1), 2-23. doi: 10.1111/j.1471-1842.2007.00701.x.

Kaplan, A. M., and Haenlein, M. (2010). Users of the world, unite! The challenges and opportunities of social media. *Business Horizons, 53*(1), 59-68.

Kardong-Edgren, S. E., Oermann, M. H., Ha, Y., Tennant, M. N., Snelson, C., Hallmark,... Hurd, D. (2009). Using a Wiki in Nursing Education and Research. *International Journal of Nursing Education Scholarship, 6*(1), 1-10. doi:10.2202/1548-923X.1787.

Kim, J. Y., Gudewicz, T. M., Dighe, A. S., and Gilbertson, J. R. (2010). The pathology informatics curriculum wiki: Harnessing the power of user-generated content. *Journal of Pathology Informatics, 1*, 10. doi:10.4103/2153-3539.65428.

Klumpp, E. (2010). Dr. Blog? *Psychiatry, 7*(8), 50-52.

Ko, H.-C., and Kuo, F.-Y. (2009). Can Blogging Enhance Subjective Well-Being Through Self-Disclosure? *CyberPsychology and Behavior, 12*(1), 75-79.

Konovalov, S., Scotch, M., Post, L., and Brandt, C. (2010). Biomedical Informatics Techniques for Processing and Analyzing Web Blogs of Military Service Members. *Journal of Medical Internet Research, 12*(4). doi:10.2196/jmir.1538.

Kovic, I., Lulic, I., and Brumini, G. (2008). Examining the Medical Blogosphere: An Online Survey of Medical Bloggers. *Journal of Medical Internet Research, 10*(3). doi:10.2196/jmir.1118.

Kubben, P. L. (2010). Introducing Neurosurgery 2.0. *Surgical Neurology International, 1*, 8. doi:10.4103/2152-7806.63900.

Lagu, T., Kaufman, E. J., Asch, D. A., and Armstrong, K. (2008). Content of Weblogs Written by Health Professionals. *Journal of General Internal Medicine, 23*(10), 1642-1646. doi:10.1007/s11606-008-0726-6.

Leuf, B. and Cunningham, W. (2001). *The wiki way: quick collaboration on the Web*. Boston: Addison-Wesley.

Matias, N. (2003). What is a wiki? [Web log post]. Retrieved from http://blogs.sitepoint.com/what-is-a-wiki/.

Naik, A. D., and Singh, H. (2010). Electronic Health Records to Coordinate Decision Making for Complex Patients: What Can We Learn from Wiki? *Medical Decision Making, 30*(6), 722-731. doi:10.1177/0272989X10385846.

Nordqvist, C., Hanberger, L., Timpka, T., and Nordfeldt, S. (2009). Health Professionals' Attitudes Towards Using a Web 2.0 Portal for Child and Adolescent Diabetes Care: Qualitative Study. *Journal of Medical Internet Research, 11*(2). doi:10.2196/jmir.1152.

O'Reilly, T. (2005). What is Web 2.0. Design patterns and business models for the next generation of software. *Communications and Strategies. 65*(1) 17-37. Retrieved from http://oreilly.com/pub/a/web2/archive/what-is-web-20.html?page=1.

Ostrow, A. (2009). Number of social networking users has doubled since 2007. *Mashable*. Retrieved from http://mashable.com/2009/07/28/social-networking-users-us/.

Randeree, E. (2009). Exploring technology impacts of Healthcare 2.0 initiatives. *Telemedicine Journal and E-Health: The Official Journal of the American Telemedicine Association, 15*(3), 255-260. doi:10.1089/tmj.2008.0093.

Schroeder, S. (2009). The Web in numbers: The rise of social media. *Mashable*. Retrieved from http://mashable.com/2009/04/17/web-in-numbers-social-media/.

Sobel, J. (2010, November 5). HOW: Technology, Traffic and Revenue - Day 3 SOTB 2010. *Technorati*. Retrieved April 25, 2011, from http://technorati.com/blogging/article/how-technology-traffic-and-revenue-day/.

Spek, V., Cuijpers, P., NyklíCek, I., Riper, H., Keyzer, J., and Pop, V. (2006). Internet-based cognitive behaviour therapy for symptoms of depression and anxiety: a meta-analysis. *Psychological Medicine, 37*(03), 319. doi:10.1017/S0033291706008944.

Swan, M. (2009). Emerging Patient-Driven Health Care Models: An Examination of Health Social Networks, Consumer Personalized Medicine and Quantified Self-Tracking. *International Journal of Environmental Research and Public Health, 6*(2), 492-525. doi:10.3390/ijerph6020492.

Takahashi, Y., Uchida, C., Miyaki, K., Sakai, M., Shimbo, T., and Nakayama, T. (2009). Potential Benefits and Harms of a Peer Support Social Network Service on the Internet for People With Depressive Tendencies: Qualitative Content Analysis and Social Network Analysis. *Journal of Medical Internet Research, 11*(3). doi:10.2196/jmir.1142.

Thrift, N. (2006). *Knowing Capitalism*. London: Sage Publications.

Tunick, R., and Mednick, L. (2008). Commentary: Electronic Communication in the Pediatric Setting–Dilemmas Associated with Patient Blogs. *Journal of Pediatric Psychology, 34*(5), 585-587. doi:10.1093/jpepsy/jsn088.

Van De Belt, T. H., Engelen, L. J., Berben, S. A., and Schoonhoven, L. (2010). Definition of Health 2.0 and Medicine 2.0: A Systematic Review. *Journal of Medical Internet Research, 12*(2). doi:10.2196/jmir.1350.

Van Voorhees, B. W., Fogel, J., Reinecke, M. A., Gladstone, T., Stuart, S., Gollan, J... Bell, C. (2009). Randomized Clinical Trial of an Internet-Based Depression Prevention Program for Adolescents (Project CATCH-IT) in Primary Care: 12-Week Outcomes. *Journal of Developmental and Behavioral Pediatrics, 30*(1), 23-37. doi:10.1097/DBP.0b013e3181966c2a.

White, A., Kavanagh, D., Stallman, H., Klein, B., Kay-Lambkin, F., Proudfoot, J., . . . Young, R. (2010). Online Alcohol Interventions: A Systematic Review. *Journal of Medical Internet Research, 12*(5). doi:10.2196/jmir.1479.

Wilson, J. A. B., Onorati, K., Mishkind, M., Reger, M. A., and Gahm, G. A. (2008). Soldier Attitudes about Technology-Based Approaches to Mental Health Care. *CyberPsychology and Behavior, 11*(6), 767-769.

Wilson, P., Petticrew, M., and Booth, A. (2009). After the gold rush? A systematic and critical review of general medical podcasts. *Journal of the Royal Society of Medicine, 102*(2), 69-74. doi:10.1258/jrsm.2008.080245.

Yellowlees, P., and Nafiz, N. (2010). The Psychiatrist-Patient Relationship of the Future: Anytime, Anywhere? *Harvard Review of Psychiatry, 18*(2), 96-102.

In: Handbook of Technology in Psychology ... ISBN: 978-1-62100-004-4
Editor: Luciano L'Abate and David A. Kasier

Chapter 7

COMPUTER-BASED COGNITIVE STIMULATION PROGRAMS TO REMEDIATE AGE-RELATED COGNITIVE DECLINE: WHAT MAKES A PROGRAM EFFECTIVE?

Peter B. Delahunt[1], Jessica B. Morton and Henry W. Mahncke
[1]University of California, Davis Medical Center, Sacramento, California, US

ABSTRACT

Cognitive functioning typically starts to decline in our thirties and is evidenced by reduced performance on a range of functions including working memory, perceptual processing speed, reasoning ability, and attentional capacity (Salthouse, 2004). Although the rate of decline is variable, for many the decline will eventually lead to a reduced ability to perform basic tasks that are important for maintaining independence including driving, managing money, shopping, management of medications, memory, and using public transit. Reduced independence negatively impacts health-related quality of life measures (Jobe, Smith, Ball, Tennstedt, Marsiske, et al., 2001), produces higher rates of depression (Marottoli, Mendes de Leon, Glass, and Williams, 1997), and leads to increased risk of nursing home care (Wolinsky, Callahan, Fitzgerald, and Johnson, 1993), which in turn leads to increased mortality rates (Wolinsky, Callahan, Fitzgerald, and Johnson, 1992).

Clearly it is desirable to preserve cognitive functioning to maintain independence and quality of life. Numerous longitudinal studies have now demonstrated an association between engagement in higher levels of cognitively stimulating activities and preservation of cognitive function (Verghese, Lipton, Katz, Hall, et al., 2003: Akbaraly, Portet, Fustinoni, Dartigues, et al., 2009). This observation has led to considerable interest in the use of cognitively stimulating activities as interventions to maintain or improve cognitive function and associated quality of life, with the phrase "use it or lose it" entering the vernacular, and popular recommendations for activities ranging from crossword puzzles to video game bowling becoming common.

In particular, computerized cognitive training programs specifically designed to maintain or improve cognitive function in older adults have become commonly available over the past 10 years (Fernandez and Goldberg, 2008). These programs vary widely in their design goals, with some designed from neuropsychological or neurological principles while others are intended principally for entertainment purposes.

Many such programs cite the longitudinal data noted above to support their use in age-related cognitive decline ("Dakim - Why It Works" n. d.: : "Reference Materials for Further Brain Training Study - Lumosity," n. d.). However, data from longitudinal studies alone cannot rigorously establish that engagement in cognitively stimulating activity causally drives the preservation of cognitive function – it can only demonstrate the association, and it has been argued that high levels of engagement in cognitively stimulating activities itself are a consequence, rather than a cause, of preserved cognitive function (Mackinnon, Christensen, Hofer, Korten, and Jorm, 2003). Given this fundamental methodological limitation, it is crucial for novel computerized cognitive training programs to independently establish their efficacy.

This chapter will lay out a framework for the rigorous clinical evaluation of computer-based cognitive training programs, review randomized controlled trials of such programs evaluated in normal aging populations, establish common attributes that appear to be important for efficacious training programs based on the published evidence, and relate these attributes to current knowledge of brain plasticity.

MEASURING EFFICACY

Approaches to Assessing Benefits

Clinical trials of cognitive training programs often include a broad spectrum of outcome measures. A useful approach to categorizing such measures is to consider where they lie on a spectrum of generalization extending from the specifics of the trained task up to real-world function.

Train to Task

An important first step is to determine if a participant is actually getting better at the task on which they are training. While this may initially sound trivial, this type of assessment serves as an important positive control for task learning – without task improvement it is unlikely that generalization to broader cognitive measures will be seen. The magnitude of train-to-the-task improvements can also be used to segment populations into responders and non-responders, and to analyze which sub-populations may be treatable or not treatable.

Neuropsychological Measures of Generalization

Improvements on task performance are useful, however it is crucial to establish that benefits from cognitive training programs generalize to untrained tasks. This can be measured through the use of standardized neuropsychological tests (provided that the tests are sufficiently distinct from the training program). The virtue of such measures is that well validated standardized tests can be used with blinded raters and other quality control measures (Lezak, 2004) to provide objective evidence of generalized cognitive improvement.

Self-Report Measures of Generalization

Beyond establishing generalization on neuropsychological measures, it is important to establish if participants themselves notice changes in their every day lives. This can be measured using well-designed and validated questionnaires, which can help establish that the magnitude of neuropsychological changes are sufficient to impact every day life.

Real-World Measures of Generalization

In addition to self-report measures, it is important to measure generalization to directly-observed real-world functions that are important to older individuals, including such skills such as mobility (e.g., driving, falls), finances (e.g., money management), and health management (e.g., medication management). Unfortunately, these real-world skills can be difficult to assess reliably in clinical trials given the challenges with reproducibility and quantification that accompany the assessment of such complex skills.

Long-Term Outcomes

Finally, the most important measure of benefit from a cognitive training program is its effects on the long-term outcome of aging, including the diagnosis of Mild Cognitive Impairment or dementia, or the movement to assisted living facility or nursing home. Such measures generally require studies with several years of follow-up.

Based on this framework, we now turn our attention to published studies of randomized controlled trials of computerized cognitive training programs in older adults.

Review of Studies

Useful Field of View (UFOV) Training

UFOV refers to the visual area over which information can be extracted at a single glance without head or eye movements (Ball, Beard, Roenker, Miller, and Griggs, 1988). UFOV generally reduces with age and has been shown to be an important predictor of a range of outcomes including automobile crashes (Owsley, Ball, Sloane, Roenker, and Bruni, 1991: Sims, McGwin, Allman, Ball, and Owsley, 2000: Edwards, Ross, Wadley, Clay, Crowe, et al., 2006: Rubin, Ng, Bandeen-Roche, Keyl, Freeman, et al., 2007), bumping into objects while walking (Broman, West, Munoz, Bandeen-Roche, Rubin, et al., 2004), and falling (Vance, Ball, Roenker, Wadley, Edwards, et al., 2006). The UFOV training exercise was developed by Drs. Karlene Ball and Daniel Roenker of the University of Alabama and Western Kentucky University, who founded Visual Awareness Incorporated. The exercise requires the participant to discriminate central stimuli while concurrently locating a peripheral target. It is adaptive with the stimulus presentation times adjusted depending on performance. The task is made more challenging as training progresses by the addition of distracters and eventually a more difficult central task. The training is generally set up for ten separate one-hour sessions. UFOV training is often referred to as 'speed of processing' or 'speed' training in the literature.

UFOV training has been evaluated in a number of studies. Two of these studies (ACTIVE and SKILL) include large participant cohorts that continue to be followed. ACTIVE (Advanced Cognitive Training in the Independent and Vital Elderly) is the largest study of cognitive interventions conducted to date. It is a multi-site double-blinded randomized controlled clinical trial which enrolled 2,832 participants aged 65 and over from 6 different geographical locations across the United States. Participants were randomized to one of four groups: memory training, reasoning training, UFOV training, and a no-contact control. All participants in the intervention groups received 10 sessions of training on separate days. The sessions lasted for 60 to 90 minutes. The memory and reasoning training programs used non-computer strategy based approaches in group settings with a facilitator. UFOV training was the only computer-based cognitive stimulation intervention in ACTIVE. Various data analyses have been and continue to be conducted on the ACTIVE cohort. Detailed information on the ACTIVE study design and methods has been published (Jobe, Smith, Ball, Tennstedt, Marsiske, et al., 2001).

UFOV training produces large train-to-task improvements. For example, 87% percent of the ACTIVE UFOV group showed reliable improvements with an average effect size of 1.46 (Ball, Berch, Helmers, et al., 2002). The training effects are long lasting, with an effect size of 0.76 measured 5 years post training (Willis et al., 2006). This general pattern of results was further confirmed in the ACCELERATE study (N = 159) (Vance, Dawson, Wadley, Edwards, Roenker, et al., 2007).

Although these strong train-to-task results were seen, initial analyses in the ACTIVE study showed that UFOV training did not show significant transfer to memory or reasoning measures (Ball, Berch, Helmers, et al., 2002). The ACCELERATE study again confirmed this pattern of results, showing no transfer to various neuropsychological measures with the exception of the Starry Night task, a visual signal detection task (Vance et al., 2007).

However, important generalization was seen from UFOV training in the ACTIVE study to multiple participant-reported measures of function. Performance on Independent Activities of Daily Living (IADLs) in the UFOV group with an additional four hours of training (the "booster" group) was significantly improved immediately after training compared with the group receiving the standard ten hours of training (K. Ball, Berch, Helmers, et al., 2002); and decline on IADLs in the overall group was slowed over a five year period when compared to the control group, with a Cohen's *d* effect size of 0.26 (Willis et al., 2006). Participants in the UFOV group were 38% less likely than the control group to report serious health-related quality of life decline measured at the two years post training and 26% less likely at five years (Wolinsky, Unverzagt, Smith, R. Jones, Wright, et al., 2006: Wolinsky, Unverzagt, Smith, Jones, Stoddard, et al., 2006). The UFOV trained group also was 30% less likely to experience serious decline in mood and worsening of depression symptoms measured five years post-training (Wolinsky, Vander Weg, et al., 2009). In a separate analysis, the UFOV group was shown to be 38% less likely to go from no depressive symptoms at baseline to showing depressive symptoms at the one year follow up (Wolinsky, Mahncke, Vander Weg, et al., 2009). Wolinsky et al. used a model based on previous work by others that relating the SF-36 health-related quality of life measure to total medical expenditures (Fleishman, Cohen, Manning, and Kosinski, 2006) to show that UFOV training was predicted to reduce total medical expenditures by 3.3% in the first year post training (Wolinsky, Mahncke, Kosinski, Unverzagt, et al., 2009).

Generalization was also demonstrated to real-world performance measures. Participants who received an additional 4 hours of UFOV training (in the "booster" group) showed significantly improved everyday speed (Ball, Berch, Helmers, et al., 2002), a directly observed composite measure of real-world functions (Owsley, M. Sloane, McGwin, and Ball, 2002), an effect that was maintained at the five year follow up (Willis et al., 2006). Furthermore, the examination of crash rates obtained from state records for the ACTIVE participants in the 5 to 6 year period following training (Ball, Edwards, Ross, and McGwin Jr., 2010) demonstrated that the UFOV trained group had a 48% reduction in at-fault crashes per mile driven compared to the control group.

Other work using real-world outcome measures has shown significant transfer to real-world measures for UFOV training. Edwards and colleagues examined the effect of UFOV training on timed instrumental activities of daily living for the SKILL cohort (Edwards, Wadley, Vance, et al., 2005). Participant's performance on a number of timed tasks was measured before and after training. The tasks included looking up a number in a phone book, counting change, looking for a food item in a cupboard, and locating and reading instructions on a medicine container. The study found that the UFOV trained group was significantly faster than the active-control group on the tasks after training (Cohen's *d* of 0.40). These results corroborated results from a previous study conducted by Edwards showing similar results (Edwards, Wadley, Myers, Roenker, et al., 2002) in a non-UFOV impaired population (Cohen's *d* of 0.41).

UFOV training has also been evaluated in the SKILL (Staying Keen In Later Life) cohort that included 895 community-dwelling adults ages 60 and over. Participants were recruited from mass-mailings to residents of Birmingham, Alabama, Bowling Green, Kentucky, and surrounding areas. Unlike the ACTIVE study, only participants with poor pre-training UFOV baseline measures received UFOV training. SKILL used an active control group who underwent internet training. Further information on the inclusion criteria can be found elsewhere (Edwards, Myers, et al., 2009).

The SKILL study was used to examine the effect of UFOV training on driver mobility (Edwards, Myers, et al., 2009). Participants completed questionnaires at the start of the study and three years later post training. The questionnaires focused on driving space (how far the participant drove from home), driving exposure (e.g., driving at night, in rush hour, left turns) and driving difficulty (e.g., difficulty rating for driving alone, when changing lanes). After training the trained group performed similarly to the high UFOV performing reference group (non-trained) participants at the three-year follow up while the active control declined significantly.

The SKILL and ACTIVE data sets were combined to examine the effect of UFOV training on driver cessation (Edwards, Delahunt, and Mahncke, 2009). Only participants with poor pre-training UFOV baseline performance were included in the experimental group. The analysis showed that the trained group was 40% less likely to cease driving in the three-year follow up period compared to the control group.

The effect of UFOV training on driving performance was examined by Roenker et al. (Roenker, Cissell, Ball, Wadley, and Edwards, 2003). They randomly assigned poor UFOV performers to either UFOV training (N=48) or traditional driver training (N=22). They used high performing UFOV participants as a low-risk reference control (N=25). All participants were 55 and over. The UFOV group trained until they met a pre-determined UFOV performance criterion. This train-to-task improvement took on average 4.5 hours training. An on-the-road driving evaluation was performed before and after training, and at the 18 months follows up. It was found that both training groups showed improvements immediately after training but only the improvements for the UFOV group persisted at the 18 month follow up. The UFOV group reduced dangerous maneuvers by 36% at the 18 month follow up while the control group and simulator trained groups showed increases. In addition, the UFOV training group improved at a complex reaction time task that the authors equate to stopping 22 feet sooner when driving at 55 mph.

Scalf et. al. (Scalf, Colcombe, McCarley, et al., 2007) used functional magnetic resonance imaging to determine if UFOV training drove functional activation changes. Their study documented robust changes in activation in a distributed network of parietal and frontal areas generally associated with attention function. This work is the first to document that brain changes in humans associated with a specific form of cognitive training.

An interesting result from this family of studies is that evaluating transfer of training benefits is not a straightforward issue. Concluding whether an exercise shows good generalizability is critically dependent on the outcome measures chosen. For example, UFOV training failed to transfer to memory and reasoning measures in the ACTIVE study (Ball, Berch, Helmers, et al., 2002) and also showed little transfer to a range of neuropsychology measures often used to assess cognitive performance (Vance et al., 2007). However, remarkable real world transfer has been shown using other outcome measures including TIADLs (Edwards, Wadley, Vance, et al., 2005: Ball, Edwards, and Ross, 2007), driving safety measures (Roenker et al., 2003), and at-fault crash incidence reduction (Ball et al., 2010). The lack of transfer across domains shows suggests that transfer might be limited across functional area (e.g., faster perceptual processing does not transfer to reasoning ability) but does transfer to tasks that are dependent on the trained function (e.g., faster perceptual processing reduces crash risk).

These results are particularly interesting and compelling because they derive from large, well designed studies with multiple outcome measures, and longitudinal follow-up. The results show that speed training is particularly plastic, with large train-to-task effects that generalize to real-world functions essential for successful aging including independent living, everyday speed and driving function. This generalization appears to occur directly by the improvement of the core speed function without accompanying improvements in other unrelated neuropsychological mechanisms (e.g., memory, reasoning). These results imply that clinical trials of cognitive training programs should employ multiple measures of generalization across the assessment ladder described above to have the best opportunity to detect such generalization regardless of the mechanism by which the generalization may be driven.

Brain Fitness Program (BFP)

BFP is a commercially available auditory computer-based cognitive stimulation program from Posit Science. It was designed to improve the speed and accuracy of auditory processing with the goal of improving verbal memory. The overall program is composed of six interrelated training exercises that in aggregate span the acoustic organization of speech from frequency-modulated sweeps through syllables, short words, sentences, and narratives. The tasks include speed of processing time-order judgments, instruction following, and verbal memory. The training difficulty adapts to the users ability level. The training is completed in 40 one-hour sessions. The scientific basis of this program has been described elsewhere (Mahncke, Bronstone, and Merzenich, 2006).

The BFP was first tested in a small randomized controlled study conducted in a classroom setting at an active retirement community (Mahncke, Bronstone, and Merzenich, 2006). Fifty-one participants aged 63-94 (mean age 79.9) were randomly assigned to one of three groups; BFP training, active control, and no contact control. The active control watched education videos on computers and the exposure time matched the BFP group. The primary outcome was a global auditory memory score calculated using the normative RBANS (Repeatable Battery for the Assessment of Neuropsychological Status) population data. The BFP group significantly improved on this outcome (effect size 0.41) while the control groups showed no significant change.

A second randomized controlled study was conducted in a larger population (N = 182) with a similar three-arm setup as in the previous study (Mahncke, Connor, et al., 2006). The BFP trained group showed significant trained to task improvements as well as a significant improvement in the RBANS global memory measure with an effect size of 0.25. The control groups did not change significantly. It was noted that more than half the participants performed highly on the RBANS tests prior to training allowing little room for improvement. A second analysis was conducted on the participants who did not show ceiling effects and the effect size increased to 0.61. A three-month follow up showed that training effects were maintained.

These two studies showed significant within group changes in the outcome measures for the BFP group. However, the studies were insufficiently powered to test for between-group differences. To address this issue and to expand the number of outcome measures, a larger randomized controlled trial called IMPACT (Improvements in Memory with Plasticity-based Adaptive Cognitive Training) was implemented (Smith, Housen, et al., 2009). IMPACT was a large multi-site double-blind clinical trial that enrolled 487 participants aged 65. Participants were randomly assigned to either the BFP training group or to an active control group who watched educational videos on a computer for an equal exposure time.

The IMPACT results showed large gains in a train to task measure with participants in the experimental group improving on the auditory speed of processing exercise by an average of 113% (Cohen's d effect size of 0.87). A battery of neuropsychological memory assessments was administered before and after the training period to quantify changes in memory. The tests included RBANS, Rey Auditory Verbal Learning Test (RAVLT), Wechsler Memory Scale (WMS-III), and the Rivermead Behavioral Memory Test (RBMT). The participants in the BFP training group improved significantly compared to the active control participants, with Cohen's *d* effect sizes ranging from 0.20 to 0.43. A self-report questionnaire measuring participants' perception of cognitive ability (Cognitive Self-Report Questionnaire or CRSQ-25) was administered pre and post-training. The CSRQ-25 consists of 25 statements about cognition and mood in everyday life over the past 2 weeks, answered using a 5-point Likert scale. The BFP group significantly improved on this measure (effect size 0.33) compared to the active control. These benefits were shown to persist over a three-month no-contact follow-up period, with some waning of effect size seen without further training (Zelinski, Spina, et al., 2011).

These results are particularly relevant because they derive from large, well-controlled studies with multiple outcome measures, and are the first to demonstrate the direct improvement of memory, which is generally the primary complaint of aging individuals. The results again show that speed training is particularly plastic, with large train-to-task effects, and show that the improvements are sufficiently substantial to be noted by participants in a self-reported outcome measure.

Other Training Programs

Several other training programs have been used in single studies with compelling results worth noting. Each of these studies represents an initial effort, typically at a single research site and with a relatively low number of participants; however such studies show promise with multiple types of cognitive training programs and show how this field is likely to expand in coming years.

Buschkuehl et al. (Buschkuehl, Jaeggi, et al., 2008) trained older participants using simple adaptive visual sequencing working memory and reaction time tasks, with a control group performing physical exercise. The cognitive training group considerably increased their performance on the trained task and benefits transferred to other visual working memory tasks. However, at the one year follow up there were no significant differences between the experimental and control groups. Similar explicit working memory training approaches have been employed by Li et. al. (Li, Schmiedek, Huxhold, Röcke, et al., 2008) and Dahlin et. al. (Dahlin, Nyberg, Bäckman, and Neely, 2008) with a broadly similar pattern of results, including some limited generalization to other working memory-related tasks (Dahlin, Bäckman, Neely, and Nyberg, 2009). It is still an open question whether this approach will generalize to broader participant-reported outcome measures or directly-observed functional measures; however given the obvious face validity of directly exercising working memory, and the importance of working memory in generalized cognitive function, further research in this area is important.

Basak et. al. (Basak, Boot, Voss, and Kramer, 2008) asked participants to train on a strategy video game (‘Rise of Nations’) for 90 minutes sessions over a period of 4 to 5 weeks giving a total training time of 23.5 hours. They improved significantly more than the control group not only on the game itself, but also on measures of executive control functioning including task switching, working memory, visual short-term memory and reasoning.

Mozolic et. al. (Mozolic, Long, Morgan, Rawley-Payne, and Laurienti, 2009) employed a creative training paradigm designed to improve the selective suppression of task irrelevant stimuli in an effort to improve processing of attended stimuli. This approach improved performance on the trained task, and also showed generalization to several other executive function measures. In addition, magnetic resonance imaging demonstrated increased resting cerebral blood flow to the pre-frontal cortex following training (Mozolic, Hayasaka, and Laurienti, 2010).

Jennings et. al. (Jennings, Webster, Kleykamp, and Dagenbach, 2005) used a novel repetition-lag procedure in which participants classified words presented in sequence as novel or familiar based on the prior exposure in study periods preceding the presentation or in the presentation itself. This training exercise yielded substantially improved task performance, which generalized to unrelated cognitive skills including N-back and digit-symbol substitution.

Berry et. al. (Berry, Zanto, Clapp, Hardy, et al., 2010) used a visual speed training exercise (Sweep Seeker, Posit Science) to examine the relationship between changes in visual evoked potentials and working memory. Trained participants showed significant changes in early visual responses, showing that changes in early visual processing were driven by the training exercise. The magnitude of the evoked potential change was significantly correlated with the observed change in visual working memory performance, indicating that the improved cognitive function derived from improved sensory performance in visual cortex.

We note briefly that several quite interesting results using computerized cognitive training approaches have been demonstrated in young adults, but to our knowledge have not yet been replicated in older adults and are hence outside of the scope of this review. In particular, action video games have been shown to improve measures of visual speed and cognitive function in young adults (Green and Bavelier, 2003), and specific adaptive working memory training program from CogMed has been shown to improve working memory measures in children with ADHD and related conditions (Klingberg, 2010), and dual N-back training has been shown to improve measures of fluid intelligence (Jaeggi, Buschkuehl, Jonides, and Perrig, 2008). All three approaches are potentially effective in older adults as well, and merit studies specifically in that population.

By far the most popular approach to cognitive training over the past few years has arisen from the Japanese video game giant Nintendo. Brain Age and its various sequels have sold more than 25 million copies worldwide (Henderson, 2008). Although the Brain Age game was designed with input from a neuroscientist, and derived from work on cognitive stimulation in Alzheimer's patients (Kawashima et al., 2005), no clinical trials have been performed specifically with Brain Age.

Owen et. al. (2010) sought to remedy this oversight by building their own cognitive exercises designed to replicate the design features of the Brain Age game, and performing a large-scale (N = 11,430) internet-based trial with these exercises in young/middle-aged adults. This novel and impressive clinical trial design and execution showed no effect of the Owen brain exercises on measures of cognitive function relative to a control group. Unfortunately, the study was marred by a very high dropout rate (more than 75% of participants enrolled did not complete the study) and a relatively low amount of cognitive training (an average of four hours of training, with some participants completing only 20 minutes over the course of the study). Nonetheless, this negative result makes a strong argument that the Owen brain exercises are ineffective at this level of training intensity. The result is quite interesting because it argues for the specificity of cognitive training exercises – finding exercises that are *ineffective* is as useful as finding exercises that are *effective*, and these results should guide future exercise development. Owen et. al. (2010) note that the study is to be repeated with older participants, and these results should be equally exciting.

SYNTHESIS OF COMMON ATTRIBUTES AND THEIR RELATIONSHIP TO BRAIN PLASTICITY

Having reviewed the existing literature above, we can now suggest three key attributes of effective computerized cognitive training programs. Interestingly, the effectiveness of these principles can be accounted for by basic mechanisms of brain plasticity. Brain plasticity refers to the brain's lifelong capacity for physical and functional change, and over the past twenty years has become a well-accepted fundamental operating principle of brain organization and function.

Speeded

An interesting common principle of the exercises examined above is their focus on improving core speed of processing. For example, UFOV training requires the user to discriminate and identify visual targets at progressively shorter presentation times, while BFP training progressively moves to more and more rapidly presented stimuli and accelerated speech. Rise of Nations as a real-time strategy game contains implicit speed training, as the user must complete sets of tasks in an increasingly rapid fashion to keep up with the computer enemy. Of course, cognitive slowing is a central feature of aging. This observation coupled with the generally large effect sizes seen in speed training suggest that the elaboration of existing programs focused on speed and the development of new such programs represents a fruitful avenue of research and clinical trials. One of the fundamental changes in the aging brain is the general decline in processing speed (Salthouse, 2004). Brain processing becomes noisier with age (Schmolesky, Wang, Pu, and Leventhal, 2000: Yu, Wang, Li, Zhou, and Leventhal, 2006). Degraded signals lead to cortical representations that are of lower fidelity and are less reliable. It is important for a cognitive training regime to include exercises designed to improve processing speed and accuracy. Training should drive the brain to make accurate discriminations along elementary and complex stimulus dimensions. This can be done by starting off with simple stimuli in elementary configurations and adjusting the complexity of the task as training progresses. Exercises that challenge the brain to process information faster and more accurately are important to reverse the decline.

Adaptive

The programs examined in this chapter all adapt exercise difficulty based on user performance. For example, UFOV training adjusts up or down in difficulty following each block of stimuli, while BFP training adapts continuously on a trial-by-trial basis, and working memory training programs adapt the span of items to be remembered. This observation is of course consistent with the well-known concept of a zone of proximal development (Vygotskiĭ and Cole, 1978), wherein optimal learning rates are driven with content that is neither too easy nor too difficult. This requirement for adaptivity is ideally suited for computerized training programs, where individuals typically train independently, allowing the difficult to be optimized on an individual basis. This approach is distinct from classroom-based cognitive training approaches, which by their nature cannot adapt to individual differences as effectively or as continuously.

Intensive

The programs discussed above all required multiple hours of training. For example the UFOV training program usually runs for 10 hours, and the BFP is designed for 40 hours of training. Effective exercises require that the participant undergo many thousands of trials during the course of training. This is required to ensure that the cortical representations of behaviorally important inputs are integrated to create robust cortical responses (Merzenich and Sameshima, 1993). This typically requires many hours of training. It is best if this training is spread over multiple sessions to reduce fatigue and to maximize consolidation of learning by allowing sleep between sessions (Karni, Tanne, Rubenstein, Askenasy, and Sagi, 1994).

In this way, effective cognitive training is similar to effective strength training or effective weight loss – all three regimens that require intensive effort on the part of people to drive real benefits. However, an interesting aspect to computerized cognitive training programs is that they should be able to benefit from the computer games revolution – despite their intensity, effective training programs should be able to be as engaging and compelling (and perhaps as addictive) as video games.

Discussion

The number of computer-based cognitive training programs targeting age-related cognitive decline has expanded rapidly in recent years. These programs vary widely in their quality of design with some based on principles of brain plasticity, others derived from fundamental neuropsychological principles, and many intended principally for entertainment purposes. Many such programs make efficacy claims based on the well-established literature relating cognitive activity levels to dementia incidence; however such data is entirely inadequate to support the claims of efficacy from any particular cognitive training program as an intervention to improve cognitive function. Rigorous clinical trials are required to make such claims, just as would be required for a new pharmaceutical or a new medical device that was designed to improve cognitive function.

Fortunately, a substantial number of high quality randomized controlled clinical trials have been performed and published. In particular, large scale multi-site trials of UFOV training and BFP training have documented improvements across a number of domains relevant to improving quality of life and real-world function in older individuals, including memory, attention, driving skills, depressive symptoms, and health-related quality of life. Initial studies with a variety of other programs, including adaptive working memory training and a real-time strategy game, show promise and merit further rigorous studies.

CONCLUSION

On the basis of this literature, we suggest that it is time to move beyond the simple-minded questions of "does cognitive training work: yes/no" (Owen, Hampshire, Grahn, Stenton, et al., 2010) to the following questions to advance the basic and applied science of cognitive training

- How can effective forms of cognitive training programs be further optimized to yield broader and deeper benefits? Benefits to executive function and social cognition are yet to be established, and represent important aspects of cognitive function for older individuals. In addition, optimal dosing and re-training procedures have yet to be established.
- Are there differential responders that should use different forms/types of cognitive training to maximize benefit? It seems highly likely that cognitive training programs could be further optimized based on the characteristics of their users, and this type of personalized training should show even greater effects than those demonstrated to date.
- What clinical trial designs and measures can we standardize to improve the comparability of future studies and their ease of design and execution to encourage excellent studies? Non-comparable studies fundamentally limit the rate of progress in this field. The NIH Toolbox effort (Gershon, Cella, Fox, Havlik, Hendrie, et al., 2010) represents an excellent first effort to standardize endpoints in the field of cognitive aging, which should look to the field of cognitive impairment in schizophrenia for a prime example of how coordinated research can be unleashed through standardization of clinical trial designs and measures (Buchanan, Davis, Goff, M. F. Green, Keefe, et al., 2005).
- Do programs that improve cognitive function delay the onset of dementia? While it seems intuitive that this would be true, it is not necessarily the case. Clinical trials addressing this issue will be large, expensive, and time-consuming; but are essential if the fundamental promise of this field is to be realized.
- How can effective programs most efficiently be delivered to individuals in need, and in doing so, help the entire population who can benefit? This issue more than the others is not a scientific question – it will require cooperation and coordination across health care payers, delivery systems, and eventually a full integration into the health care system.

As this field advances in the years to come and the above questions are addressed in a step-by-step fashion, we expect that computer-based cognitive training will enter the mainstream of aging care, and we will look back to the 20th century, when people aged without an active program of brain health, as the dark ages.

REFERENCES

Akbaraly, T. N., Portet, F., Fustinoni, S., Dartigues, J. F., Artero, S., Rouaud, O., Touchon, J., et al. (2009). Leisure activities and the risk of dementia in the elderly. *Neurology*, *73*(11), 854.

Ball, K., Edwards, J. D., Ross, L. A., and McGwin Jr., G. (2010). Cognitive training decreases motor vehicle collision involvement of older drivers. *Journal of the American Geriatrics Society*, *58*(11), 2107-2113. doi:10.1111/j.1532-5415.2010.03138.x.

Ball, K. K., Beard, B. L., Roenker, D. L., Miller, R. L., and Griggs, D. S. (1988). Age and visual search: Expanding the useful field of view. *JOSA A*, *5*(12), 2210–2219.

Ball, K., Berch, D. B., Helmers, K. F., Jobe, J. B., Leveck, M. D., Marsiske, M., Morris, J. N., et al. (2002). Effects of cognitive training interventions with older adults: a randomized controlled trial. *Journal of the American Medical Association*, *288*(18), 2271.

Ball, K., Edwards, J. D., and Ross, L. A. (2007). The impact of speed of processing training on cognitive and everyday functions. *The Journals of Gerontology Series B: Psychological Sciences and Social Sciences*, *62*(Special Issue 1), 19.

Basak, C., Boot, W. R., Voss, M. W., and Kramer, A. F. (2008). Can training in a real-time strategy video game attenuate cognitive decline in older adults?. *Psychology and Aging*, *23*(4), 765.

Berry, A. S., Zanto, T. P., Clapp, W. C., Hardy, J. L., Delahunt, P. B., Mahncke, H. W., and Gazzaley, A. (2010). The influence of perceptual training on working memory in older adults. *PloS One*, *5*(7), e11537. doi:10.1371/journal.pone.0011537.

Broman, A. T., West, S. K., Munoz, B., Bandeen-Roche, K., Rubin, G. S., and Turano, K. A. (2004). Divided visual attention as a predictor of bumping while walking: the Salisbury Eye Evaluation. *Investigative Ophthalmology and Visual Science*, *45*(9), 2955.

Buchanan, R. W., Davis, M., Goff, D., Green, M. F., Keefe, R. S. E., Leon, A. C., Nuechterlein, K. H., et al. (2005). A summary of the FDA-NIMH-MATRICS workshop on clinical trial design for neurocognitive drugs for schizophrenia. *Schizophrenia Bulletin*, *31*(1), 5-19. doi:10.1093/schbul/sbi020.

Buschkuehl, M., Jaeggi, S. M., Hutchison, S., Perrig-Chiello, P., Däpp, C., Müller, M., Breil, F., et al. (2008). Impact of working memory training on memory performance in old-old adults. *Psychology and aging*, *23*(4), 743.

Dahlin, E., Bäckman, L., Neely, A. S., and Nyberg, L. (2009). Training of the executive component of working memory: Subcortical areas mediate transfer effects. *Restorative neurology and neuroscience*, *27*(5), 405–419.

Dahlin, E., Nyberg, L., Bäckman, L., and Neely, A. S. (2008). Plasticity of executive functioning in young and older adults: Immediate training gains, transfer, and long-term maintenance. *Psychology and aging*, *23*(4), 720.

Dakim - Why It Works. (n.d.). Retrieved April 30, 2011, from http://www.dakim.com/why-it-works/scientific-studies/

Edwards, J. D., Ross, L. A., Wadley, V. G., Clay, O. J., Crowe, M., Roenker, D. L., and Ball, K. K. (2006). The useful field of view test: normative data for older adults. *Archives of Clinical Neuropsychology*, *21*(4), 275–286.

Edwards, J. D., Wadley, V. G., Vance, D. E., Wood, K., Roenker, D. L., and Ball, K. K. (2005). The impact of speed of processing training on cognitive and everyday performance. *Aging and Mental Health*, *9*(3), 262–271.

Edwards, J. D., Delahunt, P. B., and Mahncke, H. W. (2009). Cognitive speed of processing training delays driving cessation. *The Journals of Gerontology. Series A, Biological Sciences and Medical Sciences*, *64*(12), 1262-1267. doi:10.1093/gerona/glp131.

Edwards, J. D., Myers, C., Ross, L. A., Roenker, D. L., Cissell, G. M., McLaughlin, A. M., and Ball, K. K. (2009). The longitudinal impact of cognitive speed of processing training on driving mobility. *The Gerontologist*, *49*(4), 485-494. doi:10.1093/geront/gnp042.

Edwards, J. D., Wadley, V. G., Myers, R. S., Roenker, D. L., Cissell, G. M., and Ball, K. K. (2002). Transfer of a speed of processing intervention to near and far cognitive functions. *Gerontology*, *48*, 329–340.

Fernandez, A., and Goldberg, E. (2008). *The state of the brain fitness software market*. San Francisco, CA: Sharp Brains.

Fleishman, J. A., Cohen, J. W., Manning, W. G., and Kosinski, M. (2006). Using the SF-12 health status measure to improve predictions of medical expenditures. *Medical Care*, *44*(5:Suppl), I54-63.

Gershon, R. C., Cella, D., Fox, N. A., Havlik, R. J., Hendrie, H. C., and Wagster, M. V. (2010). Assessment of neurological and behavioural function: the NIH Toolbox. *The Lancet Neurology*, *9*(2), 138–139.

Green, C. S., and Bavelier, D. (2003). Action video game modifies visual selective attention. *Nature*, *423*(6939), 534–537.

Henderson, L. (2008, March 12). The 10 Highest Selling Games on the Nintendo DS. *Yahoo!* Retrieved April 29, 2011, from http://www.associatedcontent.com/article/644603/the_10_highest_selling_games_on_the.html

Jaeggi, S. M., Buschkuehl, M., Jonides, J., and Perrig, W. J. (2008). Improving fluid intelligence with training on working memory. *Proceedings of the National Academy of Sciences*, *105*(19), 6829.

Jennings, J. M., Webster, L. M., Kleykamp, B. A., and Dagenbach, D. (2005). Recollection training and transfer effects in older adults: Successful use of a repetition-lag procedure. *Aging, Neuropsychology, and Cognition*, *12*(3), 278–298.

Jobe, J. B., Smith, D. M., Ball, K., Tennstedt, S. L., Marsiske, M., Willis, S. L., Rebok, G. W., et al. (2001). ACTIVE: A cognitive intervention trial to promote independence in older adults. *Controlled Clinical Trials*, *22*(4), 453–479.

Karni, A., Tanne, D., Rubenstein, B. S., Askenasy, J. J., and Sagi, D. (1994). Dependence on REM sleep of overnight improvement of a perceptual skill. *Science*, *265*(5172), 679.

Kawashima, R., Okita, K., Yamazaki, R., Tajima, N., Yoshida, H., Taira, M., Iwata, K., et al. (2005). Reading aloud and arithmetic calculation improve frontal function of people with dementia. *The Journals of Gerontology Series A: Biological Sciences and Medical Sciences*, *60*(3), 380.

Klingberg, T. (2010). Training and plasticity of working memory. *Trends in Cognitive Sciences*, *14*(7), 317–324.

Lezak, M. D. (2004). *Neuropsychological Assessment*. Oxford University Press, USA.

Li, S. C., Schmiedek, F., Huxhold, O., Röcke, C., Smith, J., and Lindenberger, U. (2008). Working memory plasticity in old age: Practice gain, transfer, and maintenance. *Psychology and aging*, *23*(4), 731.

Mackinnon, A., Christensen, H., Hofer, S. M., Korten, A. E., and Jorm, A. F. (2003). Use It and Still Lose It? The Association Between Activity and Cognitive Performance Established Using Latent Growth Techniques in a Community Sample. *Aging, Neuropsychology, and Cognition*, *10*(3), 215. doi:10.1076/anec.10.3.215.16451.

Mahncke, H. W., Connor, B. B., Appelman, J., Ahsanuddin, O. N., Hardy, J. L., Wood, R. A., Joyce, N. M., et al. (2006). Memory enhancement in healthy older adults using a brain plasticity-based training program: a randomized, controlled study. *Proceedings of the National Academy of Sciences*, *103*(33), 12523-12528. doi:10.1073/pnas.0605194103.

Mahncke, H. W., Bronstone, A., and Merzenich, M. M. (2006). Brain plasticity and functional losses in the aged: scientific bases for a novel intervention. *Progress in Brain Research*, *157*, 81-109. doi:10.1016/S0079-6123(06)57006-2.

Marottoli, R. A., Mendes de Leon, C. F., Glass, T. A., and Williams, C. S. (1997). Driving cessation and increased depressive symptoms: Prospective evidence from the New Haven EPESE. *Journal of the American Geriatrics Society*.

Merzenich, M. M., and Sameshima, K. (1993). Cortical plasticity and memory. *Current Opinion in Neurobiology*, *3*(2), 187–196.

Mozolic, J. L., Hayasaka, S., and Laurienti, P. J. (2010). A cognitive training intervention increases resting cerebral blood flow in healthy older adults. *Frontiers in Human Neuroscience*, *4*(16), 1-10.

Mozolic, J. L., Long, A. B., Morgan, A. R., Rawley-Payne, M., and Laurienti, P. J. (2009). A cognitive training intervention improves modality-specific attention in a randomized controlled trial of healthy older adults. *Neurobiology of Aging*, *32*(4), 655-668.

Owen, A. M., Hampshire, A., Grahn, J. A., Stenton, R., Dajani, S., Burns, A. S., Howard, R. J., et al. (2010). Putting brain training to the test. *Nature*, *465*(7299), 775–778.

Owsley, C., Ball, K., Sloane, M. E., Roenker, D. L., and Bruni, J. R. (1991). Visual/cognitive correlates of vehicle accidents in older drivers. *Psychology and Aging*, *6*(3), 403.

Owsley, C., Sloane, M., McGwin, G., and Ball, K. (2002). Timed instrumental activities of daily living tasks: Relationship to cognitive function and everyday performance assessments in older adults. *Gerontology*, *48*, 254–265.

Reference Materials for Further Brain Training Study - Lumosity. (n.d.). . Retrieved April 30, 2011, from http://www.lumosity.com/references

Roenker, D. L., Cissell, G. M., Ball, K. K., Wadley, V. G., and Edwards, J. D. (2003). Speed-of-processing and driving simulator training result in improved driving performance. *Human Factors*, *45*(2), 218.

Rubin, G. S., Ng, E. S. W., Bandeen-Roche, K., Keyl, P. M., Freeman, E. E., and West, S. K. (2007). A prospective, population-based study of the role of visual impairment in motor vehicle crashes among older drivers: the SEE study. *Investigative Ophthalmology and Visual Science*, *48*(4), 1483.

Salthouse, T. A. (2004). What and when of cognitive aging. *Current Directions in Psychological Science*, *13*(4), 140.

Scalf, P. E., Colcombe, S. J., McCarley, J. S., Erickson, K. I., Alvarado, M., Kim, J. S., Wadhwa, R. P., et al. (2007). The neural correlates of an expanded functional field of view. *The Journals of Gerontology: Series B*, *62*(Special Issue 1), 32.

Schmolesky, M. T., Wang, Y., Pu, M., and Leventhal, A. G. (2000). Degradation of stimulus selectivity of visual cortical cells in senescent rhesus monkeys. *Nature Neuroscience*, *3*(4), 384–390.

Sims, R. V., McGwin, G., Allman, R. M., Ball, K., and Owsley, C. (2000). Exploratory study of incident vehicle crashes among older drivers. *Journal of Gerontology: Series A, Biological Science and Medical Science*, *55*(1), 22–27.

Smith, G. E., Housen, P., Yaffe, K., Ruff, R., Kennison, R. F., Mahncke, H. W., and Zelinski, E. M. (2009). A cognitive training program based on principles of brain plasticity: results from the Improvement in Memory with Plasticity-based Adaptive Cognitive Training (IMPACT) study. *Journal of the American Geriatrics Society*, *57*(4), 594-603. doi:10.1111/j.1532-5415.2008.02167.x.

Vance, D. E., Ball, K. K., Roenker, D. L., Wadley, V. G., Edwards, J. D., and Cissell, G. M. (2006). Predictors of falling in older Maryland drivers: a structural-equation model. *Journal of Aging and Physical Activity*, *14*(3), 254.

Vance, D., Dawson, J., Wadley, V., Edwards, J., Roenker, D., Rizzo, M., and Ball, K. (2007). The ACCELERATE study: The longitudinal effect of speed of processing training on cognitive performance of older adults. *Rehabilitation Psychology*, *52*(1), 89.

Verghese, J., Lipton, R. B., Katz, M. J., Hall, C. B., Derby, C. A., Kuslansky, G., Ambrose, A. F., et al. (2003). Leisure activities and the risk of dementia in the elderly. *New England Journal of Medicine*, *348*(25), 2508–2516.

Vygotskiĭ, L. S., and Cole, M. (1978). *Mind in society: The development of higher psychological processes*. Harvard Univ Pr.

Willis, S. L., Tennstedt, S. L., Marsiske, M., Ball, K., Elias, J., Koepke, K. M., Morris, J. N., et al. (2006). Long-term effects of cognitive training on everyday functional outcomes in older adults. *Journal of the American Medical Association*, *296*(23), 2805.

Wolinsky, F. D., Mahncke, H. W., Vander Weg, M. W., Martin, R., Unverzagt, F. W., Ball, K. K., Jones, R. N., et al. (2009). The ACTIVE cognitive training interventions and the onset of and recovery from suspected clinical depression. *The Journals of Gerontology. Series B, Psychological Sciences and Social Sciences*, *64*(5), 577-585. doi:10.1093/geronb/gbp061.

Wolinsky, F. D., Callahan, C. M., Fitzgerald, J. F., and Johnson, R. J. (1992). The risk of nursing home placement and subsequent death among older adults. *Journal of Gerontology*, *47*(4), S173.

Wolinsky, F. D., Callahan, C. M., Fitzgerald, J. F., and Johnson, R. J. (1993). Changes in functional status and the risks of subsequent nursing home placement and death. *Journal of Gerontology*, *48*(3), S94.

Wolinsky, F. D., Unverzagt, F. W., Smith, D. M., Jones, R., Stoddard, A., and Tennstedt, S. L. (2006). The ACTIVE cognitive training trial and health-related quality of life: protection that lasts for 5 years. *The Journals of Gerontology Series A: Biological Sciences and Medical Sciences*, *61*(12), 1324.

Wolinsky, F. D., Unverzagt, F. W., Smith, D. M., Jones, R., Wright, E., and Tennstedt, S. L. (2006). The effects of the ACTIVE cognitive training trial on clinically relevant declines in health-related quality of life. *The Journals of Gerontology Series B: Psychological Sciences and Social Sciences*, *61*(5), S281.

Wolinsky, F. D., Mahncke, H. W., Kosinski, M., Unverzagt, F. W., Smith, D. M., Jones, R. N., Stoddard, A., et al. (2009). The ACTIVE cognitive training trial and predicted medical expenditures. *BMC Health Services Research*, *9*, 109. doi:10.1186/1472-6963-9-109.

Wolinsky, F. D., Vander Weg, M. W., Martin, R., Unverzagt, F. W., Ball, K. K., Jones, R. N., and Tennstedt, S. L. (2009). The effect of speed-of-processing training on depressive symptoms in ACTIVE. *The Journals of Gerontology. Series A, Biological Sciences and Medical Sciences*, *64*(4), 468-472. doi:10.1093/gerona/gln044.

Yu, S., Wang, Y., Li, X., Zhou, Y., and Leventhal, A. G. (2006). Functional degradation of extrastriate visual cortex in senescent rhesus monkeys. *Neuroscience*, *140*(3), 1023–1029.

Zelinski, E. M., Spina, L. M., Yaffe, K., Ruff, R., Kennison, R. F., Mahncke, H. W., and Smith, G. E. (2011). Improvement in Memory with Plasticity-based Adaptive Cognitive Training (IMPACT): results of the 3-month follow-up. *Journal of the American Geriatrics Society*, *59*, 258-265.

In: Handbook of Technology in Psychology ... ISBN: 978-1-62100-004-4
Editor: Luciano L'Abate and David A. Kasier

Chapter 8

HARNESSING THE ACTIVE MIND: GAME-BASED LEARNING FOR DEVELOPING NAVIGATION SKILLS IN THE BLIND

Mark A. Halko, Jaime Sánchez and Lotfi B. Merabet[1]
[1]Harvard University, US

Game play has historically been intertwined with learning, though its direct causal link remains an issue of continuous discussion and debate. More recently, the growing popularity of video games has sparked new interest within the educational domain as an adjunctive form of teaching basic curriculum as well as for specialty training of skills. Early computer games such as "Where in the World is Carmen Sandiego?", " Number Munchers", "Math Blaster" and "Oregon Trail" were created around the time that personal computers (PCs) became popular fixtures in the home, office and classroom. Today, as technology continues to rapidly evolve and costs continue to decrease, the potential, flexibility and accessibility of "smart" computer devices are becoming more and more incorporated into our daily activities. This convergence of factors has proved a fertile ground for the development of specialized computer and video-game based educational strategies.

The very nature of video games provide appealing characteristics for learning that can serve as a platform to interact with subject material in novel ways (Dede, 2009). Combined with realistic virtual environments, video game-based learning can provide exposure to novel situations which may be too dangerous, unfeasible, or too expensive to carry out in real word situations. The success of flight simulators for pilot training or military combat simulations serve as a prototypical examples (Fletcher, 2009). Finally, video games can provide an enjoyable means to promote repetition and repeated exposure of material which is considered necessary to establish strong, long term memories and skill perfection.

Key to success for a viable video-game approach is the demonstration that acquired knowledge and skill can be transferred from the virtual to the real physical world. Thus, the challenge is not to create a new video game which merely simulates what is experienced in the real world, but rather to leverage the educational strengths and tailor the instructive components of a game to allow for learning which may not be as robust or as feasible in the outside world. Specifically, if gaming is tailored to develop individual skills within scenarios coupled with real life instruction, great gains can be potentially made (Fletcher, 2009) particularly for specialty populations that may show limited learning using more traditional didactic teaching methods.

This strategy has been pursued with "brain training games" but unfortunately have shown only limited skill transferability and overall, have failed to demonstrate large cognitive improvements across the general population (Owen et al., 2010). Other studies involving action video game play (particularly first-person shooter games) have demonstrated selective improvement outside of general game-play skills such as improved visual-attentive processing and visual contrast sensitivity (Dye et al., 2009; Green and Bavelier, 2003; Li et al., 2009; Li et al., 2010). Thus, interest in this arena, particularly coupled with neuroscience-based scientific research, still remains an exciting area of exploration and development.

Along these lines, how the brain continues to learn and transfer skills gained from game-based learning also remains a matter of intense debate whose repercussions could potentially influence the way education is carried out with future generations. What are the underlying brain mechanisms responsible for learning and transfer of video-game acquired skills? Further, are these mechanisms universal or unique to specific population and/or skill set being developed? Here, we investigate the neuroscientific basis for video game training in a specific population (individuals with profound blindness) and discuss future directions of population-specific neuroscience-based video-game training.

SPATIAL NAVIGATION IN THE BLIND

Maintaining functional independence is of utmost importance to a blind individual. The formal instruction of navigation skills, referred to as orientation and mobility (OandM) training, is geared at developing strategies to assist with route planning, updating information regarding one's position, and reorienting to reestablish travel as needed (Blasch et al., 1997). In order to navigate effectively, a blind individual needs to develop sensory awareness (i.e. acquiring information about the world through remaining sensory modalities), searching skills (so as to locate items or places efficiently) and an understanding of the spatial relationships that exist between oneself and the objects in the surrounding environment (Loomis et al., 2001; Rieser, 2008; Spivak et al., 1997). This representation of external space is referred to as a cognitive spatial map (Landau et al., 1981; Strelow, 1985; Tolman, 1948). Unlike the sighted, blind individuals cannot rely on visual cues to gather this information to visually order and classify their physical environment. Instead, a blind individual has to rely on other sensory channels to obtain appropriate spatial information regarding their surroundings (Thinus-Blanc and Gaunet, 1997).

Not surprisingly, the theoretical underpinnings related to navigation skills in the blind have been the subject of intense debate. Given the importance of visual cues, it has been largely assumed that blind individuals (especially children) have cognitive difficulties in representing spatial environments (Axelrod, 1959) and consequently, have impaired navigation skills. However, contradictory evidence (particularly in relation to the importance of prior visual experience) has contrasted sharply with this assumption (Loomis, et al., 2001). Certain studies have reported that there are no differences in terms of how well blind individuals are able to mentally represent and interact with spatial environments (Landau, et al., 1981; Morrongiello et al., 1995). Specifically, it has been shown in certain spatial navigation tasks that blind individuals exhibit equal (Loomis, et al., 2001) and even superior navigation performance (Fortin et al., 2008) when compared to sighted individuals.

Given these reports, one has to ask whether differences in spatial mental constructs and navigation skill are uniquely due to visual deprivation itself (and/or other developmental factors such as the timing and profoundness of vision loss) or, can they also reflect an impoverished or incomplete acquisition of necessary spatial information through other sensory channels? From a rehabilitation standpoint, it could be argued that perhaps what is missing is a better way to access, manipulate and transfer acquired information; a gap that could be potentially closed through the use of appropriate technology. The use of computer-based video games may serve to bridge that gap.

Navigating Using Audio Based Virtual Environments

With respect to navigation, information captured through sound is very important for developing a sense of spatial orientation and distance, as well as obstacle detection and avoidance (Ashmead et al., 1989; Rieser, 2008). Previous work with the blind has shown that spatial information obtained through novel computer-based approaches using sound (Ohuchi et al., 2006; Riehle et al., 2008) as well as tactile information (Johnson and Higgins, 2006; Lahav, 2006; Pissaloux et al., 2006) may prove useful for developing navigation skills. We have extended these concepts with the goal of developing audio-based virtual environments as a means to teach, motivate and develop spatial navigation skills in individuals with profound visual impairment. Specifically, by interacting with auditory cues that describe and characterize a particular environment (e.g., text to speech (TTS) to provide heading information or identifying an encountered obstacle), appropriate contextual or iconic sounds to identify a particular object (e.g. a door knock signaling the presence of a door), and the conceptual alignment of spatial features using audio-based information (e.g. using stereo spectral cues to help orient to and localize the location of an object), a user with profound blindness can learn to navigate a relatively complex route (Sanchez and Saenz, 2006). Key to this approach is the fact that auditory-based spatial information is acquired sequentially, within context, and through a highly interactive interface that greatly engages a user to actively explore a given environment and construct a cognitive spatial map effectively and efficiently.

As an early effort, Sanchez and colleagues have developed “AudioDoom”, an auditory-based computer game that serves to engage blind children in play and improve spatial navigation and problem solving skills (Sanchez and Lumbreras, 1998). The game is loosely based on a popular adventure computer video game called “Doom” in which a player navigates through a predetermined labyrinth of walls and corridors locating various items and avoiding monsters so as to find his/her way to an exit portal and start the next level. Key to succeeding in this game is to maintain an internal mental map regarding the spatial location of objects encountered and keeping track of areas explored. Briefly, the auditory version of the game (“AudioDoom”; (Sanchez and Lumbreras, 1998) works much the same way but involves the use of sound spectral cues (for example, door knocks and footsteps) as a means to acquire contextual spatial information regarding one’s surroundings during game play.

Using a keyboard, mouse or joystick, a gamer can move in any direction (stepping forward or turning right or left) and interact with the environment in a step by step fashion (i.e. through a series of sequential "encounters") so as to pass through a corridor, open a door, pick up treasure and so on.

The gaming structure organizes the level into several pre-determined corridors, dead ends, and pathways giving a sense of the entire area laid out over a three dimensional space. As the paths to be explored are constrained by the use of corridors rather than true open spaces, a player is able to maintain their sense of orientation and heading. Thus, played out in a corresponding three dimensional auditory virtual world, the user builds a spatial mental representation based on these sequential and causal encounters within a goal-directed navigation framework (Sanchez and Lumbreras, 1998). These observations reported during initial field testing of AudioDoom are important in terms of our overall discussion of navigation skill. Specifically, they demonstrate that first: auditory information can provide for accurate cues that describe spatial environments and the relationships between objects and second: users of the game who have profound blindness can generate accurate spatial cognitive maps based on auditory information using an interactive and immersive virtual environment. Furthermore, the interactive and immersive nature of the game provides for not only a strong motivating drive, but also demonstrates that spatial cognitive constructs can be learned implicitly and rather simply through causal interaction with the software.

Building upon this initial work and observations, we hypothesized that users with profound visual impairment who interact with a virtual environment that represents a real place (for example, a building in a individual's school) can not only create an accurate cognitive spatial map of that place, but may also potentially transfer this acquired spatial information to a large-scale, real-world navigation task. Key to demonstrating this premise would be to develop a flexible and modifiable software platform that leverages the advantages associated with both gaming and the contextual information provided during interactive virtual navigation. Following through with these notions, we have been investigating the feasibility and effectiveness of using an audio-based virtual navigation software called Audio-based Environment Stimulator (AbES). This software is similar to AudioDoom in terms of its audio-based navigation and interactive capabilities, but has the added feature of a floor plan editor that allows an investigator to generate virtually any physical space desired including open rooms and corridors, multiple floors as well as furniture and obstacles. The software also incorporates various data collecting methods that can be used to assess behavioral performance (e.g. reconstruction of the route travelled including the time taken to navigate to target, distance traveled and errors made).

The virtual environment is scaled so that each step is meant to represent one typical step in real physical space. Using a keyboard, a user explores the building virtually, moving through the environment and listening to appropriate spectral cues after each step taken (e.g. a knocking sound in the left stereo channel is heard as the player walks past a door on the left and walking up stairs is associated with sequential steps of increasing pitch). Orientation is based on cardinal compass headings (i.e. "north" signifies forward and gives the impression of moving forward in a three dimensional environment). The user also has a "where am I?" key that can be pressed at any time to access TTS-based information that describes their current location in the building, orientation and heading, as well as the identity of objects and obstacles in their path. As a proof of principle, pilot data from early blind test subjects (aged in their mid to late twenties at the time of study) suggests that after approximately 40 to 60 minutes of interacting with AbES, users were indeed able to survey and explore the layout of the building and locations of the target objects virtually. Furthermore, subjects was able to demonstrate a transfer of their cognitive spatial knowledge in a real world navigation task by locating objects found within a room in the actual physical building (Figure 8.1).

Another unique feature is the fact that AbES can be played in two modes; "directed navigation" or "game" mode. In directed navigation, a facilitator places the user in any location in the building and directs the individual to a target destination so as to simulate the navigation and exploration of the building. In the game mode, the user interacts with the virtual world on their own (i.e without a facilitator) with the goal of exploring the entire building in order to collect hidden gems while avoiding roving monsters that can potentially take the gems away and hide them elsewhere. Thus, in either mode, users interact with the virtual environment to gain spatial information and generate a cognitive map of the spatial surroundings. However, given the implicit nature of acquiring spatial information through gaming, we have speculated that the construction of these cognitive spatial cognitive maps may prove to be different depending on the mode of play. In other words, AbES played in game mode (i.e. "unsupervised learning") is in effect designed to promote full exploration of the building thereby maximizing creativity and encouraging the development of "higher level" spatial skills (Blasch, et al., 1997). By comparison, we hypothesized that individuals who interact with AbES in directed navigation mode (i.e. supervised learning) will generate spatial constructs that are limited to the actual routes encountered and as defined by the facilitator. This later point is of particular importance not only in terms of generating cognitive spatial maps but also with regards to safety. Indeed, early pilot work confirms this hypothesis. It would be reasonable to assume that individuals who have a more "robust" cognitive spatial map of their surroundings are more likely to be flexible in their spatial thinking and thus can come up with alternate routes for navigation when needed as opposed to relying on rote memory alone. Current work is now aimed at a large-scale investigation of these hypotheses by assessing how well individuals are able to transfer their acquired spatial information from the virtual to real physical environment and as a function of the mode of acquiring that information. (Figure 8.1)

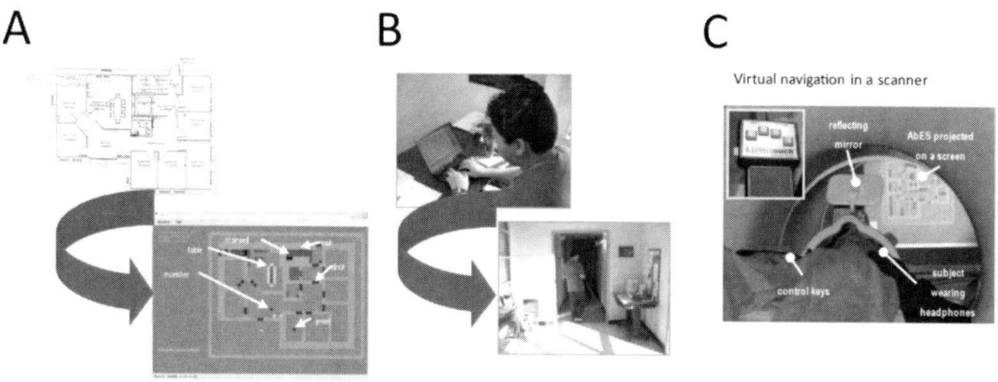

Figure 8.1. Creating and investigating the process of learning real physical environments with a video game. (A) a real physical environment is converted into a virtual representation where one "step" corresponds to one unit of space in the virtual environment. (B) Environment naive subjects are trained in the virtual environment and then tested on navigation within the real physical environment. (C) Navigation in the virtual environment is assessed within an fMRI.

The Neuroscience of Spatial Navigation

To understand how the blind learn to navigate, it worth identifying and comparing the brain structures and processes that are present not only in sighted individuals, but also what has been learned from extensive work using animal models. Considerable effort has been devoted to uncovering the brain systems involved for storage of cognitive maps of our environment (Figure 2). The most striking evidence for the housing of a spatial cognitive information comes from the discovery of place cells within the a particular area of the brain called the hippocampus (Ekstrom et al., 2003; O'Keefe and Dostrovsky, 1971). In a rodent model, these cells selectively respond to portions of the animal's environment and an entire ensemble of place cells can form a complete representation of that environment. Another cortical area of the brain, the entorhinal cortex, is believed to be intimately connected to the hippocampus and contains a population of brain cells which adjust for the relative size of the entire environment. These brain cells, termed "grid cells (Hafting et al., 2005) fire in multiple locations, in a consistent spatial pattern (hence the term "grid cells") and have also been identified in the human using functional neuro-imaging techniques (Doeller et al., 2010) (see later discussion on neuroimaging).

As an intriguing case study, London taxicab drivers, who undergo extensive mental visualization training through the streets of London, show an increase in posterior hippocampus volume that correlates with the number of years of driving experience (as compared to non-taxi driving Londoners familiar with the city) (Maguire et al., 2000; Maguire et al., 2006). In addition, the development of improved navigation abilities in one environment transfers to novel environments (Woollett and Maguire, 2010). This morphological change suggests that the brain can undergo neuroplastic modification as the result of experience in wayfinding. Not surprisingly, damage to the hippocampus results in impaired spatial navigation abilities (Astur et al., 2002).

Although the hippocampus and surrounding regions seem to be important for the acquisition and maintenance of spatial locations in memory, other non-spatial processes are also involved in navigation. The head of the caudate nucleus is strongly associated with response based strategies (Hartley et al., 2003; Iaria et al., 2003). Specifically, where subjects do not utilize spatial landmarks, but instead utilize non-spatial strategies to navigate an environment. For non-spatial strategies, one may remember a sequence of turns or name elements of a maze as opposed to relying on the location of landmarks as global indices (Iaria, et al., 2003). A structural neuroimaging study (i.e. voxel-based morphometry) has shown that the size of the hippocampus is inversely correlated with size of the caudate, within the regions selective for navigation (Iaria, et al., 2003).

Navigation is a complex cognitive process that draws upon a number of cognitive skills and functions including the construction and retrieval of a spatial cognitive map of the environment, assessment of one's current position within the map, episodic memory of past events at the same location (e.g. "last time I passed this intersection on the way to my house, I turned left"), heading planning, re-routing around obstructions, and planning future movements.

Despite careful elucidation of the role of specific areas of the brain (described above) in navigation processes, the reality is that many additional brain areas are also likely to participate in the task of spatial navigation. Previous work has identified numerous key part of the brain including: the hippocampus/parahippocampal area, posterior cingulate and precuneus, retrosplenial cortex, inferior parietal lobe, prefrontal cortex (Aguirre et al., 1996; Gron et al., 2000; Hartley, et al., 2003; Spreng et al., 2009). Some of these areas are associated with an "action planning" network, such as pre-supplemental motor area (pre-SMA) and retrosplenial cortex (Spiers and Maguire, 2006). Others, such as inferior parietal cortex, precuneus, hippocampus/parahippocampal and ventro-medial prefrontal cortex are associated with what has been termed the "default network", which is largely a memory network (Buckner et al., 2008; Raichle et al., 2001; Raichle and Snyder, 2007; Spreng, et al., 2009) implicated in future oriented thought (Andrews-Hanna et al., 2010). Thus, it is best to describe the act of navigation as implicating an entire network of brain areas involved with memory, sensation and cognition rather than a particular single structure that mediates all of navigation. Primary sensory areas provide the inputs to a complex system that must process location information, compare it to previous experience, and execute the motor movements for movement through an environment.

The Role of the Visual Cortex in the Blind

Humans rely heavily on vision in order to interact effectively with their environment. Conversely, blind individuals must learn to adapt to living in a world without the use of visual cues and thus rely heavily on other senses such as touch and hearing. To date, there is considerable evidence that adaption to blindness results in superior skill performance pertaining to the remaining senses. For example, behavioral studies have reported increased perceptual tasks within the auditory and tactile domains (Gougoux et al., 2004; Roder et al., 1999; Van Boven et al., 2000).

Intriguingly, neuroimaging studies in the blind performing tactile, auditory and verbal memory tasks have shown that the occipital cortex (normally ascribed to visual processing) is implicated in carrying out these non-visual tasks (Amedi et al., 2003; Burton et al., 2002; Kujala et al., 1995; Roder et al., 2002; Sadato et al., 1996).

What then is the role of the visual cortex in the blind? The visual cortex could serve as an extra processing unit for the compensatory abilities reported in the case of blindness. Gougoux and colleagues performed a PET study examining sound localization in blind and sighted individuals.

In blind individuals, a correlation was found between visual areas and cerebral blood flow (CBF) in response to the sound localization task in blind subjects (Gougoux et al., 2005). Amedi and cowrokers reported an fMRI study where a verbal memory task evoked responses in primary visual cortex in blind individuals (Amedi, et al., 2003). When word recall was retested after a certain time period, a positive correlation between activation and performance was observed. Specifically, subjects with the highest activation in primary visual cortex remembered the most words. Similarly, performance on the Wechsler Memory Scale correlated with primary visual cortex activation.

Causal support for the role of the visual cortex comes from clinical and experimental results. Acquired alexia (i.e. the acquired inability to read) following and occipital stroke and virtual reversible lesions caused by repetitive transcranial magnetic stimulation (rTMS) disrupt Braille reading skills.

In a rare case report, a congenitally blind, proficient Braille-reading individual had a bilateral occipital stroke which resulted in an inability to read Braille, despite normal tactile perception (Hamilton et al., 2000). Experimental evidence of a link between occipital cortex can be reliably demonstrated with rTMS, which can temporarily inactivate a region of cortex (Pascual-Leone et al., 2000).

In blind Braille readers, rTMS temporarily disrupts Braille reading (Cohen et al., 1997; Kupers et al., 2007). These findings suggest that visual cortex activation is not just a by-product of profound vision loss, but serve a functional utility for the processing of non-visual, task-relevant information.

Linking Visual Cortex to Brain Networks

Could crossmodal recruitment of the visual cortex also be implicated in navigation performance in the blind? The results of previous studies suggest that the visual cortex may be well suited for the integration of spatial and temporal information arising from non-visual inputs (Pascual-Leone and Hamilton, 2001). Thus, recruitment of visual cortex may be central to improved navigation performance (Figure 8.3):

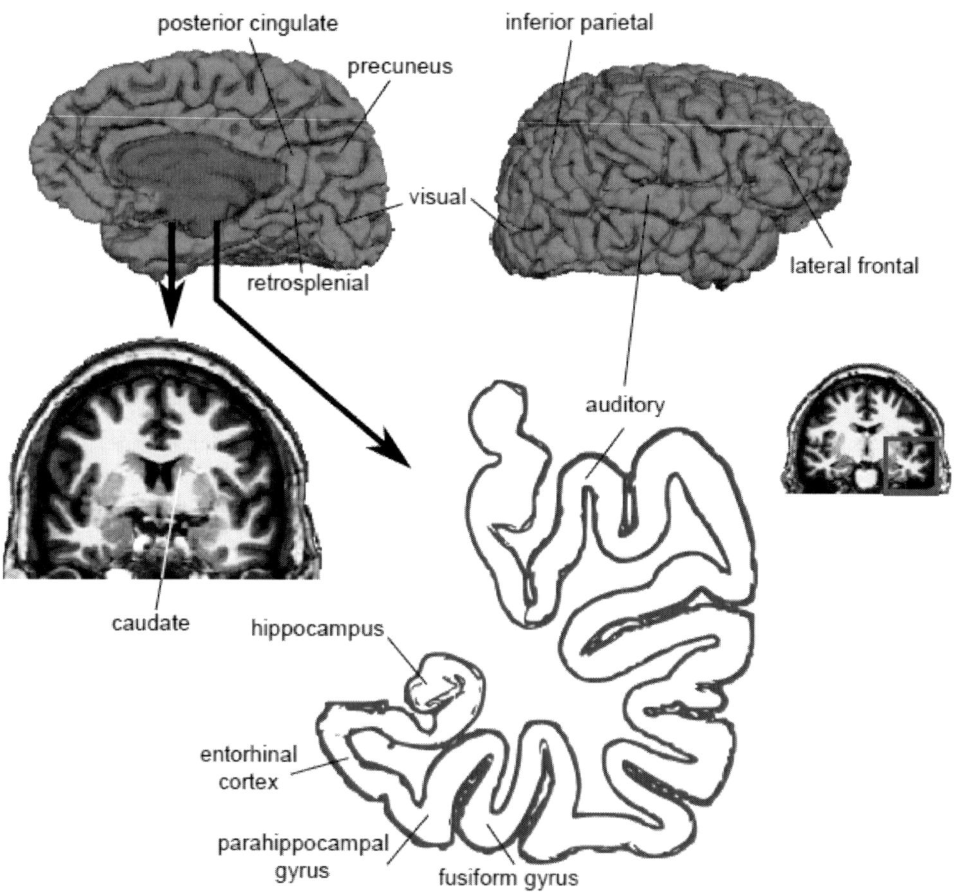

Figure 8.2. Brain areas associated with cross-modal blind navigation.

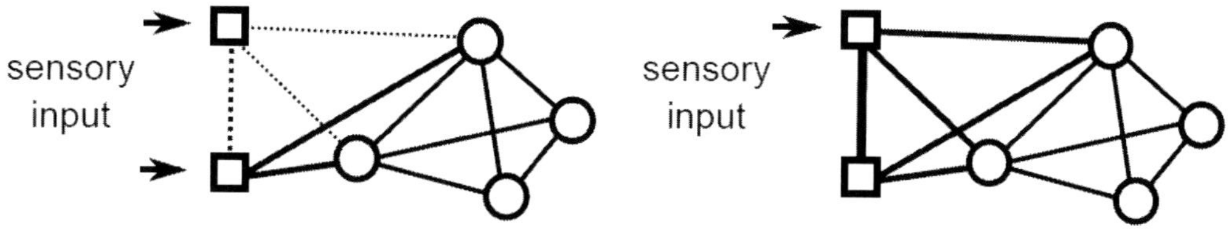

Figure 8.3. Network model of cross-modal changes to cognitive networks. Squares represent primary sensory areas, circles represent nodes in a cognitive network. (a) The dominant sensory modality supplies the primary connections to the cognitive network. (b) When the dominant sensory modality loses its input, the subordinate sensory modality utilizes existing connections with the dominant sensory system, as well as strengthens its own connections to the cognitive network.

If the visual cortex (as well as other brain areas) are implicated in navigation performance, one would need a method to assess navigation performance within the neuroimaging environment in order to address this question. Virtual environments could allow for structured investigation of navigation processes.

Furthermore, within a virtual environment, it is possible to compare brain activation on a variety of behavioral tasks as well as the strategies used to perform the tasks in question. This fact also has clinical importance within the clinical setting.

Orientation and mobility (OandM) training in the blind traditionally utilizes the instructor's clinical intuition as to which techniques and strategies are best used to proceed with training.

Each individual may respond best to different types and forms of instruction. Thus, exploiting the flexibility of virtual environments in the fMRI scanner can be used to compare, head-to-head, which learning strategies lead to optimal behavioral performance and the neural networks associated with that performance (see footnote).

Assessing Neuronavigation in the Sighted and Blind

There is no established literature regarding navigation processes in the blind. This is perhaps due to the difficulty in transitioning from the sighted virtual environment into a modality (e.g. tactile or auditory) that would allow a blind used to navigate effectively.

Other groups have investigated navigation in blind individuals with sensory substitution devices. Kupers et al (Kupers et al., 2010) utilized a "tongue display unit," which is a tactile device placed on the tongue, coupled to a camera that captures images and generates tactile spatial representations.

In their study, they found that simple navigation performance (i.e. two turn route) lead to activation patterns consistent with the navigation network of the sighted individuals. Specifically, strong activation was observed within the medial parietal and dorsolateral frontal cortices as well as the visual cortex.

In a series of on-going studies, we have adapted our audio-based virtual navigation system (AbES) for the fMRI environment. Here, a study subject (blind or sighted) can play the game solely using audio cues. We asked subjects to navigate within the virtual environment, starting from one room to another room.

In a pilot study, we have examined the brain responses of a blind individual to the brain responses of a sighted individual (blindfolded) carrying out the same task (Figure 8.4). Interestingly, activation for auditory-based navigation in the blind was similar to what would be obtained in sighted navigation and auditory stimulation. Consistent with auditory stimulation, superior temporal sulcus, angular gyrus and Wernike's area activation were all found.

The action and route planning network, consisting of pre-SMA, retrosplenial and medial/superior parietal cortex areas (Spiers and Maguire, 2006) was also activated. Visual areas near the occipital cortex and posterior lateral occipital and middle temporal gyrus were also activated. Parahippocampal and caudate areas were active, consistent with spatial navigation processes.

However, the critical comparison is comparing a blindfolded sighted subject to a blind subject. In the absence of visual input, the sighted subject recruited visual cortex, as well as within the route-planning network, but not caudate. The cortical networks recruited in the blindfolded sighted individual appear to be largely similar to the networks recruited in the blind individual. This is consistent with the hypothesis that blind individuals utilize existing brain network organization when visual cortex is recruited.

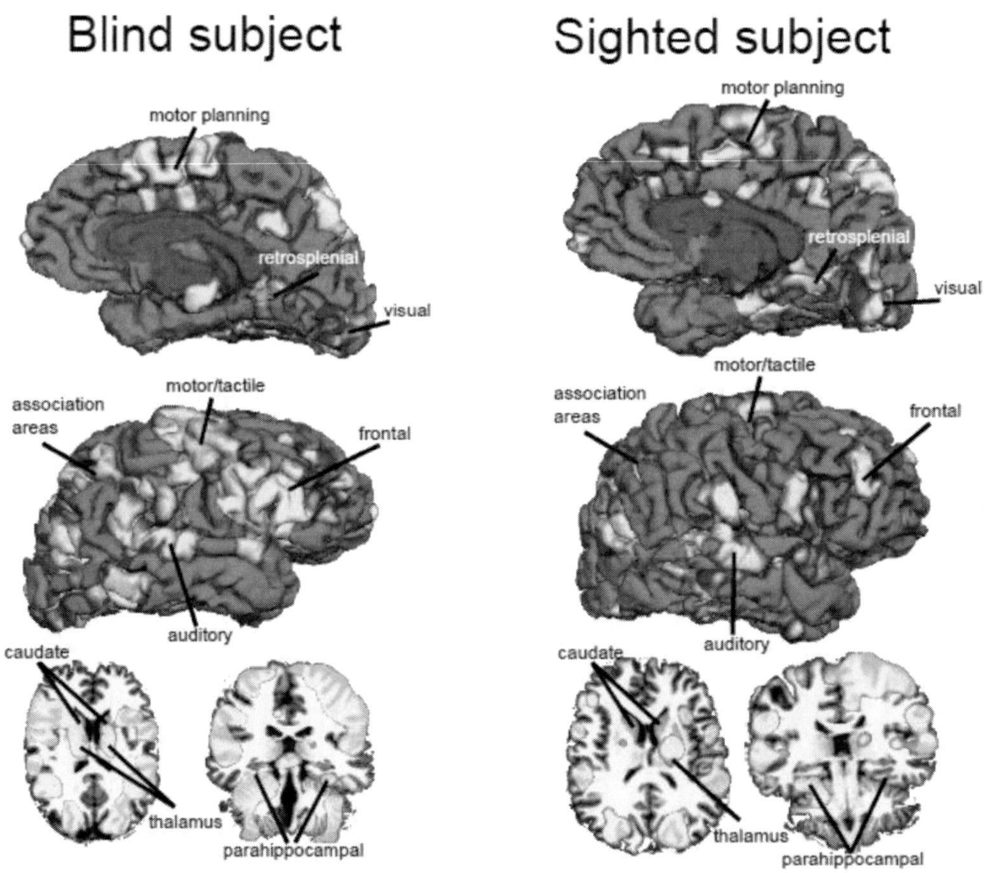

Figure 8.4. Brain activation to navigation in a sighted and blind subject using AbES, measured with fMRI. The activation shown compares activation related to navigation v.s. rest.

Open Questions

Moving around through our everyday environment is an exercise in constant multisensory bombardment. Sights, smells, sounds, touch and proprioceptive sensations serve as rich cues as to our location. In sighted individuals, we heavily rely on visual representations of our environment to successfully navigate: from turning the proper amount to make a 90 degree turn, to recognizing landmarks. For blind individuals, these visual sensations must be substituted with the remaining senses. In order to make up for the lack of rich input, blind individuals must learn to develop robust cognitive maps, as well as improved localization abilities via remaining sensation. This is supported by evidence of increased hippocampal volume in blind subjects, corresponding to the same changes observed in London taxi-drivers (Fortin, et al., 2008).

These changes in navigation strategies and skills provide a unique opportunity to better understand the neuroscience of navigation. Nearly every study on spatial navigation in the human has been performed on sighted individuals with visual stimuli. It is not clear which areas involved within the navigation network are predominately visual, or predominantly multisensory. In addition, it is not clear which pathways to the navigation network are most efficient.

It is best to think of the relationship between learning strategies and the neuroscience of navigation as circular. As we better understand how the brain represents visual environments in blind individuals, we can better adapt training strategies to improve OandM training. As we identify challenges, which have to be overcome in OandM training, we can look to the neuroscience to identify how we can utilize existing networks to solve these challenges.

FOOTNOTE: Neuroimaging techniques such as fMRI allow us to follow more closely and objectively phenomena related to behavioral performance at the level of the human brain. Unlike standard MRI images that give high quality anatomical images of the brain, functional MRI takes advantage of the fact that when a region of the brain is highly active, there is an oversupply of oxygenated blood to that region. By measuring the relative amounts of oxygenated and deoxygenated blood, it is possible to determine which regions of cortex are more active for a given task over a time scale of a few seconds. This signal is then analyzed to generate images of the brain that reflect regions of the brain implicated with the behavioral task being carried out (see Logothetis, 2008)

References

Aguirre, G. K., Detre, J. A., Alsop, D. C., and D'Esposito, M. (1996). The parahippocampus subserves topographical learning in man. *Cerebral Cortex 6*(6), 823-829.

Amedi, A., Raz, N., Pianka, P., Malach, R., and Zohary, E. (2003). Early 'visual' cortex activation correlates with superior verbal memory performance in the blind. *Nature Neuroscience, 6*, 758-766.

Andrews-Hanna, J. R., Reidler, J. S., Sepulcre, J., Poulin, R., and Buckner, R. L. (2010). Functional-anatomic fractionation of the brain's default network. *Neuron, 65*(4), 550-562.

Ashmead, D. H., Hill, E. W., and Talor, C. R. (1989). Obstacle perception by congenitally blind children. *Perception and Psychophysics, 46*, 425-433.

Astur, R. S., Taylor, L. B., Mamelak, A. N., Philpott, L., and Sutherland, R. J. (2002). Humans with hippocampus damage display severe spatial memory impairments in a virtual Morris water task. *Behavioural Brain Research, 132*, 77-84.

Axelrod, S. (1959). *Effects of early blindness; performance of blind and sighted children on tactile and auditory tasks*. New York,: American Foundation for the Blind.

Blasch, B. B., Wiener, W. R., and Welsh, R. L. (1997). *Foundations of orientation and mobility* (2nd ed.). New York: AFB Press.

Buckner, R. L., Andrews-Hanna, J. R., and Schacter, D. L. (2008). The brain's default network: anatomy, function, and relevance to disease. *Annals of the New York Academy of Sciences, 1124*, 1-38.

Burton, H., Snyder, A. Z., Conturo, T. E., Akbudak, E., Ollinger, J. M., and Raichle, M. E. (2002). Adaptive changes in early and late blind: a fMRI study of Braille reading. *Journal of Neurophysiology, 87*, 589-607.

Cohen, L. G., Celnik, P., Pascual-Leone, A., Corwell, B., Falz, L., Dambrosia, J., Honda, M., Sadato, N., Gerloff, C., Catala, M. D., and Hallett, M. (1997). Functional relevance of cross-modal plasticity in blind humans. *Nature, 389*, 180-183.

Dede, C. (2009). Immersive interfaces for engagement and learning. *Science, 323*(5910), 66-69.

Doeller, C. F., Barry, C., and Burgess, N. (2010). Evidence for grid cells in a human memory network. *Nature, 463*, 657-661.

Dye, M. W., Green, C. S., and Bavelier, D. (2009). The development of attention skills in action video game players. *Neuropsychologia, 47*, 1780-1789.

Ekstrom, A. D., Kahana, M. J., Caplan, J. B., Fields, T. A., Isham, E. A., Newman, E. L., and Fried, I. (2003). Cellular networks underlying human spatial navigation. *Nature, 425*, 184-188.

Fletcher, J. D. (2009). Education and training technology in the military. *Science, 323*, 72-75.

Fortin, M., Voss, P., Lord, C., Lassonde, M., Pruessner, J., Saint-Amour, D., Rainville, C., and Lepore, F. (2008). Wayfinding in the blind: larger hippocampal volume and supranormal spatial navigation. *Brain, 131*, 2995-3005.

Gougoux, F., Lepore, F., Lassonde, M., Voss, P., Zatorre, R. J., and Belin, P. (2004). Neuropsychology: pitch discrimination in the early blind. *Nature, 430*(6997), 309.

Gougoux, F., Zatorre, R. J., Lassonde, M., Voss, P., and Lepore, F. (2005). A functional neuroimaging study of sound localization: visual cortex activity predicts performance in early-blind individuals. *PLoS Biology, 3*, e27.

Green, C. S., and Bavelier, D. (2003). Action video game modifies visual selective attention. *Nature, 423*, 534-537.

Gron, G., Wunderlich, A. P., Spitzer, M., Tomczak, R., and Riepe, M. W. (2000). Brain activation during human navigation: gender-different neural networks as substrate of performance. *Nature Neuroscience, 3*, 404-408.

Hafting, T., Fyhn, M., Molden, S., Moser, M. B., and Moser, E. I. (2005). Microstructure of a spatial map in the entorhinal cortex. *Nature, 436*, 801-806.

Hamilton, R., Keenan, J. P., Catala, M., and Pascual-Leone, A. (2000). Alexia for Braille following bilateral occipital stroke in an early blind woman. *Neuroreport, 11*, 237-240.

Hartley, T., Maguire, E. A., Spiers, H. J., and Burgess, N. (2003). The well-worn route and the path less traveled: distinct neural bases of route following and wayfinding in humans. *Neuron, 37*, 877-888.

Iaria, G., Petrides, M., Dagher, A., Pike, B., and Bohbot, V. D. (2003). Cognitive strategies dependent on the hippocampus and caudate nucleus in human navigation: variability and change with practice. *Journal of Neuroscience 23*, 5945-5952.

Johnson, L. A., and Higgins, C. M. (2006). A navigation aid for the blind using tactile-visual sensory substitution. *Conference Proceedings - IEEE Engineering in Medicine and Biology Society, 1*, 6289-6292.

Kujala, T., Huotilainen, M., Sinkkonen, J., Ahonen, A. I., Alho, K., Hamalainen, M. S., Ilmoniemi, R. J., Kajola, M., Knuutila, J. E., Lavikainen, J., and et al. (1995). Visual cortex activation in blind humans during sound discrimination. *Neuroscience Letters, 183*, 143-146.

Kupers, R., Chebat, D. R., Madsen, K. H., Paulson, O. B., and Ptito, M. (2010). Neural correlates of virtual route recognition in congenital blindness. *Proceedings of the National Academy of Sciences of the United States of America, 107*, 12716-12721.

Kupers, R., Pappens, M., de Noordhout, A. M., Schoenen, J., Ptito, M., and Fumal, A. (2007). rTMS of the occipital cortex abolishes Braille reading and repetition priming in blind subjects. *Neurology, 68*, 691-693.

Lahav, O. (2006). Using virtual environment to improve spatial perception by people who are blind. *Cyberpsychology, Behavior, and Social Networking, 9*, 174-177.

Landau, B., Gleitman, H., and Spelke, E. (1981). Spatial knowledge and geometric representation in a child blind from birth. *Science, 213*, 1275-1278.

Li, R., Polat, U., Makous, W., and Bavelier, D. (2009). Enhancing the contrast sensitivity function through action video game training. *Nature Neuroscience, 12*(5), 549-551.

Li, R., Polat, U., Scalzo, F., and Bavelier, D. (2010). Reducing backward masking through action game training. *Journal of Vision, 10*, 1-13.

Logothetis, N. K. (2008). What we can do and what we cannot do with fMRI. *Nature, 453*, 869-878.

Loomis, J. M., Klatzky, R. L., and Golledge, R. G. (2001). Navigating without vision: basic and applied research. *Optometry and Vision Science, 78*, 282-289.

Maguire, E. A., Gadian, D. G., Johnsrude, I. S., Good, C. D., Ashburner, J., Frackowiak, R. S., and Frith, C. D. (2000). Navigation-related structural change in the hippocampi of taxi drivers. *Proceedings of the National Academy of Sciences of the United States of America, 97*, 4398-4403.

Maguire, E. A., Woollett, K., and Spiers, H. J. (2006). London taxi drivers and bus drivers: a structural MRI and neuropsychological analysis. *Hippocampus, 16*, 1091-1101.

Morrongiello, B. A., Timney, B., Humphrey, G. K., Anderson, S., and Skory, C. (1995). Spatial knowledge in blind and sighted children. *Journal of Experimental Child Psychology, 59*, 211-233.

O'Keefe, J., and Dostrovsky, J. (1971). The hippocampus as a spatial map. Preliminary evidence from unit activity in the freely-moving rat. *Brain Research, 34*(1), 171-175.

Ohuchi, M., Iwaya, Y., Suzuki, Y., and Munekata, T. (2006). *Cognitive-map Formation of Blind Persons in a Virtual Sound Environment.* Paper presented at the Proceedings of the 12th International Conference on Auditory Display, London, UK.

Owen, A. M., Hampshire, A., Grahn, J. A., Stenton, R., Dajani, S., Burns, A. S., Howard, R. J., and Ballard, C. G. (2010). Putting brain training to the test. *Nature, 465*, 775-778.

Pascual-Leone, A., and Hamilton, R. (2001). The metamodal organization of the brain. *Progress in Brain Research, 134*, 427-445.

Pascual-Leone, A., Walsh, V., and Rothwell, J. (2000). Transcranial magnetic stimulation in cognitive neuroscience–virtual lesion, chronometry, and functional connectivity. *Current Opinion in Neurobiology, 10*, 232-237.

Pissaloux, E., Maingreaud, F., Velazquez, R., and Hafez, M. (2006). Space cognitive map as a tool for navigation for visually impaired. *Conference Proceedings - IEEE Engineering in Medicine and Biology Society, 1*, 4913-4916.

Raichle, M. E., MacLeod, A. M., Snyder, A. Z., Powers, W. J., Gusnard, D. A., and Shulman, G. L. (2001). A default mode of brain function. *Proceedings of the National Academy of Sciences of the United States of America, 98*, 676-682.

Raichle, M. E., and Snyder, A. Z. (2007). A default mode of brain function: a brief history of an evolving idea. *Neuroimage, 37*(4), 1083-1090; discussion 1097-1089.

Riehle, T. H., Lichter, P., and Giudice, N. A. (2008). An indoor navigation system to support the visually impaired. *Conference Proceedings - IEEE Engineering in Medicine and Biology Society, 1*, 4435-4438.

Rieser, J. J. (2008). *Blindness and brain plasticity in navigation and object perception*. New York: Lawrence Erlbaum Associates/Taylor Francis Group.

Roder, B., Stock, O., Bien, S., Neville, H., and Rosler, F. (2002). Speech processing activates visual cortex in congenitally blind humans. *The European Journal of Neuroscience, 16*, 930-936.

Roder, B., Teder-Salejarvi, W., Sterr, A., Rosler, F., Hillyard, S. A., and Neville, H. J. (1999). Improved auditory spatial tuning in blind humans. *Nature, 400*, 162-166.

Sadato, N., Pascual-Leone, A., Grafman, J., Ibanez, V., Deiber, M. P., Dold, G., and Hallett, M. (1996). Activation of the primary visual cortex by Braille reading in blind subjects. *Nature, 380*, 526-528.

Sanchez, J., and Lumbreras, M. (1998). 3D aural interactive hyperstories for blind children. *International Journal of Virtual Reality, 4*, 20-28.

Sanchez, J., and Saenz, M. (2006). Three-dimensional virtual environments for blind children. *Cyberpsychology, Behavior, and Social Networking, 9*, 200-206.

Spiers, H. J., and Maguire, E. A. (2006). Thoughts, behaviour, and brain dynamics during navigation in the real world. *Neuroimage, 31*, 1826-1840.

Spreng, R. N., Mar, R. A., and Kim, A. S. (2009). The common neural basis of autobiographical memory, prospection, navigation, theory of mind, and the default mode: a quantitative meta-analysis. *Journal of Cognitive Neuroscience, 21*, 489-510.

Strelow, E. R. (1985). What is needed for a theory of mobility: direct perception and cognitive maps–lessons from the blind. *Psychological Review, 92*, 226-248.

Thinus-Blanc, C., and Gaunet, F. (1997). Representation of space in blind persons: vision as a spatial sense? *Psychological Bulletin, 121*, 20-42.

Tolman, E. C. (1948). Cognitive maps in rats and men. *Psychological Review, 55*(4), 189-208.

Van Boven, R. W., Hamilton, R. H., Kauffman, T., Keenan, J. P., and Pascual-Leone, A. (2000). Tactile spatial resolution in blind braille readers. *Neurology, 54*, 2230-2236.

Woollett, K., and Maguire, E. A. (2010). The effect of navigational expertise on wayfinding in new environments. *Journal of Environmental Psychology, 30*, 565-573.

In: Handbook of Technology in Psychology ...
Editor: Luciano L'Abate and David A. Kasier
ISBN: 978-1-62100-004-4

Chapter 9

The Uses of Technology for and with Children with Autism Spectrum Disorders

Olga Solomon
University of Southern California, US

> 'Throughout history and everywhere in the present, children live, learn, and play in homes, streets and other places, using the objects that come to hand and the spaces they inherit from adults".
>
> *Marta Gutman and Ning De Coninck-Smith*
> *"Designing Modern Childhoods: History, Space and the Material Culture of Children"*

Contemporary humans living in post-industrial societies are surrounded by technology and transformed by it in often-unpredictable ways. Cell phones and iPads have shifted the ways in which humans use their hands as our opposable thumbs are put to new uses in texting and video-gaming, and as our index fingers are used to move objects on touch screens of cell phones, iPads and laptops. At a popular-culture level, an IBM super-computer "Watson" competes with human experts and wins in a televised game of Jeopardy watched by millions of the program's fans. In different guises, from digital medical records to the use of robotics in surgery to streaming live video in long-distance psychological evaluations, technology has become part of our lives at work, school and home, and in health care, from diagnosis to intervention.

This chapter reviews the technological advances relevant to the diagnosis and interventions for Autism Spectrum Disorders (ASD) that have been developed through the collaborations of researchers, clinicians, families and the industry, and that are currently used by practitioners, teachers, and families. The goal of this chapter is to inform the reader about the technologies used across clinical disciplines and across home, community, clinic, and school settings, so that practitioners, parents and teachers can gain access to information on ASD-relevant technologies. This chapter builds upon the recent reviews of technological advances specific to clinical psychology (Harwood and L'Abate, 2010) and low-cost approaches to promote physical and mental health (L'Abate, 2007). It also draws upon the author's involvement in an Innovative Technology for Autism initiative originally launched by the Cure Autism Now Foundation and continued by the Autism Speaks Foundation, the largest parent advocacy non-profit foundation in the United States to support autism research (see Goodwin, 2008).

There are two related domains of technological innovation designed to address the complex and varied needs of children with Autism Spectrum Disorders (ASD) and their families. The first domain is related to the use of technology to facilitate timely and appropriate identification, assessment and diagnosis of ASD, especially when the families' access to experienced professionals is limited because of scheduling, geographical or socio-economic constraints; or because family members may find it difficult to describe clinically relevant autistic symptomatology when speaking with professionals during clinical visits (Goodwin, 2008).

The second area in which the use of technology has shown promise is the support of communication, participation and learning for children with ASD. For typically developing children and children with special needs alike, the use of technology has altered how children communicate, learn, play and engage in everyday activities. A 9 year old 'texting' her friends, a 10 year old in the middle a busy airport speaking on the cell phone, or a 13 year old with a Face Book or MySpace account have become familiar signs of contemporary childhood and adolescence. While these uses of technology may now seem unremarkable and taken for granted, they offer unique new resources for participation and engagement of children diagnosed with ASD who have an interest in technology and a documented proclivity for the use technological devices (e.g. Gillette, 2003).

ASD: Diagnostic Challenges in the Midst of a Public Health Crisis

Autism Spectrum Disorders (ASD) have become an urgent public health concern: an estimated 1.5 million individuals in the U.S. and tens of millions worldwide are affected by autism. U.S. government statistics suggest the prevalence of autism is increasing 10-17 percent annually. ASD are currently estimated to affect approximately 1%, or one in 110 children in the United States. The prevalence of ASD is 4 times higher in boys than in girls and approximately 40% of those affected have an intellectual disability. It is estimated that about 730,000 individuals between the ages of 0 to 21 have an ASD (Centers for Disease Control and Prevention, 2009).

The increase in prevalence has been attributed to many factors including changes in diagnostic criteria and migration from other clinical diagnoses (Byrd, 2003; Fombonne, 2001), however, there is disagreement in the research community about whether the size of increase in prevalence can be fully accounted for by this migration (e.g. Croen et al., 2002; Grinker, 2008). Most evidence points to changes in diagnostic practices and improved recognition and awareness (Fombonne, 2005; Mandell, Stahmer and Brodkin, 2008)

ASD consist of three DSM-IV diagnoses: Autistic Disorder, Asperger's Disorder, and Pervasive Developmental Disorder-Not Otherwise Specified (PDD-NOS). At the time of this writing, while the preparation of the next Diagnostic Statistical Manual, 5th Edition of the American Psychiatric Association is in progress with several important changes expected, these three Autism Spectrum Disorders belong to a group of Pervasive Developmental Disorders that additionally include Childhood Disintegrative Disorder and Rett's Disorder.

ASD are expected to manifest in childhood in the first years of life through such symptoms as impaired social reciprocity, delayed or absent spoken language, and proclivity for restricted behaviors, interests and activities (APA, 2000). The diagnosis of Autistic Disorder is made based upon the following criteria:

Table 1. DSM-IV Diagnostic Criteria for Autistic Disorder 299.00 (APA, 2000)

A. A total of six (or more) items from (1), (2), and (3), with at least two from (1), and one each from (2) and (3):
   (1) qualitative impairment in social interaction, as manifested by at least two of the following:
      (a) marked impairment in the use of multiple nonverbal behaviors such as eye-to-eye gaze, facial expression, body posture, and gestures to regulate social interaction
      (b) failure to develop peer relationships appropriate to developmental level
      (c) a lack of spontaneous seeking to share enjoyment, interests, or achievements with other people, (e.g., by a lack of showing, bringing, or pointing out objects of interest to other people)
      (d) lack of social or emotional reciprocity
   (2) qualitative impairments in communication as manifested by at least one of the following:
      (a) delay in, or total lack of, the development of spoken language (not accompanied by an attempt to compensate through alternative modes of communication such as gesture or mime)
      (b) in individuals with adequate speech, marked impairment in the ability to initiate or sustain a conversation with others
      (c) stereotyped and repetitive use of language or idiosyncratic language
      (d) lack of varied, spontaneous make-believe play or social imitative play appropriate to developmental level
   (3) restricted repetitive and stereotyped patterns of behavior, interests and activities, as manifested by at least two of the following:
      (a) encompassing preoccupation with one or more stereotyped and restricted patterns of interest that is abnormal either in intensity or focus
      (b) apparently inflexible adherence to specific, nonfunctional routines or rituals
      (c) stereotyped and repetitive motor mannerisms (e.g hand or finger flapping or twisting, or complex whole-body movements)

B. persistent preoccupation with parts of objects Delays or abnormal functioning in at least one of the following areas, with onset prior to age 3 years: (1) social interaction, (2) language as used in social communication, or (3) symbolic or imaginative play.

C. The disturbance is not better accounted for by Rett's Disorder or Childhood Disintegrative Disorder

Characterized by heterogeneity in the range of symptoms and degree of their severity across affected population, manifestations of ASD also change over the course of an individual's lifespan development (Wing, 1996). A popular saying in the autism community and among some researchers, "If you have met one person with autism, you have met one person with autism", attributed to an adult with ASD, author Stephen Shore, highlights the uniqueness of each affected person's manifestations and experience of ASD. The heterogeneity of ASD creates challenges in planning clinical interventions because understanding a child's unique individual profile is as important as understanding ASD as a clinical syndrome (Greenspan and Weeder,1999). Tomchek et al. (2010) provide an overview (see Table 2) of potentially challenging areas of occupation for individuals with ASD. In the field of occupational science and occupational therapy, occupations are defined as meaningful activities that support a person's health, well-being, and development (AOTA, 2008).

Table 2. Potentially Challenging Areas of Occupation (Tomchek et al., 2010: S126)

Areas of Occupation	Examples of Occupations (Daily Life Activities) That May Be Challenging for People With an ASD
Activities of daily living (ADLs)	Bathing, showering, dressing, eating, toilet hygiene, personal hygiene and grooming, sexual activity
Instrumental activities of daily living (IADLs)	Communication management, community mobility, meal preparation, fiscal management, health management and maintenance, home establishment and management, meal preparation and cleanup, safety and emergency maintenance, shopping
Rest and sleep	Achieving a calm state, sleep preparation, developing routines that support sleep, sleep participation
Education	Formal and informal education participation; written language skills; computer use; assistive technology; skills needed to access the curriculum; participation and transition within school environments such as classroom, cafeteria, playground, hallways, auditorium; and other school activities
Work	Employment interests and pursuits, employment seeking and acquisition, job performance, retirement preparation and adjustment, volunteer exploration, volunteer participation
Play and leisure	Play and leisure, exploration and participation
Social participation	Appropriate interactions with others (e.g., family, peers and friends, community), social competencies and management of novel situations, behavior management

Currently, no pathognomonic signs or biomarkers have been specifically associated with ASD and diagnosis is made based upon the clinical interpretation of a broad spectrum of observed behavior manifestations using standardized methods of evaluation (Lord and Spence, 2006). Asperger's Disorder and PDD-NOS are milder forms of Autistic Disorder. In Asperger's Disorder, language development proceeds without delay although the degree to which language develops in a normative fashion has been questioned (e.g. Landa, 2000). The diagnosis of PDD-NOS is given when there are insufficient criteria for a diagnosis of Autistic Disorder or Asperger's Disorder but many of the symptoms are present.

Evaluation by an experienced clinician remains the best 'instrument' for diagnosis of autism in infants and toddlers (Cox et al., 1999; Lord, 1995). Research suggests that such expert diagnosis is highly stable and accurate in children as young as 2 years of age (Charman and Baird, 2002; Lord et al., 2006). Access to expert diagnosis, however, presents a significant challenge to families.

Filipek et al. (2000) identified significant challenges to early diagnosis for children with ASD. In a survey of 1,300 families, although most parents had serious concerns about their children's development by 18 months and sought medical evaluation by age 2, the average age at diagnosis of autism was about 6 years. Fewer than 10% of the children in the survey were diagnosed at initial visit; another 10% were either assured that their child "would grow out of it" or advised to return if their worries persisted. Over 80% of the surveyed families were referred to another clinician when their children's mean age was 40 months. At this second visit, 25% were told "not to worry," 40% were given a formal diagnosis, and yet another 25% were referred to a third or fourth professional. Approximately 20% of the surveyed families reported that they paid privately for an evaluation or had to exert considerable pressure to obtain a referral. Over 30% of the surveyed parents who were referred for subsequent third and fourth evaluations reported that the practitioners who diagnosed their children did not offer any help or information regarding education, therapy, or referrals to parent support groups. Only about 10% of the parents reported that a clinician explained their child's impairments linked to ASD. Almost 50% of the surveyed families reported that other parents and the school system, rather than the health care professionals, were the major source of information and assistance over time (Filipek et al., 2000).

Although an ASD diagnosis may be made by an experienced professional based upon DSM-IV criteria, a reliable diagnosis ideally involves two instruments representing the best practice for ASD diagnosis: the Autism Diagnostic Observation Schedule (ADOS), a four-module interactive assessment based upon the level of language development that the child has achieved (Lord et al., 1989) and the Autism Diagnostic Interview, Revised (ADI-R) a detailed interview with the caregiver that focuses on details of the child's development between the ages of 3 and 4 years (Lord, Rutter and Le Couteur, 1994; Rutter, Le Couteur and Lord, 2003). While a number of diagnostic instruments, characterized by varied requirements for training and time of administration, have been developed to identify ASD, it is usually the case that the instruments requiring more training and certification, and taking longer to administer such as ADOS and ADI-R, have better reliability and validity than briefer instruments requiring less training. The Childhood Autism Rating Scale (CARS, Schopler, 1980) for example, is a brief checklist that can be filled out by a clinician using different kinds of data, from direct observations and parent interviews to medical records. While CARS does not take a long time to administer nor does it require extensive training and experience, it also over-identifies children with intellectual disabilities and under-identifies children with PDD-NOS. Another checklist instrument, the Gilliam Autism Rating Scale (GARS; Gilliam 1995), is completed by a caregiver, is brief and easy to score and also requires minimal training of the clinician. Empirical studies on GARS, however, found poor reliability and validity and under-identification of children with ASD (Lecavalier, 2005; South et al., 2002; Williams, Atkins and Soles, 2009). Comparative analysis of three widely used screening instruments for ASD diagnosis in toddlers at high risk indicated that no instrument or individual item showed satisfying power in discriminating ASD from non-ASD (Oosterling et al., 2009).

Recent efforts are under way to overcome this challenge by designing rapid and reliable assessment paradigms that would obviate clinic-based, time-intensive and high-cost behavioral assessments. Lee et al. (2010) report on the reliability and validity of an Internet-based Interactive Autism Network (IAN)-implemented parent survey involving assessments of verbal children ages 4-17 with an existing diagnosis of ASD. Although there were several caveats that weaken the generalizability of this study's results, statistical analysis of accuracy of an Internet – based parental report within families participating in an ASD-specific autism registry suggested that it has a high concordance rate with clinic-based best practice assessments (ADI-R and ADOS). The picture is complex because *whether and how* diagnostic instruments are used vary across clinical settings with different eligibility requirements, as well as across individual clinicians. Although best practice guidelines in ASD diagnosis have been developed (Filipek et al., 2000; Johnson and Myers, 2007; Lord and Bishop, 2010), and American Academy of Pediatrics recently recommended that all 18- and 24-month-olds be screened for autism spectrum disorders (Zwaigenbaum et al., 2009), most psychologists assessing children for the Department of Developmental Services (DDS) eligibility who were surveyed in the Hering (2005) study, reported not using best practice guidelines. The result of this lack of consistency is that different diagnostic practices produce often-conflicting evaluations, with the rate of agreement by different evaluators for individual children in one study reported to be only 45% (Williams, Atkins and Stoles, 2009).

Thus, the situation 'on the ground' is often a far cry from the recommended best practice guidelines and parents may receive conflicting diagnoses from different professionals, a traumatic and confusing experience that adds stress and uncertainty to the lives of children and families. When the racial and ethnic backgrounds and socio-economic status of the families enter the picture, it gains an additional level of complexity. For example, population-level studies strongly indicate an unprecedented scale of health and service disparities in ASD diagnosis for African American children. Demographic and epidemiological data about the children's age at diagnosis and the duration of the diagnostic process indicate that African American children are 2.6 times less likely than Caucasian children to receive an ASD diagnosis on their first specialty care visit and to have a much higher probability of being misdiagnosed with adjustment disorder, conduct disorder or ADHD. A national study of age of diagnosis correlates in Medicaid-enrolled children with ASD found that socio-demographic characteristics as well as local healthcare resources and state policies contribute to disparities in the age of diagnosis (Mandell et al., 2009). Caucasian children entered the mental health system at an earlier age (6.0 versus 7.1 years, $p = .005$); however, after adjusting for age, sex, and time eligible for Medicaid, African American children required more time in treatment before receiving the diagnosis: African American children receiving Medicaid were diagnosed on average at 7.9 years of age, 18 months later than Caucasian children on Medicaid who were diagnosed at 6.3 years (Mandell et al., 2002, 2003, 2006, 2007; Mandell, Stahmer and Brodkin, 2008; Stahmer and Mandell, 2007). The picture that emerges from these studies is of systematic delays in diagnosis and challenges to secure appropriate services once the diagnosis is received. In summary, there is a shortage of professionals trained in ASD who have the experience and the resources to follow best practice guidelines in diagnosis and interventions. Because of this shortage, families often have to wait weeks and even months for an appointment. Even when families are able to secure an appointment with an experienced clinician, they may have to travel if they live far from major cities (Oberleitner, 2006). This is an area of urgent need where tele-health and medical information technologies, combined with sophisticated digital video-capturing, storage and streaming, offer unique opportunities to families who otherwise may not have access to a high quality diagnostic evaluation and assessment.

USE OF TELE-HEALTH IN ASD ASSESSMENT AND INTERVENTION

Tele-health is a technique that uses telecommunication and information technologies to transfer basic patient information in order to deliver health care services at a distance, including diagnosis and treatment (Loane and Wootton, 2003). Tele-health technologies, also called 'telemedicine', have been used since the 1990s to increase access to health care in medically-underserved areas and populations (Office for the Advancement of Telehealth, 2003).

For example, distance writing, an Internet-based computer-mediated psychological intervention has been used in mental health disciplines for over a decade as a complimentary, cost-effective way to augment face-to-face 'talk therapy', as well as to reach those previously unreachable through more traditional therapeutic approaches (see L'Abate, 2001). Tele-health technology has been used to ameliorate geographic and scheduling limitations and bring health care professionals together with children and families when there are concerns about a child's development. Two kinds of tele-health technologies are currently available: 1) Real-time, synchronous tele-health includes video-conferencing in real-time between a provider and a patient, or a real-time consultation between providers; 2) Asynchronous, or 'store-and - forward' tele-health is capturing medical information electronically and then forwarding it to a provider (Stamm, 1998). "Store-and-forward' has been a tele-health technology of choice for ASD because it is often difficult to accurately sample symptom-relevant behaviors such as self-injury, seizures, tantrums or aggression in real time during a tele-health session with a provider.

Accurate presentation of these behaviors through 'store-and-forward' video-imaging modalities improves providers' understanding of the child's challenges in the natural environments and creates an extensive video-based medical record for diagnosis or follow-up evaluations, for example, when a medication is prescribed. Grady (2002) found that this type of patient-provider communication reduced overall stress in the family, and saved time and financial resources (see also Oberleitner, 2006; Reischl and Oberleitner, 2009).

Figure 1 demonstrates a flow diagram of a strategy developed by Talk Autism, a communication service shared by organizations to access a common database of resources and distance learning library (Oberleitner et al., 2006, p. 231). In this diagram, a caregiver video-records an episode of concern or interest; the video is either mailed conventionally or electronically transferred to an appropriate health care provider; the provider reviews the video; the provider responds to the caregiver with recommendations via telephone, letter or e-mail; the video-recorded episode is archived for future reference.

The use of tele-health for ASD holds promise in a number of clinical disciplines, such as counseling, neurology, psychiatry, social work, speech and language, occupational therapy, and physical therapy. The integration of technology and health care, however, has been slow to develop in mainstream clinical practice, hindered in part by legal, financial, and regulatory barriers, including ambiguities in third-party reimbursement coverage of tele-health applications, and concerns about the confidentiality, privacy and security of health information.

Figure 1. Flow Diagram of the Use of Tele-health for ASD (Oberleitner et al. 2006, p. 231)

To overcome the challenges in provider acceptance, third-party payer reimbursement and liability, the tele-health community must demonstrate this technology's efficacy and produce cost effectiveness data through high-quality, peer-reviewed clinical studies. A report by the Office of Technology Policy (2004) recommends that to overcome these challenges, "tele-health suppliers (manufacturers and services firms), providers (clinics and clinicians), payers (third party insurers), and other stakeholders must be prepared to work together to address a wide array of needs, issues and opportunities" (p.10). Additionally, rigorous cost-benefit and business case analyses are necessary to justify public funding for developing mainstream applications for tele-health.

USE OF UBIQUITOUS COMPUTING FOR BEHAVIOR IMAGING IN ASD

Behavior Imaging® technology has emerged to meet the needs of families and practitioners for an accurate and systematic integration of behavior information with health records in the ASD diagnosis and intervention. The goal of this field is to make an impact on ASD-related health care industry similar to other 'imaging' techniques such as X-Rays, MRIs and CT scans (Oberleitner et al., 2010). Parents and other caregivers often report the need to record, store, and analyze video data about their children with ASD to make it available to health care professionals (Hayes et al., 2004).

The usefulness of behavior imaging technologies depends not only on the clinical skills and expertise of the practitioners but also upon the availability of high quality, extensive video data that captures clinically relevant behavior. Moreover, personal health records that document clinical history and that are synchronized with the behavioral video data must be made available to clinicians to assist in long-distance functional assessment. Finally, these synchronized data streams should be available for marking and sharing among individual clinicians and their interdisciplinary teams. Such marking and sharing may include 'tagging', or classifying / categorizing certain segments or points in the video, assembling them in clinically meaningful sub-corpora or collections, and having long-distance, internet-based, secure access in order to work collaboratively on the interpretation of these data.

But how are such streams of data created? The technological complexity of this process ranges from a digital video-recording of a child's behavior carried out by a family member to a sophisticated, automated capture system that is part of a built-in home environment. Hayes et al. (2004) report on the development of a technologically sophisticated automated capture system used for collecting data on activities that are part of interventions for children with ASD and for keeping records about these activities. An important part of this process was defining what kinds of intervention data will be captured because different individuals in different locations, both professionals and family members, may be administering and monitoring interventions. The automated capture system allowed to document which team member was providing which intervention, how these professionals or family members communicated with one another, and what kinds of records or assessments they were administering. Essentially, the capture system made it possible for family members and professionals to visually *show* what they did with the child with ASD, and how the child responded, rather than trying to *describe* it. Hayes et al. (*ibid)* identified three discreet types of information: 1) duration, or how long the child was engaged in an activity, and whether the activity was appropriate; 2) performance, or how often the child responded to a question or a direction in an appropriate way; and 3) written narrative description of the child's behavior. An example of the automated capture and access system (Hayes, 2004) is *Abaris* designed at the Georgia Institute of Technology by Gregory Abowd, the Distinguished Professor at the School of Interactive Computing and Executive Director of the Health Systems Institute. The purpose of *Abaris* is to document Discreet Trial Training (DTT) intervention where a team of therapists takes turns working with a child for approximately 2 to 3 hours a day every day, which involves completing hundreds of trials per session. Before the automated capture system was available, in the end of the session each therapist was expected to manually add up the data, calculate the percentages of successfully completed trials, create graphs that reflect progress and write narrative notes, a task that was both difficult and time-consuming. *Abaris* automated this process by capturing and integrating the individual therapist's data with the video of the session. Rather than using a pen and pencil, therapists use a tablet PC and specially designed electronic forms (called CareLogs) to enter the data. Summary statistics are automatically calculated and then are available for graphing.

Reflecting its dual function, capture and access, *Abaris* consists of two parts located on the same computer: one part for capture, or recording of data, and another for its access and analysis. Other components include a web cam for capturing video and audio, a wireless microphone for voice recognition, and a digital pen for writing on specially printed paper. The structure of *Abaris* is reflected in Figure 2:

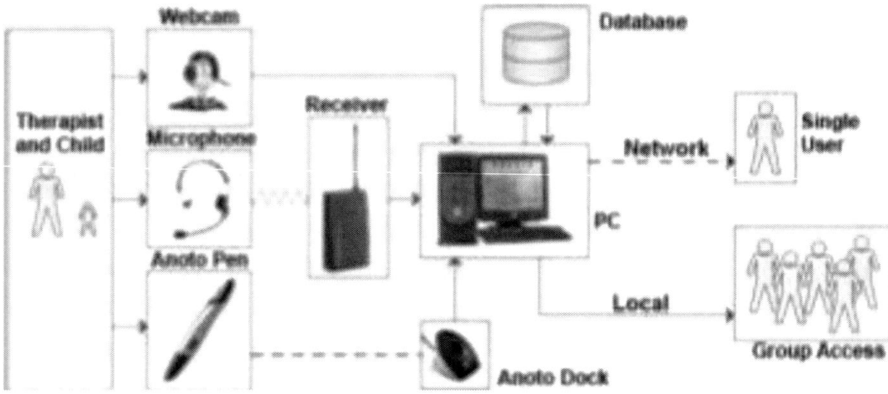

Figure 2. Components of the Abaris system (Klientz et al., 2005)

To make automated capture more extensive and continuous, whole home environments have been adapted to integrate data capture systems. Georgia Tech's Aware Home project (Klentz et al., 2008), part of the university's Aware Home research initiative, is a three story, 5040 square feet home that functions as a laboratory for the interdisciplinary design of technology, its development and evaluation. Automated capture and access technologies like *Abaris* can be useful in the early detection of autism, enhancing collaborations among families, health care professionals and teachers, and making systematically collected, high quality data available for research on ASD (Klentz et al., 2007). One of Aware Home project's most recent innovations is "Social Mirrors", an environment-embedded social networking system to support adults with ASD wanting to live independently (Gregory Abowd, personal communication, March 29, 2011).

USE OF TECHNOLOGY TO SUPPORT COMMUNICATION, PARTICIPATION AND LEARNING OF CHILDREN WITH ASD

There is growing interest in technology-based interventions for children with ASD. Practitioners working with children with ASD and their families will be more clinically effective when they have a deep understanding of the technologies used by their clients and the meaning these technologies have for them. In the second part of this chapter the discussion will turn to how technology is used to enhance communication, learning and participation of children with ASD. This is an area much discussed in the popular media; however, the analytic intersection of 'technology', 'childhood' and 'autism' present both challenges and opportunities. The challenges lie, on one hand, in romanticizing technology as a magically omnipotent, uniformly accessible tool that identifies and alleviates impairments imposed by autism and by doing so fundamentally transforms the lives of children and their families.

On the other hand, there is a danger in not appreciating this powerfully transformative tool enough, especially in post-industrial "knowledge societies" of the Internet Age where being computer- and digitally-literate is on par with traditionally defined literacy (Organization for Economic Co-operation and Development, 2001). The solution to this quandary lies perhaps in paying careful attention to how technology, childhood and autism intersect in a realm of what anthropologist George Marcus (1995) calls "the imaginary", an orientation to imagining futures that arises out of technology and scientific practice being both imaginative and visionary (p. 4).

Especially in clinical settings, where practitioners and family members collaborate in carrying out interventions that 'emplot' desirable, hoped-for futures (Mattingly, 1998), there is tension between technology's potential and the reality of people's everyday lives. The tension lies between technology, especially its Internet-based varieties, as a tool for preparing young people for participation in the 'knowledge economy' (Hargreaves, 2003) and technology as a mediator of universal literacy in the context of globalization. As Moss (2008) writes, "Even in our discussions about computer technologies and digital literacies, the ideal norm and what is real can be far apart. In the ideal, the Internet, specifically the World Wide Web, has provided a means for cultures to participate in crosscultural communications. Digital technologies bring the global world to the most remote villages, making the world smaller. Presumably, this global world has an impact on community or vernacular literacies as this contact with the wider world increases. Yet, unequal access resulting in a digital divide is still a reality. Material resources, or lack thereof, continue to contribute to who has power and who does not. Also, while digital technologies expand our definitions of writing, there is still a tension about what kind of writing will prepare our young people to be able to succeed in a technologically sophisticated, global society. What kinds of writing and other literacies should be valued is still a question" (p. 562-563).

Considering the barriers to how information is created, exchanged and used via various technologies in communities around the world, Hawisher and Selfe (2000) decry the myth of a 'global village' perpetuated mostly in the U.S. and in other post-industrial societies by the view of the Internet and World Wide Web as a "global literacy environment in which peoples from all over the world can communicate with one another without significant barriers posed by geopolitical location, language, culture, and everyday practices and attitudes" (p. 2), or as in case of autism, by all of the above plus the communicative challenges posed by a developmental disorder that impacts the very ability to communicate (see also Warschauer, 2004). Another challenge lies in recognizing the practices of identifying and understanding autism itself: a highly heterogeneous, developmentally variable phenomenon that is both clinical and socio-cultural (Solomon, 2010). In considering how technology is used for and with children, including children diagnosed with ASD, there is a need to acknowledge that institutional setting in which the children are participants are organized by ideologies regarding the uses of technology as well as by local notions of competence, achievement, and development (e.g. Mehan, 1996). These institutional ideologies are reproduced and ratified through situated practices, whether these institutions are hospitals and clinics, schools, after-school care programs or online communities. Another analytic hurdle is a representation of children in post-industrial societies as passive users and consumers of technology. There is a lament about "a new type of child" who is a "technologically astute, information-loaded and brand-literate product of advanced consumer culture" (Lindstrome and Seybold, 2003, cf Clarke, 2008). A less pessimistic view is that children and families incorporate technologies into their lives and by doing so, transform them in sometimes unpredictable and paradoxical ways. In a related discussion about mass media as exemplified by Disney cultural products, and its interpretive, assimilating consumption (Certeau, 1984) by children and families, Mattingly (2006) writes:

"The power of mass media to shape cultural identities has also been a topic of central concern within contemporary culture theory. Stuart Hall, for example, has argued that mass media is a primary vehicle for ideological production, a means through which groups construct images of their lives (Hall 1997). Mass media might be colonizing but ethnographic studies of the actual practices of cultural production and reception help to complicate the picture of media as ideology machine. Such studies reveal the way that the global is made "local" in specific contexts and show that the meaning of a media text is not given by the text itself but by the processes through which it is taken up and consumed by particular interpretive communities (Abu-Lughod 1991; Metcalf 2001; Rapp and Ginsburg 2001). Reception theories in media studies have been at pains to argue that meaning is a complex and local practice of negotiation between the world of the text and the world of the audience. It is a historically particular and local invention (see Spitulnik 1993 for a review of this literature). In other words, the audience does not merely consume, it "poaches," as Michel de Certeau (1984) puts it, helping to construct meaning through its practices of consumption" (p. 498).

Thus, there is a need for an ethnographically informed, person-centered examination of the situated practices and ideologies of the uses of technology for and with children with ASD, as well as its geo-political, economic and user-centered accessibility. In this sense, the 'social life' of technology, to paraphrase Appadurai (1986), has to be part of the analytic picture. Similarly to Georg Simmel's conceptualization of 'economic objects' (1907/1978, cf Appadurai 1986), the 'social life' of technology for children with ASD involves understanding its value not as its inherent property but rather a property assigned to it by subjects - be they the children themselves, their family members, teachers, or clinicians - who are guided by their own subjectivities in a space between desire and enjoyment, "which is a distance that can be overcome" (Appadurai 1986: 3). The tension that exists in this space is worth considering because it tells us about the practical logics that guide how technology is used for and with children with ASD in the first place, what the underlying practical reasoning is behind its use, and what possible and imagined futures are "emplotted" (Mattinlgy, 1998) in the process. In the very least, this space can tell us about the newly visible potentialities and competencies that the uses of technology afford to children, families, clinicians who are part of the children's lifeworlds.

The uses of technology for and with children with ASD also foreground the materiality of human bodies in social life. Philosopher Judith Butler (1993) interrogates the processes of 'construction' of a gendered body, asking questions that may be relevant to understanding the place of technology in the lives of children with ASD and their families: "What are we to make of constructions without which we would not be able to think, to live, to make sense of it all, those that have acquired for us a kind of necessity? Are certain constructions of the body constitutive in this sense: that we could not operate without them, that without them there would be no "I", no "we"? Thinking of the body as constructed demands a re-thinking of the meaning of construction itself". (p. xi) These questions suggest a view of the uses of technology as constructing a certain kind of a human being where materiality of the body and the materiality of technology exist in a constitutive relationship, making an 'unintelligible' and 'unlivable' body 'intelligible' to others and 'livable' for self. There are several example of such ontological flexibility afforded by technology. In the field of Socially Assistive Robotics, socially intelligent robots engage in contingent social interaction with children with ASD and their family members, thus supporting communication among human participants (Dautenhahn, 2003; Feil-Seifer and Mataric, 2005, 2008). In the field of Virtual Reality,

Embodied Conversational Agents also called "Virtual Peers" engage children with autism in interpersonal relationships through collaborative storytelling (Cassell, 2001; Tartaro and Cassell, 2006, 2008).

In less philosophical terms, technology helps realize the potentialities of children with ASD as human beings with rich subjectivities, communicative intentions and abilities, and inner worlds: It quite literally helps 'construct' a different kind of being than the child has been until technology was introduced into his or her lifeworld. The possibility that technology can have this effect disrupts and problematizes the notion of agency as residing within an individual, suggesting a more complex agentive relationship between human beings and their material and semiotic worlds (Barad, 1996, 2007). Some of these themes can be heard in parental narratives about how their children with ASD engage with technology. Consider, for example, a story told around Christmas time by a mother of a four-year old boy diagnosed with ASD; at the time this story was told, the boy's verbal communication has been limited to only a few words[1]. The narrative is shared with a group of family members of children with ASD that meets as part of an ethnographic study on health and service disparities in ASD diagnosis for African American children. The group is mediated by two researchers.

Mother: He knows that Santa is coming and is going to bring him something. And the funny thing, we were asking, what do you want, what do you want and he says, "Ipad, Ipad." (laughs) Because he plays with my husband's Ipad and he can-, he's great with computers you know. He may not be able to speak really well, but he can navigate the Ipad, he can download, he knows how to go to websites to download the games and applications that he wants and he can stay on it for hours, so that- I mean-, his preschool teacher sent us an article about a child with autism who found their niche with an Ipad, and I said well I can relate to that because my son can spend hours upon hours on the Ipad. Like this is mine, I have it. And I bought some computer software for him, so he plays this Blue's Clues computer game for hours. Like last Saturday, it was just the two of us and I was kind of, you know, making some preparations, so I had the laptop and he stayed on it for 4 hours straight and when I tried to pull him away, like, "ok it's too long, let's take a break, let's eat," he doesn't want to eat or anything, he likes technology, so- I think that's good. I mean in some respects, I don't want him to have too much, but then again, 10,000 hours is the uh, expert level, so, with anything. I read The Outliers, and that was one of the things that they talk about. Once you reach 10,000 hours doing anything, you become an expert at it. Yeah, Bill Gates, among other people, had 10,000 hours built in, professional athletes, 10,000 hours. So that seems to be benchmark for mastery of something.

Researcher: So are you keeping track of hours?

Mother: I'm keeping track of his hours, I was like, "ok, we got 4 today, we're on our way!" (Laughs)

Researcher: Were you surprised with the Ipad?

Mother: Yes! I mean it was almost instant. It took him maybe 20 minutes before he figured out, "Ok, if I slide my finger across it makes pages move. What if I push this button?"

---

[1]This example was selected from a digital video-and audio- data corpus collected for an ethnographic, interdisciplinary project 'Autism in Urban Context: Linking Heterogeneity with Health and Service Disparities' (1 R01 MH089474-01; NIH / NIMH; Olga Solomon, PI) that examines health and service disparities in ASD diagnosis for African American children living in urban Los Angeles.

And he, he figured it out. Meanwhile, I'm still sitting there trying to get onto the Internet and he's like, "Mommy, let me show you."

Such narratives about families' experience with technology offer a glimpse of the ways in which it is used for and with children with ASD. These narratives allow to see how impairments imposed by ASD are experienced, interpreted and mediated; what theories of competence and disability are at play; and what socio-cultural practices are improvised and routinely carried out to manifest certain kinds of technologically expressed selves and mediated identities.

In this sense, the kinds of technology and the practices of its use offer a unique picture of ASD from the perspectives of the children, their families and clinicians who work with them. Such examination illustrates that competence may be as much an attribute of the socio-cultural and socio-interactional context as of an individual's mind or body, i.e. the inclusion of technology in educational and family settings supports, mediates and affords children's communicative competencies that otherwise may not be visible.

TYPES OF TECHNOLOGY USED FOR AND WITH CHILDREN WITH ASD

Technological devices used for and with children with ASD may be categorized by the following six domains of engagement in social interaction that technology mediates (see Table 3 for examples):

1) *Medium of communication.* Communication *via speech* is enhanced by voice output keyboards; communication *via text* is enhanced through the use of computer keyboards, texting on the cell phone, iPads and other text-generating devices.
2) *Social ontologies.* Ontological properties of interlocutors can be manipulated and configured favorably, for example, minimizing facial information, in Socially Assistive Robotics; and by designing Virtual Peers in Virtual Reality environments. Individuals with ASD also have opportunities of altering their own social identities through the choice of avatars in Virtual Reality environments, e.g. Second Life.
3) *Affect recognition and affect communication.* This domain of functioning is technologically enhanced through affective computing and wireless bio-sensing[2] where individuals with autism wear biosensors that make visible their physiological states such as stress and anxiety.
4) *Enhancing participation through auditory and tactile prompting, and scheduling.* Experience of participating in activities with others is enhanced though auditory and tactile prompting devices that keep an individual with ASD focused and on task; scheduling devices enhance a sense of predictability of social environments, promote confidence and decrease anxiety.
5) *Turn-taking of social interaction.* This domain of functioning can be technologically enhanced via the use of DiamondTouch™ StoryTable, a multi-user touch and gesture, drag-and-drop device to support small group collaboration, where children

[2]Developed by Affectiva, a company created through the collaboration of researchers at MIT Media Lab and the technology industry.

with autism have to coordinate their actions in order to participate in an enjoyable play activity.

6) *Safety-related skills.* Several Virtual Reality applications have been developed to teach safety-related skills such as what to do in the event of a fire, or how to cross a busy street.

Table 3. Examples of technologies and supporting research across six domains

Domain of technological intervention	Technology and Research
1) Medium of Communication Voice	Voice out-put communication aids (VOCA) [3] (Gillette, 2003; Mirenda, 2001, 2003)
Writing	Computer keyboards (Heimann, et al., 1995); e-mail (Burke, and Kraut and Williams, 2010; Benford, 2008) Cell phones (Durkin et a., 2010; Goldsmith and LeBlank, 2004) iPads and other tablet PCs
2) Social ontologies "Other"	Socially Assistive Robotics (Dautenhahn, 2003Feil-Seifer and Mataric, 2005; 2008) Virtual Peers (Cassell, 2001; Tartaro and Cassell, 2006, 2008) CosmoBot™[4] (Brisben, Lockerd, and Lathan,2004)
"Self"	Second Life (Boulos, Hetherington and Wheeler, 2007)
3) Affect recognition and affect communication	Affective Computing, Wireless Bio-sensing Technology (el Kaliouby et al., 2006; Picard, 2009; Picard and Goodwin, 2008) Virtual Reality for training eye gaze skills (Trepagnier et al., 1998)
4) Enhancing participation through auditory and tactile prompting	Tactile and auditory prompting devices[5] (Coyle and Cole, 2004; Shabani et al., 2002; Taylor and Levin, 1998; Taylor, Huges, Richard, Hoch, and Coello, 2004)
through scheduling	Microsoft PowerPoint (Rehfeldt, Kinney, Root, and Stromer, 2004); Video enhanced activity schedules (Dauphin, Kinney, and Stromer, 2004); Computer based activity schedules (Stromer et al., 2006)
5) Turn-taking of social interaction	StoryTable system based on DiamondTouch (Bauminger et al., 2007)
6) Safety-related skills	Do2Learn Virtual reality application (Strickland, 1997; Strickland et al, 1996; Rizzo, Strickland and Bouchard, 2004)

Such a domain-based approach contributes to the existing reviews on the uses of technology for individuals with ASD. Goldsmith and LeBlank (2004), for example, review five types of technology used as a temporary instructional aid: 1) tactile and auditory

[3]Developed by Archimedes Access and Research and Technology, Inc.

[4]Developed by Anthrotronics, is a patented interactive robotic system designed to enhance development of children with special needs including ASD.

[5]Developed by Follow Through Inc., 2003; JTECH Communications Inc., 2004

prompting devices; 2) video-based instruction and feedback, 3) computer-aided instruction, 4) virtual reality, and 5) robotics. Other categories that differentiate the uses of technology for ASD include assistive technologies; alternative and augmentative communication systems, designed either to supplement (i.e., augment) children's existing speech or to act as a primary (i.e., alternative) method of expressive communication; voice output communication aides (VOCA); interactive games; educational software; affective computing; online communities; ubiquitous computing; wireless bio-sensing; user-centered collaborative design processes; and product assessment (see also Gillette, 2003; Gillette et al., 2007; Goldsmith and LeBlank, 2004; Goodwin, 2008 for reviews).

CHALLENGES AND PROMISES OF TECHNOLOGY IN CLINICAL PRACTICE

While there is a sense of excitement about the possibilities of the use of technology for and with children with ASD, there remain serious challenges to bridging development, research and practice. Kimbal and Smith (2007) review the challenges related to computer technology use and conclude that work needs to be done in crossing the bridge from research to practice and in transforming promising and sound experimental findings into effective, affordable, and readily available technological products. Until the bridge from research to practice is crossed, the task of integrating the technology into clinical or educational practice rests primarily on the shoulders of families and practitioners rather than developers and researchers. Even though there is strong empirical support for the use of some technologies with children with ASD, there are not enough 'ready to use' applications that provide logistic feasibility for the use in clinical, educational settings (McConnell, 2002).

On the other hand, there may be an advantage to the absence of ready-made prescriptions on how to use the multiplicity of technological devices presently on the market. As Goldsmith and LeBlanc (2004) point out, PDA's, cell phones, laptops, and MP3 players have entered mainstream society and are becoming more and more affordable. A child with ASD using an iPad or a cell phone for prompting or scheduling will look indistinguishable from the neurotypical others using these devices. The task of the clinician is to find ways to adapt these devices into a technologically enhanced clinical practice.

ACKNOWLEDGMENT

The author wishes to thank the following individuals who have been members of the Innovative Technology for Autism Advisory Board: Gregory Abowd, Katarina Boser Danielle Gillette, Matthew Goodwin, and Albert (Skip) Rizzo. A special thank you to Portia Iversen who first initiated this initiative as part of the Cure Autism Now foundation's area of research. Others whose research has contributed to the author's on-going commitment to the study of technology for and with individuals with ASD are: Justine Cassell, Sharon Cermak, David Feil-Seifer, Gillian Hayes, Julie Kienz, Mary Lawlor, Maja Mataric, and Elinor Ochs.

REFERENCES

Abu-Lughod, L. (1991). Writing against culture. In R. G. Fox (Ed.), *Recapturing anthropology: Working in the present* (pp. 137-162). Santa Fe, NM: School of American Research Press.

American Occupational Therapy Association. (2008). Occupational therapy practice framework: Domain and process (2nd ed.). *American Journal of Occupational Therapy, 62*, 625–683.

American Psychiatric Association. (2000). Diagnostic and statistical manual of mental disorders (4th ed., text. rev.). Washington, DC: American Psychiatric Association.

Appadurai, A. (1986). *The social life of things: Commodities in cultural perspective*. Cambridge, MA: Cambridge University Press.

Barad, K. (1996). Meeting the universe halfway: Realism and social constructivism without contradiction. In L. H. Nelson and J. Nelson (Eds.), *Feminism, science, and the philosophy of science* (pp. 161-194). London: Kluwer.

Barad, K. (2007). *Meeting the universe halfway: Quantum physics and the entanglement of matter and meaning*. Durham, NC: Duke University Press.

Bauminger, N., Gal, N., Goren-Bar, D., Kupersmitt, J., Pianesi, F., Stock, O., Weiss, P.L.T., Yifit, R., and Zancanaro, M. (2007). Enhancing social communication in high-functioning children with autism through a co-located interface. Proceedings of the 6th International Workshop on Social Intelligence Design, July 2-4 2007, Trento, Italy.

Benford, P. (2008). The use of internet-based communication by people with autism (Doctoral dissertation). University of Nottingham.

Boulos, M. N. K., Hetherington, L., and Wheeler, S. (2007). Second life: An overview of the potential of 3-D virtual worlds in medical and health education. *Health Information and Libraries Journal, 24*(4), 233-245. doi: 10.1111/j.1471-1842.2007.00733.x.

Brisben, A.J., Lockerd, A.D., and Lathan, C. (2004). Design evolution of an interactive robot for therapy. *Telemedicine Journal and e-Health. 10*(2), 252-259.

Burke, M., Kraut, R., and Williams, D. (2010). *Social use of computer-mediated communication by adults on the autism spectrum*. Paper presented at the Proceedings of the 2010 ACM Conference on Computer Supported Cooperative Work, Savannah, Georgia, USA.

Butler, J. (1993). *Bodies that matter: On the discursive limits of "sex"*. New York, NY: Routledge.

Byrd, R. S., University of California, Davis., and M.I.N.D. Institute. (2003). Report to the legislature on the principal findings from the epidemiology of autism in California: A comprehensive pilot study. Davis, CA: M.I.N.D. Institute.

Cassell, J. (2001). Embodied conversational agents: Representation and intelligence in user interfaces. *AI Mag., 22*(4), 67-83.

Centers for Disease Control and Prevention. (2009). Autism spectrum disorders (ASDs): Data and statistics. Retrieved March 30, 2011, from http://www.cdc.gov/ncbddd/autism/data.html

Certeau, M. (1984). *The practice of everyday life*. Berkeley, CA: University of California Press.

Charman, T., and Baird, G. (2002). Practitioner review: Diagnosis of autism spectrum disorder in 2- and 3-year-old children. *Journal of Child Psychology and Psychiatry, 43*(3), 289-305. doi: 10.1111/1469-7610.00022.

Cox, A., Klein, K., Charman, T., Baird, G., Baron-Cohen, S., Swettenham, J., Drew, A., Wheelwright, S. (1999). Autism spectrum disorders at 20 and 42 months of age: Stability of clinical and ADI-R diagnosis. *The Journal of Child Psychology and Psychiatry and Allied Disciplines, 40*(5), 719-732.

Coyle, C., and Cole, P. (2004). A videotaped self-modeling and self-monitoring treatment program to decrease off-task behavior in children with autism. *Journal of Intellectual and Developmental Disability, 29,* 3-15.

Croen, L. A., Grether, J. K., Hoogstrate, J., and Selvin, S. (2002). The changing prevalence of autism in California. *Journal of Autism and Developmental Disorders, 32*(3), 207-215. doi: 10.1023/a:1015453830880.

Dauphin, M., Kinney, E. M., Keogel., R.L., and Stromer R. (2004). Using video enhanced activity schedules and matrix training to teach socio-dramatic play to a child with autism. *Journal of Positive Behavior Interventions, 6*(4):238–250.

Dautenhahn, K. (2003). Roles and functions of robots in human society: Implications from research in autism therapy. *Robotica, 21*(04), 443-452. doi:10.1017/S0263574703004922.

Durkin, K., Whitehouse, A., Jaquet, E., Ziatas, K., and Walker, A. J. (2010). Cell phone use by adolescents with Asperger Syndrome. *Research in Autism Spectrum Disorders, 4*(2), 314-318. doi: 10.1016/j.rasd.2009.09.017.

el Kaliouby, R., Picard, R., and Baron-Cohen, S. (2006). Affective computing and autism. *Annals of the New York Academy of Sciences, 1093*(1), 228-248. doi: 10.1196/ annals.1382.016.

Feil-Seifer, D. and Mataric, M. (2005). *Defining socially assistive robotics*. Proceedings of the 2005 IEEE, 9th International Conference on Rehabilitation Robotics, Chicago, IL.

Feil-Seifer, D., and Mataric, M. (2008). *Robot-assisted therapy for children with autism spectrum disorders*. Proceedings of the 7th International Conference on Interaction Design and Children, Chicago, IL.

Filipek, P. A., Accardo, P. J., Ashwal, S., Baranek, G. T., Cook, E. H., Dawson, G., Gordon, B., Gravel, J.S., Johnson, C.P., Kallen, R.J., Levy, S.E., Minchew, N.J., Ozonoff, S., Prizant, B.M., Rapin, I., Rogers, S.J., Stone, W.L., Teplin, S.W., Tuchman, R.F., and Volkmar, F. R. (2000). Practice parameter: Screening and diagnosis of autism. *Neurology, 55*(4), 468-479.

Gillette, D. (2003). Technology's role in encouraging social and cognitive development in children with autism. *Journal of Developmental and Learning Disabilities, 35,* 97-109.

Gillette, D., Hayes, G., Abowd, G., Cassell, J., el Kaliuby, R., Strickland, D., and Weiss, P.T. (2007). *Interactive technologies for autism*. Proceedings of the CHI 2007 Extended Abstracts on Human Factors in Computing Systems, doi:10.1145/1240866.1240960.

Gilliam, J. E. (1995). *Gilliam Autism Rating Scale (GARS).* Austin, TX: Pro-Ed.

Goldsmith, T. R., and LeBlanc, L. A. (2004). Use of technology in interventions for children with autism. *Journal of Early and Intensive Behavior, 1*(2), 166-178.

Goodwin, M. S. (2008). Enhancing and accelerating the pace of autism research and treatment. *Focus on Autism and Other Developmental Disabilities, 23*(2), 125-128. doi: 10.1177/ 1088357608316678.

Grady, B. J. (2002). A comparative cost analysis of an integrated military telemental health-care service. *Telemedicine Journal and e-Health, 8*(3), 293-300. doi: 10.1089/15305620260353180.

Greenspan, S., and Wieder, S. (2006). *Engaging autism: Using the Floortime approach to help children relate, communicate, and think*. Cambridge, MA: Da Capo Press.

Grinker, R. R. (2008). *Unstrange Minds*: remapping the World of Autism. New York, NY: Basic Books.

Gutman, M. and Coninck-Smith, N. D. (2008). *Designing Modern Childhoods: History, Space and the Material Culture of Children.* New Brunswick, NJ: Rutgers University Press.

Hall, S. (1997). The local and the global: Globalization and ethnicity. In A. McClintock, A. Mufti and E. Shohat (Eds.), *Dangerous liaisons: Gender, nation, and postcolonial perspectives* (pp. 173-187). Minneapolis, MN: University of Minnesota Press.

Hargreaves, A. (2003). *Teching in the knowledge society: Education in the age of insecurity.* New York, NY: Teachers College Press.

Harwood, T. M., and L'Abate, L. (2010). *Self help in mental health: A critical evaluation.* New York, NY: Springer-Science.

Hawisher, G. E., and Selfe, C. L. (2000). *Global literacies and the world-wide web.* New York, NY: Routledge.

Hayes, G. R., Kientz, J. A., Truong, K. N., White, D. R., Abowd, G. D., and Pering, T. (2004). Designing capture applications to support the education of children with autism. In N. Davies, E. Mynatt and I. Siio (Eds.), *UbiComp 2004: Ubiquitous computing* (Vol. 3205, pp. 161-178). Berlin, Germany: Springer.

Heimann, M., Nelson, K., Tjus, T., and Gillberg, C. (1995). Increasing reading and communication skills in children with autism through an interactive multimedia computer program. *Journal of Autism and Developmental Disorders, 25*(5), 459-480. doi: 10.1007/bf02178294.

Hering, M. M. (2005). *A survey of psychologists' diagnostic practices in the assessment of children with autistic disorder* (Doctoral dissertation). Pepperdine University.

Johnson, C. P., Myers, S. M., and The Council on Children With Disabilities. (2007). Identification and evaluation of children with autism spectrum disorders. *Pediatrics, 120*(5), 1183-1215. doi: 10.1542/peds.2007-2361.

Kientz, J. A., Patel, S. N., Jones, B., Price, E., Mynatt, E. D., and Abowd, G. D. (2008). The Georgia Tech Aware Home. *Proceedings of the CHI 2008*, Florence, Italy.

Kientz, J. A., Boring, S., Abowd, G. D., and Hayes, G. R. (2005). Abaris: Evaluating automated capture applied to structured autism interventions. In M. Beigl, S. Intille, J. Rekimoto and H. Tokuda (Eds.), *UbiComp 2005: Ubiquitous Computing* (Vol. 3660, pp. 903-903). Berlin, Germany: Springer.

Kientz, J. A., Hayes, G. R., Westeyn, T. L., Starner, T., and Abowd, G. D. (2007). Pervasive computing and autism: Assisting caregivers of children with special needs. *IEEE Pervasive Computing, 6*, 28-35. doi: 10.1109/MPRV.2007.18.

Kimball, J. W., and Smith, K. (2007). Crossing the Bridge. *Focus on Autism and Other Developmental Disabilities, 22*(2), 131-134. doi: 10.1177/10883576070220020501.

L'Abate, L. (Ed.). (2001*). Distance writing and computer-assisted interventions in psychiatry and mental health.* Westport, CT: Ablex.

L'Abate, L. (Ed.). (2007). *Low-cost approaches to promote physical and mental health*. New York, NY: Springer-Science.

Landa, R. (2000). Social language use in Asperger syndrome and high-functioning autism. In A. Klin, F. R. Volkmar and S. S. Sparrow (Eds.), *Asperger syndrome* (pp. 125-155). New York, NY: Guilford Press.

Lecavalier, L. (2005). An evaluation of the Gilliam Autism Rating Scale. *Journal of Autism and Developmental Disorders, 35*(6), 795-805. doi: 10.1007/s10803-005-0025-6.

Lee, H., Marvin, A. R., Watson, T., Piggot, J., Law, J. K., Law, P. A.,... Nelson, S. F. (2010). Accuracy of phenotyping of autistic children based on internet implemented parent report. *Am J Med Genet Part B, 153B*(6), 1119-1126. doi: 10.1002/ajmg.b.31103.

Lindstrome, M., and Seybold, P. (2003). *Brainchild: Remarkable insights into the minds of today's global kids and understanding their relationship with brands.* Sterling, VA: Kogan Page.

Loane, M., and Wootton, R. (2002). A review of guidelines and standards for telemedicine. *J. Telemed Telecare, 8*(2), 63-71. doi: 10.1258/1357633021937479.

Lord, C. (1995). Follow-up of two-year-olds referred for possible autism. *Journal of Child Psychology and Psychiatry, 36*(8), 1365-1382. doi: 10.1111/j.1469-7610.1995.tb01669.x.

Lord, C., and Bishop, S. L. (2010). Autism spectrum disorders: Diagnosis, prevalence, and services for children and families. *Sharing Child and Youth Development Knowledge, 24*(2), 1-27.

Lord, C., Risi, S., DiLavore, P. S., Shulman, C., Thurm, A., and Pickles, A. (2006). Autism from 2 to 9 years of age. *Arch Gen Psychiatry, 63*(6), 694-701. doi: 10.1001/archpsyc.63.6.694.

Lord, C., Rutter, M., and Couteur, A. (1994). Autism diagnostic interview-revised: A revised version of a diagnostic interview for caregivers of individuals with possible pervasive developmental disorders. *Journal of Autism and Developmental Disorders, 24*(5), 659-685. doi: 10.1007/bf02172145.

Lord, C., Rutter, M., Goode, S., Heemsbergen, J., Jordan, H., Mawhood, L., and Schopler, E. (1989). Austism diagnostic observation schedule: A standardized observation of communicative and social behavior. *Journal of Autism and Developmental Disorders, 19*(2), 185-212. doi: 10.1007/bf02211841.

Lord, C., and Spence, S. (2006). Autism spectrum disorders: Phenotype and diagnosis. In S. O. Moldin and J. L. R. Rubenstein (Eds.), *Understanding autism: From basic neuroscience to treatment* (pp. 1-24). Boca Raton, FL: Taylor and Francis.

Mandell, D. S., Cao, J., Ittenbach, R., and Pinto-Martin, J. (2006). Medicaid expenditures for children with autistic spectrum disorders: 1994 to 1999. *Journal of Autism and Developmental Disorders, 36*(4), 475-485. doi: 10.1007/s10803-006-0088-z.

Mandell, D. S., Ittenbach, R., Levy, S., and Pinto-Martin, J. (2007). Disparities in diagnoses received prior to a diagnosis of autism spectrum disorder. *Journal of Autism and Developmental Disorders, 37*(9), 1795-1802. doi: 10.1007/s10803-006-0314-8.

Mandell, D. S., Listerud, J., Levy, S. E., and Pinto-Martin, J. A. (2002). Race differences in the age at diagnosis among medicaid-eligible children with autism. *Journal of the American Academy of Child and Adolescent Psychiatry, 41*(12), 1447-1453.

Mandell, D. S., Stahmer, A. C., and Brodkin, E. S. (2008). Autism spectrum disorders in childhood. In T. P. Gullotta and G. M. Blau (Eds.), *Handbook of childhood behavioral issues: Evidence based approaches to prevention and treatment* (pp. 345-369). New York, NY: Taylor and Francis Group.

Mandell, D. S., Wiggins, L. D., Carpenter, L. A., Daniels, J., DiGuiseppi, C., Durkin, M. S., Giarelli, E., Morrier, M.J. Nicholas, J.S., Pinto-Martin, J. A., Shattuck, P.T., Thomas, K.C., Yeargin-Allsopp, M., and Kirby, R. S. (2009). Racial/ethnic disparities in the identification of children with autism spectrum disorders. *Am J Public Health, 99*(3), 493-498. doi: 10.2105/ajph.2007.131243.

Marcus, G. (1995). *Technoscientific imaginaries: Conversations, profiles, and memoirs.* Chicago, IL: University of Chicago Press.

Mattingly, C. (1998). *Healing dramas, clinical plots.* Cambridge, MA: Cambridge University Press.

Mattingly, C. (2006). Pocahontas goes to the clinic: Popular culture as lingua franca in a cultural borderland. *American Anthropologist, 108*(3), 494-501. doi: 10.1525/aa.2006.108.3.494.

McConnell, S. R. (2002). Interventions to facilitate social interaction for young children with autism: Review of available research and recommendations for educational intervention and future research. *Journal of Autism and Developmental Disorders, 32*(5), 351-372. doi: 10.1023/a:1020537805154.

Mehan, H., Villanueva, I., Hubbard, L., and Lintz, A. (1996). *Constructing school success: The consequences of untracking low-achieving students.* Cambridge, MA: Cambridge University Press.

Metcalf, P. (2001). Global "disjunctive" and the "sites" of anthropology. *Cultural Anthropology, 16*(2), 165-182. doi: 10.1525/can.2001.16.2.165.

Mirenda, P. (2001). Autism, augmentative communication, and assistive technology. *Focus on Autism and Other Developmental Disabilities, 16*(3), 141-151. doi: 10.1177/1088357 60101600302.

Mirenda, P. (2003). Toward functional augmentative and alternative communication for students with autism: Manual signs, graphic symbols, and voice output communication aids. *Lang Speech Hear Serv Sch, 34*(3), 203-216. doi: 10.1044/0161-1461(2003/017).

Moss, B. J. (2008). Writing in the wider community. In R. Beard, D. Myhill, J. Riley and M. Nystrand (Eds.), *The SAGE handbook of writing development* (pp. 561-573). Thousand Oaks, CA: SAGE Publications Ltd.

Oberleitner, R., Ball, J., Gillette, D., Naseef, R., and Stamm, B. H. (2006). Technologies to lessen the distress of autism. In J. Garrick and M. B. Williams (Eds.), *Trauma treatment techniques: Innovative trends* (pp. 221-242). Binghamton, NY: Haworth Press.

Oberleitner, R., Reischl, U., Lacy, T., Goodwin, M., and Spitalnick, J. S. (2011). Emerging use of behavior imaging for autism and beyond. *Commun Med Care Compunetics, 1*, 93–104 doi: 10.1007/8754_2010_5

Office for the Advancement of Telehealth. (2003). Office for the Advancement of Telehealth Welcome Page. Retrieved March 25, 2011, from http://www.hrsa.gov/ruralhealth/about/telehealth.

Oosterling, I., Swinkels, S., van der Gaag, R., Visser, J., Dietz, C., and Buitelaar, J. (2009). Comparative analysis of three screening instruments for autism spectrum disorder in toddlers at high risk. *Journal of Autism and Developmental Disorders, 39*(6), 897-909. doi: 10.1007/s10803-009-0692-9.

Picard, R. (2009). Future affective technology for autism and emotion communication. *Phil. Transactions of the Royal Society, 364*: 3575–3584.

Rapp, R., and Ginsburg, F. (2001). Enabling disability: Rewriting kinship, reimagining citizenship. *Public Culture, 13*(3), 533-556. doi: 10.1215/08992363-13-3-533.

Rehfeldt, R. A., Kinney, E. M., Root, S., and Stromer, R. (2004). Creating activity schedules using Microsoft Powerpoint. *Journal of Applied Behavior Analysis, 37,* 115-128.

Reischl, U., and Oberleitner, R. (2009). Development of a telemedicine platform for the management of children with autism. *German Journal for Young Researchers*. Retrieved March 27, 2011, from http://www.nachwuchswissenschaftler.org/2009/1/3.

Rizzo, A. A., Strickland, D., and Bouchard, S. (2004). The challenge of using virtual reality in telerehabilitation. *Telemedicine Journal and e-Health, 10*(2), 184-195. doi: doi:10.1089/tmj.2004.10.184.

Rutter, M., LeCouteur, A., and Lord, C. (2003). Autism diagnostic interview-revised manual. Los Angeles, CA: Western Psychological Services.

Schopler, E., Reichler, R., DeVellis, R., and Daly, K. (1980). Toward objective classification of childhood autism: Childhood Autism Rating Scale (CARS). *Journal of Autism and Developmental Disorders, 10*(1), 91-103. doi: 10.1007/bf02408436.

Shabani, D. B., Katz, R. C., Wilder, D. A., Beauchamp, K., Taylor, C. R., and Fischer, K. J. (2002). Increasing social initiations in children with autism: Effects of a tactile prompt. *Journal of Applied Behavior Analysis, 35,* 79-83.

Solomon, O. (2010). Sense and the senses: Anthropology and the study of autism. *Annual Review of Anthropology, 39*(1), 241-259. doi:10.1146/annurev.anthro.012809.105012.

South, M., Williams, B. J., McMahon, W. M., Owley, T., Filipek, P. A., Shernoff, E., and Ozonoff, S. (2002). Utility of the Gilliam Autism Rating Scale in research and clinical populations. *Journal of Autism and Developmental Disorders, 32*(6), 593-599. doi: 10.1023/a:1021211232023.

Spitulnik, D. (1993). Anthropology and mass media. *Annual Review of Anthropology, 22*(1), 293-315. doi: doi:10.1146/annurev.an.22.100193.001453.

Stahmer, A., and Mandell, D. S. (2007). State infant/toddler program policies for eligibility and services provision for young children with autism. *Administration and Policy in Mental Health and Mental Health Services Research, 34*(1), 29-37. doi: 10.1007/s10488-006-0060-4.

Stamm, B. H. (1998). Clinical applications of telehealth in mental health care. *Professional Psychology: Research and Practice, 29*(6), 536-542.

Strickland, D. (1997). Virtual reality for the treatment of autism. In R. Giuseppe (Ed.), *Virtual reality in neuro-psycho-physiology: Cognitive, clinical and methodological issues in assessment and rehabilitation* (pp. 81-86). Burke, VA: IOS Press.

Strickland, D., Marcus, L., Mesibov, G., and Hogan, K. (1996). Brief report: Two case studies using virtual reality as a learning tool for autistic children. *Journal of Autism and Developmental Disorders, 26*(6), 651-659. doi: 10.1007/bf02172354.

Stromer, R., Kimball, J.W., Kinney, E.M., and Taylor, B.A. (2006) Activity Schedules, Computer Technology, and Teaching Children With Autism Spectrum Disorders, *Focus on Autism and other Developmental Disabilities, 21*(1), 14–24.

Tartaro, A., and Cassell, J. (2006). *Authorable virtual peers for autism spectrum disorders.* Paper presented at The combined workshop on language-enabled educational technology and development and evaluation of robust spoken dialogue systems at the 17th European Conference on Artificial Intelligence, Riva del Garda, Italy.

Tartaro, A., and Cassell, J. (2008). Playing with virtual peers: Bootstrapping contingent discourse in children with autism. Proceedings of the 8th International Conference for the Learning Sciences. Retrieved February 10, 2011, from http://www.gerrystahl.net/proceedings/icls2008/ papers/paper290.pdf

Taylor, B. A., Hughes, C. E., Richard, E., Hoch, H., and Coello, A. R. (2004). Teaching teenagers with autism to seek assistance when lost. *Journal of Applied Behavior Analysis, 37,* 79-82.

Taylor, B. A., and Levin, L. (1998). Teaching a student with autism to make verbal initiations: Effects of a tactile prompt. *Journal of Applied Behavior Analysis, 31,* 651-654.

Tomchek, S., LaVesser, P., Watling, R., and DeLany, J. (2010). The scope of occupational therapy services for individuals with an autism spectrum disorder across the life course. *American Journal of Occupational Therapy, 64(6) Suppl*(1), S125-S136.

Trepagnier, C., Rosen, M. J., and Lathan, C. (1999). Telerehabilitation and virtual reality technology for rehabilitation: Preliminary results. *Proceedings of the 14th Annual International Conference, Technology and Persons with Disabilities,* Los Angeles, CA.

Warschauer, M. (2004). *Technology and social inclusion: Rethinking the digital divide.* Cambridge, MA: MIT Press.

Williams, M., Atkins, M., and Soles, T. (2009). Assessment of autism in community settings: Discrepancies in classification. *Journal of Autism and Developmental Disorders, 39*(4), 660-669. doi: 10.1007/s10803-008-0668-1.

Wing, L. (1996) Autistic spectrum disorders. *British Medical Journal, 312,* 327-328.

Zwaigenbaum, L., Bryson, S., Lord, C., Rogers, S., Carter, A., Carver, L., ... Yirmiya, N. (2009). Clinical assessment and management of toddlers with suspected autism spectrum disorder: Insights from studies of high-risk infants. *Pediatrics, 123*(5), 1383-1391. doi: 10.1542/peds.2008-1606.

In: Handbook of Technology in Psychology ...
Editor: Luciano L'Abate and David A. Kasier
ISBN: 978-1-62100-004-4

Chapter 10

TECHNOLOGICAL INNOVATIONS IN COMPARATIVE PSYCHOLOGY: FROM THE PROBLEM BOX TO THE 'RUMBAUGHX'

David A. Washburn, Michael J. Beran, Theodore A. Evans, Megan L. Hoffman and Timothy M. Flemming
Georgia State University, US

> By experience in working with various animals and with pathological human subjects, I am convinced of the urgent need of attention to our methods of recording reactions. We, at present, allow the experimenter too great range and place upon him over-great responsibility. As observer, he is liable both to influence the subject in his attempts to get data on reaction and, in turn, to be influenced, in his descriptions of what he sees, by his unescapable tendencies to interpret. Quite evidently, the ideal experiment is one in which the subject provides us a detailed photographic record (or other form of graphic record) of its response. It is the writer's belief that we should make systematic and persistent attempts to develop recording devices which shall free us from the observational imperfections of the experimenter. This means that our apparatus for us in Comparative Psychology must be largely automatic or self-controlling over considerable periods of time, not only with respect to the objective situation or setting in which the subject reacts, but also with respect to the recording of the several important aspects of response. We should devise types of recording mechanism which shall either operate automatically or be operated by the subject rather than by the experimenter. This would mean not the elimination of the observer but the freeing of his attention for those aspects of the total experiment which most urgently demand control (Yerkes, 1915, pg. 258).

As the other contributions to this volume doubtless attest, the impact of technology on psychology is certainly not limited to its use in studies of animal behavior. Whether the technology in question is the apparatus that provides our scientific measures, or the platform through which we intervene in behavior (e.g., in the form of training, therapy, or experimental manipulation for research), it is clear that the development of psychology as a discipline is interwoven with innovations and applications of technology. That said, it is undeniable that technology is *at least* as important for the study of nonhuman animals (hereafter "animals") as it is for the understanding and change of human behavior. Indeed, it is reasonable to suggest that—given that animals cannot be queried about their behavior through language—technology plays a more central role in comparative psychology than it does in any other subdiscipline of the field, save perhaps neuropsychology. In any event, one would be hard-pressed to find a clearer and more compelling statement about the importance of technology for any other area of psychology than the charge from Robert Yerkes (1915) to comparative psychologists. Written almost a century ago during the formative years of comparative psychology, Yerkes warned of the dangers of inadvertent cuing of animal subjects and of the perils of subtle (or not so subtle) bias in the observation, recording, description, and interpretation of animal behavior. Those cautions are equally appropriate for contemporary comparative inquiry, where the surging popularity and acceptance of comparative cognition seems in some cases to have lowered scientists' guards against old enemies to the discipline such as the "Clever Hans" phenomenon, uncritical anthropomorphism, biases in the reporting and interpreting of behavior, and violations of parsimony and conservatism. For Yerkes, one strategy for avoiding these pitfalls was the application of technology—specifically automation—in the scientific process. Whereas automation alone doesn't eliminate all the threats to sound science, it has certainly been the case that advances in comparative psychology have frequently been linked to apparatus developments that permit scientists to ask new questions or to ask familiar questions in better ways. Many of the most influential scholars in the history of comparative psychology are themselves closely identified with methodological, and frequently technological, innovations that similarly impacted the discipline. In the present chapter, we will review some of the ways that technology has been used in the study of animal behavior and cognitive competencies, and the ways that our understanding of animals has in turn been shaped and constrained by these dominant technologies and paradigms.

CAVEATS

Before reviewing the symbiotic relation between technological innovation and scientific progress in comparative psychology, we should acknowledge several concessions that we have made in this review. First, we recognize that "technology" is not limited to apparatus, much less to apparatus that runs on electricity. Whereas the present review largely, but not exclusively, highlights innovations for animal research that encourage automation, no claim is implied that one necessarily improves the quality of science by plugging an apparatus into a wall! Many of the most significant studies in the history of the discipline were conducted with low-tech apparatus; conversely, the use of sophisticated technology provides no guarantee that the data produced will be free of weaknesses or flaws.

"Technology" is generically defined as the practical application of knowledge. For the present discussion then, technology involves application of knowledge about how to ask fair questions of animals and how to record their behavioral responses in ways that maximize reliability and validity. Thus, Small's (1900, 1901) mazes were technology. Köhler's (1928) wooden crates, sticks, and string were technology. Harlow's (1949) Wisconsin General Test Apparatus—even in its most rudimentary manual forms—was technology. Clayton and

Dickinson (e.g., 1999) showed that an ice-cube tray can be technology. In our own laboratory, nonhuman primates are routinely tested with state-of-the-art computer hardware, but we also use opaque cups, small rocks, and similar technologies to promote objectivity and to control for inadvertent cuing (e.g., Beran, 2001; 2004, Brosnan and Beran, 2009).

In this sense, there have been a vast number of technological innovations in comparative psychology. Countless advances in knowledge about animal behavior and cognition are the direct result of clever application of researchers' knowledge in the design of apparatus and experimental procedures. Therefore, the present review is necessarily selective. We have chosen to highlight several influential technologies that seemed to have transformed and defined the field, and that serve to illustrate important themes or lessons that emerge across the history of comparative psychology. In other words, the pages that follow contain our subjective and presentist selection of just a few of the important technologies that have been used by comparative psychologists to promote objective and unbiased measurement of animal behavior.

Finally, we acknowledge that the present review is an extension of the Washburn, Rumbaugh, and Putney (1994) paper, titled "Apparatus as Milestones in the History of Comparative Psychology" (see also Washburn, Rumbaugh, and Richardson, 1988). In the 17 years since that paper was published, there have been some interesting technological developments to discuss—but of course our consideration of the technologies used in earlier comparative investigations will largely follow that original discussion. Like that 1994 paper, the present chapter also benefits greatly from the detailed reviews published by other scholars, particularly those in Sidowski's (1966) *Experimental Methods and Instrumentation in Psychology*.

TECHNOLOGICAL MILESTONES IN THE HISTORY OF COMPARATIVE PSYCHOLOGY

Comparative psychology as a subdiscipline is distinguished from other areas of psychology by a methodology: the method of comparison. Specifically, comparative psychology is defined as the comparison of behavior and cognition across animal (including human) species. The justification for this methodology, as compared to cognitive and behavioral studies that are focused within a single species, was well stated by Nissen (1951) in his classic chapter on "phylogenetic comparison" in Stevens's *Handbook of Experimental Psychology*:

> "Within that one species it is often impossible to abstract what is essential from what is incidental. But, when the process is observed in various contexts (i.e., species), the irreducible minimum, the essence of the process under consideration, gradually becomes clear" (pg. 351)."

In other words, examining behavior across species provides converging evidence that helps to define the essence of the processes being studied.

The history of comparative psychology is thus a story about attempts to provide such convergence (and divergence) through empirical methods. Many excellent reviews of this history have been published (for recent examples, see Church, 2001; Dewsbury, 1994). The history of the field has been told as a series of revolutionary findings and compelling ideas. It has been told as a chronological tour through the pantheon of great scholars in our history. It has been told as a survey of animal subjects that have been the focus of investigation. Irrespective of the strategy for organizing the history of the discipline, one must recognize the methodological challenge of comparing across species in ways that are valid for the psychological processes under investigation and that are fair for the animals being studied. Very often, this challenge was met by technological innovation.

PROBLEM BOX

The Thorndike Problem Box arguably holds the title of "Most Familiar Least Used Apparatus in Comparative Psychology." That is, Thorndike's research on cats in problem (or puzzle) boxes is familiar content in textbooks on introductory psychology, psychology of learning, history of psychology, and comparative psychology. However, when Thorndike (1898) published his careful empirical studies of cats' learning to escape problem boxes, it does not appear that the field suddenly adopted Thorndike's apparatus as the standard testing instrument or paradigm for psychological research on animals, with other scientists locking animals into escapable boxes in other laboratories. Rather, the renown of the problem box stems principally from the significance of Thorndike's own experimental findings with the apparatus, and more specifically from the impact of his interpretations of this research. Although Thorndike was certainly not the first comparative psychologist, his careful experimental work and compelling—if not entirely accurate—conclusions (i.e., that animals learn incrementally by trial-and-error and by virtue of the effects or outcomes of their behaviors) set the standard for empirical analysis of animals' psychological competencies. Although these findings did not send researchers worldwide into the alleys to collect their own cats for study, Thorndike's articulation of the Law of Effect that governed trial-and-error learning by animals did inspire thousands of studies across many years. Accordingly, Bitterman (1966) labeled many apparatus that have been used to study the incremental learning that results from the consequences of behavior as "Thorndikian" descendents of the problem box, in that they mimic the learning procedure even if they do not resemble the specific technological innovation.

Thorndike's technological innovation was designed to control the tendency to over-interpret relatively rare anecdotes that suggested intelligent behavior by animals. To provide this control, Thorndike built enclosures that would serve as novel (i.e., never before experienced, such that the animal's entire learning history would be known to the experimenter) problems for the animals to solve. Capitalizing on the animals' motivation to escape confinement, Thorndike locked cats into boxes in which there were multiple potential ways to open an escape door. The animals could free themselves, but had to learn which response (e.g., stepping on a treadle, pulling a chain, moving a lever) would open the door. For each individual cat, one solution would be effective on every trial; but across cats, the different escape options could be counterbalanced. Thorndike simply observed the cats' behavior and recorded the time it required each cat to escape the box across consecutive trials. He found that time-to-success and the number of attempts declined with experience. His conclusions that the cats learned gradually by trial and error rather than suddenly by insight were consistent with his data. That is, the successful behavior became more probable across trials, as the various unsuccessful attempts declined in frequency. Thorndike noted that a behavior becomes more likely when it results in “a satisfying state of affairs” (e.g., the favorable consequence of escaping confinement) and that this, rather than some reasoned insightful process, describes problem solving by animals.

Of course, animals in Thorndike’s problem box *had* to learn by trials and error, if they were to learn at all. That is, the cats were placed in a situation in which trial-and-error learning was the only possible solution to the problem. The device was designed to preclude the possibility that the animals could perceive all the information necessary to show evidence of insight. Although this fact does not undermine Thorndike’s contention that animals do learn by trial-and-error, it also doesn’t compel the conclusion that animals *only* learn in this way. Imagine that you work in an office building with six different elevators. You might learn through experience (trial and error) that one elevator is much faster than the others—but you might also learn this without riding a single elevator, but rather by stepping aside and watching them all go up and down for a few cycles. Your capacity to learn incrementally by trial and error does not obviate your capacity under other circumstances to learn rapidly and by insight. Indeed, this was the argument by Köhler (1925), who found convincing evidence for insightful problem solving by chimpanzees, but only under conditions in which the apes had access and familiarity with all the technology that was relevant to the problem its solution.

But Thorndike’s finding that cats did learn by trial and error to escape puzzle boxes remains one of the most influential results in the history of comparative psychology. Directly or indirectly, the study stimulated more than 100 years of research on the way behavior is affected by its outcomes. The puzzle box also serves as a continuing reminder to contemporary comparative psychologists of the need to guard against the over-interpretation of anecdotal evidence. At the same time, the puzzle box reminds us that it is unfair to expect intelligent behavior by animals unless problems are presented in a way that allows the animals to respond intelligently.

Discrimination Box

Apparently, Robert Yerkes believed in practicing what he preached (or perhaps it was preaching what he practiced). That is, his recognition of the importance of technology in comparative research (as in the 1915 quote in this chapter) manifested itself in numerous apparatus innovations that supported empirical research with a wide range of animal species. The Yerkes Discrimination Box was one product of his efforts, and will be discussed here as an example of a class of apparatus that facilitated research on choice behavior by comparative psychologists.

Bitterman (1966) described the discrimination box in the context of discussing runways, T-mazes, and elevated mazes. When Small (1900, 1901) examined “the mental processes of the rat” using a small version of the famous Hampton Court maze, he established an experimental paradigm in comparative psychology that would remain popular into current times. Mazes represent a milestone technology for comparative psychology in their own right, and perhaps should also be linked to research on activity and locomotion either in other enclosed spaces or in open fields (e.g., Howells, 1932). Mazes have been used to test a wide range of species and to investigate many different aspects of psychological functioning (e.g., learning, memory, wayfinding, numerical cognition; see for example Babb and Crystal, 2005; Cook, Brown, and Riley, 1985; Davis, Mackenzie, and Morrison, 1989; Reed and Richards, 1996). Some of the field's most important findings have been produced by maze-based research, including those that led Tolman (1932) to conclude that behavior is purposive, and that humans and other animals can learn the layout of a maze (a mental or "cognitive map" to guide behavior) even in the absence of the kind of reward-driven trial-and-error learning championed by Thorndike (i.e., latent learning).

But mazes were also important as a precursor to the kinds of choice apparatus that would become popular in comparative investigations. Consider a simple T-maze: The animal traverses an alley to reach a choice point, and then must learn whether to turn left or right to obtain reward. Alternatively, some discriminative cues could be used instead of left/right as the to-be-learned stimulus. For example, one alley could be light and the other dark, with the relative positions of these darkness cues changed across trials. Or one alley could have glass flooring and the other could have cloth flooring. Or the entrance to one alley could be marked with a big Y and the entrance to the other labeled with a big N. In each of these examples, the animal's task remains to determine what cue (e.g., position, appearance, color, texture, smell) is salient, and then to learn which specific stimuli in this dimension was reliably associated with the baited or rewarded alley in the maze.

This is the innovation made by Yerkes (1908) for his research with the dancing mouse, and in variations for other species across his career. The Yerkes Discrimination Box was essentially a modification of a simple T-maze (or more accurately, a Y-maze because the alleys were adjacent to one another, but separated by a partition). From the choice point, the animal could see the two different discriminative stimuli that labeled each of the choices. The animal would make its choice by moving down one alley or the other, which would open into a goal compartment that either was baited with food (if the positive or S+ stimulus had been selected) or was not rewarded (if the negative or S- stimulus had been approached). A guillotine door could then be opened to return the animal to the starting compartment, whereupon the next trial could be administered.

The advantage of the discrimination box over other apparatus like mazes and problem boxes is that new learning situations could be presented repeatedly to the same animal or animals, because the subjects are actually learning about the discriminative stimuli rather than about the maze per se. That is, a mouse might be tested for learning in which the positive choice was marked with a square and the negative choice was marked with a circle. After mastering this problem, that same animal could be tested in which the S+ was marked with an X and S- was marked with a V; and then a problem in which S+ had a mesh floor and S- had a solid floor, and so forth.

Lashley (1930) modified the discrimination box for use with rats, replacing the runways with a jumping stand. The animals could view the discriminative stimuli from the stand, and would leap from the platform toward one or the other stimulus card. If the animal chose correctly, the card would swing away and the animal (typically a rat) would land on a feeding platform. If the animal chose incorrectly, the card would not swing away and the animal would land in a net below the stimulus displays. The Lashley Jumping Stand was a significant modification of the discrimination box, because with it Lashley demonstrated that rats learned better when they actually made contact with the discriminative stimuli than when their responses were spatially discontiguous with the stimuli. (Stimulus-response spatial discontiguity will be discussed again later in this chapter.)

Harlow (1949) also produced a significant modification to the discrimination box. For testing nonhuman primates, Harlow eliminated the locomotion requirements altogether and simply allowed the animals to reach out and touch one or the other discriminative stimulus. Each stimulus covered a small well on the apparatus, and when the S+ stimulus was displaced the animal could retrieve the reward item in the well. To keep the animal from displacing both stimuli, the stimuli were located on trays that could be slid toward the animal for selection, and the tray that was not selected could immediately be withdrawn out of the animal's reach. To keep the animal from observing which well had been baited, a screen could be pulled down when both trays were in the withdrawn position. Thus, the experimental procedure consisted of retracting both trays, lowering the blind, baiting one of the wells, placing the S+ stimulus (e.g., a red wooden block) over the baited well and the S- stimulus (e.g., a tennis ball) over the unbaited well, raising the blind, sliding the trays toward the animal, observing the response and withdrawing the unselected tray, and repeating the procedure. By raising the blind only high enough for the animal to see the stimuli and the observer to see the response, but not so high that the animal and the observer could see one another, opportunities for inadvertent cuing were minimized.

Harlow (1949) named this technological innovation the Wisconsin General Test Apparatus (WGTA). This variation on the discrimination-box paradigm became the standard test apparatus for use with nonhuman primates for about four decades (e.g., Hicks, 1956; McClearn and Harlow, 1954; Murphy and Miller, 1958; Schrier, Stollnitz, and Green, 1963), and many researchers still use the WGTA for studying discrimination learning and other aspects of animal behavior (e.g., Olthof, Iden, and Roberts, 1997; Terrell and Thomas, 1990). As with the Yerkes Discrimination Box, this paradigm allowed each animal to be tested on multiple learning problems, because each novel pair of discriminative stimuli represented a new opportunity for the animal to learn which stimulus would be rewarded and which should be avoided. Although this two-choice discrimination learning procedure required trial-and-error learning (as in Thorndike's problem box), Harlow observed an interesting "learning to learn" phenomenon that is not directly explained by the Law of Effect. In two-choice discrimination learning with a WGTA, animals (rhesus monkeys in the Harlow, 1949, study) must guess on Trial-1 of each problem, as there is no way to know which stimulus is the S+ and which is the S-. Beginning with Trial-2 of each problem, the consequences of previous choices (i.e., whether or not the previous selection yielded a reward) can inform the animal's behavior. The systematic increase in the probability of selecting the S+ rather than the S- trials across problems is perfectly consistent with Thorndike's finding of incremental trial-and-error learning through the Law of Effect. *Across problems* however, the rate of incremental learning increased, so that after a few hundred problems monkeys were almost perfect on Trial-2 of each problem. That is, monkeys gradually learned which stimulus to select on the first few WGTA-presented discrimination-learning problems, but after extensive experience on this paradigm they were essentially as good on Trial-2 as they ever would become on a problem. After even a single guess, the monkeys had learned a "win-stay, lose-shift" rule such that they would continue to select the stimulus they chose on the previous trial if it resulted in food, but would switch to the other stimulus if their previous response was unrewarded. The animals had to learn this strategy, which was never specifically trained or reinforced within a problem. Harlow called this learning-how-to-learn phenomenon "learning set" because the animals seemed to acquire through experience a mental set or strategy for performing the task.

Using the discrimination box and its variations, researchers have also been able to investigate aspects of animal cognition other than two-choice discrimination learning. Rumbaugh's Transfer Index (e.g., Rumbaugh, 1969; Rumbaugh and Pate, 1984; Rumbaugh, Savage-Rumbaugh, and Washburn, 1997) is a test of intelligence (capacity for learning and transfer) that facilitates comparisons across species, and was developed using this class of technology. Matching-to-sample, delayed matching, oddity, and other paradigms that became the standard test procedures in comparative psychology were also developed using these apparatus (see reviews by Fobes and King, 1982; Rumbaugh, Washburn, and Hillix, 1996).

The Yerkes Discrimination Box and similar apparatus reminds us of two important lessons that are still relevant for comparative psychology. As the problem box showed how apparatus can constrain the type of behavior that is observed, so the discrimination box underscores the value of multiple methods and measures of behavior. Yerkes underestimated animals' capacity for discrimination learning based on visual stimuli because of the stimulus-response spatial discontiguity problem that was only revealed by Lashley's jumping stand apparatus. Even the best technology gives a limited and operationalized assessment of behavior and, by inference, of cognitive competencies—a "behavior as measured by X" picture. Converging evidence from multiple paradigms and technologies yields confidence in one's conclusions.

Second, the learning-to-learn phenomenon observed by Harlow (1949) provides an important reminder that although animal subjects are partners with researchers in the scientific enterprise, they are typically blind to the specific questions we are trying to investigate. That is, animals do not know what we as experimenters are trying to study. They are doing what they do, which is to try to make sense of the world, to find regular and predictable patterns in experience, and to satisfy basic physical and psychological needs. In the research setting, animals may behave exactly in the way that the observer anticipates, but may also learn something very different or much more than the focus of the study. The technology that empowers the scientist to ask new questions may also produce new, unanticipated answers—just as Harlow's study of trial-and-error two-choice discrimination learning within problems yielded evidence of emergent learning-to-learn processes across problems.

Obstruction Box and Shuttle Box

Two other variations on the Yerkes Discrimination Box idea merit mention. Whereas Yerkes, Lashley, Harlow, and others used the discrimination box for various studies of learning, Columbia University's Carl J. Warden modified the technology so as to study motivation. Warden was an influential scholar in the history of comparative psychology, and wrote extensively on the subject, including the three-volume *Comparative Psychology: A Comprehensive Treatise* reviews published in the 1930s. In 1927, Warden recruited a hard-working graduate student named Henry W. Nissen (who would later work with Yerkes and Lashley, and eventually become director of the Yerkes Primate Research Center) to collaborate on studies of animal motivation. Noting that it is difficult to infer the strength of animal needs or motives by observing their behavior, because it is necessary also to know the factors that might hinder the animal in pursuing these motivations, Warden developed an apparatus that allowed simultaneous and parametric manipulation of incentive and obstruction. This apparatus, the Columbia Obstruction Box, was essentially a discrimination box like the one described above, but with electric grids in each of the choice alleys. By varying the intensity of the electric current that an animal must cross to obtain the incentive, and by varying the strength of the incentive, Warden and Nissen were able to quantify the relative potency of various motivations. This quote from Nissen's (1930) study of the maternal motivation of rats illustrates this program of research: "The intensity of the maternal drive is slightly greater than that of maximal hunger and thirst drives and greater than that of the sex drive; and that this intensity decreases with age and especially with the age of the litter, and also with previous long separation from litter" (pg. 377).

The shuttle box (Warner, 1932) is basically the converse of an obstruction box. Rather than requiring animals to cross an electric grid to determine the strength of the animals' motivation, shuttle boxes are constructed so animals can avoid electric shock by moving from compartment to compartment in the apparatus (see Bitterman, 1966). Shuttle boxes have been used with a wide range of species (e.g., rats, dogs, fish) to study classical conditioning, or the learning about a conditioned stimulus (e.g., illumination of a lamp) that reliably predicts the onset of an unconditioned stimulus (shock). In a situation where shock can be avoided by shuttling from compartment to compartment, the conditioned stimulus can come to produce the escape response prior to administration of the shock. If you have ever taken a shower at a house with old plumbing, where a sudden drop in water pressure (as when someone flushes a toilet in another bathroom) signals an imminent increase in water temperature, you have experienced something of the shuttle-box paradigm. After a few pairings of reduced-pressure (conditioned stimulus) → scalding (unconditioned stimulus), you learned quickly to move out of the shower stream as soon as the drop in pressure was perceived—probably without conscious consideration of the decision.

The shuttle box was neither the most familiar nor the most widely employed technology for studying classical conditioning, though. The most familiar apparatus was certainly Pavlov's original technology for recording salivation by dogs in response to reliable tone→food pairings. The most widely employed technology for studying classical conditioning was probably the variations of the eye-blink apparatus used to puff air at the eyes of humans and other animals (see review by Gormezano, 1966). Both the shuttle box and the obstruction box serve, however, to remind contemporary comparative psychologists of the crucial role that animal motivation plays in the quality and quantity of data obtained. It is obviously important that animal subjects be motivated to perform in the studies being conducted, and a variety of methods have been used to promote motivation (e.g., through deprivation) and to increase incentive (e.g., through reward). Under some conditions, however, these manipulations cause animals to become so focused on reducing drive/obtaining incentive that they fail to perceive other salient relations that are available in the situations.

Skinner Box

One can debate the relative familiarity, influence, and significance of the various technologies used to study Pavlovian conditioning. There is little debate over the most familiar, influential, and significant apparatus used to study operant conditioning. Even the problem box and its many variations and descendents used to study Thorndikian learning pale in comparison to the contributions from the standard operant conditioning chamber, commonly known (despite B. F. Skinner's objections) as the "Skinner box." Of course, this claim is meaningless if one considers the operant chamber to *be* a problem box, as was suggested by Skinner (1932) and Bitterman (1966).

How does one simplify the learning situation even more than was done by Yerkes, Lashley, Harlow, and others for the two-choice discrimination learning procedure? One simply eliminates choice by providing only one response option! For studying learning in its most basic form, Skinner stripped away as much of the normal rich context as possible. The basic unit of Skinner's theoretical framework was "stimulus control" (i.e., the change in probability of particular behaviors in the presence of specific stimuli), but the operant chamber was really a technology that allowed unprecedented amounts of *experimental* control. In this research, animals (typically rats or pigeons) with specified learning histories were studied. The behavior of interest was usually limited to a single type of response (e.g., lever pressing by rats, pecking by pigeons), and the response rate was the primary measure. The stimuli that might provide the context for this behavior were also strictly limited and tightly controlled, such that a specific stimulus (e.g., a light) might be associated with conditions in which a specific consequence (e.g., food rewards, which of course are also a stimulus) might be available. To ensure that the animal is motivated to respond, it is typically deprived of the reward item (food or fluid), for example by decreasing the animal in body weight to a safe percentage of its free-feeding weight prior to testing. Thus, a hungry pigeon (for example) might be tested in a Skinner box equipped with a disk that could be pecked, and a light that could be turned on or off. Responses to the disk when the light is on would automatically open a grain hopper for a specified period of time. Response to the disk when the light is off did not affect the door to the hopper. All responses are automatically tabulated on a cumulative record, which also indicates when the hopper door was opened/closed and when the light was turned on/off. With a Skinner box, every aspect of this experimental situation—except for the bird's behavior itself—could be automated, such that the animal could work with minimal or no experimenter intervention. In this way, Skinner and hundreds of other psychologists were able to study the relations between stimuli and consequences on the rate of responding, which could conveniently and objectively be recorded automatically as a cumulative record of behavior.

The advantage of simplifying the experimental situation is that very clear and replicable descriptions of the factors that influence response rate could be provided. The disadvantage of such a simple stimulus-response methodology is that it yields a very simple stimulus-response psychology. That is, the paradigm excludes by design the complex information and response options that constitute more cognitive forms of behavior. Skinner and other proponents of the experimental analysis of behavior (as researchers who study operant conditioning using this apparatus like to call themselves) argued that these more complex behavioral forms were explained with long chains of simple operant associations. However, this requires a considerable logical leap, from "a hungry animal learning what it must do in a highly constrained situation in order to obtain food" to "all instances of seemingly intelligent behavior, regardless of species and context."

In fairness, this depiction of the Skinner-box paradigm is a bit of a caricature. Although there are indeed thousands of published studies that fit this simple description, creative comparative psychologists have found many ways to study more complex forms of animal behavior using operant chambers. For example, Wright and colleagues (1985) have used a variation of the Skinner box to study memory in various species, by presenting sample stimuli that must be remembered in the stimulus window, and then providing multiple response options so that the animal could choose whether a target stimulus was present or absent. Other researchers have taken advantage of the powerful automated features of the operant chamber to study concept learning, attention, numerical cognition, and other cognitive constructs (e.g., Wasserman, Kiedinger, and Bhatt, 1988; Zentall and Urcuioli, 1993; Xia, Emmerton, Siemann, and Delius, 2001). Of course in each of these examples, the distinction blurs between the Skinner box and automated instantiations of the discrimination box.

In many ways, the operant chamber was the ideal innovation in response to Yerkes's charge for "*types of recording mechanism which shall either operate automatically or be operated by the subject rather than by the experimenter*" *(Yerkes, 1915, pg. 258).* It satisfied the requirement championed by Watson, Skinner, and others for psychology to be methodologically behaviorist, by providing automated and objective assessment of an animal's behavior under specified experimental conditions, freeing the researcher's attention for those aspects of the study that demand her/his control, to rephrase more of Yerkes's words. The weakness of the Skinner box is in the philosophical or theoretical (rather than the methodological) behaviorism to which it is typically—but not inextricably—linked. In this sense, the criticisms of Skinner's radical behaviorism become the criticisms of Skinner's revolutionary box. In emphasizing overt behavior, the operant chamber may ignore important cognitive influences. As Hershberger (1988) noted,

> Skinner's error, of course, is the empty organism, not the overt behavior. This error is not in attending too much to overt behavior, but rather in attending to too little overt behavior. He has overlooked the type of overt behavior that implies a nonempty organism (namely, purposive, self-controlled input). Therefore cognitive science cannot expect to rectify Skinner's oversight by ignoring overt behavior. On the contrary, if cognitive science is to stand on Skinner's shoulders, as opposed to his face, it must consider fully all types of overt behavior, purposive included. In short, psychology must be a conative, as well as cognitive, science" (pg. 823).

In the remainder of this chapter, we turn our attention to technology innovations that promise to satisfy this standard. These computer-based apparatus provide automatic and objective recording of overt behavior, but are not limited to behaviors that can be associated simply with observable stimuli in the environment. Rather, these technologies support paradigms that permit scientist to make inferences about internal, cognitive processes that influence responses, as reflected in self-controlled, purposive, and rational behavior.

LEXIGRAM KEYBOARD

The computer has become more convenient for use in the scientific laboratory since the 1960's (Church, 2001) with its rapid and continual decrease in size since the production of the first "minicomputer" (PDP-8, Digital Equipment Corp., 1965). As one source of evidence for this change, note that the Psychonomic Society journal *Behavior Research Methods and Instrumentation* changed its name to *Behavior Research Methods, Instruments and*

Computers in 1984. Computers had become a key apparatus in psychology, and reports of computer applications provided the cutting-edge methodological innovations for our discipline. The growth of computers as a research apparatus was also fueled by development of user-friendly programming languages, experiment-generation platforms, stimulus design, and data collection software (e.g., see Cohen, MacWhinney, Flatt, M. and Provost, 1993; Stahl, 2006; Washburn, 1990). By 2005, computers were ubiquitous in psychological research; consequently, the journal changed it name again to simply *Behavior Research Methods*. These advances have allowed psychologists to create, implement, and report experiments in less time and with much less effort in comparison to the use of manual devices (Church, 2001; Washburn and Rumbaugh, 1992).

If you do a PsycINFO search with keywords "chimpanzee" and "computer" the first eight references in which chimpanzees are actually using a computer (i.e., excluding those in which a computer was used to analyze data on chimpanzee behavior) pertain to research by Duane Rumbaugh. Beginning with the 1973 paper called "Exploring the Language Skills of Lana Chimpanzee," these publications reflect the initial outcomes of the Language Analog (LANA) Project. Six of these eight initial papers represent reports of data from the LANA project, and the other two are critiques of this work. Rumbaugh's ape-language research was noteworthy in the history of comparative psychology for many reasons, but for the present chapter we focus on the innovative use of technology in the generation of a computer-based keyboard as the primary medium for communications by and with the chimpanzee named Lana. Unlike previous researchers who studied chimpanzees' communicative competencies using sign language, Rumbaugh and his team designed a keyboard-based language. With respect to this decision to use a technological solution to answer the question, "Is language uniquely human?" Rumbaugh and colleagues wrote:

> "The approach and methods have been designed (a) to enhance the objectivity and efficiency of inquiry into the language-relevant behaviors of ape Ss, and (b) to develop a technology that will allow for systematic investigation of the parameters influencing the acquisition of these behaviors. To serve these ends, a computer-controlled environment was designed and constructed within which the ape Ss can come to control increasingly the events of the 24-h day" (Rumbaugh et al., 1973b, p. 385).

If not directly inspired by Yerkes (1915), this statement certainly addresses the criteria specified by the quotation at the beginning of this chapter. Whereas attempts to teach chimpanzees to communicate vocally or through sign language had informed the field, these paradigms placed considerable interpretative burden on the language-competent human experimenters. With an electronic keyboard, there was no ambiguity about which symbol a chimpanzee selected.

Lana may or may not have understood what she was "saying" with the keyboard (although we think the data show definitively that she did use the symbols meaningfully and communicatively), or the key she pressed may or may not have been the one she intended to select. However, there was no confusing what her behavior actually *was*: that is, her responses were discrete, empirically verifiable, objectively recorded utterances. Commenting on the objectivity of their technology for ape-language research, Rumbaugh et al. (1973b) noted,

> "It should be clear that in this communication system it is always the computer that analyzes the linguistic input from the Ss and decides whether or not that input conforms to the grammar of Yerkish. Any possibility of a human O's bias with regard to the Ss' linguistic competence is therefore eliminated a priori, because competence is the perfect match of the input with the results of the programmed grammar" (pg. 389).

Rumbaugh and colleagues (Rumbaugh et al., 1973b; Rumbaugh, 1977) described how the keyboard system came into being, and the rationale for its development. The project made use of geometric symbols called "lexigrams" as abstract and arbitrary word symbols. Each lexigram was embossed on a translucent key, which would be illuminated when the lexigram was depressed by Lana or an experimenter. A Digital Equipment Corporation PDP-8/ECA computer registered each response, illuminated the corresponding keys, recorded the data on a teletype, and operated the various dispensers or other devices that Lana could control. The system was situated inside and outside the room in which Lana lived, and human caregivers communicated with her through the computerized keyboard (see Hillix and Rumbaugh, 2004).

Lana was taught to string together sequences of lexigrams (e.g., PLEASE MACHINE GIVE JUICE) from her earliest training. Rumbaugh and Ernst von Glasersfeld, a linguist associated with the project, named Lana's language "Yerkish" in honor of the Yerkes primate laboratory. Although the lexigrams did not resemble the items they represented (i.e., they were composed of overlapping simple geometric elements, so that the symbol for "box" for example was a filled triangle+a diamond+a vertical line), they were designed to fit within specific categories on the basis of their semanticity and physical appearance in an effort to provide a logical consistency. Each design element corresponded to some grammatical category. Colors of the lexigrams corresponded to rough semantic classes; for example, any violet lexigram corresponded to an autonomous actor, and any red lexigram represented something edible. Different shapes were presented in various combinations to make discriminable lexigrams. This allowed any lexigrams that Lana touched to be presented in a row of projected symbols presented on a display above the computer so that Lana could see what she, or an experimenter, had said (see Figure 1). By the second half of 1973—the year the lexigram keyboard was constructed for Lana (Hillix and Rumbaugh, 2004)—data were being published on Lana's use of the keyboard for expression and comprehension.

Figure 1. Lana and the lexigram keyboard.

Across the years, other apes have been introduced into language research using the lexigram keyboard. The keyboard itself has undergone tremendous changes in response to technological advances, but the basic elements of the paradigm (lexigrams that corresponded to people, places, actions, and objects attached to discrete keys for objective responses) have been constant across instantiations. The PDP-8 may have been a revolutionary microcomputer for its day, but was still just a 12-bit machine that weighed nearly 100 pounds. Smaller, lighter, more powerful processors allowed the lexigram keyboards to become mobile. In the 1980s, speech output was added to the keyboard—first as synthesized English, and subsequently as high-quality digitized speech (Hillix and Rumbaugh, 2004; Savage-Rumbaugh, Murphy, Sevcik, Brakke, Williams, and Rumbaugh, 1993). Commercial augmentative and alternative communication (AAC) devices began to appear on the market, and some (like the ALLTALK, a 9-pound touch-sensitive 128-key keyboard with human-quality voice output, manufactured by ACS Techologies of Rapid River, MI) were used at the LRC. These AAC devices were fitted with lexigram symbols for use both with language-trained apes and also in a parallel intervention project with nonspeaking children with mental retardation (see Romski and Savage-Rumbaugh, 1986; Romski and Sevcik, 1988, 1996). Blurring the distinction between the lexigram keyboard and the Rumbaughx-type technology, communicative symbols can also be selected from a computer screen using joystick or touchscreen input (e.g., Beran, Pate, Richardson, and Rumbaugh, 2000).

Computer-based lexigram keyboards continue to be used for communication by/with apes, including Lana, Sherman and Panzee (Beran and Washburn, 2002); Kanzi and Panbanisha (Lyn, 2010), Ai (Matsuzawa, 2009), and others. Meanwhile, the AAC field continues to flourish as scholars use similar keyboard-based communication systems to help human children and adults with language impairments (e.g., Snell et al., 2010).

Rumbaughx

Duane Rumbaugh did not invent the computer or the joystick. He was not even the first scientist to put a joystick in front of a nonhuman primate. King (1961) had demonstrated that a capuchin monkey could manipulate a joystick so as to perform a compensatory tracking task, and Miall, Wier and Stein (1986, 1988) had reported similar tracking performance by rhesus monkeys. In an Air Force study of the effects of radiation on performance, rhesus monkeys manipulated a joystick so as to control balance of a primate equilibrium platform (Yochmowitz, Patrick, Jaeger, and Barnes, 1977). In each of these early studies, the animals were using the joystick as a simple operant manipulandum, rather than as a way to "play a game" in the way that is commonly seen in cognitive, developmental, and comparative research today. According to Professor Rumbaugh (see also Savage-Rumbaugh, 1986), it was Sue Savage-Rumbaugh who first demonstrated the use of a commercial video-game system to chimpanzees Sherman and Austin. These language-trained chimpanzees reportedly were able to play these computer games after watching Savage-Rumbaugh model use of the joystick for just a few trials. Based on this event, original computer software was written to conduct a wide range of "video-task" experi-ments with the apes at Georgia State University's Language Research Center (LRC). (Note that the activities one engaged in arcades and home-gaming systems like Atari and Commodore were called "video games" at the time, so scientific questions that were instated as game-like tests were called "video tasks.")

Rumbaugh and his colleagues purchased dedicated computers specifically for comparative testing. A first-year graduate student (Washburn) was recruited in 1984 and assigned exclusively to the task of translating researchers' ideas into game-like computer tasks. Because of this student's familiarity with a particularly affordable new "home computer" option, the LRC apes quickly became a significant sector of Commodore 64 computer users during this period (!), although the first publication from this work would not appear in print until 1989 (Hopkins and Morris, 1989; Hopkins, Washburn and Rumbaugh, 1989; Rumbaugh, Washburn, and Savage-Rumbaugh, 1989).

In 1986, the National Aeronautics and Space Administration (NASA) approached Duane Rumbaugh for assistance with its research on the effects of spaceflight on rhesus monkeys' physical and psychological well-being. NASA scientists had a problem. Already several years into a multi-national, multi-disciplinary research program on the effects of microgravity and other spaceflight-related variables on the morphology and function of various physiological systems, these scientists had committed to a nonhuman primate model (the rhesus monkey) for its studies. Rhesus macaques were ideal in many ways for this type of research: They are a hearty and robust primate with well-known physiology and established utility as a surrogate for human biology. The team of scientists, veterinarians, and animal-care staff assembled by NASA for the aptly named "Rhesus Project" were highly qualified to promote the physical health of these monkeys while studying the effects of spaceflight on musculoskeletal, neurovestibular, circulatory, regulatory, and other physiological systems. However, the Animal Welfare Act of 1966, which governs the use of many types of animal for research (among other things), had been amended by Congress in 1985 with new language specifically pertaining to research with rhesus monkeys and other nonhuman primates. Scientists studying nonhuman primates were suddenly required not only to care for the physical well-being of their research animals, but additionally to generate and follow a plan to promote the *psychological* well-being of monkeys and apes maintained as research subjects. Although the Rhesus Project team was quite familiar with the physical needs of the macaques they studied, expertise from a psychologist was required to address the issues of psychological well-being. Primate expert Duane Rumbaugh was contacted to provide consultation on this issue.

Based on decades of experience in working with nonhuman primates, Rumbaugh obtained pilot funding from NASA to explore whether a computer-based system similar to the one he was using successfully with chimpanzees could be developed for rhesus monkeys. Such a system might be useful as environmental enrichment to promote and to assess the monkeys' psychological well-being. This was a risky proposal, however. Despite the apes' success in mastering the causal connections between joystick manipulations and video-task stimuli, it was not at all certain that rhesus monkeys could succeed. Not only was there little, if any, evidence in the literature to suggest that rhesus monkeys could come to understand and to respond to complex computer-generated stimuli, but there was also a substantial literature indicating that these monkeys in particular would struggle to master such skills (see review by Rumbaugh et al., 1989). Many researchers had attempted to train monkeys to respond to stimuli that were spatially discontiguous with response or reward loci; however, learning seemed inevitably to suffer when the source of the salient stimulus information was even a few centimeters away from the location where a response would be made or where a reward would be distributed (e.g., McClearn and Harlow, 1954; Meyer, Polidora, and McConnell, 1961; Murphy and Miller, 1955, 1958, 1959; Polidora and Fletcher, 1964; Schrier et al. 1963; Stollnitz and Schrier, 1962). If a monkey were required to use a joystick or similar input device (trackball, mouse, keyboard, button box, etc.) to respond to computer-graphic stimuli, there would certainly be more than a few centimeters separating the hand executing each response from the screen where the monkey needed to attend to salient stimuli. Thus, it seemed unlikely that the rhesus monkeys would actually learn, at least without painstaking shaping, to manipulate a joystick so as to perform computer-based tasks that would be enriching and informative with respect to the monkeys' psychological state. However, there were tremendous potential benefits if this risky enterprise *did* succeed, as the computer-based test system would provide an ideal platform for assessing cognitive functioning (e.g., the effects of spaceflight-relevant variables on learning, memory, attention, and so forth) as well as for giving the monkeys enjoyable activities in which to engage while their biological systems were being studied.

IBM's personal computer (PC), running on the revolutionary Microsoft Disk Operating System (MS-DOS), was selected as the platform for this pilot work. The platform had been introduced to the world several years earlier, but had become relatively affordable (albeit still about triple the price of a Commodore 64!) around this time. With Professor Kirk Richardson providing programming support (in QuickBasic) and expertise on the experimental analysis of behavior, and with two graduate students to assist with day-to-day operations, Dr. Rumbaugh designed a training protocol to teach a naïve monkey to move a joystick so as to control the movements of a computer-graphic cursor. The idea was to use autoshaping to train the monkeys to associate the sound of the dispenser with the location of food rewards. Then, the task software would be introduced such that any movement of the joystick produced movements of a cursor on the screen. The stimuli on the screen were completely static when the joystick was in its resting position. The initial training task (called "SIDE") required subjects to manipulate the cursor so as to direct the cursor into contact with a computer-graphic rectangle positioned on the outer border of the screen. Early in training, all four borders of the screen were covered with target rectangles; thus, movements of the cursor in any direction would result in contact between the cursor and target. As the subject successfully completed trials, one randomly selected rectangle would be removed, so that the monkey was not rewarded for moving the joystick in *any* direction, but rather had to direct the cursor into contact with one of the remaining sides of the screen (i.e., one that did have a target rectangle on the border). As the animal became more proficient at this task, the number of randomly selected walls to be covered with target rectangles would be systematically decreased (from 4, to 3, to 2, to 1). Subsequently, the size of the single-target rectangle would also be decreased (from the full border of the screen, to half the border, to a small rectangle). If the monkeys were unable to complete a trial or otherwise performed poorly, the task would titrate in the opposite direction (toward 4 full-border rectangles) making the trials easier to complete. In this way, the SIDE task might automatically shape the monkeys to attend to stimulus movements on the screen, to associate those movements with hand movements of the joystick, and to become increasingly skilled at discerning the location of target stimuli and at manipulating the joystick so as to bring the cursor into contact with those targets.

Two rhesus monkeys (Abel and Baker) were obtained by the LRC from NASA for this project. In parallel with this effort, Dr. James King of the University of Arizona began training rhesus monkeys to respond via touchscreen to similar computer-based tasks—just in case the macaques at LRC were unable to use the joystick. However, this companion project was quickly abandoned (joysticks being a much more practical option than touchscreens, which were quite expensive at the time), because Abel and Baker quickly demonstrated that they were able to manipulate the joystick so as to control the cursor on the SIDE task, and indeed could bring the cursor into contact with a target stimulus that moved on the screen (the CHASE task). These results were reported to NASA Rhesus Project scientists in 1987, and appeared in print in 1989 (Rumbaugh et al., 1989).

Several good decisions likely contributed to this rapid success. The decision to make the stimuli static whenever the joystick was stationary, and to accompany cursor movement with a sound that was produced from the general direction of the screen (rather than the hand or the reward cup) probably helped the monkeys to recognize the salience of the on-screen images. The SIDE task was written with an option so experimenters could manually shape the monkeys to touch the joystick. Some manual shaping did occur with Abel and Baker, but the training of subsequent animals indicates that it is not necessary use manual reinforcement of systematic approximations to the joystick to bring a naïve monkey to a point of mastery over cursor movements (although putting a grape or raisin on the joystick handle certainly can reduce the time required for the monkey to discover that the joystick is there!). For Abel and Baker (and every monkey trained subsequently at the LRC), we oriented the joystick so that the handle extended horizontally toward the monkey, such that cursor movements were isomorphic to joystick movements (e.g., deflection of the joystick handle up and to the right resulted in cursor movements toward the top-left of the computer screen). By surrounding the cursor with target images, it seems that we took advantage of the rhesus monkeys' natural curiosity—in essence rewarding the animals for trying to break the joystick, until such time that they became more interested in how to obtain the rewards than how to destroy the apparatus.

To date, we have used this same procedure to train more than 100 rhesus monkeys at the LRC or at NASA-associated facilities in California and Moscow, Russia. Once the macaques acquire the basic cursor-control skills, they rapidly learn a host of other computerized tasks, and thus have contributed data to a wide range of studies. This same configuration of apparatus and training curriculum has also been used successfully with many other primate species (e.g., Andrews, 1993; Andrews and Rosenblum, 1994; Evans, Beran, Chan, Klein, and Menzel, 2008; Leighty and Fragaszy, 2003; Torou et al., 2004; Vauclair and Fagot, 1993) and even with rats (Washburn, Rulon and Gulledge, 2004). At one point, more than three dozen laboratories worldwide were training and testing nonhuman primates using variations of the hardware and software developed at the LRC. Duane Rumbaugh named this test apparatus the Language Research Center's Computerized Test System (LRC-CTS). NASA called the version of the system developed for spaceflight the Psychomotor Test System (PTS). However, in light of the technology-naming tradition within comparative psychology—with Thorndike's Puzzle Box, Yerkes's Discrimination Box, the Skinner Box, and the like—perhaps the best name for the computer-task paradigm would be the "*Rumbaughx*" (pronounced "rum-box")!

Of course, the computerized test system has also evolved subsequently in other laboratories and for other purposes (i.e., not directly descendant from Rumbaugh's innovations). Today, experiments in which animals respond with joysticks or touch-screens to computer-generated stimuli are common, just as is the case in studies of human adults and children. We are making no claims about the relative independence or superiority of these various configurations of apparatus, software, and training/testing procedures across comparative psychology.

Figure 2. Typical configuration of the Rumbaughx.

We *are* however drawing attention to the tremendous paradigm shift that has taken place in the last few decades, in which the notion of "computer-using animals" that was so novel in 1989 is now a standard research tool in the study of cognition across species. This new testing technology has become the "operant conditioning chamber" of its era. What the Skinner box is to radical behaviorism and the experimental analysis of behavior, the Rumbaughx and its variations are to rational behaviorism (Rumbaugh, 2002; Rumbaugh and Washburn, 2003; Washburn, 2007; see also Naour, 2010) and comparative cognitive science.

What makes the Rumbaughx and similar test systems particularly useful for comparative research is the diversity of tasks and studies that can be administered using the computer. Computerized versions of many of the most important tests in cognitive psychology, behavioral neuroscience and neuropsychology, clinical assessment, and developmental psychology have been developed and administered to animals. At the same time, computerized test systems allow administration of classic testing paradigms in comparative psychology, including computerized maze learning, simple operant and respondent conditioning procedures, two-choice discrimination learning and reversal, relative numerousness judgments, matching-to-sample and delayed matching, and sequence learning. These tests and others were developed initially using physical stimuli and manual apparatus, but with the computerized-task paradigm become "*largely automatic or self-controlling over considerable periods of time, not only with respect to the objective situation or setting in which the subject reacts, but also with respect to the recording of the several important aspects of response,*" to reiterate Yerkes's (1915) words. At the LRC, scientists schedule various tasks to be available to different rhesus monkeys throughout each day. Rhesus monkeys have continuous access to dedicated Rumbaughx systems in their home cages. These monkeys engage the tasks whenever they want, moving between tasks or studies seamlessly throughout each day, and rest or engage in other activities whenever they want. The animals are not deprived of food or fluid, or reduced in body weight for purposes of testing. They do receive fruit-flavored chow pellets for completing computer-task trials, but are fully provisioned with food whether or not they work on the tasks. In this way, scientists obtain data from the monkeys at those times that the animals are most highly motivated to work—for reasons (e.g., for well-being needs like challenge or control) that may be unrelated to hunger (e.g., Washburn and Rumbaugh, 1992). These scientists include researchers physically present at the LRC of course, but also include collaborating scholars at universities across the globe who obtain data from these monkeys electronically and remotely, without ever being in the same building (or even state!) as the macaques. In a typical day, we obtain more than 1,000 completed trials from each rhesus monkey in our laboratory—all with minimal human intervention (e.g., starting computer tasks, filling pellet dispensers). Reminiscent again of the words from Yerkes (1915), the Rumbaughx allows animals to test themselves rather than being tested by the experimenters. In some studies, we have even given the monkeys menus with icons representing each of the tasks or experiments currently being administered, and allowed the monkeys to choose the task on which he will work or even the reward that will be dispensed for this work (Washburn, Hopkins, and Rumbaugh, 1991).

Capuchin monkeys and chimpanzees at the LRC live in social groups and separate voluntarily several times each day for computer-task testing. In other laboratories, the computer-test paradigm has been modified so as to facilitate testing of social colonies. Andrews (1994) described a system for social testing of nonhuman primates in which a tiny microchip is injected subcutaneously into each animal's arm (or arms, as in Andrews and Rosenblum, 1994). When the animals reached out of the caging to manipulate the joystick, the microchip was detected by a sensor. This sensor then communicated the animal's identity to the computer, which in turn could tailor task demands appropriately. In recent years, Fagot and colleagues have implemented this procedure with a social colony of baboons, allowing these animals to test themselves at computerized test stations throughout the day (Fagot and Paleressompoulle, 2009; Fagot and Bonte, 2010). This system, which allows socially housed animals to separate themselves briefly and to test themselves, yields benefits for the investigation of individual differences and the social influences on cognition.

As touchscreens have become more reliable and more affordable, many researchers have opted for them rather than joysticks as the response manipulandum of preference. Animals typically learn to respond by touchscreen faster than by joystick, as they do not have to solve the stimulus-response spatial discontiguity problem. Additionally, animals can typically respond faster with a touchscreen than with a joystick, leading for example to the impressive performance levels at startling response speeds that the chimpanzee Ai and other animals have produced on numerous computerized tasks (e.g., Biro and Matsuzawa, 1999; Inoue and Matsuzawa, 2009; see also Cantlon and Brannon, 2006; Terrace, Son and Brannon, 2003).

The foregoing discussion of the Rumbaughx and similar technologies has focused on the benefits in terms of research design. Computer-based testing allows humans, nonhuman primates, and some other animal species to be tested on identical tasks under comparable conditions. This testing paradigm yields high volumes of data. Because data collection is automated, opportunities for inadvertent cuing of animals are attenuated and all measures should be objectified. This testing paradigm also makes it possible to ask some questions that would be difficult to address otherwise. For example, studies of uncertainty monitoring (Beran, Smith, Redford, and Washburn, 2006; Smith, 2009; Smith, Redford, Beran, and Washburn, 2010; Smith, Shields, Schull, and Washburn, 1997; Smith, Shields, and Washburn, 2003; Washburn, Gulledge, Smith, and Beran, 2010; Washburn, Smith, and Shields, 2006) require dynamic adjustment of stimulus characteristics or trial difficulty and high trial counts to map continuous changes in uncertainty across decision space. These studies would be prohibitive without the computer-test system. Finally, it appears that animals may in some cases demonstrate cognitive competencies with the computer-task paradigm that they failed to demonstrate with manual apparatus. The aforementioned issue that rhesus monkeys have with stimulus-response spatial discontiguity provides one example of this competence that emerges with computer-task experience (see Beran, Rumbaugh, and Washburn, 2007; Rumbaugh et al., 1989). Macaques' performance on tests of relational learning, their ability to predict target movements, and their acquisition of knowledge about the meanings of numeric symbols are other examples of competencies that appear in computer-task testing (and that may have emerged as a result of computer-task experience) but that had not been reported for monkeys tested with manual apparatus (Rumbaugh and Washburn, 2003). It seems that we are still in the relatively early years of another significant shift in the history of comparative psychology, a shift to comparative cognition research that is in many ways revolutionizing our understanding of animal behavior and animal minds.

CONCLUSION: THE FUTURE OF TECHNOLOGY IN COMPARATIVE PSYCHOLOGY

From this review of technological innovations in the past, what predications can be made about the future of apparatus in comparative psychology? Other than the safe prediction that there *will be* technological innovations—that our history of change will certainly lead into a future marked by change—it is impossible to anticipate what new applications of knowledge will improve research on animal minds and behavior. The current authors entered the field at a time in which every monkey and ape at our laboratory would have access to a dedicated computer for research. It wasn't that long ago that each of the *scientists* might not have a dedicated computer for research. Commercial gaming technology has advanced impressively in the years since the Commodore 64 and the IBM PC. The software written for research with nonhuman animals still looks more like Pong and Pac-Man than like contemporary games—but computer-based tasks written for cognitive and neuropsychological research with humans are not much more sophisticated than the tasks being administered to apes, monkeys, pigeons, and other animals.

Given the priority within the discipline (and the funding agencies) of establishing brain-behavior relations, it seems likely that technological innovations will make it increasingly easy to understand the biological correlates of behavior and cognition. For the most part, there remains a chasm in the cognitive neuroscience literature between studies of healthy humans on the one hand, which tend to employ noninvasive neuroimaging techniques such as magnetic resonance imaging, and studies of nonhuman animals on the other, which tend to employ invasive procedures such as single-cell recording and ablation paradigms. Creative integration of neuroimaging procedures with animal-research methods will fill this gap. Some progress on this front is already beginning to appear in the literature (e.g., Molfese and Morse, 1991; Srihasam, Sullivan, Savage, and Livingstone, 2010; Taglialatela, Russell, Schaeffer, and Hopkins, 2009; Washburn et al., 2010).

Whatever the technological innovations the future brings, it seems likely that the lessons of our past will remain important for the study of comparative psychology. More than a century after Clever Hans, comparative psychology continues to embarrass itself periodically by failing to control for cuing, by celebrated examples of bias in data selection and interpretation, and with debates over anthropomorphism. It is somewhat chilling that Yerkes's (1915) concerns about the range and responsibility placed on the experimenter remain just as pertinent ten decades later. Then as now, the challenge for the comparative psychologist is to apply knowledge about experimental methods and animal behavior in a way that yields the most accurate, reliable, valid, and efficient data. We are confident that tomorrow's history will show that the scholars who are most successful in applying technology to this challenge will be recognized among the women and men who have contributed the most to scientific progress in comparative psychology.

ACKNOWLEDGMENTS

This review was supported by grants from NICHD (HD-060563 and HD-061455) and NSF (BCS 0924811 and BCS 096993). Additional support was provided by the College of Arts and Sciences of Georgia State University. Timothy M. Flemming is currently affiliated with the Université de Provence and Centre National de la Recherche Scientifique (CNRS) in Rousset, France.

REFERENCES

Andrews, M. W. (1993). Video-task paradigm extended to *Saimiri. Perceptual* and *Motor Skills,* 76, 183-191.

Andrews, M. W. (1994). An automated identification system for use with computer-controlled tasks. *Behavior Research Methods, Instruments and Computers, 26,* 32-34.

Andrews, M. W., and Rosenblum, L. A. (1994). Automated recording of individual performance and hand preference during joystick-task acquisition in group-living bonnet macaques (*Macaca radiata*). *Journal of Comparative Psychology*, *108,* 358-362.

Babb, S. J., and Crystal, J. D. (2005). Discrimination of what, when, and where: Implications for episodic-like memory in rats. *Learning and Motivation, 36*, 177-189.

Beran, M. J., Pate, J. L., Richardson, W. K., and Rumbaugh, D. M. (2000). A chimpanzee's (*Pan troglodytes*) long-term retention of lexigrams. *Animal Learning and Behavior*, *28*, 201-207.

Beran, M. J., Rumbaugh, D. M., and Washburn, D. A. (2007). Rhesus monkeys (*Macaca mulatta*) maintain learning set despite second-order stimulus-response spatial discontiguity. *Psychological Record, 57*, 9-22.

Beran, M. J., Smith, J. D., Redford, J. S., and Washburn, D. A. (2006). Rhesus macaques (*Macaca mulatta*) monitor uncertainty during numerosity judgments. *Journal of Experimental Psychology: Animal Behavior Processes, 32*, 111-119.

Beran, M. J., and Washburn, D. A. (2002). Chimpanzee responding during matching-to-sample: Control by exclusion. *Journal of the Experimental Analysis of Behavior*, *78*, 497-508.

Biro, D., and Matsuzawa, T. (1988). Numerical ordering in a chimpanzee (*Pan troglodytes*): Planning, executing, and monitoring. *Journal of Comparative Psychology, 113*, 178-185.

Bitterman, M. E. (1966). Animal learning. In J. B. Sidowski (Ed.), *Experimental methods and instrumentation in psychology* (pp. 451-484). New York: McGraw-Hill.

Cantlon, J. F., and Brannon, E. M. (2006). Shared system for ordering small and large numbers in monkeys and humans. *Psychological Science, 17*, 401-406.

Church, R. M. (2001). Animal cognition: 1900-2000. *Behavioural Processes, 54*, 53-63.

Cohen, J. D., MacWhinney, B., Flatt, M., Provost, J. (1993). PsyScope: An interactive graphic system for designing and controlling experiments in the psychology laboratory using Macintosh computers. *Behavior Research Methods, Instruments and Computers, 25*, 257-271.

Cook, R. G., Brown, M. F., and Riley, D. A. (1985). Flexible memory processing by rats: Use of prospective and retrospective information in tha radial maze. *Journal of Experimental Psychology: Animal Behavior Processes, 11*, 453-469.

Davis, H., MacKenzie, K. A., and Morrison, S. (1989). Numerical discrimination by rats (*Ratus norvegicus*) using body and vibrissal touch. *Journal of Comparative Psychology, 103*, 45-53.

Evans, T. A., Beran, M. J., Chan, B., Klein, E. D., and Menzel, C. R. (2008). An efficient computerized testing method for the capuchin monkey (*Cebus apella*): Adaptation of the LRC-CTS to a socially housed nonhuman primate species. *Behavior Research Methods, 40*, 590-596.

Dewsbury, D. A. (1984). *Comparative psychology in the Twentieth Century.* Stroudsburg, PA: Hutchinson Ross.

Fagot, J., and Bonté, E. (2010). Automated testing of cognitive performance in monkeys: Use of a battery of computerized test systems by a troop of semi-free-ranging baboons (*Papio papio*). *Behaivor Research Methods, 42*, 507-516.

Fagot, J., and Paleressompoulle, D. (2009). Automatic testing of cognitive performance in baboons maintained in social groups. *Behavior Research Methods, 41,* 396-404.

Fobes, J. L., and. King,J. E. (1982). Measuring primate learning abilities. In J. L. King and J. E. King (Eds.), *Primate behavior* (pp. 289-326). New York: Academic Press.

Gormezano, I. (1966). Classical conditioning. In J. B. Sidowski (Ed.), *Experimental methods and instrumentation in psychology*, 385-420. New York: McGraw-Hill.

Harlow, H. F. (1949). The formation oflearning sets. *Psychological Review,* 56, 51-65.

Hershberger, W. A. (1988). Psychology as a conative science. *American Psychologist, 43,* 823-824.

Hicks, L. H. (1956). An analysis of number-concept formation in the rhesus monkey. *Journal of Comparative and Physiological Psychology, 49*, 212-218.

Hillix, W. A., and Rumbaugh, D. M. (2004). *Animal bodies, human minds: Ape, dolphin, and parrot language studies.* New York: Kluwer Academic.

Hopkins, W. D., and Morris, R. D. (1989), Laterality for visual-spatial processing in two language-trained chimpanzees. *Behavioral Neuroscience, 103*, 227-234.

Hopkins, W. D., Washburn, D. A., and Rumbaugh, D. M. (1989). Note on hand use in the manipulation of joysticks by rhesus monkeys (*Macaca mulatta*) and chimpanzees (*Pan troglodytes*). *Journal of Comparative Psychology, 103*, 91-94.

Howells, T. H. (1932). Apparatus for measuring activity. *Journal of Experimental Psychology, 15,* 226-233.

Inoue, S., and Matsuzawa, T. (2009). Acquisition and memory of sequence order in young and adult chimpanzees (*Pan troglodytes*). *Animal Cognition, 12*, S59-S69.

King, W. J. (1961). Continuous compensatory tracking by a Cebus monkey. *Science, 134,* 947-948.

Kluver, H. (1933). *Behavior mechanisms in monkeys.* Chicago: University of Chicago Press.

Köhler, W. (1925). *The mentality of apes.* New York: Harcourt, Brace.

Lashley, K. S. (1930). The mechanism of vision: I. A method for rapid analysis of pattern-Vision in the rat. *Journal of Genetic Psychology,*37, 453-460.

Leighty, K. A., and Fragaszy, D. M. (2003). Primates in cyberspace: Using interactive computer tasks to study perception and action in nonhuman animals. *Animal Cognition, 6*, 137-139.

Lyn, H. (2010). Environment, methodology, and the object choice task in apes: Evidence for declarative comprehension and implications for the evolution of language. *Journal of Evolutionary Psychology, 8*, 333-349.

Matsuzawa, T. (2009). Comparative cognitive science: The 30 years of Ai project. *Japanese Journal of Animal Psychology, 59*, 135-160.

McClearn, G. E., and Harlow, H. F. (1954). The effect of spatial contiguity on discrimination learning by rhesus monkeys. *Journal of Comparative and Physiological Psychology, 45,* 391-394.

Meyer, D. M., Polidora, V. J., and McConnell, D. G. (1961). Effects of spatial S-R contiguity and response delay upon discriminative performances by monkeys. *Journal of Comparative and Physiological Psychology, 54*, 175-177.

Miall, R. C., Weir, D. J., and Stein, J. F. (1986). Manual tracking of visual targets by trained monkeys. *Behavioural Brain Research, 20*, 185-201.

Miall, R. C., Weir, D. J., and Stein, J. F. (1988). Planning of movement parameters in a visuo-motor tracking task. *Behavioural Brain Research, 27*, 1-8.

Molfese, D. L., and Morse, P. A. (1991). Developmental changes in nonhuman primate patterns of brain lateralization for the perception of speech cues: Neuroelectrical correlates. In N. A. Krasnegor, D. M. Rumbaugh, R. L. Schiefelbusch, and M. Studdert-Kennedy (Eds.), *Biological and Behavioral Determinants of Language Development*, (189-207). Hillsdale, NJ: Erlbaum.

Murphy, J. V., and Miller, R. E. (1955). The effect of spatial contiguity of cue and reward in the object-quality learning of rhesus monkeys. *Journal of Comparative and Physiological Psychology, 48*, 221-224.

Murphy, J. V., and Miller, R. E. (1958). Effect of the spatial relationship between cue, reward, and response in simple discrimination learning. *Journal of Experimental Psychology, 56*, 26-31.

Murphy, J. V., and Miller, R. E. (1959). Spatial contiguity of cue, reward, and response in discrimination learning by children. *Journal of Experimental Psychology, 58*, 485-489.

Naour, P. (2010). E. O. Wilson and B. F. Skinner: A dialogue between sociobiology and radical behaviorism. New York: Springer.

Nissen, H. W. (1930). A study of maternal behavior in the white rat by means of the obstruction method. *The Pedagogical Seminary and Journal of Genetic Psychology, 37,* 377-393.

Nissen, H. W. (1951). Phylogenetic comparison. In S. S. Stevens (Ed.), *Handbook of Experimetnal Psychology* (347-386). New York: Wiley and Sons.

Olthof, A., Iden, C. M., and Roberts, W. A. (1997). Judgments of ordinality and summation of number symbols by squirrel monkeys (*Saimiri sciureus*). *Journal of Experimental Psychology: Animal Behavior Processes, 23*, 325-339.

Polidora, V. J., and Fletcher, H. J. (1964). An analysis of the importance of S-R spatial contiguity for proficient primate discrimination performance. *Journal of Comparative and Physiological Psychology, 57*, 224-230.

Reed, P., and Richards, A. (1996). The von Restorff effect in rats (*Rattus norvegicus*). *Journal of Comparative Psychology, 110*, 193-198.

Romski, M. A., and Savage-Rumbaugh, E. S. (1986). Implications for language intervention research: A nonhuman primate model. In E. S. Savage-Rumbaugh (Ed.), *Ape language, from conditioned response to symbol*, New York: Columbia University Press.

Romski, M. A., and Sevcik, R. A. (1988). Augmentative and alternative communication systems: Considerations for individuals with severe intellectual disabilities. *AAC: Augmentative and Alternative Communication, 4*, 83-93.

Romski, M. A., and Sevcik, R. A. (1996). *Breaking the speech barrier: Language development through augmented means*. Baltimore: Paul H. Brookes Publishing.

Rumbaugh, D. M. (1969). The Transfer Index: An alternative measure of learning set *Proceedings of the 2nd International Congress of Primatology* (Vol. 1, pp. 267-273). New York: Karger.

Rumbaugh, D. M. (Ed.) (1977). *Language learning by a chimpanzee: The Lana project*. New York: Academic Press.

Rumbaugh, D. M. (2002). Emergents and rational behaviorism. *Journal of Cognitive Educatoin and Psychology, 2*, 163-171.

Rumbaugh, D. M., Gill, T. V., Brown, J. V., von Glaserfeld, E. C., Pisani, P., Warner, H., and Bell, C. L. (1973a). Exploring the language skills of Lana chimpanzee. *International Journal of Symbology, 4*, 1-9.

Rumbaugh, D. M., Gill, T. V., Brown, J. V., von Glaserfeld, E. C., Pisani, P., Warner, H., and Bell, C. L. (1973b). A computer-controlled language training system for investigating the language skills of young apes. *Behavior Research Methods and Insturments, 5*, 385-392.

Rumbaugh, D. M., and Pate, J. L. (1984). The evolution of cognition in primates: A comparative perspective. In H. L. Roitblat, T. G. Bever and H. S. Terrace (Eds.), *Animal cognition* (pp. 569-587). Hillsdale, NJ: Erlbaum.

Rumbaugh, D. M., Richardson, W. K., Washburn, D. A., Savage-Rumbaugh, E. S., and Hopkins, W. D. (1989). Rhesus monkeys (*Macaca mulatta*), video tasks, and implications for stimulus-response spatial contiguity. *Journal of Comparative Psychology, 103*, 32-38.

Rumbaugh, D. M., Savage-Rumbaugh, E. S., and Washburn, D. A. (1996). Toward a new outlook on primate learning and behavior: Complex learning and emergent processes in comparative perspective. *Japanese Psychological Research, 38*, 113-125.

Rumbaugh, D. M., and Washburn, D. A. (2003). *Intelligence of apes and other rational beings*. New Haven, CT: Yale University Press.

Rumbaugh, D. M., Washburn, D. A., and Hillix, W. A. (1996). Respondents, operants, and emergents: Toward an integrated perspective on behavior. In K. Pribram and J. King (Eds.), *Learning as a self-organizing process* (pp. 57-73). Hillsdale, NJ: Erlbaum.

Rumbaugh, D. M., Washburn, D. A., and Savage-Rumbaugh, E. S. (1989). On the care of captive chimpanzees: Methods of enrichment. In E. F. Segal (Ed.), *Housing, care and psychological wellbeing of captive and laboratory primates* (357-375), Park Ridge, NJ: Noyes Publications.

Savage-Rumbaugh, E. S. (1986). *Ape language: From conditioned response to symbol*. New York: Columbia University Press.

Savage-Rumbuagh, E. S., Murphy, J., Sevcik, R. A., Brakke, K. E., Williams, S. L., and Rumbaugh, D. M. (1993). Language comprehension in ape and child, *Monographs of the Society for Research on Child Development, 58*, 1-256.

Schrier, A. M., Stollnitz, F., and Green, K. F. (1963). Titration of spatial S-R separation in discrimination by monkeys (*Macaca mulatta*). *Journal of Comparative and Physiological Psychology, 56*, 848-851.

Schrier, A. M., Stollnitz, F., and Green, K. F. (1963). Titration of spatial S-R separation in discrimination by monkeys (*Macaca mulatta*). *Journal of Comparative and Physiological Psychology, 56*, 848-851.

Sidowski, J. B. (Ed.), *Experimental methods and instrumentation in psychology*. New York: McGraw-Hill.

Skinner, B. F. (1932). On the rate offormation of a conditioned reflex. *Journal of General Psychology*, 7, 274-285.

Small, W. S. (1900). An experimental study of the mental processes of the rat. *American Journal of Psychology*, 11, 135-165.

Small, W. S. (1901). Experimental study of the mental processes of the rat II. *American Journal of Psychology*, 12, 206-239.

Smith, J. D. (2009). The study of animal metacognition. *Trends in Cognitive Sciences, 13*, 389-396.

Smith, J. D., Redford, J. S., Beran, M. J., and Washburn, D. A. (2010). Rhesus monkeys (*Macaca mulatta*) adaptively monitor uncertainty while multi-tasking. *Animal Cognition, 13*, 93-101.

Smith, J. D., Shields, W. E., Schull, J., and Washburn, D. A. (1997). The uncertain response in humans and animals. *Cognition, 62*, 75-97.

Smith, J. D., Shields, W. E., and Washburn, D. A. (2003). A comparative approach to metacognition and uncertainty monitoring. *Behavioral and Brain Sciences, 26*, 317-339.

Snell, M. E., Brady, N., McLean, L., Ogletree, B. T., Siegel, E., Sylvester, L., Mineo, B., Paul, D., Romski, M. A., Sevcik, R. (2010). Twenty years of communication intervention research with individuals who have severe intellectual and developmental disabilities, *American Journal on Intellectual and Developmental Disabilities, 115*, 364-380.

Srihasam, K., Sullivan, K., Savage, T., and Livingstone, M. S. (2010). Noninvasive functional mri in alert monkeys. *NeuroImage*, Jan 29, 2010.

Stollnitz, F., and Schrier, A. M. (1962). Discrimination learning by monkeys with spatial separation of cue and response. *Journal of Comparative and Physiological Psychology, 55*, 876-881.

Taglialatela, J. P., Russell, J. L., Schaeffer, J. A., and Hopkins, W. D. (2009). Visualizing vocal perception in the chimpanzee brain, *Cerebral Cortex, 19*, 1151-1157.

Terrace, H. S., Son, L. K., and Brannon, E. M. (2003). Serial expertise in rhesus macaques. *Psychological Science, 14*, 66-73.

Terrell, D. F., and Thomas, R. K. (1990). Number-related discrimination and summation by squirrel monkeys (*Saimiri sciureus sciureus* and *S. boliviensus boliviensus*) on the basis of the number of sides of polygons. *Journal of Comparative Psychology, 104*, 238-247.

Thorndike, E. L. (1898). Animal intelligence: An experimental study of the associative processes in animals. *Psychological Review, Monograph Supplement*, *2*, 8.

Tolman, E. C. (1932). *Purposive behavior in animals and men.* New York: Century.

Torou, L. R., Kahar, C. W., Adcock, D., Bloomsmith, M. A., and Maple, T. L. (2004). Computer-assisted enrichment for zoo-housed orangutans (*Pongo pygmaeus*). *Animal Welfare, 13*, 445-453.

Vauclair, J., and Fagot, J. (1993). Manual and hemispheric specialization in the manipulation of a joystick by baboons (*Papio papio*). *Behavioral Neuroscience, 107*, 210-214.

Warner, L. H. (1932). The association span of the white rat. *Journal of Genetic Psychology, 41,* 57-89.

Washburn, D. A. (1990). PC-compatible computer-generated stimuli for video-task testing. *Behavior Research Methods, Instruments, and Computers, 22*, 132-135.

Washburn, D. A. (2007). The perception of emergents. In D. A. Washburn (Ed.), *Primate perspectives on behavior and cognition* (109-123). Washington, DC: American Psychological Association.

Washburn, D. A., Gulledge, J. P., Beran, M. J., and Smith, J. D. (2010). With his memory magnetically erased, a monkey knows he is uncertain. *Biology Letters, 6*, 160-162.

Washburn, D. A., Hopkins, W. D., and Rumbaugh, D. M. (1991). Perceived control in rhesus monkeys (*Macaca mulatta*): Enhanced video-task performance. *Journal of Experimental Psychology: Animal Behavior Processes, 17*, 123-127.

Washburn, D. A., Rulon, M. J., and *Gulledge*, J. P. (2004). A new breed of computer users: Cursor control via joystick manipulation by rats. *Behavior Research Methods, Instruments and Computers, 36*, 173-179.

Washburn, D. A., and Rumbaugh, D. M. (1992). Investigations of rhesus monkey video-task performance: Evidence of enrichment. *Contemporary Topics in Laboratory Animal Science, 31*, 6-10.

Washburn, D. A., Rumbaugh, D. M., and Putney, R. T. (1994). Apparatus as milestones in the history of comparative psychology. *Behavior Research Methods, Instruments, and Computers, 26*, 231-235.

Washburn, D. A., Rumbaugh, D. M., and Richardson, W. K. (1988). Apparatus in comparative psychology. In G. Greenberg and M. Haraway (Eds.) *Comparative Psychology: A Handbook* (pp. 221-225). New York: Garland Publishing.

Washburn, D. A., Smith, J. D., and Shields, W. E. (2006). Rhesus monkeys (*Macaca mulatta*) immediately generalize the *Uncertain* response. *Journal of Experimental Psychology: Animal Behavior Processes, 132*, 185-189.

Wasserman, E. A., Kiedinger, R. E., and Bhatt, R. S. (1988). Conceptual behavior in pigeons: Categories, subcategories, and pseudocategories. *Journal of Experimental Psychology: Animal Behavior Processes, 14,* 235-246.

Wright, A. A., Santiago, H. C., Sands, S. F., Kendrick, D. F., and Cook, R. G. (1985). Memory processing of serial lists by pigeons, monkeys, and people. *Science, 229,* 287-289.

Xia, L., Emmerton, J., Siemann, M., and Delius, J. D. (2001). Pigeons (*Columba livia*) learn to link numerositites with symbols. *Journal of Comparative Psychology*, *115*, 83-91.

Yochmowitz, M. Z., Patrick, R., Jaeger, R. J., and Barnes, D. (1977). Protracted radiation-stressed primate performance. *Aviation, Space and Environmental Medicine, 48,* 598-606.

Yerkes, R. M. (1907). The *dancing mouse*. New York: Macmillan.

Yerkes, R. M. (1915). Notes: The role of the experimenter in comparative psychology. *Journal of Animal Behavior, 5,* 258.

Zentall, T. R., and Urcuioli, P. J. (1993). Emergent relations in the formation of stimulus classes by pigeons. *Psychological Record, 43*, 795-810.

Section III. Advances in Assessment

In: Handbook of Technology in Psychology …
Editor: Luciano L'Abate and David A. Kasier
ISBN: 978-1-62100-004-4

Chapter 11

MOBILE TECHNOLOGIES IN EDUCATION AND HEALTHCARE

Robert K. Atkinson*[1], *Andre Denham and John Quick
[1]Arizona State University, Tempe, Arizona, US

Historically man has sought to develop tools that would allow for the mobile recording and transmission of information. The creation of paper, the book, and the Guttenberg press all serve as early examples of man's attempt to develop mobile devices. In the late 1800s, Elisha Gray successfully filed a patent for a device that sought to leverage existing telegraph technology in order to send and receive handwritten notes. The Rand Company is credited with developing the first digital tablet in 1964. Due to financial and technological restraints, the tablet market was fairly silent until the development of the Wang Freestyle Tablet and stylus in the late 1980s, and the Apple Newton in the early 1990s. In the 1990s and early 2000s, the line of Palm PDAs (Personal Digital Assistant) followed these tablets. Microsoft made the first major foray into tablet computing in 2000 with its tablet PCs, which allowed for the input of information through either a stylus or through a keyboard. Tablet PCs did not receive mainstream usage and only found a home within the few industries, healthcare being one (Kolakowski, 2011).

Smartphones ushered in a new era of mobile devices in the 2000s. These devices combined the affordances of a PDA with a low-powered tablet PC and a telephone within one device. Smartphones helped to bridge the gap between consumers and enterprise mobile device users. While the smartphone was a major step forward for mobile devices, they were still plagued by a variety of issues, such as high cost, a lack of applications, no centralized application store, a lack of processing power, and frequent crashing. These contributed to a lack of widespread adoption by consumers. The 2007 Apple iPhone addressed many of the issues associated with previous smartphones and simultaneously introduced a robust multi-touch user interface (Kolakowski, 2011).

The Apple iPhone has contributed to the rapid increase of smartphone usage in the past few years. In a 2009 report published by Portio Research, smartphone ownership was predicted to achieve a Compound Annual Growth Rate of 18.5 percent from 2009 to 2014 (Portio, 2009). Smartphones are expected to encompass 43 percent of mobile phone ownership by 2015 (RBR, 2011). This is compared to its current market share of 31 percent.

The rapid growth in smartphone usage is being closely mirrored by the growth of the tablet market.

The current leader in the tablet market, Apple, sold 4.9 million units of their latest tablet, the iPad 2, during the second fiscal quarter of 2011 (Apple, 2011). The limiting factor impeding the sale of iPads was Apple's inability to produce quantities sufficient to meet consumer demand. The year 2011 has also seen the release of tablets by other companies such as Blackberry, Samsung, Motorola, and Hewlett Packard. As more manufacturers introduce tablets to consumers, the portion of the personal computing market usually associated with low-end laptops and netbooks will be siphoned off. Prognosticators have recently lowered the projected amount of laptops and netbooks to be shipped in 2011 by 10.5% based on this current trend (Gonsalves, 2011).

eReaders have also carved out a significant share of the market device market. eReaders are devices that allow for users to transport and purchase digital books. Barnes and Nobles' Nook and Amazon's Kindle are just two examples of eReaders similar in form factor and human-computer interface as tablets, that have been recently introduced to the computing market. The recent commercial success of eReaders is based on their ability to cater to those who are interested in a device dedicated entirely to the reading and purchasing of literature (Koz, 2010).

These newly introduced mobile devices combined with the trends in their usage have piqued the attention of two specific communities: education and healthcare. Educational technologists are interested in the potential of these devices to address long-standing pedagogical issues. For example, practitioners are interested in the ability of Global Positioning System (GPS) hardware, commonly found in tablets, to provide location based instructional (Gayeski, 2002). Those in the healthcare industry are interested in the use of mobile devices in the training of medical support personnel, as a tool for the mobile conducting of administrative tasks and an aid in the day-to-day tasks of physicians. Their excitement is warranted as the technological capabilities of these devices have provided affordances that would be deemed impossible only a decade ago.

In the book *Learning Unplugged*, Gayeski (2002) outlines several compelling arguments for the integration of mobile technologies into educational and training settings. The first is the ability of mobile devices to provide training anytime, and anywhere through the use of Wi-Fi and cellular network connectivity. This allows educators and training departments to provide instruction that goes beyond Web-Based Training, which requires the user to be tethered to either a desktop or a laptop. Gayeski's second argument is related to the use of mobile devices as pocket-sized repositories of knowledge. Information that would usually be contained in reference books or job aids can now travel easily with students, trainees, and medical personnel. Finally, Gayeski points to the leveraging of GPS hardware to provide location-based instruction, navigational information, and track the location of personnel or equipment in order to efficiently assign resources.

The 2011 Horizon Report suggests to practitioners that the combination of features within a mobile device is an impetus to employ them within educational settings. The ability to interact with video, audio, the Internet, and applications specifically designed for educational use within one device makes their use appealing. The Horizon Report also points to the ease of deployment within settings as an additional advantage of mobile devices (Johnson, Smith, Willis, Levine, and Haywood (2011).

CONTEXT OF USE

Educators have already begun to see the benefit of integrating mobile devices in their practice. Mobile devices provide educators access to informal learning spaces. Costabile, De Angeli, Lanzilotti, Ardito, Bouno, and Pederson (2008) describe the use of a mobile device to support a middle school excursion to an archeological park. Thorton and Houser (2005) describe studies in which they used mobile phones to assist teaching English to Japanese students on the university level. Students were sent emails containing vocabulary words to their phones on a regular schedule. The conclusion of this study was that students that received emails on a mobile device performed better on their recall of English words than students who were not sent emails and just encouraged to study on their own.

Those in industry are also looking at the iPad as a means to improve their business practices. For instance, in the effort to move pilots beyond the use of paper maps for navigation and while en-route, the executives of a few major commercial and cargo airlines have begun the process of petitioning the Federal Aviation Administration (FAA) to allow the use of iPads in-flight. Delta and Alaska are two of the many airline companies that have expressed serious interest in the adoption of iPads (Rosenberg, 2011). Executive Jet Management, a charter jet company, has already received permission to use iPads for pre-flight planning and other activities. In additon there are airlines interested in looking to the iPad as a potential new means of in-flight entertainment.

Airlines are not alone in the looking at iPads to revolutionize how they do business. National Football League (NFL) teams are planning on adopting the iPad to replace the large and complicated play calling sheets and playbooks that are used by their coaching staffs and players (Terdiman, 2011). For a player, a tablet could be used to learn the team playbook by allowing for the ability to easily sort through formations and personnel packages. Players and coaches would be able to watch animations of plays while also being able to diagram new ones and rapidly share them with the entire team. This would also allow NFL teams to save costs by eliminating the need to print "as much as 5000 pages of printouts per game" (Terdiman, 2011). In a regular season in which each of the 32 NFL teams play 16 games, adoption of the iPad would eliminate the need to printout approxiamaterly 2,560,000 sheets of paper.

Mobile devices also have potential to impact the way physicians practice medicine. Fischer, Stewart, Mehta, Wax, and Lapinsky (2003) performed a comprehensive review of the possibility and feasibility of handheld computing within medicine. Currently, the use of handhelds within the medical community is limited to PDAs. Nevertheless, PDAs have shown to be beneficial to physicians as a means of referencing medical literature, tracking patients, accessing billing, and writing prescriptions digitally. Burdette, Herchline, and Oehler (2008) point to the comparable ability of smartphones, which combine the functionality of multiple devices within a single device, as a means to assist physicians to be more connected to medical and patient information. Speaking directly to physicians who specialize in infectious diseases, the authors point to availability of several commonly used applications such as (medical calculators, the Sanford guide, Hopkins Antibiotic Guide, etc.) as potential reasons to use mobile devices.

OBJECTIVES

It is our intent to provide a comprehensive overview of recent applications of mobile device usage within educational, training, and medical settings. We also intend to forecast the future of mobile device usage in the aforementioned settings and supply some direction on how to get there. Lastly, we intend to provide an overview of the strengths and weaknesses of mobile devices. Specifically, our analysis will focus on eReaders, tablets, and smartphones. This overview will also include information on design guidelines for mobile applications. This chapter will serve to provide those in education and healthcare with a snapshot of the current trends, devices and practices native to mobile technology. This would assist in the integration of these devices within their respective disciplines.

RECENT APPLICATIONS

eReaders

Electronic book readers have become extremely popular amongst consumers over the past few years and a proliferation of devices have emerged. These thin and lightweight devices have screens that are capable of presenting text in a way that closely approximates the feeling of reading from paper. Users are able to read novels, textbooks, and other types of documents. Many eReaders now feature wireless connectivity that allows for the downloading and off-device storage of content. The most popular eReaders today are the Amazon Kindle, Barnes and Noble Nook, and Sony Reader. Hardware prices typically range from \$150 to \$300. Little software is available for eReaders beyond those used to read, and in some cases annotate, text and most applications are built into the devices. Independent development of e-reader software is not widespread yet and the devices generally lack the hardware necessary to implement substantial applications.

In education, eReaders are commonly looked at with hopes that hard copies of textbooks can be entirely replaced and that student bodies will carry electronic books (e-books) instead (Johnson, Smith, Willis, Levine, and Haywood, 2011). To date, this has proved unfeasible. While eReaders and e-books are portable, inexpensive, and multifunctional, a lack of available textbooks, usability and accessibility difficulties, and insufficient features necessary for scholarly work have obstructed their adoption at the university level (Blumenstein, 2010; Gerlich, Browning, and Westermann, 2010; Johnson et al., 2011).

Further, recent studies and pilot programs, at institutions such Ohio State University, Arizona State University, and Princeton University to name a few have shown that student opinions of eReaders as educational tools are largely negative. These universities and others have experimented with e-reader pilot programs with little success (Blumenstein, 2010; Cheng, 2010; Damast, 2010; Gerlich et al., 2010; Marmarelli and Ringle, 2010; Roscorla, 2011). Nevertheless, eReaders are still considered a viable solution in education (Johnson et al., 2011). Therefore, the adoption of eReaders into the classroom seems likely. The same is expected in the industry of healthcare. EReaders are ideal for the aggregation and simplification of reading materials, such as reference books, patient records, and medical journals (Robertson, Miles, and Bloor, 2010).

Tablets

The preceding generations of tablets were more like a full-fledged computer, running traditional desktop operating systems, such as Windows XP. With the same, flat form factor, medical professionals used a pen stylus directly on the screen to input information. These tablets were used by healthcare employees for tasks such as patient record management and evaluations (Garson and Adams, 2008; Martins and Jones, 2005; Richter et al., 2008).

A new breed of tablet has recently emerged, beginning with the release of the Apple iPad in 2010. In contrast to the preceding generation of tablets, these devices are thinner, lighter, use touchscreen interfaces, and have their own specialized operating systems and software applications. These tablets commonly have screens that range in size from seven to 10 inches. The most common operating systems in use today are Apple's iOS and Google's Android. Though the Apple iPad is the most popular tablet today (Keizer, 2011), alternatives are beginning to surface such as the Samsung Galaxy, Motorola Xoom, and Blackberry Playbook. Software applications for tablets, often referred to as "apps," are commonly distributed through a dedicated online store featured on the device itself or accessible through the user's desktop computer. Applications are abundant in both the education and health fields. For instance, Apple's App Store for iOS devices has dedicated education and medical categories, each containing thousands of applications (Apple Computer Inc., 2011; Husain, 2011). Furthermore, the creation of apps for devices running the iOS or Android operating systems is possible thanks to software development kits (SDKs), which allow for the independent creation and distribution of software. Tablets tend to range in price from $500 to $1,000 and their apps cost $1 to $5 each in most cases.

Many educators see tablets as an all-encompassing toolkit for education. Unlike eReaders, which are purely for reading, tablets can be used for a wide variety of activities, such as Internet browsing, playing learning games, writing, and drawing. Several higher education institutions, including Notre Dame University, University of Minnesota, and Oklahoma State University have put iPad pilot programs into place that provide incoming students with the devices and subsequently monitor their experiences (McCombs and Liu, 2011; Pepperdine University, 2011; Perkins, Hamm, Pamplin, Morris, and McKelvain, 2011; Yeung and Cheng, 2011). Many of these pilot studies are just getting underway, but early findings suggest benefits of student engagement, information aggregation, multipurpose applications, and time management (Angst and Malinowski, 2010). Furthermore, some institutions have started iPad programs for use by faculty and administrators (Dickerson, Winslow, Lee, and Geer, 2011; Penny, 2011). With the recent addition of cameras to tablets, new serious games and augmented reality applications are also anticipated (Carmigniani et al., 2011). The versatile nature of tablets makes them a promising learning technology that is sure to be further examined in upcoming years.

Tablets can also benefit health professionals by helping them efficiently organize and access learning materials (Robertson et al., 2010). Additionally, tablets have been proposed for use in assessing patient risk, handling patient records, and viewing and annotating radiological images (Robertson et al., 2010; Rubin, Rodriguez, Shah, and Beaulieu, 2008; Zhange et al., 2011). In fact, administrators have successfully incorporated the new generation of tablets into hospitals and clinics. For example, the iPad has been used as a reference during surgery (Sugimoto, 2010; Volonte, Robert, Ratib, and Triponez, 2011; Wodajo, 2011). Additionally, many publishers of human anatomy reference books are making their guides available as apps. The iMedicalApps website (www.imedicalapps.com) is a useful resource that provides news and reviews of apps relevant to healthcare professionals. These applications and more are proof that iPads have more of a practical application that just for entertainment.

Smartphones

Smartphones, like modern tablets, are nearly full-fledged computers that are contained in personal, portable form factors. They come in various shapes and sizes with screens that typically measure in the 2.5 to five inch range. In contrast to traditional mobile phones, which are primarily designed to make calls, smartphones incorporate many additional features such as web browsing, GPS location technology, accelerometers and gyroscopes, high-quality cameras, office software, and time management applications. Smartphone adoption worldwide has grown tremendously, with current consumer ownership ranging from 31%-36% based on recent market research (The Neilsen Company, 2011; Olswang LLC., 2011). Smartphones tend to make use of specialized operating systems, rather than using those of their desktop counterparts. The most widespread operating systems today are Android, Symbian, iOS, Blackberry OS, and Windows Phone (Canalys, 2011). The commercial trend is to distribute software, as apps through a centralized marketplace, such as Apple's App Store or Android's Market. All of the aforementioned smartphone operating systems provide SDKs to independent developers, although iOS and Android are currently the most popular development platforms. Some operating systems are open source (Android, Symbian), whereas others are not (iOS, Blackberry OS, Windows Phone). Due to the openness of smartphone operating systems to third-party developers, an abundance of apps can be found on nearly all of these platforms, though iOS and Android currently feature the largest selections (Lookout Inc., 2011). Thousands of health and education apps are currently available through smartphone application marketplaces (Apple Computer Inc., 2011; Husain, 2011).

In education, smartphones have piqued the interest of practitioners and researchers alike. Mobile learning via smartphones has been examined in terms of pedagogical strategies and benefits across several levels of education, including informal learning (Chen, Seow, So, Toh, and Looi, 2010; Cochrane and Bateman, 2010; Cochrane, Bateman, and Flitta, 2009; Kukulska-Hulme and Pettit, 2007). Explorations into pedagogical smartphone use by academic faculty have also begun (Herrington, 2008; Herrington et al., 2010). The iPhone itself has been targeted as a platform for content-specific applications in math, computer science, and language learning (O'Rourke, MacDonald, and Goldschmidt, 2010; Tabata, Yin, Ogata, and Yano, 2010; Yuan, Chae, Nantwi, Natriello, and Garg, 2010). Additional smartphone topics of interest include assessment and augmented reality applications (Backer, 2010; Pence, 2011).

Likewise, smartphones are already being researched and applied in healthcare. For instance, some hospitals are using smartphones for patient monitoring, virtual visits, and self-assessment (Armstrong, Nugent, Moore, and Finlay, 2010; Busis, 2010; Lam, Wong, Wong, Wong, and Mow, 2009). Smartphones have also found a home in medical education (Phillippi and Wyatt, 2010; Trelease, 2008). Other application areas include medical records, diagnostic tools, reference material, and imaging (Sarasohn-Kahn, 2010). In one sense, smartphones can be viewed as convergent successors to previously separated clinical technologies, such as pagers, mobile phones, and personal digital assistants (Burdette, Herchline, and Oehler, 2008). Smartphones' increasing capability to offer desktop-caliber applications, multipurpose technologies in a single package, and a portable form factor, make them attractive one-stop solutions for healthcare professionals.

FUTURE APPLICATIONS

The future of mobile devices in the fields of psychology, psychiatry, and neurology will soon replace less developed technologies and supersede previous expectations. A few of the developing and upcoming trends in mobile technology are highlighted here.

Just in Time Reference

Just in time learning is the concept of accessing critical knowledge if and when it is most needed to accomplish the task at hand (Gee, 2007). In a learning context, just in time information can enhance a student's ability to solve problems and make connections between the real world and learning content. In a healthcare settings, just in time information can help nurses, doctors, and emergency response personnel make rapid and appropriate decisions that are critical to patients' wellbeing. Mobile devices are a tremendous benefit to just in time delivery systems, because they can travel with users and be accessed virtually anytime, anywhere that they are needed. A few examples of mobile applications that deliver just in time information are Human Body Anatomy, Nurse's Pocket Drug Guide, and Pocket Universe.

Serious Games

Serious games, a term that refers to games designed primarily for purposes other than entertainment, is an emerging field that uses digital and non-digital games to engage users and allow them to acquire and practice new skills in authentic environments. In education, serious games are being used to enhance student learning in a wide array of topics in science, math, language, and social studies.

Examples of serious games include Quest Atlantis, Do I Have a Right?, and Life Preservers. In the health sector, games are being created to help educate patients about their medical conditions and involve them in self-regulation. Furthermore, serious games are being used to train practitioners on complex tasks, such as surgeries and patient interactions. Examples of serious games related to health care include Re-Mission, Killer Flu, and Zero Hour: America's Medic.

Many of today's games are delivered via an Internet browser, desktop computer, or gaming console. However, the increasing technical capabilities and proliferation of mobile devices is likely to lead to the development of mobile serious games as well.

Gesture and Motion-Based Computing

An explosion of interest in gesture and motion-based computing has arrived thanks to recent commercial products, such as the Nintendo Wii and Microsoft Xbox Kinect. These technologies have begun to replace traditional keyboard, joystick and mouse interfaces with more intuitive, natural, and aesthetically pleasing ones. As seen in the Apple iPhone, touchscreens have quickly become the hardware standard for smartphones and tablets. As a result, many mobile devices are starting to feature embedded accelerometers and gyroscopes. However, software that utilizes the affordances of these advancements is just beginning to be explored. In both health and education, motion-based and gesture computing can be used to increase the authenticity with which actions are performed and the accessibility of computing systems to individuals of varied capabilities. Current mobile application examples include OsiriX HD and Blausen Human Atlas in health and Star Walk in education.

Augmented Reality Applications

Augmented reality refers to allowing users to "see the real world, with virtual objects superimposed upon or composited with the real world" (Azuma, 1997, p. 356). Recently, thanks to the inclusion of one or more cameras in nearly all cellphones and smartphones, augmented reality applications have begun to emerge. In particular, iOS and Android applications, such as Layar Reality Browser, Vernier Video Physics, and Meal Snap, are augmenting users' everyday worlds with information about the objects and events that they encounter. Augmented reality has great potential to improve health and education. For example, students could use augmented reality to explore historical world locations or to witness the math and science underlying everyday interactions. Further, doctors, nurses, and other health professionals could use augmented reality to easily view patient information in real time and even to help diagnose and develop treatment plans for patients based on the additional information provided through augmentation.

Wearable Computing

Sensors, monitors, and full-fledged computers that can be worn on the body are on the horizon. These devices are capable of unobtrusively collecting vast amounts of data about users' mental and physical states. For instance, wearable sensors are being used to monitor the physiological states of people with Autism to better understand the gap between emotion that is outwardly portrayed and inwardly felt by an individual (Picard, 2009). Additionally, commercial companies, such as BodyMedia, have released systems that use sensors to track users' physical activity and report it back to them online or via mobile devices. Furthermore, brainwave and eye-tracking technologies from companies like Emotiv and Tobii are being employed by researchers to better understand medical interventions and entertainment applications (Campbell et al., 2010; Faro, Giordano, Spampinato, De Tommaso, and Ullo, 2010; Koutepova, Liu, Lan, and Jeong, 2010; Trojano, Moretta, Estraneo, and Santoro, 2010). Systems such as these are expected to rapidly become smaller, more powerful, and incorporated into a wide variety of everyday devices. As data mining techniques are refined and general computing power increases, it is predicted that unprecedented amounts and types of information about the human mind and body will be collected in coming years. Yet, as technical capabilities rapidly develop, it will also be necessary for education and health professionals to determine appropriate ways to interpret and employ this new information towards beneficent ends.

TECHNICAL INFORMATION

Pros and Cons

All technologies have strengths and weaknesses. Mobile devices are no exception. The perfect device may be elusive, but often the right technology for the time, place, and situation is readily available. The following tables depict the relative pros and cons of eReaders, tablets, and smartphones, as applied to practice in the fields of education and healthcare. These can be used to assist in aligning personal, professional, and institutional needs to the capabilities of different mobile devices.

Table 1. Pros and cons of eReaders

eReaders	
Pros	Cons
• More portable than tablets • Most comfortable and efficient choice for reading large amounts of text • Able to customize viewing options, such as text size • Able to read text aloud as audio • Can store and organize entire document collections • Wireless download of content • Content often cheaper than hard copies	• Less portable than smartphones • Text annotation is inefficient • Web browsing is inefficient • Lack multipurpose functions found in smartphones and tablets • Price may not justify lack of features relative to other mobile devices • Tend to use proprietary e-book formats that are incompatible with other devices • Content is licensed, not owned, by purchaser

Table 2. Pros and cons of tablets

Tablets	
Pros	Cons
• Larger screen size and more power than smartphones and eReaders • Multipurpose (web, email, reading, photos, games, calendars, etc.) • Good ratio of features to price • Many apps available • Apps often free or low cost • Wireless download of content • SDKs allow for third-party development • Some include cameras • Strong battery life	• Less portable than smartphones and eReaders • Cost more than smartphones and eReaders • May pay for features that are not used • Not capable of everything that a standard computer can do • Data input less efficient than standard keyboard • Inefficient file management • Usually non-replaceable battery

Table 3. Pros and cons of smartphones

Smartphones	
Pros	Cons
• More portable than eReaders and Tablets • Device costs often low due to telecom provider subsidies • Multipurpose (web, email, reading, photos, games, calendars, etc.) • Good ratio of features to price • Many apps available • Apps often free or low cost • Wireless download of content • SDKs allow for third-party development • Often include cameras, GPS, motion sensors, phone, and text messaging	• Screen size limits range of applications and reading comfort • Phone/data subscription fees lead to high long-term costs • Can be fragile and susceptible to wear • Usually non-replaceable battery • Inefficient data input • Inefficient file management

Naturally, all mobile devices have their respective advantages and disadvantages. For example, some include an impressive amount of features in a small package, but simultaneously have a higher cost. Others focus on doing one specific thing well, but miss out on the potential to satisfy multiple user needs. In addition to the pros and cons of mobile hardware itself, the quality and abundance of software is an important consideration when selecting from an array of potential technologies. Fortunately, design guidelines for mobile applications have begun to emerge. Many of these heuristics are derived from good design techniques found in other content domains, such as web, instructional, and usability design. Mobile design guidelines can be used to assess the quality of existing applications and to assist in the development of new ones.

Design Guidelines

Mobile device manufacturers such as Apple (Apple, 2011), Google (Android, 2011), and Microsoft (Microsoft, 2010) have all developed a robust set user-interface (UI) guidelines. These guidelines are intended to assist mobile application developers with creating a rich user-interface experience. While these guidelines speak to application development for a variety of devices, there are consistencies among these guidelines that can be used to develop a list of UI principles for mobile application development. These UI guidelines are presented in the form of a list, but it is not our intent to dictate any order or priority. All of these UI guidelines are interrelated and when applied correctly contribute to development of applications that are intuitive in their functionality and user-centered.

Be Consistent. Users of mobile devices have an expectation of how mobile applications should function. In order to meet these expectations mobile applications must be consistent in their functionality. Apple, Microsoft, and Google, all provide within their SDKs a standard set of icons, buttons, and menu bars. This helps to create a sense of consistency from one application to another. Through repeated use of a mobile operating system and applications that adhere to UI guidelines, mobile device users are easily able to navigate through applications, regardless of the applications intended functionality. The Microsoft UI guidelines refer to the terms "consistently" or "consistent" in nineteen different sections while Apple's UI refers to those two terms thirty-eight different times. An emphasis on consistency helps to eliminate user confusion and emphasizes a user-centered approach to application development.

Handle Changes to Screen Orientation. Most mobile devices currently being manufactured have the ability to detect changes in their orientation. This is accomplished through the use of an accelerometer and/or gyroscope. Mobile application developers are able to take advantage of this functionality by changing the layout of their apps. A developer can have different layouts for each position. Mobile device users are familiar with this functionality as well. It is important for application developers to ensure that their application can be viewed in all possible device orientations. Apple (Apple, 2011) encourages this approach to UI design, but also cautions developers to remember that the content should not take a back seat when viewed in orientations outside of the primary orientation. Google also encourages developers to take advantage of ability of devices to change screen orientation but allow for the development of applications that only work in one orientation. The Windows Mobile 7 UI guidelines require applications to support landscape mode. In addition both left and right landscape views must also be supported. Windows Mobile 7 applications that require text input are also required to include a landscape keyboard.

Be Conscious of Touch Target Area. A mobile device with a multi-touch display has become the industry standard. Most mobile devices eschew the use of physical keyboards, and limit the number of buttons on their products. Human touch has become the accepted means of human-computer interaction. One thing that has not changed is the use of a Graphical User Interface (GUI) approach to launch and interact with applications. With that in mind, it is important that developers are conscious of the size of items within their mobile applications and make all touch "target" areas the size of a fingertip. Apple (2011) notes that though we vary in size as humans, average finger size does not vary. Apple and Microsoft recommends specific dimensions for the size of touch target areas based on the touch control's intended function. Creating touch target areas that are too small distract from the UI experience, increase the number of errors made, and require users to focus more on the interface than on the application itself. Microsoft encourages developers to go to the lengths necessary to manipulate "size, spacing, location, and visuals to reduce the difficulty in acquiring a target through finger touch" (Microsoft, 2011, p. 75). Failure to practice this principle could easily result in users discontinuing the use of an application in favor of another.

Use Touch Gestures Effectively. Along with being conscious of touch target area sizes, developers should also ensure that they make appropriate use of touch gestures. Apple, Microsoft, and Google all advise against the use of a gesture that goes counter to the standard use of that gesture. For example the UI guideline for Microsoft outlines six gestures: tap, double tap, pan, flick, touch and hold, and pinch and stretch. The intended use of the pinch gesture is to zoom in on content and should not be used to execute an action. That is the intend purpose of the tap gesture. The gestures that are used within an application should be simple and intuitive. The majority of interactions should be limited to gestures that only require the use of one finger. This will allow the content to remain primary and not the UI experience. As stated earlier the content is of foremost concern. Within the Windows Phone

Developer Tools is native support for touch gesture development. This support helps to maintain consistency within applications and the phone operating system, while also allowing for the creation of gestures that custom touch controls (Microsoft, 2010).

Provide Tactile or Visual Feedback. Many detractors of multi-touch mobile devices bemoaned the lack of tactile feedback. While tactile feedback is very important within human-computer interactions, providing visual feedback serves as an excellent alternative. An additional means of feedback is haptic feedback. Haptic feedback serves as an equivalent to visual feedback by using vibrations to indicate acceptance of a touch command. Recommendations related to feedback can be found throughout the UI guidelines of mobile device manufacturers. For example the Android operating system provides access to haptic feedback effects through code. Apple even lists feedback among its six human interface principles. Through the use of animation, sound, haptic feedback or some visible change in the screens appearance, users should be aware that they're intended (or unintended) interactions have been registered by the device.

Make Use of Metaphors. Associations between real world interactions with objects, and the interactions with virtual objects or GUI's can be powerful if harnessed correctly. By combining the graphical capabilities of mobile devices and gestures, mobile application developers can design interfaces that intuitive for users. Page turning within iBooks, Apple's e-reader application is an excellent example of using metaphors to create natural interaction within an application. Users are able to turn the page with the use of a tap or by metaphorically dragging the page. The application itself recognizes the drag as a command to turn the page and provides visual feedback through an animation that shows the page turning. The user is able to control the speed of the page turning animation. This animation works both when turning to a new page or returning to a previously read page. This connection to the feedback that one would experience in a real life interaction with book and the UI experience within iBooks speaks to proper use of a metaphor.

User Control is Paramount. All of the guidelines mentioned speak to the notion of user control. A consistent focus across all of the UI guidelines of major mobile device manufacturers is the importance of creating an experience that puts the user in control of all actions that take place within a mobile application. Users are given control when they are responsible for initiating the opening and closing of an application. User control also encompasses the ability to recover from errors in an efficient manner. Consistency in UI along with feedback, metaphors, contribute greatly to granting users control over their experience, while maintaining the aesthetic integrity of the application.

CONCLUSION

Mobile devices provide educators and healthcare practitioners with a tool that has tremendous potential. For educators and practitioners, mobile devices have the potential to provide access to informal learning settings, which have been historically difficult to access. Mobile devices also provide educators the ability to leverage GPS, Wi-Fi, and cellular network connectivity, to push content to learners based on their location. Mobile devices allow for user-centered and ubiquitous learning that is not limited to brick and mortar facilities.

In healthcare, mobile devices may revolutionize the way that medicine is practiced. For example, consider a psychologist who has prescribed a regiment to a patient suffering from agoraphobia (the irrational fear of open spaces). The psychologist asks the patient to steadily increase the amount of time that he spends in a busy shopping mall from thirty minutes a day to two hours a day. Through the use of a mobile device with Foursquare (a location-based application), the patient can log when he arrives at the mall, the stores he visited, and when he departed the mall. This psychologist (with the permission of the patient) could track the progress of the patient through the Foursquare website between office visits. In another example, a physical therapist could make use of the Nike Plus running application to track the physical activity of a patient. Moreover, a dietician can develop an application that allows clients to log meal-by-meal caloric intake and track weight variations. Further still, a neurologist could conduct checkups through the use of video conferencing software on mobile devices. Lastly, an ER physician could use a tablet to analyze MRI results moments after the procedure is complete. These are just a few of ways that mobile devices can be integrated into the fields of education and healthcare.

Although mobile devices have great potential to transform fields, practitioners should exercise caution before investing resources in new technologies. Within education there currently exists no unified theory of mobile learning. Sharples, Taylor, and Vavoula (2010) began the process of defining a theory of learning with mobile devices by providing a comprehensive litmus test for which any theory should be validated against. They postulate that any theory of mobile learning should stand in stark contrast against theories that have been developed to address learning in traditional contexts (classroom, job training, etc.). A theory of mobile learning should contain language that ties the ability to be in motion as an aid to learning. Mobile learning should not and is not repurposed e-learning. It is mobile learning is not truly taking place is the learning is not required to be mobile. The authors also question any theory that does not cover the full spectrum of learning from informal to formal settings. Mobile devices can be seamlessly integrated within each context and a theory of mobile learning should account for this. The author's assert that learning is constructed by the learner and constitutes a social process. Therefore learning that is assisted by the use of a mobile device should be user centered and situated within the proper context.

Another reason to exercise caution when integrating mobile devices into one's discipline is that no well-developed authoring tools currently exists. E-learning tools, such as Articulate and Captivate, allow subject matter experts to design and develop online training efficiently and consistently. These authoring tools help to significantly reduce the development cost and time associated with creating training. There are several companies that currently purport to have developed authoring environments that create mobile applications but they all have their strengths and weakness. If one could create the ideal mobile authoring tool, it would allow for the publishing of applications to a variety of mobile operating systems. This authoring tool would also allow for offline and online content delivery. Laslty, the tool would allow for both code-based and What You See Is What You Get (WYSIWYG) development.

On a cautionary note, the use of mobile devices to administer learning instruction and healthcare treatment has the potential to infringe on the privacy of students and patients. Mobile devices store a tremendous amount of personal information about each user. While mobile devices provide practitioners with ubiquitous access to clientele, care must be taken to ensure that the rights of end users are respected and protected. This is a matter of law within the medical community. The Health Insurance Portability and Accountability Act (HIPAA) of

1996, was put into law to protect the privacy of medial patients. The educational equivalent to HIPAA is the Family Educational Rights and Privacy Act (FERPA). This federal law protects the privacy of students and regulates how, where, and when student education records can be accessed and disseminated.

Notwithstanding the aforementioned challenges, mobile devices are powerful tools for enhancing learning and modern medicine. Mobile technologies are still in their infancy and the integration of mobile devices within these education and healthcare has yet to be empirically tested. In education, considerable work still needs to be done to develop a unified theory of mobile learning. This would go a long way in identifying the pedagogical benefits of mobile devices. Within the medical community, efficiency and viability testing should be conducted to determine how mobile devices best serve practitioners and patients alike.

REFERENCES

Angst, C. and Malinowski, E. (2010). *Findings from e-reader project, phase 1: Use of iPads in MGT40700, Project Management.* Retrieved from http://www.nd.edu/~cangst/NotreDame_iPad_Report_01-06-11.pdf

Android (2011). *User Interface Guidelines.* Retrieved from http://developer.android.com/guide/practices/ui_guidelines/index.html

Apple Computer, Inc. (2011). *Thousands of apps. Endless potential.* Retrieved from http://www.apple.com/education/apps/

Apple Computer, Inc. (2011). *iOs Human Interface Guidelines: User Experience.* Retrieved from http://developer.apple.com/library/ios/documentation/userexperience/conceptual/mobilehig/MobileHIG.pdf

Apple Computer, Inc. (2011). *Q2 FYll Earnings Release.* Retrieved from http://investor.apple.com/

Armstrong, N., Nugent, C., Moore, G., and Finlay, D. (2009). Using smartphones to address the needs of persons with Alzheimer's disease. *Annals of Telecommunications, 65*(9-10), 485-95.

Azuma, R. T. (1997). A survey of augmented reality. *Presence: Teleoperators and Virtual Environments, 6*(4), 355-85.

Backer, E. (2010). Using smartphones and Facebook in a major assessment: The student experience. *e-Journal of Business Education and Scholarship of Teaching, 4*(1), 19-31.

Blumenstein, L. (2010). Kindle accessibility suit settled with universities. *Library Journal, 135*(4), 18-20.

Burdette, S. D., Herchline, T. E., and Oehler, R. (2008). Practicing medicine in a technological age: Using smartphones in clinical practice. *Clinical Infectious Diseases, 47*(1), 117-22.

Busis, N. (2010). Mobile phones to improve the practice of neurology. *Neurologic Clinics, 28*(2), 395-410.

Campbell, A., Choudhury, T., Hu, S., Lu, H., Mukerjee, M.K., Rabbi, M., and Raizada, R.D.S. (2010). NeuroPhone: Brain-mobile phone interface using a wireless EEG headset. *Proceedings of the second ACM SIGCOMM Workshop on Networking, Systems, and Applications on Mobile Handhelds, USA*, 3-8.

Canalys. (2011). *Google's Android becomes world's leading smart phone platform.* Retrieved from http://www.canalys.com/pr/2011/r2011013.pdf.

Carmigniani, J., Furht, B., Anisetti, M., Ceravolo, P., Damiani, E., and Ivkovic, M. (2011). Augmented reality technologies, systems, and applications. *Multimedia Tools and Applications, 51*(1), 341-77.

Chen, W., Seow, P., So, H. J., Toh, Y., and Looi, C. K. (2010). Extending students' learning spaces: Technology-supported seamless learning. *Proceedings of the 9th International Conference of the Learning Sciences, USA,* 484-91.

Cheung, J. (2010). *Business school ditches Kindle DX after trial run.* Retrieved from http:// arstechnica.com/gadgets/news/2010/05/business-school-students-not-fond-of-kindle-for-academics.ars.

Cochrane, T., and Bateman, R. (2010). *Smartphones give you wings: Pedagogical affordances of mobile web 2.0.* Paper presented at the 26th ASCILITE Conference, ASCILITE 2009, Auckland, New Zealand. Retrieved from http://web.me.com/thom _cochrane/MobileWeb2/Research_Outputs/ConferenceProceedings/Ascilite2009.pdf.

Cochrane, T., Bateman, R., and Flitta, I. (2009). *Integrating mobile web 2.0 with tertiary education.* Paper presented at m-ICTE2009, Lisbon, Portugal. Retrieved from http://web.me.com/thom_cochrane/MobileWeb2/Research_Outputs/ConferenceProceedin gs/mICTe2009.pdf.

Costabile, M. F., De Angeli, A., Lanzilotti, R., Ardito, C., Buono, P., and Pederson, T. (2008). Explore! Possibilities and Challenges of Mobile Learning. *Proceedings of the Twenty Sixth Annual SIGCHI Conference on Human Factors in Computing Systems,* (145-154).

Damast, J. (2010). *E-book readers bomb on college campuses.* Retrieved from http://www. businessweek.com/bschools/content/jun2010/bs20100610_200335.htm.

Dickerson, J., Winslow, J., Lee, C. Y., and Geer, C. (2011). iPrincipals: School administrator iPad utilization – part 1. In *Proceedings of Society for Information Technology and Teacher Education International Conference 2011* (pp. 2952-3). Chesapeake, VA: AACE.

Faro, A., Giordano, D., Spampinato, C., De Tommaso D., and Ullo, S. (2010). An interactive interface for remote administration of clinical tests based on eye tracking. *Proceedings of the 2010 Symposium on Eye-Tracking Research and Applications, USA*, 69-72.

Fischer S, Stewart T.E., Mehta, S., Wax R., and Lapinsky S.E. (2003) Handheld computing in medicine. *Journal of American Medical Information Association, 10*(2):139-149.

Garson, K., and Adams, C. (2008). Security and privacy system architecture for an e-hospital environment. *Proceedings of the 7th symposium on identity and trust on the Internet IDtrus 08, USA, 283,* 122-30.

Gayeski, D. (2002). *Learning Unplugged: Using Mobile Technologies for Organizational Training and Performance Improvement.* New York: AMACOM.

Gee, J. P. (2007). *Good video games and good learning.* New York: Peter Lang.

Gerlich, R. N., Browning, L., and Westermann, L. (2010). E-readers v. traditional print media on campus: Gender differences and challenges to be overcome. *Proceedings of the Academy of Educational Leadership, USA, 15,* 47-51.

Gonsalves, A. (2011, March 4). *Tablet Growth Project to Slow PC Demand.* Retrieved from http://www.informationweek.com/news/hardware/desktop/229300180.

Herrington, A. (2008). *Adult educators' authentic use of smartphones to create digital teaching resources*. In *Hello! Where are you in the landscape of educational technology? Proceedings of ascilite Melbourne 2008, Australia,* 414-418. Retrieved from http://www.ascilite.org.au/conferences/melbourne08/procs/herrington-a.pdf.

Herrington, J., McKenzie, S., Pascoe, R., Woods-McConney, A., MacCallum, J. and Wrign, P. (2010). A whole-school approach to technological literacy: Mobile learning and the iPhone. In Z. Abas et al. (Eds.), *Proceedings from Global Learn Asia Pacific 2010* (pp. 181-186). Chesapeake, VA: AACE.

Husain, I. (2011). *Mac app store presents possibilities for medical community due to low barrier of entry*. Retrieved from http://www.imedicalapps.com/2011/01/mac-app-store-medical-community/

Johnson, L., Smith, R., Willis, H., Levine, A., and Haywood, K. (2011). *The 2011 horizon report*. Austin, Texas: The New Media Consortium.

Keizer, G. (2011). *iPad lead shrinks as tablet sales grow*. Retrieved from http://www.computerworld.com/s/article/9207319/iPad_lead_shrinks_as_tablet_sales_grow.

Koutepova, T., Liu, Y., Lan, X., and Jeong, J. (2010). Enhancing video games in real time with biofeedback data. *Proceedings of the 3rd ACM SIGGRAPH Conference* and *Exhibition on Computer Graphics and Interactive Technologies in Asia, Korea*, n.p.

Koz, M. (2010, May 17). *A Brief History of eBooks*. Retrieved from http://goodereader.com/blog/electronic-readers/a-brief-history-of-ebooks/.

Kolakowski, N. (2011, April 25). *Enterprise Mobility: From Teleautograph to Apple iPad: The Table PC's First 123 years*. Retrieved from http://www.eweek.com/c/a/Mobile-and-Wireless/From-Telautograph-to-Apple-iPad-The-Tablet-PCs-First-123-Years-519437/

Kukulska-Hulme, A., and Petit, J. (2007). Self-service education: Smartphones as a catalyst for informal collective and individual learning. *Proceedings of Mlearn '07 Mobile Learning Conference, Australia,* 16-19.

Lam, S., Wong, K., Wong, K., Wong, W., and Mow, W. (2009). A smartphone-centric platform for personal health monitoring using wireless wearable biosensors. *Proceedings of 7th International Conference on Information, Communications and Signal Processing, China,* 1-7.

Lookout, Inc. (2011). *App genome report – February 2011*. Retrieved from https://www.mylookout.com/appgenome.

Looking at the Future of Smartphone Ownership. (2011, February, 10). Retrieved from http://www.rbr.com/media-news/research/looking-at-the-future-of-smartphone-ownership.html.

Marmarelli, T., and Ringle, M. (2010). *The Reed College Kindle study*. Retrieved from http://web.reed.edu/cis/about/kindle_pilot/Reed_Kindle_report.pdf.

Martins, H., and Jones, M. R. (2005). What's so different about mobile information communication technologies (MICTs) for clinical work practices? A review of selected pilot studies. *Health Informatics Journal, 11*(2), 123-34.

Microsoft. (2010). *UI design Interaction Guide for Windows Phone 7.* Retrieved from http://msdn.microsoft.com/en-us/library/bb158602.aspx.

Neilsen Company, The. (2011). *State of the media 2010: U.S. audiences and devices.* Retrieved from http://blog.nielsen.com/nielsenwire/wp-content/uploads/2011/01/nielsen-media-fact-sheet-jan-11.pdf.

Olswang, LLC. (2011). *Does it add up? Convergence survey 2011 executive summary.* Retrieved from http://www.olswang.com/convergence2011/pdfs/Executive_Summary_-_Convergence_Report_2011.pdf.

O'Rourke, J., MacDonald, I., and Goldschmidt, D. (2010). Learning computer science concepts using iPhone applications. *Journal of Computing Sciences in Colleges, 25*(6), 121-8.

Pence, H. E. (2011). Smartphones, smart objects, and augmented reality. *The Reference Librarian, 52*(1), 136-45.

Penny, C. (2011). Slide to unlock: An iPad pilot project for higher education faculty. In *Proceedings of Society for Information Technology and Teacher Education International Conference 2011* (pp. 1931-2). Chesapeake, VA: AACE.

Pepperdine University. (2011). *Technology and learning: iPad studies at other institutions.* Retrieved from http://services.pepperdine.edu/techlearn/tools/ipad/similarstudies.htm

Perkins, S., Hamm, S., Pamplin, K., Morris, J., and McKelvain, R. (2011). Exploring learning with the iPad: ACU connected and the future of digital texts. In *Proceedings of Society for Information Technology and Teacher Education International Conference 2011* (pp. 1640-2). Chesapeake, VA: AACE.

Phillippi, J. C., and Wyatt, T. H. (in press). Smartphones in nursing education. *CIN: Computers, Informatics, Nursing*. doi: 10.1097/NCN.0b013e3181fc411f.

Picard, R. W. (2009). Future affective technology for Autism and emotion communication. *Philosophical Transactions of the Royal Society B: Biological Sciences, 364*(1535), 3575-3584.

Portio Research Inc. (2009). *Smartphone Futures 2009-2014: Worldwide Market Analysis and Competitor Positioning in the High-End Handset Market in 2009 and Beyond (PDF).* Retrieved from http://www.portioresearch.com/Smartphone_brochure_Nov09.pdf.

Richter, J. G., Becker, A., Koch, T., Nixdorf, M., Willers, R., Monser, R.,...Schnelder, M. (2008). Self-assessments of patients via tablet pc in routine patient care: Comparison with standardised paper questionnaires. *Ann. Rheum. Dis, 67*(12), 1739-41.

Robertson, I., Miles, E., and Bloor, J. (2010). The iPad and medicine. *BMJ Careers, 342.* Retrieved from http://careers.bmjknowledge.org/careers/advice/view-article. html?id=20001584.

Robles, G. (2009). "Virtual Reality: Touch/Haptics". In Goldstein B (Ed.). *Encyclopedia of Perception.* Vol. 2, pp. 1036-1038. Sage Publications: Thousand Oaks CA (PDF). Retrieved 04-29-2011.

Roscorla, T. (2011). *CSU system shares e-book pilot results.* Retrieved from http://www.convergemag.com/classtech/California-State-University-Shares-E-book-Pilot-Results.html.

Rosenberg. A. (2011, March 7). *Delta Air Lines, Alaska Airlines push for iPads as In-Cockpit Flight Aids.* Retrieved from http://www.digitaltrends.com/mobile/devices/delta-air-lines-alaska-airlines-push-for-ipads-as-in-cockpit-flight-aids/.

Rubin, D. L., Rodriguez, C., Shah, P., and Beaulieu, C. (2008). iPad: Semantic annotation and markup of radiological images. *AMIA Annual Symposium Proceedings 2008, USA,* 626-30.

Sarasohn-Kahn, J. (2010). *How smartphones are changing health care for consumers and providers.* Retrieved from http://www.chcf.org/~/media/Files/PDF/H/PDF%20HowSmartphonesChangingHealthCare.pdf.

Sharples, M., Taylor, J., and Vavoula, G. (2007). A Theory of Learning for the Mobile Age. In Richard Andrews, Caroline Haythornwaite (Eds.), *The SAGE Handbook of E-Learning Research* (87-99). Los Angeles : SAGE.

Sugimoto, M. (2010). *World 1st iPad surgery: OsriX 3D navigation* [Video file]. Retrieved from http://www.youtube.com/watch?v=n5nbNIpqdAY.

Tabata, Y., Yin, C., Ogata, H., and Yano, Y. (2010). An iPhone quiz system for learning foreign languages. *Proceedings of 2nd International Asia Conference on Informatics in Control, Automation and Robotics, China,* 299-302.

Terdiman, D. (2011, February 4). *NFL Teams May Replace Playbooks with iPads.* Retrieved from http://news.cnet.com/8301-13772_3-20030638-52.html.

Thornton, P., Houser, C. (2005), Using mobile phones in English education in Japan, *Journal of Compute Assisted Learning*, 21, 217-228.

Trojano, L., Moretta, P., Estraneo, A., and Santoro, L. (2010). Neuropsychologic assessment and cognitive rehabilitation in a patient with locked-in syndrome and left neglect. *Archives of Physical Medicine and Rehabilitation, 91*(3), 498-502.

Trelease, R. B. (2008). Diffusion of innovations: smartphones and wireless anatomy learning resources. *Anatomical Sciences Education, 1*(6), 233-9.

Volonte, F., Robert, J. H., Ratib, O., and Triponez, F. (in press). A lung segmentectomy performed with 3D reconstruction images available on the operating table with an iPad. *Interactive Cardiovascular and Thoracic Surgery.* Retrieved from http://icvts.ctsnetjournals.org/cgi/rapidpdf/icvts.2010.261073v1.pdf.

Wodajo, F. M. (2010). *Column: The iPad in the hospital and operating room.* Retrieved from http://www.surgisphere.com/SurgRad/issues/volume-2/1-january-2011–pages-1-112/152-column-the-ipad-in-the-hospital-and-operating-room.html

Yeung, M., and Cheung, H. (2011). iPEP talk: Pedagogical conversations from the iPad exploration project. In *Proceedings of Society for Information Technology and Teacher Education International Conference 2011* (pp. 3036-41). Chesapeake, VA: AACE.

Yuan, T., Chae, H. S., Nantwi, G., and Garg, P. (2010). An examination of 300 iPhone math applications: Implications for the development of an educational math game. In *Proceedings of World Conference on Educational Multimedia, Hypermedia, and Telecommunications 2010* (pp. 4083-88). Chesapeake, VA: AACE.

Zhang, P., Recia, A., Willis, G., Smitham, P., Gilbert, L., Santer, M., and Grange, S. (2011). Catch before a fall – an iPad application for osteoporosis risk assessment. In *IADIS e-Society, Spain,* 10-13.

In: Handbook of Technology in Psychology … ISBN: 978-1-62100-004-4
Editor: Luciano L'Abate and David A. Kasier

Chapter 12

NEUROCOGNITIVE TECHNOLOGY: INFORMATION PROCESSING AND EVENT RELATED POTENTIALS

Jason S. Moser and Tim P. Moran
Michigan State University, US

The past several decades have seen event-related brain potentials (ERPs) become a useful tool for studying neural processes in real time. The historical roots of ERPs can be traced back to some of the earliest neurophysiological research done on human participants. Over the course of the past half-century, ERP research has become a mainstay in the field of cognitive neuroscience. Clinical applications of such techniques were also obvious to early neurophysiologists, and in recent years clinical psychophysiology has emerged as an independent and productive field in its own right. This chapter is aimed at introducing the ERP technique and describing its clinical applications.

A BIT OF ANCIENT HISTORY

Modern ERP work owes its beginnings to Hans Berger who recorded the first electrical potentials from the scalp of human participants (Berger, 1929). While recording resting electroencephalogram (EEG; i.e. scalp potentials recorded while participants are asked to simply sit still), Berger discovered alpha waves (brain activity oscillating at approximately 10Hz and characterizing a wakeful relaxed state). In subsequent work, Berger took the first EEG recordings during a seizure (Berger, 1933) thus introducing possible clinical applications of scalp recordings. The neurophysiologists of Berger's day were very skeptical of his work, as some thought his findings were simply an electrical artifact that would not replicate. However, several independent researchers subsequently replicated Berger's findings over the course of the next few years (Adrian and Matthews, 1934; Jasper and Carmichael, 1935; Gibbs, Davis, and Lennox, 1935). During this time, Gibbs et al. (1935) and Gibbs, Lennox, and Gibbs (1936) further elucidated the pathological EEG patterns indicative of seizure activity, marking the beginning of clinical electroencephalography. In 1939 physiologists recorded the first event-related potentials (ERPs; i.e. electrical potentials recorded in response to specific events) from humans listening to auditory stimuli (Davis, Davis, Loomey, Harvey and Hobart). The discovery that brain potentials could be modulated by external stimuli marked the beginning of experimental electrophysiology. The next several decades saw dramatic changes in ERP methodology, the most important of which was the increased use of computers in the 1960s and 1970s. In more recent years, researchers have begun applying the ERP technique to elucidating the mechanisms underlying behavioral/psychiatric disorders as well as neurological disorders.

BIOLOGICAL ORIGINS OF EEG AND ERPS

EEG consists of electrical potentials recorded at the scalp that typically originate in large populations of the pyramidal neurons of cortical structures. Specifically, EEG is the sum of post-synaptic potentials elicited by sensory, motor and cognitive processes (among others), the activation of which then passively propagates to the scalp via volume-conduction (cf. Luck, 2005). When neurotransmitters are released from pre-synaptic dendrites, current flows into the apical dendrites of the post-synaptic cell causing a net charge at the apical dendrites (whether this post-synaptic potential is negative or positive[1] depends on whether the neurotransmitter was excitatory or inhibitory, respectively). As a consequence of this, current of the opposite polarity will flow from the basal dendrites and cell body in order to complete the circuit. This results in the cell creating a small dipole (in this case a dipole is a cell with a positive end and a negative end). This single-cell dipole is far too small to be detected by a scalp electrode. Instead, it is necessary for multiple dipoles to summate in order for a scalp recording to be possible. Animal research shows that such summation or synchronization occurs within networks of neurons and therefore can propagate a large enough signal to the scalp (Llinas, 1988; Pizzagalli, 2007). A standard scalp electrode is approximately 10mm in diameter, which is considerably larger than a single neuron, which is approximately 20μm long. The area of a scalp electrode therefore covers thousands of neurons (Baillet, Mosher, and Leahy, 2001; Pizzagalli, 2007). Scalp-recorded potentials thus reflect the summation of voltages associated with a wide array of simultaneously active neurons and, therefore, manifold processes. However, it is possible to extract potentials associated with specific processes via a simple averaging procedure. The on-going EEG activity is time-locked, or anchored, to a specific, relevant stimulus and segmented for a given duration following the stimulus. Trials with the same stimuli are averaged together. Neural activity consistently following the stimulus will remain following the averaging procedure whereas unrelated, or random, neural activity will be attenuated or eliminated by the averaging procedure (Luck and Girelli, 1998). This EEG activity related to a specific event is referred to as an event-related potential (ERP). ERPs consist of several components – i.e., a series of positive and negative peaks following stimulus onset – that are defined by when in time and at what scalp positions they occur. It is possible to measure the effects of experimental manipulations on these

[1]The use of the terms "positive" and "negative" are often a point of confusion for those unfamiliar with ERP research. Positive and negative describe physical characteristics of electrical potentials and should be thought of as similar to describing the poles of magnets as "north" and "south;" it does not describe a good-bad dimension or affective valences. Furthermore, the absolute magnitude of the measured voltages does not confer useful information. It must be compared to some baseline. For example, a positive voltage may be described as a negativity if it is less positive (i.e. more negative) than the baseline condition.

components and thereby assess the cognitive procedures involved in the processing of events as well as their time course (Luck and Hillyard, 1994).

RECORDING AND ANALYZING EEG/ERPS

There are a variety of systems and techniques available to record and analyze electrical potentials. Some of the more common systems are briefly discussed here: BioSemi, Electro Geodesics International (EGI) and Advanced Neuro Technology (ANT) all produce a variety of recording systems including the state-of-the-art dense-array montages. Dense-array recording systems utilize between 64 and 256 electrodes placed around the scalp (around the entire head as the number of sensors increases) within approximately 2 cm of each other. Most companies also offer smaller arrays made up of 16 - 32 electrodes that are more affordable and still provide a wealth of functional brain data. Large numbers of sensors, however, allow researchers to estimate which cortical structure(s) give rise to the ERP(s) under investigation[2]. Although localization techniques can be useful, high-density systems do come with some drawbacks. As the number of sensors increases, the time required to setup each participant increases as well. One way companies like BioSemi and EGI have attempted to address the problem of long setup time for dense array montages is to develop high-impedance systems that require less setup time per electrode. Electrical impedance can be thought of as the extent to which a given material opposes the passage of electrical current. To deal with impedance, researchers often abrade participants' skin during setup, which has the effect of removing dead skin and other particulates on the epidermis thereby establishing a better connection between the sensor and the scalp. High-impedance systems, on the other hand, are capable of operating with minimal abrading and thus shorter setup time. Caution should be noted, however, as high-impedance systems have the potential to distort EEG signals. Moreover, Luck (2005) noted that the potential for human error also increases with the number of sensors. Even an advanced system will provide poor results if the quality of the setup and data collection is not monitored by a keen eye. Another company, Neuroscan, should be noted for its versatility. Neuroscan has developed an EEG system capable of taking scalp recordings from within an operating fMRI scanner. This allows for high temporal (i.e., EEG) and spatial (i.e., fMRI) resolution data to be recorded simultaneously.

Although there are some significant differences between these EEG systems, the data extracted from them all must go through many of the same processing steps. All EEG systems record scalp-potentials using electrodes that are composed of a conductive substance attached to a wire. The conductive substance is most commonly a silver-silverchloride (Ag-AgCl) compound although tin is common as well. These electrodes are typically affixed to participants' scalps via either a spandex or stretch-lycra cap. A conductive gel is used to bridge the gap between the sensor that is embedded in the cap and the scalp. Finally, before the signal is stored on a computer for further analyses, the signal must be digitized. EEG is

---

[2] While electroencephalography is not generally used to image specific brain regions, mathematical procedures have been developed to attempt to localize the neural source of the scalp-recorded activity. These findings should be interpreted carefully and always in light of functional magnetic resonance imaging (fMRI) findings (see Luck, 2005 for a more detailed treatment). It also should be noted that dense-array systems suffer from diminishing returns. The spatial resolution gained by adding additional sensors decreases greatly beyond approximately 60 sensors.

initially recorded as a continuous, unbroken wave. In order to be quantified, this wave must be digitized into discrete sampling points. This usually occurs at a rate between 256 and 1,024Hz (256-1,024 discrete samples per second). Although a higher sampling rate is generally more representative of the data, the Nyquist-Shannon Theorem states that the sampling rate only needs to be twice the frequency of the waveform under investigation (e.g., if examining theta, which ranges from 4-7 Hz, one would need to sample at least 14 times per second). Further processing occurs on specialized analysis software after data collection, or 'offline' as EEG/ERP researchers refer to it. Most EEG systems are sold with analysis software that is compatible with the hardware. ANT, for example, offers its own software (ASA) for offline analyses. Like most specialized analysis programs, ASA utilizes a simple, user-friendly interface for data analysis. Although this type of program is quite easy to use and requires little training, the interface tends to constrain the range of analyses that can be performed. In terms of stand-alone analysis packages, BrainVision Analyzer (BVA) is a powerful program that is compatible with a number of EEG/ERP data files and offers a number of signal analysis options. For the more programming-oriented researcher, more flexible programs, such as MatLab, are also available. MatLab is a versatile analysis tool utilizing a high-level programming language. Its versatility allows it to handle almost any analytic need, however, the programming language requires considerably more training and skill.

The first step in analysis is filtering. Human neural activity generally oscillates at a frequency less than 40Hz. High-frequency activity (typically greater than 100Hz) is generated by muscle activity and is therefore not of interest to most ERP researchers. It is also important to attenuate 60Hz (50Hz in Europe) activity that is generated by the electronic devices being used to deliver imperative stimuli to subjects during data collection. It is therefore common to set a low-pass filter (a filter that reduces the influence of frequencies above a given cut-off and "passes" frequencies below that cutoff) between 30 and 40Hz. It is also important to attenuate very low-frequency activity resulting from skin potentials such as those generated by sweat. Setting a high-pass filter (the opposite of a low-pass filter) between 0.01 and 1 Hz will take care of that. Other artifactual activity, such as voltage deflections caused by ocular activity, also needs to be removed from the recorded data. Like post-synaptic potentials, eyes can be thought of as dipoles. During eye-movements and blinks the orientation of the dipole shifts which causes measurable voltage changes across the scalp. These changes are generally much larger than brain activity and can obscure the activity of interest. There are several methods available to correct for ocular artifacts that utilize regression methods; one of the most commonly used was developed by Gratton, Coles and Donchin (1983). It is also fairly common to simply remove trials during which an ocular artifact occurred, however, a major problem that arises from this method is that data is being completely discarded and will ultimately limit the number of trials available for averaging and statistical analysis. Artifactual activity can also arise from random sources and this is handled by artifact rejection methods. The specific criteria for removing these artifacts vary from lab to lab but these methods tend to have several features in common: a cut-off is set for the maximal voltage change that will be allowed within a given period of time, for the minimum voltage change that will be allowed in a given period of time and for the maximum voltage change that will be allowed between contiguous sampling points. Finally the data are segmented with reference to the event (e.g., stimulus onset or the occurrence of a particular response) of interest and then averaged. The exact timing of each event is recorded along with the EEG data. Segments (or epochs) of a given duration are created, aligned based on the timing of the stimulus. The resulting stimulus-locked (or response-locked) brain activity is then averaged for each experimental condition. It is also common to perform a baseline-correction procedure. The absolute voltage of a waveform can be influenced by ongoing activity preceding stimulus onset. For example, if the ongoing activity preceding a stimulus is consistently a negative value then the stimulus-locked activity will be more negative (or less positive) than it otherwise would have been. Thus, the average activity for a given time-period preceding the stimulus is averaged and subtracted from the waveform to reduce the influence of ongoing activity.

Figure 1 depicts several commonly studied ERP components that result after the above-described processing procedures are performed. This waveform was computed by averaging the brain activity following the presentation of a stimulus across several like trials. Most early components (those occurring shortly after stimulus presentation) are named for their polarity and ordinal position. For example, the "P1" is the first positive peak[3] and the "N2" is the second negative peak occurring in the waveform.

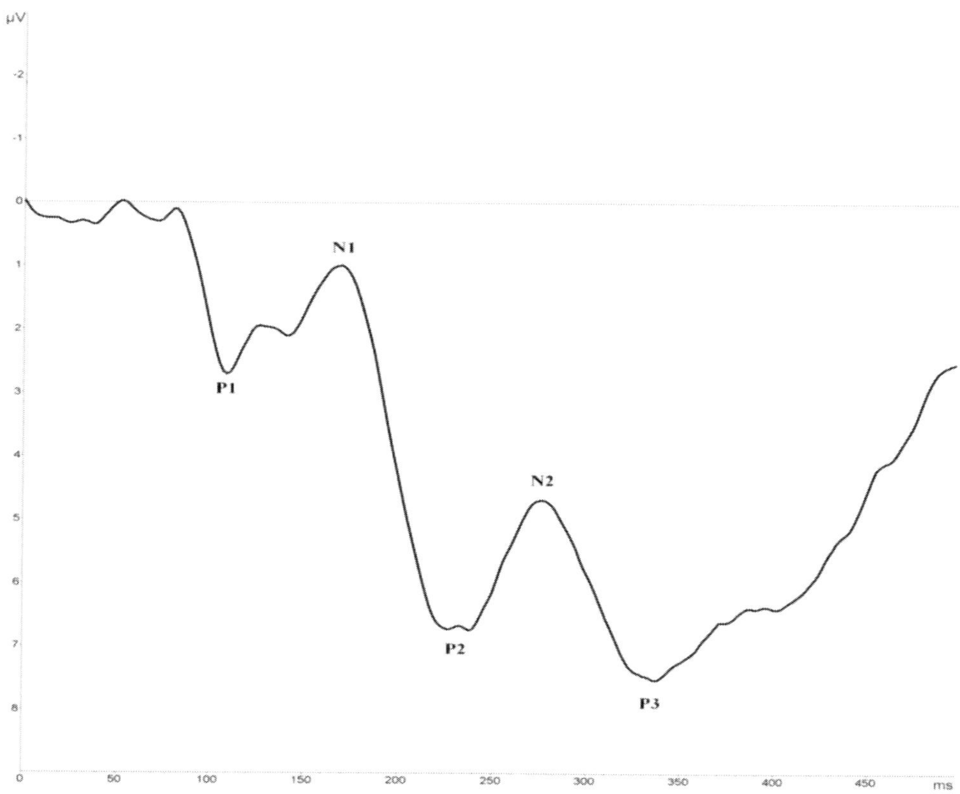

Figure 1. An averaged waveform elicited by the presentation of a stimulus. Positive and negative peaks (labeled) are considered to be separate components representing different sensory and cognitive processes. It should be noted that ERPs are plotted upside-down such that negative values are plotted above 0 and positive values below 0.

[3] All disciplines have their idiosyncrasies; radiologists, for example, display the left side of images on the right and vice versa. ERP researchers are no different. It is customary in ERP research to plot negative above the x-axis and positive below the x-axis. Thus, a positive-going deflection, such as the P1, will be plotted as a downward deflection.

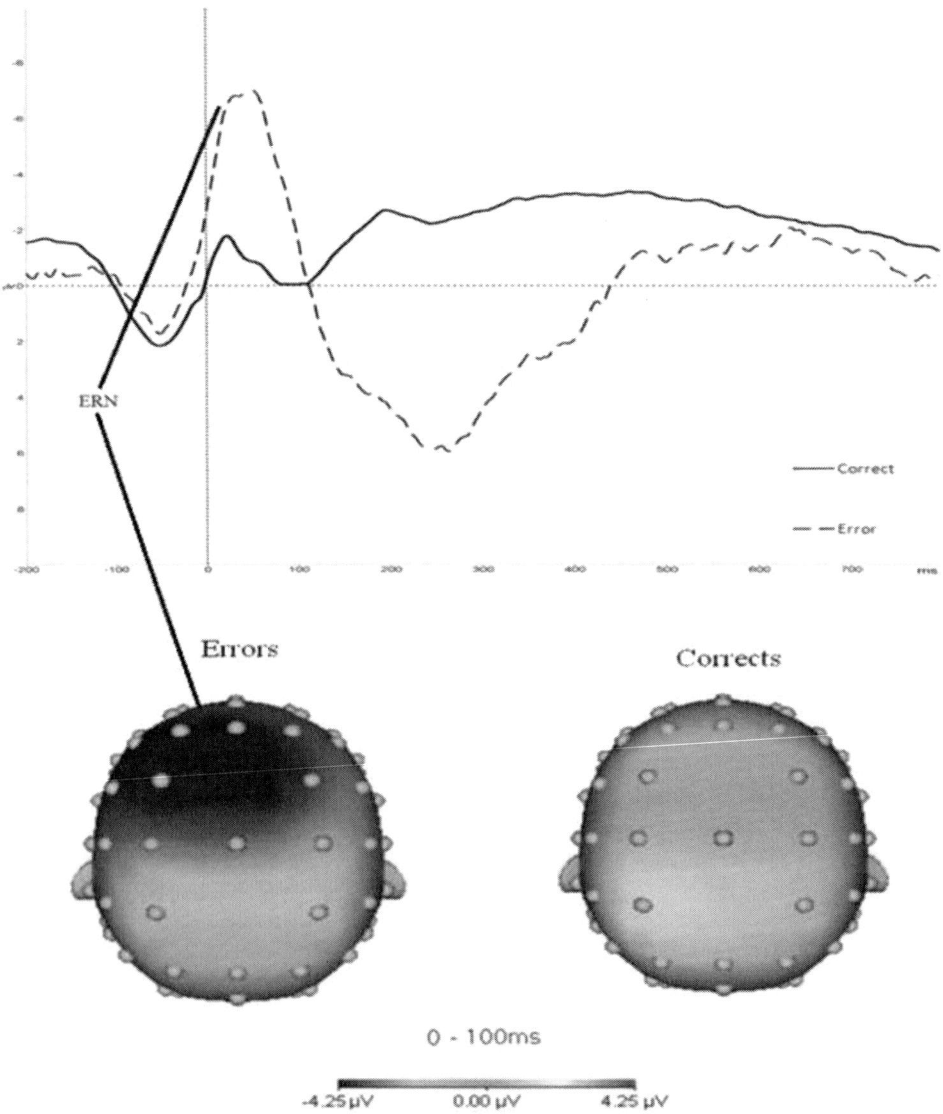

Figure 2. The error-related negativity (ERN), a negative-going deflection peaking approximately 50ms following the commission of an erroneous response, is depicted in the top panel by the broken line. The scalp topography of the ERN (bottom left) is observed as a broad negativity occurring over fronto-central electrode sites consistent with findings implicating the anterior cingulate cortex as the neural generator of the ERN. Conversely, the scalp topography following correct responses (bottom right) displays relatively little activity.

There are a few exceptions to this general naming-scheme, however. Some components, like the N170 (not shown), are named for their polarity and latency following the stimulus. Other components, such as the Error-Related Negativity (ERN; depicted in Figure 2) are named for their functional significance.

Many components are defined by visual inspection of the waveform and identification of their onset and offsets. This method, however, is fraught with subjectivity for obvious reasons. To avoid complete subjectivity, it is helpful to reference existing literature to aid component identification. Another method has been used to define components by their functional significance. To do this, researchers engage in what some have dubbed "ERPology", that is, the study of the ERPs themselves (Luck, 2005). By determining what manipulations affect a component, investigators are able to determine which cognitive processes are represented in the recorded activity. Doing this has taught us a great deal about human neural activity, but it should be met with some caution. As noted earlier, potentials recorded at the scalp reflect the summated activity of a large number of neurons and therefore are not representative of unitary processes. Furthermore, contextual factors can also influence the waveform. For example, many of the early components are not modality-independent; the processes reflected in components elicited by visual stimuli are not necessarily the same as those elicited by auditory stimuli. Finally, some researchers apply mathematical procedures such as Principal Components Analysis (PCA) to isolate components.

Whether identifying components by visual inspection, ERPology, or PCA, the final step in ERP processing is quantification. Components are typically quantified by 'peak picking' – taking the maximum voltage in a given time window – or averaging – taking the average activity in a given time window – procedures. Luck (2005) has outlined the advantages and disadvantages of peak picking versus averaging and favors averaging, as do we.

WHY USE ERPS?

Information processing models of adaptive and maladaptive psychological functioning have had a major impact on our understanding of human nature. For example, Broadbent's (1958) Filter Model of selective attention and Atkinson and Shiffrin's (1968) Modal Model of memory have shaped the way we understand and talk about cognitive processes, even in our everyday vernacular.

Building on early theories of basic cognitive processes, Foa and Kozak (1986) proposed an information processing model of fear and anxiety that has significantly influenced the way researchers and clinicians alike conceptualize and treat anxiety disorders. All of these theories share a focus on the passage of information through simple to more complex stages that are instantiated in the brain.

However, these theories were proposed prior to the mainstream use of neuroimaging techniques and therefore relied mostly on behavioral (indirect) measures to substantiate their claims. The mainstream use of neuroimaging technology has since aided in directly testing the primary tenets of such theories, leading to a more refined understanding of ourselves. Thus, neuroimaging technology has enabled us to begin opening up the 'black box' of the mind/brain and acquire direct measures of information processing. The growing trends in agency funding further underscore the importance of directly measuring the brain at work.

Given the dominance of information processing models as a way to understand normal and abnormal behavior and the focus on direct (brain) measures, ERPs represent an ideal means by which to contribute to progress in psychological science. Specifically, ERPs are particularly well suited to directly study information processing for at least three reasons: 1) ERPs are direct measures of online neural activity, 2) ERPs are characterized by excellent temporal resolution, and 3) as a corollary to #2, the ERP waveform can be decomposed into specific 'components' that allow for the examination of the sequence of constituent operations involved in stimulus- and response- processing on the order of milliseconds. Therefore, ERPs provide a more precise measure of information processing by separating out processes involved in stimulus evaluation from those of response execution that are blurred by other measures such as reaction time and fMRI. Moreover, ERPs are less expensive and invasive to record than other neuroimaging technologies like fMRI and positron emission tomography (PET) and therefore have a greater potential to be utilized in various populations and contexts.

ERPs have been utilized in a number of areas of research. Most relevant to the current chapter, ERPs have been used by researchers to contribute greatly to information processing models of attention (for a review see Hopfinger, Luck, and Hillyard, 2004), language (for a review see Federmeier and Kutas, 2001), and emotion (for a reviews see Eimer and Holmes, 2007 and Olofsson, Nordin, Sequeira, and Polich, 2008). Thus, ERPs already have a longstanding presence in psychological science and a solid base of evidence around which to build wide-reaching applications.

With regard to their utility in clinical research, ERPs have several attractive attributes. That ERPs provide temporally sensitive measures of information processing is of great import to testing information processing theories of psychopathology that dominate the literature. For example, Eysenck and colleagues have proposed an 'Attentional Control Theory' of anxiety (Eysenck, Derakshan, Santos, and Calvo, 2007) that posits that anxiety is associated with impairments in the efficiency of particular executive functions. Eysenck and Derakshan specifically say that it is important to "...ensure that the differences in processing between high-anxious and low-anxious individuals can be identified as precisely as possible" (p.5) by measures with excellent temporal resolution, the purpose for which ERPs are principally designed. 'Vigilance-avoidance' models of anxiety (e.g., Mogg and Bradley, 1998) suggest that anxious individuals initially allocate preferential attention to threatening stimuli and subsequently avoid these same stimuli. ERPs are ideally suited for testing such models, as they can detect covert shifts in attention over time with millisecond precision. Recent studies have begun using ERPs in this way and provide preliminary evidence for vigilance-avoidance dynamics in anxiety (Adenauer et al., 2010; Holmes, Nielson, and Green, 2008; Mueller et al., 2009; Moser, Huppert, Duval, and Simons, 2008). In general, ERPs may be especially sensitive to detecting the presence of processing abnormalities in psychopathological groups. Several studies have demonstrated ERP differences between negative affective (anxious and depressed) and control groups in the face of comparable behavioral performance (Fallgatter et al., 2004; Gehring, Himle, and Nisenson, 2000; Hajcak, McDonald, and Simons, 2004; Moser, Hajcak, Huppert, Foa, and Simons, 2008; Weinberg, Olvet, and Hajcak, 2010).

Finally, that ERPs are less expensive and invasive than other neuroimaging techniques is important for reaching a broad range of clinical populations. The relative lower cost of ERP data collection is important to collecting large samples that will ultimately contribute to determining the generalizability of findings and to linking ERP findings to theories derived from large-scale interview and questionnaire studies (e.g., Krueger, 1999). Being less invasive than other neuroimaging techniques has clear advantages for clinical patients who, for example, may not feel comfortable in tight places or exposed to loud noises, as is the case in fMRI. The promise of ERPs to have an important impact on the study and practice of psychological clinical science is therefore strong.

WHO BENEFITS FROM ERP RESEARCH?

ERPs have been recorded in humans as young as four months old (e.g., Reynolds, Courage, and Richards, 2010) and as old as 80 years old (e.g., Polich, Howard, and Starr, 1985). Thus, ERPs have been and can be used to illuminate information processing abilities across the lifespan. Although much of the ERP research relevant to the study of psychopathology has focused on individuals in early- to middle-adulthood, there is a growing body of ERP evidence in childhood psychopathology. For instance, ERPs have even been used in temperamentally difficult infants as young as nine months old (Marshall, Reeb, and Fox, 2009).

As with many other areas of clinical psychology and psychiatry (and psychology in general), the trend in ERP research is to: 1) begin with convenient samples such as college students chosen high or low on a particular measure of psychopathology, 2) extend to adult clinical patients, and 3) finally, downward extend the paradigms to younger populations. This progression makes fine intuitive and practical sense. The use of convenient samples allows for the collection of large amounts of data in order to uncover and replicate candidate ERP correlates of psychopathology for later use in (the more difficult task of extending to) patients and children. We review ERP studies of psychopathology more thoroughly in the next section.

Unfortunately, given the preponderance of females in college psychology courses and psychological clinics, most ERP studies are underpowered to test sex differences in information processing and, most definitely, to test sex differences in ERP-psychopathology associations. Some ERP studies that have directly tested sex differences in information processing have demonstrated differences between males and females. For example, Hoffman and Polich (1999) found larger P300 amplitudes – an index of attention processing – in females than males. Some have shown that sex differences in the P300 are attributable to seasonal variations, perhaps coinciding with hormonal changes (Deldin, Duncan, and Miller, 1994). In general, however, sex differences in ERPs are understudied. More importantly, there are very few studies examining sex differences in the relationships between ERPs and psychopathology despite overwhelming evidence indicating sex differences in the nature and expression of mental health problems (e.g., Hartung and Widiger, 1998). The role of sex as a moderator of relationships between ERPs and psychopathology represents an exciting area for future research.

With regard to the use of ERPs to study specific clinical diagnoses, there have been a number of applications. ERPs have been used in anxiety (e.g., Gehring et al.,, 2000) and depression (e.g., Holmes and Pizzagalli, 2008), schizophrenia (e.g., Ford, White, Kim, and Pfefferbaum, 1994), substance use problems (e.g., Polich, Pollock, and Bloom, 1994), personality disorders (e.g., De Bruijn et al., 2006), attention deficit-hyperactivity disorder (e.g., Albrecht et al., 2008), traumatic brain injury (e.g., Perlstein, Larson, Dotson, and Kelly, 2006), as well as neurological conditions, most notably, paralysis (e.g., Birbaumer et al., 1999). ERPs have thus far contributed to elucidating the pathophysiology of a number of clinical problems.

In the following section, we focus largely on the ERN and P300 (or P3) as the respective relationships these components bear to psychopathology are some of the most extensively studied. For a more exhaustive review, please see Hansenne (2006). As depicted in Figure 2, the ERN is a negative-going deflection peaking approximately 50ms following the commission of an error in an arbitrarily defined task. The ERN is thought to reflect a basic action-monitoring process generated in the anterior cingulate cortex (ACC; Falkenstein, Hohnsbein, Hoorman, and Blanke, 1991; Gehring, Coles, Meyer, and Donchin, 1990; Figure 2 also depicts the scalp topography of voltages associated with the commission of an error). The P300 is a slow, positive-going wave that peaks approximately 300ms following the presentation of task-relevant stimuli and is thought to be an index of attention, orienting and memory updating (Donchin, 1981; Donchin and Coles, 1988; see Figure 1).

Clinical Investigations Using ERPs

Much of the early research on the exact function the ERN focused solely on cognitive mechanisms. Some believe that the ERN is a negative reinforcement signal following an erroneous response (Holroyd and Coles, 2002) whereas others believe it reflects a response-conflict signal, indexing the coactivation of mutually exclusive responses (Yeung, Botvinick, and Cohen, 2004). Recent evidence has also revealed that the ERN is sensitive to psychopathological and affective variables.

The relationship between the ERN and anxiety has been well established. In 2000, Gehring, Himle, and Nisenson showed that outpatients diagnosed with obsessive-compulsive disorder (OCD) displayed a potentiated ERN when compared with healthy controls. Subsequent research by Hajcak and Simons (2002) extended these findings to college students who displayed obsessive-compulsive characteristics. Hajcak, Franklin, Foa, and Simons (2008) also demonstrated that the enhanced ERN observed in OCD is not diminished by successful treatment, suggesting that enhanced ERN is a trait marker of OCD. fMRI studies showing that hyperactivation of the ACC – the area of the brain responsible for the generation of the ERN (Posner and Rothbart, 1998; Dehaene, Posner, Tucker, 1994) – is related to OC symptoms (Breiter, et. al., 1996) further highlight the role of aberrant response monitoring in OCD. Enhanced ERN is not confined to participants experiencing OCD or OCD-like symptoms, however. Chronic worriers (Hajcak, McDonald, and Simons, 2003) and individuals diagnosed with generalized anxiety disorder (GAD; Weinberg, et al., 2010) also demonstrate enhanced ERN amplitudes. Enhanced ERN only seems to characterize individuals with trait anxiety, as inducing transient fear states in phobic individuals does not increase the magnitude of the ERN (Moser, Hajcak, and Simons, 2005). We have previously suggested that the enhanced ERN in anxiety may therefore reflect a biological marker of inflated concerns about/overvaluation of mistakes that are often reported by anxious individuals (Hajcak, Moser, Yeung, and Simons, 2005).

A potentiated ERN is not limited to anxiety disorders; it has also been observed in depressed individuals. Holmes and Pizzagalli (2008) and Chiu and Deldin (2007), for instance, found increased ERN amplitude in depressed patients. Similarly, fMRI data suggest increased ACC activity in depression (Steele and Lawrie, 2004). Some have proposed that these findings reflect the negative biases associated with depression that are thought to play an important role in the etiology and maintenance of depressive symptoms (Beck, 1967; Ingram, Miranda, and Segal, 2006).

Due to the high degree of symptom overlap and comorbidity between anxiety and depression, Clark and Watson (1991)'s tripartite model proposed that both anxiety and depression are characterized by high negative affect. Thus, a parsimonious explanation may be that an enhanced ERN is a characteristic of the negative affect underlying both anxiety and depression. This possibility has found some empirical support; to date, two studies have found that a potentiated ERN is related to self-report measures of negative affect (Luu, Collins, and Tucker, 2000; Hajcak et. al, 2004).

Apart from the negative affective/internalizing problems, some researchers have also examined relations between the ERN and 'externalizing' problems (Krueger, 1999; Krueger, McGue, and Iacono, 2001) such as substance use disorders and antisocial personality. Franken and colleagues (2007) found an attenuated ERN in patients being treated for cocaine dependency, a finding consistent with other imaging studies showing reduced ACC activation in individuals with a number of different substance use problems (e.g., Goldstein et al., 2007). Importantly, Hall, Bernat, and Patrick (2007) found that a reduced ERN was associated with increased externalizing scores on a composite self-report measure representing various aspects of the externalizing construct. Individuals with externalizing problems are often characterized by impulsivity and lack of behavioral constraint (Moeller, et. al., 2001) as well as diminished sensitivity to punishment (Corr, 2002). Consistent with this, Dikman and Allen (2000) found a reduced ERN in participants scoring in the bottom 3% of Gough's (1960) socialization scale during a punishment condition. Studies using other measures of disinhibitory personality such as impulsivity scales have found consistent results (Potts, George, Martin, and Barratt 2006; Ruchsow, Spitzer, Gron, Grothe, and Kiefe, 2005). Thus, externalizers appear to suffer from the opposite problem of internalizing individuals in that they are under-responsive to mistakes.

Given the relationships between the ERN and psychopathology, Olvet and Hajcak (2008) recently proposed that the ERN might represent an endophenotype for various forms of psychopathology. Endophenotypes are unobservable traits that mediate the relationship between genes and phenotypic characteristics such as overt behaviors (Gottesman and Gould, 2003). Pathological presentations have proven to be complex and heterogeneous and are very likely multigenic in origin (Chakravarti and Little, 2003). Endophenotypes are helpful in this regard as they generally reflect simpler, more homogenous processes and are determined by a smaller set of genes. Olvet and Hajcak (2008) proposed that the ERN may be a useful endophenotype for studying internalizing and externalizing psychopathology insofar as it reflects action-monitoring and self-regulation abnormalities that may mediate the relationship between genetic predispositions and disease.

The P300 has also been a valuable tool in understanding psychopathology. Levit, Sutton, and Zubin (1973) were some of the first to show attenuated P3 amplitude in depressed participants during an auditory oddball task in which a unique tone is presented within a series of like tones. However, a number of studies that followed failed to replicate this finding (Bruder et. al., 1991; Have, Kolbeinson, and Pétursson, 1991). Other studies have found slower reaction times in depression but no differences in P3 amplitude or latency (Gangadhar, Ancy, Janakiramaiah, and Umapathy, 1993; Bolz and Giedke, 1981) suggesting abnormalities in response selection and preparation rather than stimulus evaluation. Thus, it is currently unclear to what extent the P3 reveals altered information processing in depression.

Research focusing on the relationship between the P3 and anxiety disorders has been somewhat more extensive than that focusing on depression. One of the earliest studies found a reduced P3 to visual stimuli in OCD patients (Beech, Ciesielski, and Gordon, 1983). Towey et. al. (1993) later replicated the P3 abnormalities in OCD using auditory stimuli. Taken together, these findings suggest abnormal attentional control in OCD. This is further supported by findings that a reduced P3 amplitude is predictive of attentional control deficits (Kim et. al., 2003). Studies investigating the P3 in panic disorder and PTSD have, to date, found inconsistent results. Clark, McFarlane, Weber, and Battersby (1996) found that patients suffering from panic disorder exhibited a potentiated P3 in response to infrequent tones. This was interpreted as evidence for enhanced automatic processing or perhaps an inability to filter stimuli that would otherwise not reach conscious awareness. This is consistent with clinical conceptualizations of panic disorder which characterize patients as generally unable to integrate complex exogenous stimuli and are distracted by extraneous stimuli. However, several subsequent studies have failed to find this enhanced P3 (Iwanami, Isono, Okajima, and Kamijima, 1997; Wang, et al., 2003). Several studies, on the other hand, have shown attenuated P3 amplitudes in PTSD patients (Mcfarlane, Weber, and Clark, 1993; Felmingham, Bryant, Kendall, and Gordon, 2002). However, subsequent studies were unable to reproduce these findings (Kimble, Kaloupek, Kaufman, and Deldin, 2000; Neylan, et. al., 2003). More research is needed in this area to clarify these contradictory findings.

Aberrations in P3 amplitude have also been fairly consistently observed within the externalizing spectrum. Male participants with antisocial personality disorder (APD) exhibit an attenuated P3 (Bauer, 1997) and the reduction in P3 amplitude was predicted by symptom severity (Costa, et. al., 2000). Similar results have also been observed in adolescents with conduct disorder, APD's developmental antecedent (Bauer and Hesselbrock, 1999). Importantly, reduced P3 has also been related to the latent externalizing construct underlying these psychopathologies (Gilmore, Malone, Bernat, and Iacono, 2010). In fact, evidence from adolescent studies suggests reduced P300 may be an endophenotype of externalizing psychopathology (Hicks et al., 2007).

Although the P3 has been used to index purely cognitive deficits in psychopathological groups, there is a growing body of research showing the utility of the P3 in revealing biases in processing emotionally relevant stimuli. For instance, enhanced P3 amplitudes to fear-relevant stimuli have been demonstrated in socially anxious (Moser et al., 2008) and spider phobic (e.g., Kolassa, Musial, Mohr, Trippe, and Miltner, 2005) individuals. Enhanced P3 amplitude has also been shown in alcoholics while viewing alcohol cues (Hermann, Weijers, Wiesbeck, Böning, and Fallgatter, 2001). Thus, although the P3 has not revealed as many consistent results as some my have hoped, the strongest data show a relationship between reduced P3 and externalizing problems. Hansenne (2006) has proposed that many of the inconsistent findings may owe partially to heterogeneity in symptom criteria and methodologies. Thus, one explanation as to these disparate findings is that abnormalities in the P3 are related to specific subtypes or symptom-clusters. For example, Pierson et al. (1996) found the largest modulation of the P3 in patients with symptoms relating to blunted affect.

DIRECTIONS FOR FUTURE RESEARCH

To this point, we have provided a background on what ERPs are, what they are good for, and how they have already contributed to our understanding of normal and abnormal information processing. In the current section we focus our sights on the future and address the question: Where do we go from here? While we recognize that there are innumerable ways to address this question, we hone our thoughts on what we consider some of the most important prospects for advancing ERP studies of psychopathology in particular (and individual differences more broadly). We begin by highlighting the importance of being conceptually rigorous and follow with concrete ideas for practical applications of clinical electrophysiology.

Perhaps most importantly, ERP researchers should strive to develop thoughtful theoretical models that clearly articulate what important and specific biopsychosocial processes ERPs reflect. The ERN is a good example of an ERP component that has attracted a substantial amount of attention in clinical/individual differences research, but suffers from the lack of an integrative theory of its functional significance across different groups of individuals. For instance, an enhanced ERN characterizes anxious and depressed individuals (cf. Olvet and Hajcak, 2008). However, an enhanced ERN has also been found in individuals with higher school achievement (Hirsh and Inzlicht, 2010) and those who demonstrate greater daily stress regulation (Compton et al., 2008). On the other hand, reduced ERNs have been reported in people with a strong belief in god (Inzlicht, McGregor, Hirsh, and Nash, 2009) and those reporting higher levels of satisfaction with life (Larson, Good, and Fair, 2010) as well as externalizers (Hall et al., 2007) and borderline personality disorder patients (de Bruijn et al., 2006). Thus, the magnitude of the ERN does not discriminate 'healthy' from 'unhealthy' individuals very well. Having a relatively large ERN could mean you are good at regulating daily stress or chronically bad at it. This begs at least two questions: 1) what does the magnitude of the ERN reflect, and 2) how useful will the ERN be as a diagnostic tool? Cognitive neuroscientists have debated the ERN's functional significance, with some arguing that it reflects the detection of conflict between correct and erroneous responses (Yeung et al., 2004) whereas others argue that it reflects a reinforcement-learning signal (Holroyd and Coles, 2002). Unfortunately, researchers from the clinical/individual differences subfield have failed to develop an integrative model that draws on existing computational models of the ERN (Yeung et al., 2004) or on existing comprehensive models of particular psychological problems (e.g., Attentional Control Theory of Anxiety; Eysenck et al., 2007). Thus, no one to date has accounted for the disparate ERN findings in different groups. As with most other cognitive abnormalities, the ERN will have to be interpreted in the context of other relevant data (e.g., behavior) to fully appreciate its role in healthy and unhealthy information processing. Researchers should be more cognizant of the multitude of findings that already exist and work from extant theories to develop thoughtful predictions when applying such ERPs to new groups of individuals or in novel contexts.

Practically, our view is that ERP researchers should adopt methods utilized with other sources of data (e.g., self-report). One of the biggest limitations to the clinically relevant ERP findings to date is the fact that many were revealed in small samples. It is not that we doubt the validity of the extant evidence, rather small samples greatly restrict the ability to examine potentially important moderating variables such as sex and ethnicity that have been shown to affect the nature and expression of psychopathology (Hartung and Widiger, 1998) and ERP morphology (e.g., Deldin et al., 1994). Moreover, because of the small, between-groups designs typically employed in ERP studies of psychopathology, the analysis strategies that can be applied to the data are limited. A discussion of concrete examples for future directions follows.

The internalizing-externalizing model described above (cf. Krueger, 1999) was developed by submitting large questionnaire and interview data to factor analyses. This model draws on decades of research from personality psychology suggesting that covariance between seemingly distinct psychological problems arises from shared underlying processes (e.g., Millon, 1969). In line with these findings, research within diagnostic categories has also shown that mental disorders are better conceptualized as representing dimensions along a continuum from mild to extreme rather than discrete categories (e.g., Ruscio, Ruscio, and Keane, 2002). A major implication of this research is a primary motivation for collecting larger ERP datasets: that shared, core psychopathological processes likely represent more viable targets for mapping onto biological processes than highly overlapping categorical disorders.

Future studies should therefore marry methods from research on ERP indicators of disease that has historically been based on small-sample experimental designs with methods from research on the structure of psychopathology that has historically been based on large-sample self-report and interview correlational designs. Others have called for such a marriage between methods in the startle eye blink literature (e.g., Vaidyanathan, Patrick, and Cuthbert, 2009). The implications of this approach are that researchers can begin to evaluate convergence between units of analysis (i.e., subjective report and biological assay) and can use such findings to feed back into models of psychopathology in order to bring into focus more valid conceptualizations and biological markers of disease for use in diagnostics, treatment, and prevention. Findings from these sorts of large-scale studies can also feed back into models of the psychological processes reflected in ERPs and thus aid in construct validation efforts. To realize such goals, future research should involve the collection of ERP measures in large and diverse samples of individuals representing a range of psychopathological symptoms and employ statistical modeling techniques aimed at deriving meaningful, parsimonious structure (e.g., factor analysis).

We are aware of only one study (resulting in two publications) mapping ERP measures onto dimensions of psychopathology in adults. As described above, Hall et al. (2007) showed that higher scores on a self-report measure of externalizing problems was associated with reduced magnitude of the ERN in a sample of college students. Nelson, Patrick, and Bernat (in press) used the same dataset and showed that reductions in both the ERN and P300 were associated with increased self-reported externalizing problems. Based on the superior fit of dimensional models of psychopathology and the initial work that has demonstrated relationships between the externalizing dimension and ERPs, additional studies in this area are surely warranted.

Another specific example of a future direction for ERP research is to adopt a multivariate approach. Again, we discuss the application of this approach to psychopathology/individual differences research, however, the multivariate approach could also be employed for the sake of better understand the construct(s) reflected by certain ERPs. Unlike behavioral studies of psychopathology that commonly use multiple measures for assessing constructs, psychophysiological studies routinely use a single task to elicit a few components – akin to subscales of a single behavioral instrument. This is most certainly the case for ERP studies of psychopathology and personality. To our knowledge, there is only one ERP study of adult psychopathology or personality that has recorded ERP components from multiple tasks. Nelson et al. (in press) showed that an ERP composite made up of the ERN measured in a flanker task, the P300 measured in a flanker task, and the P300 measured in a feedback task was a better predictor of externalizing psychopathology than any of the individual ERP scores.

It is disappointing that only one study to date has employed this multivariate approach, as a single ERP component is limited in its ability to significantly increment current clinical methods. As Hansenne (2006) remarked in his review of ERPs in psychopathology research, "grouping two or more different ERP components [across different tasks] would greatly improve the clinical usefulness of brain potentials" (p. 24). As demonstrated by Nelson et al. (in press), the multivariate approach capitalizes on the increasing power of combining measures. Ultimately this approach will strengthen diagnostic and treatment methods and aid in the search for target genes, as these candidate endophenotypes – the P300 and ERN – lie closer to gene action. More ERP researchers should also consider employing multiple time-point designs. Longitudinal studies in which ERPs are used to predict psychological outcomes at future time-points as Compton et al. (2008) did – using the ERN to predict subjects' ability to regulate daily stressors over a week – will help establish the prognostic validity of ERPs. ERPs could be used to help predict the onset and course of psychological problems. For instance, ERPs recorded shortly after or before a traumatic experience could be used to predict the onset and course of posttraumatic stress symptoms. ERPs can also be used to index response to psychological or psychopharmacological treatment. One recent study showed that P300-related activity changed with successful treatment of adult spider phobia (Leutgeb, Schäfer, and Schienle, 2009) whereas another recent study showed no change in the ERN after successful treatment of pediatric obsessive-compulsive disorder, which suggests that the ERN is a trait, but not state indicator, of anxiety (Hajcak, Franklin, Foa, and Simons, 2008). Future studies using ERPs as indicators of treatment response will be important for better understanding the mechanisms of change in common psychotherapies as well as to further illuminate the role of ERPs in the pathogenesis of psychological conditions.

An interesting avenue for future research that hasn't been studied to date is to use ERPs as indices of cognitive change in cognitive training paradigms. For example, a large body of evidence demonstrates biased attention to threat in anxiety (Mathews and MacLeod, 2005). Recent studies have shown that this threat bias can be re-trained away from threat, and that this un-training of the threat bias is associated with significant decreases in anxiety symptoms (Amir, Beard, Burns, and Bomyea, 2009; Amir et al., 2009; Schmidt, Richey, Buckner, and Timpano, 2009). To date, however, the assessment of change in attention bias by cognitive training programs has been restricted to reaction time. Future studies should also include ERP measures in cognitive training programs, as ERPs can provide a more refined picture of changes in specific attention processes and have often proved to be more sensitive measures of cognitive abnormalities (e.g., Falgatter et al., 2004). ERPs have already been shown to reveal attention biases in anxiety (e.g., Moser et al., 2008), and thus the logical next step would be to include ERPs in cognitive training studies.

Considering Incorporating ERPs into your Research Program?

Above we have attempted to make the case that ERPs are an important biological tool for studying normal and abnormal information processing, and we also provided some possibilities for moving the field forward. Convinced? If you might be interested in incorporating ERPs into your research program, what next? There are undoubtedly numerous paths to take, however, we sketch some important considerations and steps one could take toward integrating ERPs into one's work.

The researcher should start by doing some reading so as to better appreciate the cost-benefit ratio for adopting ERPs. Obviously, ERPs are a more costly methodology to adopt than paper and pencil questionnaires and computer tasks and thus the first item on one's agenda should be to determine whether he/she (and his/her host institution) could afford the expense. As mentioned above, ERP systems range from basic (16-32 electrodes) to more sophisticated (64-256 electrodes). Perusing the websites of the companies referenced above would be a good start.

Nonetheless, many of the most widely studied ERPs such as the P300 and ERN can be recorded with basic setups. Another consideration is space. Recording ERPs is usually achieved by having two adjacent rooms: a subject running room and a control room. Thus, the researcher would have to have access to such a space. Also, some researchers install metal shielding – which adds additional cost to setup – to protect against electrical noise interference, however, many systems (e.g., BioSemi) do not require such shielding for acceptable recordings.

In terms of evaluating the utility of ERPs, we have attempted to provide a very convincing case for why ERPs are useful in the study of normal and abnormal information processing. However, one should ask him/herself whether ERPs are best suited to answer his/her most pressing research questions. If the timing and function of particular cognitive and affective processes is central to one's research, then ERPs are most definitely worth investing in. Also, don't do it alone. The researcher should establish and maintain relationships with investigators who already use ERPs. This could be achieved by simply discussing the method/technology over coffee with a colleague, should you work with someone who already uses the technique. Writing to established researchers in the field is also a good way to get connected to the method, and if you should be so lucky to reside close to one you could make visits to his/her laboratory for instruction and demonstrations and likewise have him/her visit your lab to help troubleshoot. Finally, attending conferences where ERP research is presented (Society for Psychophysiological Research's annual meeting; see http://sprweb.org) and specialized ERP workshops (e.g., Steven Luck's ERP Bootcamp; see http://erpinfo.org) will provide perhaps some of the best ways to gain knowledge of and access to those who use the methodology. Maintaining such relationships will be important to one's long-term success with the method. As a final practical note, many, if not all, of the systems referenced above are manufactured and distributed by thoughtful and helpful individuals. Maintaining contact and healthy relations with your manufacturer/distributor will also aid in your long-term success in applying ERPs to whatever strikes your fancy.

CONCLUSION

ERPs are direct measures of neural activity characterized by excellent temporal resolution, revealing cognitive and affective processing stages on the order of milliseconds. ERPs are relatively non-invasive and inexpensive measures of neural activity compared to other neuroimaging techniques such as fMRI. They are principally useful for scientists studying and applying information processing models in which the identification of effects on discrete neural functions is of primary import. Numerous successful applications of the ERP technique to individuals of varying age, sex, and clinical syndrome across a variety of contexts have already been documented. Future investigations should be built on this solid base and be focused on testing and refining integrative theoretical models and take cues from methods employed with other sources of data (e.g., self-report) in terms of analysis and application. Practically, aspiring 'ERPers' should think critically about how ERPs could increment their existing research program and if they are worth the time, effort and expense. Surely, we think they are, and we hope that ERPs continue to be used to advance psychological science for years to come.

REFERENCES

Adenauer, H., Pinosch, S., Catani, C., Gola, H., Keil, J., Kiβler, J., and Neuner, F. (2010). Early processing of threat cues in posttraumatic stress disorder- Evidence for a cortical vigilance – Avoidance reaction. *Biological Psychiatry*, *68*, 451-458.

Adrian, E.D., and Matthews, B.H.C. (1934). The Berger rhythm: Potential changes from the occipital lobes in man. *Brain*, *57*, 355-385.

Albrecht, B., Brandeis, D., Uebel, H., Heinrich, H., Mueller, U. C., Hasselhorn, M., et al. (2008). Action monitoring in boys with attention-deficit/hyperactivity disorder, their nonaffected siblings, and normal control subjects: Evidence for an endophenotype. *Biological Psychiatry, 64*, 615-625.

Amir, N., Beard, C., Burns, M., and Bomyea, J. (2009). Attention modification program in individuals with generalized anxiety disorder. *Journal of Abnormal Psychology*, *118*, 28-33.

Amir, N., Beard, C., Taylor, C.T., Klumpp, H., Elias, J., Burns, M., & Chen, X. (2009). Attention training in individuals with generalized social phobia: A randomized controlled trial. *Journal of Consulting and Clinical Psychology, 77(5)*, 961-973.

Atkinson, R. C., and Shiffrin, R. M. (1968). Human memory: A proposed system and its control processes. In K. W. Spence (Ed.), *The psychology of learning and motivation: Advances in research and theory* (pp. 89-195). New York: Academic Press.

Baillet, S., Mosher, J.C. and Leahy, R.M. (2001). Electromagnetic brain mapping. *IEEE Signal Processing Magazine*, *18*, 14-30.

Bauer, L.O. (1997). Frontal P300 decrements, childhood conduct disorder, family history, and the prediction of relapse among abstinent cocaine abusers. *Drug and Alcohol Dependence 44*, 1–10.

Bauer, L.O., and Hesselbrock, V.M., (1999). P300 decrements in teenagers with conduct problems: Implications for substance abuse risk and brain development. *Biological Psychiatry, 46*, 263–272.

Beech, H.R., Ciesielski, K.T., and Gordon, P.K. (1983). Further observations of evoked potential in obsessional patients. *British Journal of Psychiatry*, *142*, 605-609.

Beck, A. (1967). *Depression: Clinical, Experimental, and Theoretical Aspects.* New York: Harper Row.

Berger, H. (1929). Ueber das Elektrenkphalogramm des Menschen. *Archives fur Psychiatrie Nervenkrankheiten, 87*, 527-570.

Berger, H. (1933). Uber das Electrenkephalogramm des Menschen. *Archives fur Psychiatrie Nervenkrankheiten, 100*, 301–320.

Birbaumer, N., Ghanayim, N., Hinterberger, T., Iversen, I., Kotchoubey, B., Kubler, A., et al. (1999). A spelling device for the paralysed. *Scientific Correspondence, 398*, 297-298.

Bolz, J., and Giedke, H. (1981). Controllability of an adverse stimulus in depressed patients and healthy controls: a study using slow brain potentials. *Biological Psychiatry, 16*, 441-452.

Breiter, H.C., Rauch, S.L., Kwong, K.K., Baker, J.R., Weisskoff, R.M., and Kennedy, D.N. et al., (1996). Functional magnetic resonance imaging of symptom provocation in obsessive–compulsive disorder. *Archives of General Psychiatry, 53*, 595–606.

Broadbent, D. E. (1958). *Perception and communication*. London: Pergamom University Press.

Bruder, G.E., Towey, J.P., Stewart, J.W., Friedman, D., Tenke, C., and Quitkin, F.M. (1991). Event-related potentials in depression: influence of task, stimulus hemifield and clinical features on P3 latency. *Biological Psychiatry, 30*, 233-246.

Chakravarti, A., and Little, P. (2003). Nature, nurture, and human disease. *Nature, 421,* 412-414.

Chiu. P.H., and Deldin, P.J. (2007). Neural evidence for enhanced error detection in major depressive disorder. *American Journal of Psychiatry, 164*, 608–616.

Clark, C.R., McFarlane, A.C., Weber, D. L., and Battersby, M. (1996). Enlarged frontal P300 to stimulus change in panic disorder. *Biological Psychiatry, 39*, 845-856.

Clark, L.A., and Watson, D. (1991). Tripartite model of anxiety and depression: Psychometric evidence and taxonomic implications. *Journal of Abnormal Psychology, 100,* 316-336.

Compton, R. J., Robinson, M. D., Ode, S., Quandt, L. C., Fineman, S. L, and Carp, J. (2008). Error-monitoring ability predicts daily stress regulation. *Psychological Science, 19*, 702-708.

Corr, P.J. (2002). J. A. Gray's reinforcement sensitivity theory: Tests of the joint subsystems hypothesis of anxiety and impulsivity. *Personality and Individual Differences, 33*, 511–532.

Costa, L., Bauer, L., Kuperman, S., Porjesz, B., O'Connor, S., Hesselbrock, V., Rohrbaugh, J., and et al. (2000). Frontal P300 decrements, alcohol dependence, and antisocial personality disorder. *Biological Psychiatry 47*, 1064–1071.

Davis H., Davis, P.A., Loomis, A. L., Harvey, E. N., and Hobart, G. (1939). Electrical reactions of the human brain to auditory stimuli during sleep. *Journal of Neurophysiology, 2*, 500-514.

De Bruijn, E. R. A., Grootens K. P., Verkes, R. J., Buchholz, V., Hummelen, J. W., and Hulstijn, W. (2006). Neural correlates of impulsive responding in borderline personality disorder: ERP evidence for reduced action monitoring. *Journal of Psychiatric Research, 40*, 428-437.

Dehaene, S., Posner, M.I., and Tucker, D.M. (1994). Localization of a neural system for error detection and compensation. *Psychological Science, 5*, 303-305.

Deldin, P. J., Duncan, C. C., and Miller, G. A. (1994). Season, gender, and P300. *Biological Psychology, 39*, 15-28.

Dikman, Z. V., and Allen, J. J. (2000). Error monitoring during reward and avoidance learning in high- and low-socialized individuals. *Psychophysiology, 37,* 43–54.

Donchin, E. (1981). Surprise!...surprise? *Psychophysiology, 18*, 493–513.

Donchin, E., and Coles, M.G.H. (1988). Is the P300 component a manifestation of contextual updating? *Behavioral and Brain Sciences, 11*, 357–427.

Eimer, M., and Holmes, A. (2007). Event-related brain potential correlates of emotional face processing. *Neuropsychologia, 45*, 15-31.

Eysenck, M.W., Derakshan, N., Santos, R., and Calvo, M. G. (2007). Anxiety and cognitive performance: Attentional control theory. *Emotion, 7*, 336-353.

Falgatter, A. J., Herrmann, M. J., Roemmler, J., Ehlis, A. C., Wagener, A., Heidrich, A., et al. (2004). Allelic variation of serotonin transporter function modulates the brain electrical response for error processing. *Neuropsychopharmacology, 29,* 1506–1511.

Falkenstein, M., Hohnsbein, J., Hoormann, J., and Blanke, L. (1991). Effects of cross-modal divided attention on late ERP components: II. Error processing in choice reaction tasks. *Electroencephalography and Clinical Neurophysiology, 78*, 447–455.

Federmeier, K.D., and Kutas, M. (2001). Meaning and modality: Influences of context, semantic memory organization, and perceptual predictability on picture processing. *Journal of Experimental Psychology, 27*, 202-224.

Felmingham, K.L., Bryant, R.A., Kendall, C., and Gordon, E. (2002). Event-related potential dysfunction in posttraumatic stress disorder: the role of numbing. *Psychiatry Research, 109*, 171-179.

Foa, E. B., and Kozak, M. J. (1986). Emotional processing of fear: Exposure to corrective information. *Psychological Bulletin, 99*, 20-35.

Ford, J. M., White, P., Lim, K. O., and Pfefferbaum, A. (1994). Schizophrenics have fewer and smaller P300s: A single- trial analysis. *Biological Psychiatry, 35*, 96-103.

Franken, I. H. A., van Strien, J. W., Franzek, E. J., and van de Wetering, B. J. (2007). Error-processing deficits in patients with cocaine dependence. *Biological Psychology, 75*, 45-51.

Gangadhar, B.N., Ancy, J., Janakiramaiah, N. and Umapathy, C. (1993). P300 amplitude in non-bipolar, melancholic depression. *Journal of Affective Disorders, 28*, 57-60.

Gehring, W.J., Coles, M.G.H., Meyer, D.E., and Donchin, E. (1990). The error-related negativity: an event-related brain potential accompanying errors. *Psychophysiology, 27,* S34.

Gehring, W.J., Himle, J., and Nisenson, L.G. (2000). Action-monitoring dysfunction in obsessive–compulsive disorder. *Psychological Science, 11*, 1–6.

Gibbs, F.A., Davis, H., and Lennox, W.G. (1935). The electroencephalogram in epilepsy and in conditions of impaired consciousness. *Archives of Neurology and Psychiatry, 34,* 1133–1148.

Gibbs, F.A., Lennox W.G., and Gibbs, E.L. (1936) The electroencephalogram in diagnosis and in localization of epileptic seizures. *Arch. Neurol. Psychiatry, 36*, 1225–1235.

Gilmore, C.S., Malone, S.M., Bernat, E.M. and Iacono, W.G. (2010). Relationship between the P3 event-related potential, its associated time-frequency components, and externalizing psychopathology. *Psychophysiology, 47*, 123-132.

Goldstein, R.Z., Tomasi, D., Rajaram, S., Cottone, L.A., Zhang, L., and Maloney, T. (2007). Role of the anterior cingulate and medial orbitofrontal cortex in processing drug cues in cocaine addiction. *Neuroscience, 144*, 1153–1159.

Gottesman I.I, and Gould, T.D. (2003). The endophenotype concept in psychiatry: Etymology and strategic intentions. *American Journal of Psychiatry*, *160*, 636–645.

Gough, H. (1960). Theory and measurement of socialization. *Journal of Consulting Psychology, 24*, 23-30.

Gratton, G., Coles, M. G. H., and Donchin, E. (1983). A new method for off-line removal of ocular artifact. *Electroencephalography and Clinical Neurophysiology*, *55*, 468–484.

Hajcak G., Franklin M.E., Foa E.B., and Simons R.F. (2008). Increased Error-Related Brain Activity in Pediatric Obsessive-Compulsive Disorder Before and After Treatment. *American Journal of Psychiatry*, *165*, 116-123.

Hajcak, G., McDonald, N., and Simons, R.F. (2003) Anxiety and error-related brain activity. *Biological Psychology*, *64*, 77-90.

Hajcak, G., McDonald, N., and Simons, R.F. (2004). Error-related psychophysiology and negative affect. *Brain and Cognition*, *56*, 189–197.

Hajcak,G., Moser, J.S., Yeung, N., and Simons, R.F. (2005). On the ERN and the significance of errors. *Psychophysiology, 42,* 151-160.

Hajcak, G., and Simons, R.F. (2002). Error-related brain activity in obsessive-compulsive undergraduates. *Psychiatry Research*, *110*, 63-72.

Hall, J.R., Bernat, E.M., and Patrick, C.J. (2007). Externalizing psychopathology and the error-related negativity. *Psychological Science*, *18*, 326–333.

Hansenne, M. (2006). Event-related brain potentials in psychopathology: Clinical and cognitive perspectives. *Psychologica Belgica. Special Issue: Experimental Psychopathology, 46*, 5-36.

Hartung, C. M., and Widiger, T. A. (1998). *Psychological Bulletin, 123*, 260-278.

Have, G., Kolbeinson, H., and Pétursson, H. (1991). Dementia and depression in old age: psychophysiological aspects. *Acta Psychiatrica Scandinavia*, *83*, 329-333.

Herrman, M. J., Weijers, H. –G., Wiesbeck, G. A., Böning, and Fallgatter, A. J. (2001). Alcohol cue-reactivity in heavy and light social drinkers as revealed by event-related potentials. *Alcohol and Alcoholism, 36*, 588-593.

Hicks, B., Bernat, E., Malone, S.M., Iacono, W.G., Patrick, C.J., Krueger, R.F., and McGue, M. (2007). Genes mediate the association between P3 amplitude and externalizing disorders. *Psychophysiology, 44*, 98-105.

Hirsh, J.B., and Inzlicht, M. (2010). Error-related negativity predicts academic performance. *Psychophysiology, 47*, 192-196.

Hoffman, L. D., and Polich, J. (1999). P300, handedness, and corpus callosal size: gender, modality, and task. *International Journal of Psychophysiology, 31*, 163-174.

Holmes, A., Nielsen, M. K., and Green, S. (2008). Effects of anxiety on the processing of fearful and happy faces: An event-related potential study. *Biological Psychology*, *77*, 159-173.

Holmes, A.J., and Pizzagalli, D.A. (2008). Spatiotemporal dynamics of error processing dysfunctions in major depressive disorder. *Archives of General Psychiatry*, *65*,179–188.

Holroyd, C.B., and Coles, M.G.H. (2002). The Neural Basis of Human Error Processing: Reinforcement Learning, Dopamine, and the Error-Related Negativity. *Psychological Review*, *109*, 679-709.

Hopfinger, J. B., Luck, S. J., and Hillyard, S. A. (2004). Selective attention: Electrophysiological and neuromagnetic studies. In M. S. Gazzaniga (Eds.), *The cognitive neurosciences* (3rd ed., pp.5610-574). Cambridge, MA, US: MIT Press.

Ingram R., Miranda J., and Segal, Z. (2006). Cognitive vulnerability to depression. In L. B. Alloy and J. H. Riskind (Eds.), *Cognitive Vulnerability to Emotional Disorders* (pp. 63-91). Hillsdale, NJ: Lawrence Erlbaum Associates.

Inzlicht, M., McGregor, I., Hirsh, J.B., and Nash, K. (2009). Neural markers of religious conviction. *Psychological Science, 20*, 385-392.

Iwanami, A., Isono, H., Okajima, Y., and Kamijima, K. (1997). Auditory event-related potentials in panic disorder. *European Archives Psychiatry Clinical Neuroscience, 247*, 107-111.

Jasper, H.H., and Carmichael, L. (1935). Electrical potentials from the intact human brain. *Science, 81*, 51-53.

Kim, M.S., Kang, S.S., Youn, T., Kang, D.H, Kim, J.J., and Kwon, J.S. (2003). Neuropsychological correlates of P300 abnormalities in patients with schizophrenia and obsessive-compulsive disorder. *Psychiatry Research, 123*, 109-123.

Kimble, M., Kaloupek, D., Kaufman, M., and Deldin, P. (2000). Stimulus novelty differentially affects attentional allocation in PTSD. *Biological Psychiatry, 47*, 880-890.

Kolassa, I. -T., Musial, F., Mohr, A., Trippe, R. H., and Miltner, W. H. R. (2005). Electrophysiological correlates of threat processing in spider phobics. *Psychophysiology, 42*, 520-530.

Krueger, R.F. (1999). The structure of common mental disorders. *Archives of General Psychiatry, 56*, 921–926.

Krueger, R. F., McGue, M., and Iacono, W. F. (2001). The higher order structure of common DSM mental disorders: Internalization, externalization, and their connections to personality. *Personality and Individual Differences, 30*, 1245-1259.

Larson, M.J., Good, D. A., and Fair., J. E. (2010). The relationship between performance monitoring, satisfaction with life, and positive personality traits. *Biological Psychology, 83*, 222-228.

Leutgeb, V., Schäfer, A., and Schienle, A. (2009). An event-related potential study on exposure therapy for patients suffering from spider phobia. *Biological Physiology, 82*, 293-300.

Levit, R.A., Sutton, S., and Zubin, J. (1973). Evoked potential correlates of information processing in psychiatric patients. *Psychological Medicine, 3*, 487-494.

Llinas, R.R. (1988). The Intrinsic Electrophysiological Properties of Mammalian Neurons: Insights into Central Nervous System Function. *Science, 242*, 1654-1664.

Luck, S.J. (2005). *An Introduction to the Event-Related Potential Technique* (pp. 27-32). MIT Press.

Luck, S.J. and Girelli, M. (1998). Electrophysiological Approaches to the Study of Selective Attention in the Human Brain. In R. Parasuraman (Ed.), *The Attentive Brain* (pp. 71-94). MIT Press.

Luck, S.J., & Hillyard, S.A. (1994). Spatial filtering during visual search: Evidence from human electrophysiology. *Journal of Experimental Psychology: Human Perception and Performance, 20(5)*, 1000-1014.

Luu, P., Collins, P., and Tucker, D. M. (2000). Mood, personality, and self-monitoring: Negative affect and emotionality in relation to frontal lobe mechanisms of error monitoring. *Journal of Experimental Psychology: General, 129*, 43–60.

Marshall, P. J., Reeb, B. C., and Fox, N. A (2009). Electrophysiological responses to auditory novelty in temperamentally different 9-month-old infants. *Developmental Science, 12*, 568-582.

Mathews, A., and MacLeod, C. (2005). Cognitive vulnerability to emotional disorders. *Annual Review Clinical Psychology, 1*, 167-195.

McFarlane, A.C., Weber, D.L., and Clark, C.R. (1993). Abnormal stimulus processing in posttraumatic stress disorder. *Biological Psychiatry, 34*, 311-320.

Millon, T. (1969). *Modern Psychopathology: A Biosocial Approach to Maladaptive Learning and Functioning*. Philadelphia: Saunders.

Moeller, F.G., Dougherty, D.M., Barrat, E.S., Schmitz, J.M., Swann, A.C., and Grabowski, J. (2001). The impact of impulsivity on cocaine use and retention in treatment. *Journal of Substance Abuse Treatment, 21*, 193–198.

Mogg, K., and Bradley, B. P. (1998). A cognitive-motivational analysis of anxiety. *Behavior Research and Therapy, 36*, 809-848.

Moser, J. S., Hajcak, G., Huppert, J. D., Foa, E. B., and Simons, R. F. (2008). Interpretation bias in social anxiety as detected by event-related brain potentials. *Emotion, 8*, 693-700.

Moser J.S., Hajcak G., and Simons R.F. (2005). The effects of fear on performance monitoring and attentional allocation. *Psychophysiology, 42*, 261–268.

Moser, J.S., Huppert, J.D., Duval, E., and Simons, R.F. (2008). Face processing biases in social anxiety: An electrophysiological study. *Biological Psychology, 78*, 93-103.

Mueller, E. M., Hofmann, S. G., Santesso, D. L., Meuret, A. E., Bitran, S., and Pizzagalli, D. A. (2009). Electrophysiological evidence of attentional biases in social anxiety disorder. *Psychological Medicine: A Journal of Research in Psychiatry and the Allied Sciences. 39*, 1141-1152.

Nelson, L.D., Patrick, C. J., and Bernat, E. M. (in press). Operationalizing proneness to externalizing psychopathology as a multivariate psychophysiological phenotype. *Psychophysiology*.

Neylan, T.C., Jasiukaitis, P.A., Lenoci, M., Scott, J.C., Metzler, T.J., Weiss, D.S., Schoenfeld, F.B., and Marmar, C.R. (2003). Temporal instability of auditory and visual event-related potentials in posttraumatic stress disorder. *Biological Psychiatry, 53*, 216-225.

Olofsson, J.K., Nordin, S., Sequeira, H., and Polich, J. (2008). Affective picture processing: An integrative review of ERP findings. *Biological Psychology, 77*, 247-265.

Olvet, D.M., and Hajcak, G. (2008). The error-related negativity (ERN) and psychopathology: towards an Endophenotype. *Clinical Psychology Review, 28*, 1343 – 1354.

Perlstein, W. M., Larson, M. J., Dotson, V. M., and Kelly, K. G. (2006). Temporal dissociation of components of cognitive control dysfunction in severe TBI: ERPs and the cued-Stroop task. *Neuropsychologia, 44,* 260-274.

Pierson, A., Ragot, R., van Hooff, J., Partiot, A., Renault, B., and Jouvent, R. (1996). Heterogeneity of information-processing alterations according to dimensions of depression: An event-related potentials study. *Biological Psychiatry, 40*, 98-115.

Pizzagalli, D. A (2007). Electroencephalography and high-density electrophysiological source localization. In J.T. Cacioppo, L.G. Tassinary and G.G. Berntson (Eds.), *Handbook of Psychophysiology* (3rd ed., pp. 56-84). Cambridge, UK: Cambridge University Press.

Polich, J., Howard, L., and Starr, A. (1985). Stimulus frequency and masking as determinants of P300 latency in event-related potentials from auditory stimuli. *Biological Psychology, 21*, 309-318.

Polich, J., Pollock, V. E., and Bloom, F. E. (1994). Meta-analysis of P300 amplitude from males at risk for alcoholism. *Psychological Bulletin, 115*, 55-73.

Posner, M.I., and Rothbart, M.K. (1998). Attention, self-regulation and consciousness. *Philosophical Transactions of the Royal Society of London B, 353*, 1915-1927.

Potts, G.F., George, M.R., Martin, L.E. and Barratt, E.S. (2006). Reduced punishment sensitivity in neural systems of behavior monitoring in impulsive individuals. *Neuroscience Letters, 397*, 130–134.

Reynolds, G. D., Courage, M. L., and Richards, J. E. (2010). Infant attention and visual preferences: Converging evidence from behavior, event-related potentials, and cortical source location. *Developmental Psychology, 46*, 886-904.

Ruchsow, M., Spitzer, M., Gron, G., Grothe, J., and Kiefe, M. (2005). Error processing and impulsiveness in normals: evidence from event-related potentials. *Cognitive Brain Research, 24*, 317–325.

Ruscio, A.M., Ruscio, J., and Keane, T.M. (2002). The latent structure of posttraumatic stress disorder: A taxometric investigation of reactions to extreme stress. *Journal of Abnormal Psychology, 111*, 290-301.

Schmidt, N.B., Richey, A., Buckner, J.D., and Timpano, K.R. (2009). Attention training for generalized social anxiety disorder. *Journal of Abnormal Psychology, 118*, 5-14.

Steele, J. D., and Lawrie, S. M. (2004) Segregation of cognitive and emotional function in the prefrontal cortex: A stereotactic meta-analysis. *NeuroImage, 21*, 868-875.

Towey, J.P., Bruder, G.E., Tenke, C.E., Leite, P., DeCaria, C., Friedman, D., and Hollander, E. (1993). Event-related potential and clinical correlates of neurodysfunction in obsessive-compulsive disorder. *Psychiatry Research, 49*, 167-181.

Vaidyanathan, U., Patrick, C. J., and Cuthbert, B. N. (2009). Linking dimensional models of internalizing psychopathology to neurobiological systems: Affect-modulated startle as an indicator of fear and distress disorders and affiliated traits. *Psychological Bulletin, 135*, 909-942.

Wang, J., Miyazato, H., Randall, M., Hokama, H., Hiramatsu, K., and Ogura C. (2003). The N200 abnormalities of auditory event-related potentials in patients with panic disorder. *Progress in Neuropsychopharmacology and Biological Psychiatry, 27*, 1013-1021.

Weinberg, A., Olvet, D. M., and Hajcak, G. (2010). Increased error-related brain activity in generalized anxiety disorder. *Biological Psychology, 85*, 472-480.

Yeung, N., Botvinick, M.W., and Cohen, J.D. (2004). The neural basis of error detection: Conflict monitoring and the error-related negativity. *Psychological Review, 111*, 931-959.

In: Handbook of Technology in Psychology …
Editor: Luciano L'Abate and David A. Kasier
ISBN: 978-1-62100-004-4

Chapter 13

DIFFUSION TENSOR IMAGING: ANALYSIS METHODS FOR GROUP COMPARISON IN NEUROLOGY

Hans-Peter Müller*[1], *Alexander Unrath, Axel Riecker and Jan Kassubek
[1]Tromsø University, Breivika, Norway

Keywords: Diffusion tensor imaging (DTI), fractional anisotropy (FA) mapping, fiber tracking (FT), magnetic resonance imaging (MRI)

The white matter tracts of the central nervous system consist of densely packed axons in addition to various types of neuroglia and other small populations of cells. The axonal membrane as well as the well-aligned protein fibers within an axon restrict water diffusion perpendicular to the fiber orientation, leading to anisotropic water diffusion in brain white matter (Moseley et al., 1990). Myelin sheaths around the axons may also contribute to the anisotropy for both intra- and extracellular water (Mori and van Zijl, 2002). A detailed study of these contributions was performed by Beaulieu (Beaulieu 2002).

The quantitative description of this anisotropy could be detected by a special diffusion-weighted magnetic resonance imaging (MRI) technique, i.e. diffusion tensor imaging (DTI) (Basser et al., 1994a, Basser et al., 1994b). DTI produces images of tissues weighted with the local microstructural characteristics of water diffusion. The image-intensities at each position are attenuated, depending on the strength (*b*-value) and direction of the so-called magnetic diffusion gradient, as well as on the local microstructure in which the water molecules diffuse. The more attenuated the image is at a given position, the more diffusion exists in the direction of the diffusion gradient. In order to measure the complete diffusion profile of the tissue, one needs to repeat the MR scans, applying different directions (and possibly strengths) of the diffusion gradient for each scan. Therefore, DTI scans derive neural tract directional information based on at least six gradient directions, sufficient to compute a diffusion tensor (LeBihan et al., 2001, Matiello et al., 1994, Matiello et al., 1997). Recently, studies were performed at 1.5 as well as at 3.0 Tesla scanners with an increased number of gradient directions (up to 128) in humans to improve data quality (Bueltmann et al., 2008, Rossi et al., 2008, Boss et al., 2007, Rossi et al., 2007, Ni et al., 2006).

In typical DTI measurements, the voxel dimensions are in the order of millimeters. Thus, a voxel always contains the averaged information of the water molecules inside the detected volume that usually covers several axons as well as the surrounding water molecules (Mori and van Zijl, 2002). Despite this multi-directional environment, DTI is sensitive to the orientation of the largest principal axis which aligns to the predominant axonal direction, i.e. the axonal contribution dominates the measured signal (Mori and van Zijl, 2002). As initially shown by Basser and co-workers (Basser et al., 1994a), DTI provides two types of information about the property of water diffusion: first, the orientation-independent extent of diffusion anisotropy (Pierpaoli and Basser, 1996, Uluğ and van Zijl, 1999) and second, the predominant direction of water diffusion in image voxels, i.e. the diffusion orientation (Douek et al., 1991, Pajevic and Pierpaoli, 1999).

Since there are several challenges in displaying tensor data, the concept of diffusion ellipsoids has been proposed (Basser et al., 1994a, Basser et al., 1994b). These ellipsoids could be set up by the Eigenvectors and the Eigenvalues of the diffusion tensor. These Eigendiffusivities represent the unidimensional diffusion coefficients in the main direction of diffusivities of the medium. Therefore, the main axis of the ellipsoid represents the main diffusion direction in the voxel which coincides with the direction of the fibers, while the eccentricity of the ellipsoid provides information about the degree of anisotropy and its symmetry – isotropic diffusion would be visualized as a sphere whereas a "cigar shape" would represent anisotropic diffusion. Therefore, from the diffusion tensor, diffusion anisotropy measures such as the fractional anisotropy (FA) and the mean diffusivity (MD) were defined. MD is introduced to get an overall evaluation of the diffusion independent of the direction, whereas FA is a measure of directional variance of the diffusion amplitude. Further anisotropy indices are the relative anisotropy or the volume ratio (LeBihan et al., 2001).

An additional access is to use the principal direction of the diffusion tensor to infer the white matter connectivity of the brain, corresponding to the tractography approach which has the intention to investigate which part of the brain is connected to which other part. Assuming that the orientation of the major component of the diagonalized diffusion tensor represents the orientation of the dominant axonal tracts, a 3-D vector field is provided in which each vector presents the fiber orientation.

Currently, there are several different approaches to reconstruct white matter tracts which could be divided into two types: the first category is based on line propagation algorithms using the local tensor information for each step of the fiber tract propagation (cf. (Mori and van Zijl, 2002, Conturo et al., 1999, Lori et al., 2002, Lazar et al., 2003)). The second category is based on global energy minimization to find the energetically most favorable path between two white matter regions, resulting in the approach of tract-based spatial statistics (TBSS) (Behrens et al., 2007, Smith et al., 2006, Smith et al., 2007, Johansen-Berg et al., 2006, Kochunov et al., 2007).

In this chapter, an overview of the DTI technique and the analysis and quantification of diffusion properties is presented. After the transformation into a stereotaxic standard space, averaging of single subject diffusion properties becomes feasible, thus allowing for subject averaging to increase the signal-to-noise ratio (SNR) and also for the comparison of subject groups.

Supplementary results could be provided by combined analysis of advanced MRI-based techniques of the computational neuroanatomy, i.e. the combination of DTI with e.g. intensity-based analysis of 3-D T1-weighted MRI in standard space (Mueller et al., 2008, Kolind et al., 2008, Reich et al., 2007).

B. DTI Acquisition

For DTI scanning at least 6 independent gradient directions have to be detected. Studies with 30, 64, or 128 gradient directions (e.g. (Ni et al., 2006)) increase the SNR and allow for a more exact detection of the diffusion tensor. The price for the recording of many gradient directions is a prolonged acquisition time which might cause problems especially when patients are investigated. That way, a good compromise between SNR and the total acquisition time has to be found, depending on the aims of each study, e.g. if a connectivity analysis is to be performed in healthy (young) subjects or if alterations of the white matter networks are to be analyzed in specific patient populations with a certain disease.

Thus, typical DTI acquisitions consist of gradient-echo recorded 13 volumes (12 for the gradient directions and one with b=0) and b-values of 800-1000 s/mm^2. In order to improve the tensor accuracy, the number of gradient directions could be increased to 32, 128, or even 256 with the drawback of an increased acquisition time.

C. Quantification of Diffusion Properties of single Subject Data – DTI Theory

The diffusion of water molecules in the presence of a strong magnetic gradient results in a proton-MRI signal loss as a result of the dephasing of spin coherence. DTI is one advanced application of MRI-based diffusion weighting which describes the application of a pair of strong gradients to detect differences in the diffusivity of water molecules among various biological tissues (LeBihan et al., 2001, LeBihan et al., 1998, Turner et al., 1990). In diffusion weighted imaging (DWI), the signal intensity in each voxel is influenced by the *b*-value (diffusion sensitation) which is related to the gradient strength. Diffusion is described using a single scalar parameter, the diffusion coefficient *D*. The effect of diffusion on the MRI signal is an attenuation *A* which depends on *D* and the *b*-value:

$$A = e^{-Db} \tag{1}$$

Thus, the apparent diffusion coefficient (ADC) for each image voxel can be calculated. However, in the presence of anisotropy in white matter, diffusion can no longer be characterized by a single scalar coefficient, but requires a tensor which in first approximation describes molecular mobility along each direction and correlation between these directions (Neeman et al., 1990, Matiello et al., 1994, Matiello et al., 1997, Basser et al., 2002). Diffusion anisotropy is mainly caused by the orientation of fiber tracts in white matter and is influenced by its micro- and macrostructural features (Figure 1). Of the microstructural features, intraaxonal organization appears to be of greatest influence on diffusion anisotropy.

Other features include the density of fiber and cell packing, degree of myelination, and individual fiber diameter.

On a macroscopic scale, the variability in the orientation of all white matter tracts in an imaging voxel influences the degree of anisotropy assigned to the respective voxel (Turner et al., 1990, Chenevert et al., 1990, Pierpaoli and Basser, 1996).

The elements of the symmetric tensor can be measured by diffusion gradients along at least six non-collinear and non-coplanar directions so that *b* (Equation 1) has become a tensor, resulting in a signal attenuation

$$\ln(A) = -(b_{xx}D_{xx} + 2b_{xy}D_{xy} + 2b_{xz}D_{xz} + 2b_{yz}D_{yz} + b_{yy}D_{yy} + b_{zz}D_{zz}) \tag{2}$$

This equation requires the need to account for possible interactions between imaging and diffusion gradients that are applied in orthogonal directions (cross terms) and even between imaging gradients that are applied in orthogonal directions (Neeman et al., 1990, Matiello et al., 1994, Matiello et al., 1997).

The second-rank diffusion tensor can always be diagonalized leaving only three non-zero elements along the main diagonal of the tensor, the Eigenvalues ($\lambda_1, \lambda_2, \lambda_3$). The Eigenvalues reflect the shape or configuration of the ellipsoid. The mathematical relationship between the principal coordinates of the ellipsoid and the laboratory frame is described by the Eigenvectors ($\vec{v}_1, \vec{v}_2, \vec{v}_3$) (Figure 2).

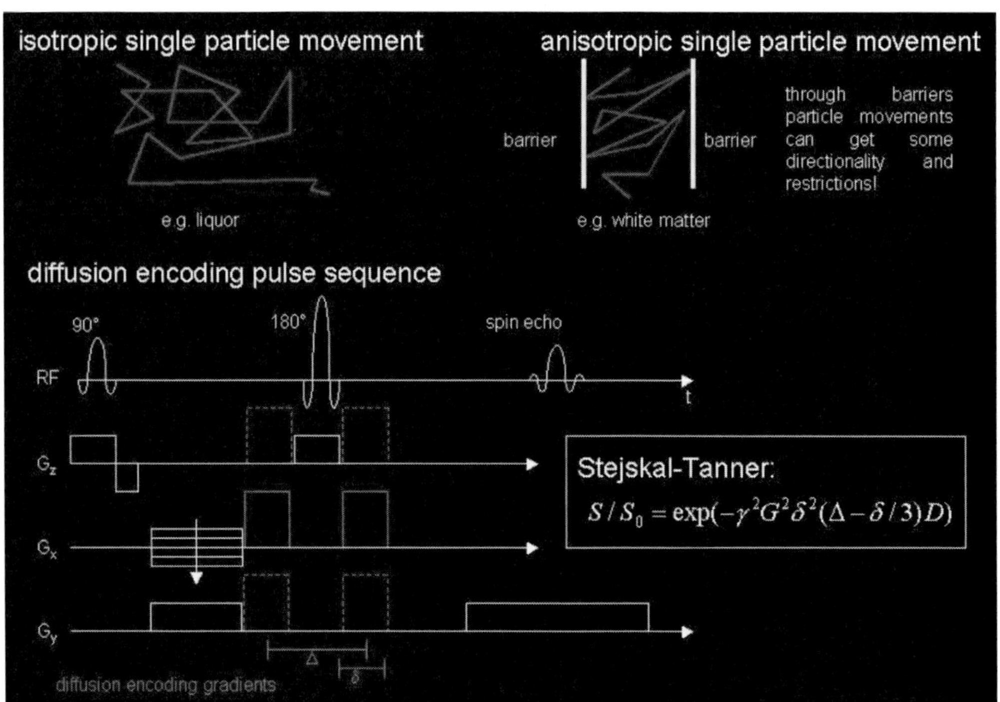

Figure 1. encoding of isotropic and anisotropic particle movement by diffusion weighted imaging.

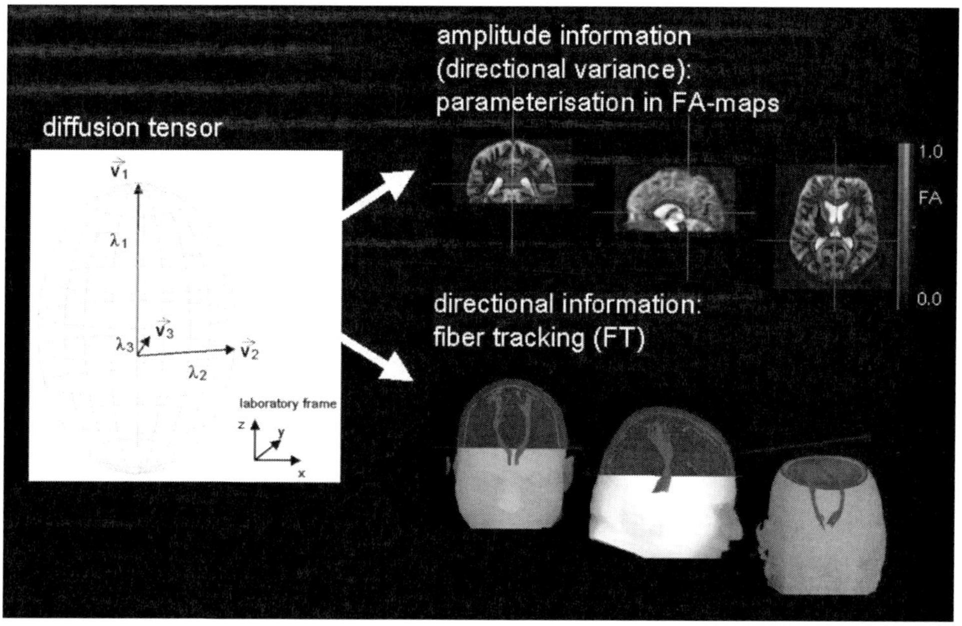

Figure 2. The diffusion tensor is used to obtain information about the diffusion amplitude via parameterization into FA-maps (intensity is related to the FA amplitude and the color-coding was red for major Eigenvector mainly in left-right direction, blue for major Eigenvector mainly in inferior-superior direction, and green for major Eigenvector mainly in posterior-anterior direction) and the diffusion orientation via directional information visualized in fiber tracts (FT), e.g. in the pyramidal tracts (the images are projections of the FTs in coronar, sagittal, and axial direction).

The diffusion anisotropy can be quantified by several indices (Pierpaoli and Basser, 1996, Wu et al., 2004), each of which distinguishes between a prolonged, "cigar shape" ellipsoid (which would be expected in oriented fiber bundles of white matter) and a sphere (which represents isotropic diffusion) (Pierpaoli and Basser, 1996). Out of these, the mean diffusivity MD, reflecting the overall diffusion amplitude independent of the direction

$$M_D = \lambda_1 + \lambda_2 + \lambda_3 \tag{3}$$

and the fractional anisotropy FA which is a measure of the directional variance of the diffusion amplitude

$$F_f = \sqrt{\frac{3}{2}\frac{(\lambda_1 - \bar{\lambda})^2 + (\lambda_2 - \bar{\lambda})^2 + (\lambda_3 - \bar{\lambda})^2}{\lambda_1^2 + \lambda_2^2 + \lambda_3^2}} \tag{4}$$

can be calculated (Figure 2). is the average of all Eigenvalues.

The extension of the DTI technique to DTI-based fiber tracking (FT) offers new possibilities to map the interconnectivity between different brain regions by the underlying white matter networks. This orientational information can be the basis for the reconstruction of the pathways of the fibers. Various techniques addressing this topic have been published (Basser et al., 2000, Conturo et al., 1999, Mori et al., 1999, Lazar et al., 2003, Jones et al.,

1999, Behrens et al., 2003a). Most of them focus on qualitative imaging of the FT, 3-D visualization and evaluation by experienced operators. There have also been various efforts in using diffusion anisotropy as a marker for white matter tract integrity (Pagani et al., 2005, Wilson et al., 2003, Kanaan et al., 2006). In these approaches, a quantitative analysis has been performed by use of the underlying FA maps for selective statistics. Smith and co-workers have developed an algorithm for an alignment-invariant tract representation to overcome normalization problems; this technique is referred to as tract based spatial statistics (TBSS) (Behrens et al., 2007, Smith et al., 2006, Smith et al., 2007, Johansen-Berg et al., 2006, Kochunov et al., 2007).

D. Standard Data Processing

1. Eddy Current Correction

Large discontinuities in bulk magnetic susceptibility produce local magnetic field gradients that notoriously degrade and distort DTI data, particularly during the use of echo-planar imaging (Basser and Jones, 2002). These eddy current-induced geometric distortions vary with the magnitude and direction of the diffusion sensitizing gradients. For the correction of this distortion, the method proposed by Shen et al. (Shen et al., 2004) relies on collecting pairs of images with reversed diffusion sensitizing gradients – these paired images are distorted with eddy currents in opposite directions. A columnwise correction in the image domain along the phase encoding direction (anterior – posterior) can be performed by searching for the maximum value of the cross-correlation between two corresponding columns (of two paired volumes) while one is shifted and scaled (fitting routine could be e.g. simplex method (Press et al., 1992)). Each column can then be corrected by applying opposite shifts and scales equal to half of the correction. Alternative techniques for the eddy current correction have been described (Morgan et al., 2004, Andersson et al., 2003).

2. Transformation to Iso-Voxels and Smoothing

As the recording technique provided voxels with non-isotropic size (usually the slice thickness was larger than the in-plane voxel size), the DTI data sets could be transformed into an isotropic grid with an isotropic voxel size (e.g. 1.0 x 1.0 x 1.0 mm^3) in the first step. The transformation chosen could be a linear nearest neighbor transformation with

$$I_{t\arg et}(i,j,k) = \sum_{v=1}^{8} a_v I_v(l,m,n) \qquad (5)$$

where was the voxel intensity at the new grid coordinates *i,j,k* and *l,m,n* were the original voxel coordinates in *x,y,z* direction, respectively. The factors were the 8 weighting factors for the interpolation.

Interpolation and smoothing are classical image processing problems for which a variety of solutions exist (Lehmann et al., 1999, Mishra et al., 2006, Mishra et al., 2007). Mishra et al. (Mishra et al., 2006) have proposed the idea of an anisotropic image interpolation method.

In that approach, the kernel for interpolation is weighted by a factor depending on the local gradients. Thus, the sharpness of the image remains preserved. This kernel was weighted with the local gradients by

$$I_{t\arg et}(i,j,k) = \frac{\sum_{v=1}^{N} I_v(l,m,n)/(r_v + g_{t\arg et})}{\sum_{v=1}^{N} 1/(r_v + g_{t\arg et})} \quad (6)$$

where was the distance between and I_v, and was the absolute value of the gradient at position *(i,j,k)*. In this way, the local gradients weighted the interpolation kernel with a sharpness dependency.

For smoothing, it had to be considered that the filter size influences the results of DTI data analysis (Jones et al., 2005). The matched filter theorem states that the width of the filter used to process the data should be tailored to the size of the difference one expects to see (Rosenfeld and Kak, 1982). In general, filter sizes for smoothing range from 4mm up to 12mm (FWHM) with a Gaussian kernel depending of a good balance between sensitivity and specificity.

3. Transformation into a Stereotaxic Standard Space

Spatial normalization allowed for arithmetic averaging of the results obtained from different subjects in order to improve the SNR and to perform finally a comparison of groups of patients and controls in case that the study aims at an analysis of the computational pathoanatomy of a specific disorder, e.g. a neurodegenerative disease which is associated with the affectation of a specific brain system. The prerequisite of statistical analysis at group level and arithmetic averaging of subject data is the normalization to a standardized stereotaxic space. Talairach and Tournoux (Talairach and Tournoux, 1988) suggested a transformation algorithm to a standard atlas involving the identification of various brain landmarks and piecemeal scaling of brain quadrants. Nowadays, most of the advanced MRI data analysis packages use normalization to the Montreal Neurological Institute (MNI) stereotaxic space. Here, a new standard brain was defined by using a large series of MRI scans of normal controls, resulting in the MNI atlas (Brett et al., 2002).

For the transformation into MNI space, a semiautomatic spatial normalization utilizing study specific templates could be performed. A schematic illustration of a possible normalization process is displayed in Figure 3. The templates used for the normalization of the DTI data sets are created from the (*b*=0) data sets of all the subjects who participated in the respective study. An iterative algorithm has previously been described (Mueller et al., 2007a) in which three templates are generated by arithmetic averaging of the data sets after affine and non-affine transformations. After creation of the templates, all data sets were consecutively transformed by affine and non-affine normalization steps into MNI space. In order to obtain the study specific templates necessary for the non-affine normalization process, an FA-template was defined by averaging all individually derived FA-maps of patients and controls. After having created the FA-template, two normalization steps were performed: First, an affine normalization of the *b0*-images (applying consequently the identical transformation to the *b1* – *b12*-images) and second, a non-affine normalization to the FA-template as it has been suggested by Smith et al. (Smith et al., 2006). The parameters of these normalization steps have to be stored so that they can be considered during the fiber tracking process – see below. The non-affine normalization process to the FA-template follows the basic ideas of the work from Ashburner and Friston (Ashburner and Friston, 1999). The non-affine registration to an FA-template has the advantage that it provides more contrast in comparison to normalization to a b0-template (Smith et al., 2006), All these analysis steps can be applied to MD analyses by use of an MD-template in the same way. An alternative approach is to use automated brain registration algorithms (Collins et al., 1994, Friston et al., 1995). During the normalization process the directional information has to be preserved. Basically, a complete non-linear MNI normalization consists of 3 deformation components (DC):

- DC 1: A rigid brain transformation to align the basic coordinate frames. The rotation angles have to be stored in a rotation matrix $\vec{\vec{R}}$.
- DC 2: An affine deformation according to landmarks. The 6 stretching parameters for the different brain regions have to be stored in a vector $\vec{S}$.
- DC 3: A non-affine normalization equalizing non-linear brain shape differences. The 3-D vector shifts are different for each voxel resulting in a 6-D matrix (a 3-D vector for each voxel of the 3-D matrix) $\vec{\vec{\vec{T}}}$.

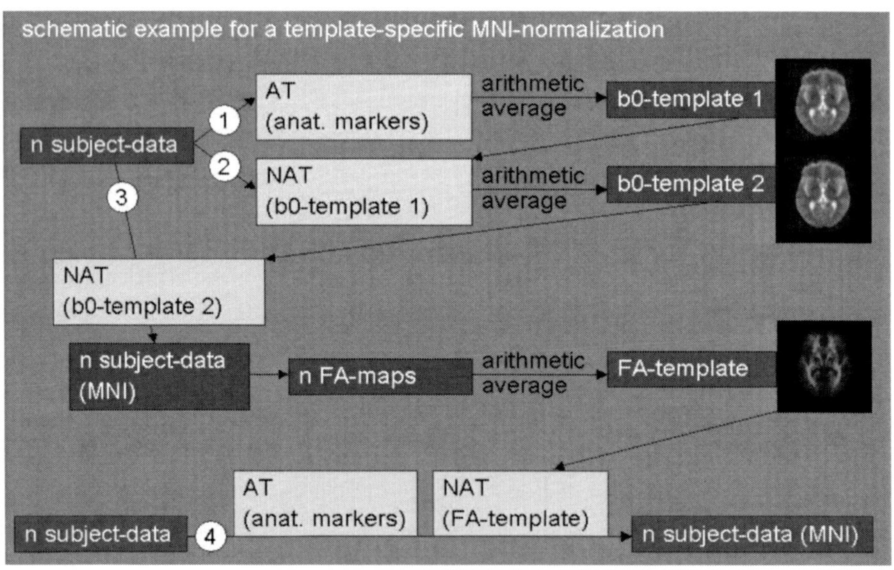

Figure 3. Scheme for an normalization procedure: after n subject data were transformed according to landmarks by an affine transformation, template 1 could be created by arithmetic averaging. By use of this template 1 in a second normalization step by a non-affine transformation, template 2 could be created by arithmetic averaging. This template could be used to perform a non-affine transformation resulting in an FA map for each subject data set. That way, the templates were created that were needed for the final transformation which was consisting of consecutive affine transformation (anatomical markers), and non-affine transformation (FA-template).

Consequently, the resulting diffusion tensor of each voxel has to be rotated according to all the rotations listed above (Figure 4).

- Rotation resulting from the aligning to the basic coordinate frame (corresponding to DC 1)
-

$$\vec{\vec{D}}_i' = \vec{\vec{R}} \cdot \vec{\vec{D}}_i \,. \tag{7}$$

- Simple trigonometry gives a rotation matrix (for each voxel independently), resulting from the 3-D vector shifts following the basic ideas of Alexander et al. (Alexander et al., 2001). The dilation matrices were used for the alignment of the tensor of each voxel to the surrounding voxels (corresponding to DC 3).

$$\vec{\vec{D}}_i'' = \vec{\vec{t}}_i \cdot \vec{\vec{D}}_i' \tag{8}$$

where are the components of $\vec{\vec{\vec{T}}}$.

- The components of the Eigenvectors ($\vec{v}_1, \vec{v}_2, \vec{v}_3$) have to be stretched according to the 6 stretching parameters of vector (dependent on the brain region s_a, $a = 1 \ldots 6$) of the affine deformation (corresponding to DC 2).

$$v_{w,j}''' = s_a v_{w,j}''' \tag{9}$$

with and $j = x, y, z$. After stretching, the Eigenvectors have to be re-normalized.

These fine-corrections of the tensor are essential for a correct FT, and the corresponding parameters have to be stored for each subject's data set independently (Mueller et al., 2007b, Mueller et al., 2009).

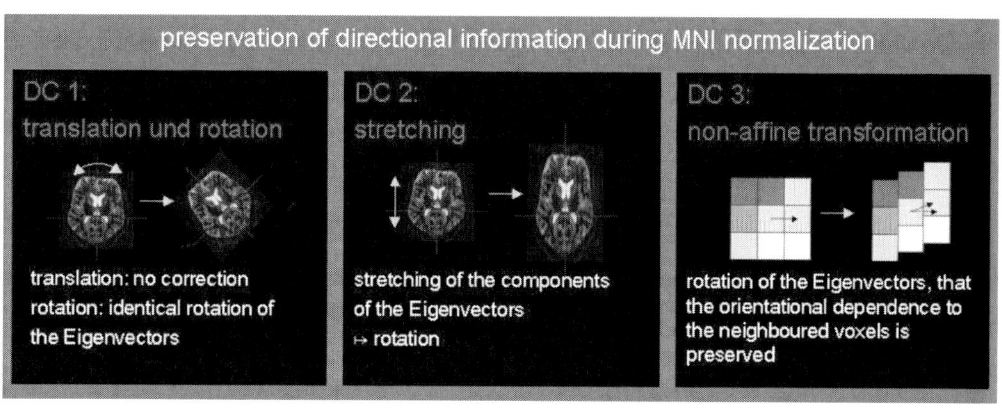

Figure 4. schematical display of principal responses on deformations to preserve directional information during normalization.

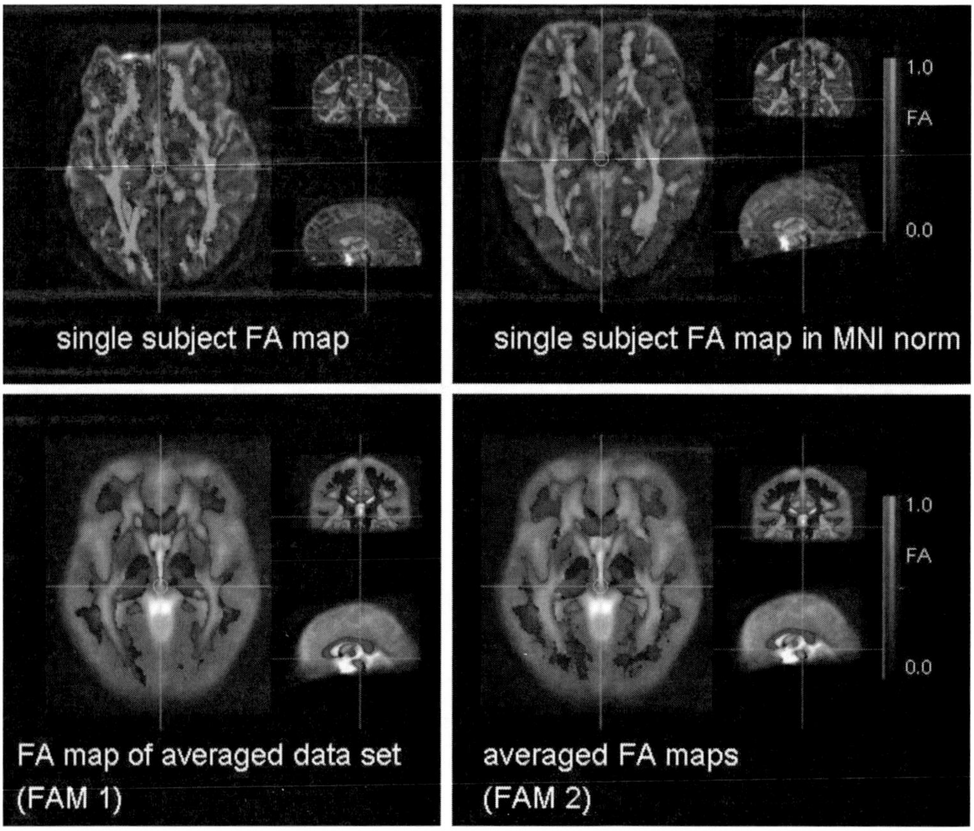

Figure 5. upper panel left: FA map calculated from data before MNI normalization, upper panel right: the same single subject data set after MNI normalization (coronar, sagittal, axial view). FA display threshold was 0.2, display background was *b0* data. lower panel left: FA map created as the arithmetic average of 10 control FA maps (FAM1), lower panel right: FA map calculated from a DTI data set which is the arithmetic average from 10 control data sets. FA display threshold was 0.2, display background was *b0* data.

In order to display the results on a morphological background, the corresponding 3-D T1-weighted data sets can be normalized to the same stereotaxic space (MNI) and arithmetically averaged; this normalization procedure could be performed in analogy to the DTI data sets by use of a customized template (Mueller et al., 2007b).

In order to show that normalization to the MNI standard space preserves specific diffusion features, selected regions of interest (ROIs) in the FA maps before and after the normalization were compared indicating deviations of less than 5%. A detailed analysis of the preservation has been reported in (Mueller et al., 2007a). Figure 5 (upper panel) shows the comparison between FA maps calculated from data before MNI normalization and the same single subject data set after MNI normalization.

4. Averaging of Single Subject Data and Group Comparison

Like in other advanced MRI methods, DTI- and FT-based studies in a clinical context pursue the ultimate goal to categorize individual patient's brain morphology in order to facilitate the diagnostic process based on some discrimination metric. In the present study, the standardization of technical approaches to group-based analyses of certain brain pathologies can be considered the prerequisite for the identification of pathological patterns of altered brain anatomy at group level.

Studies at the group level may be important if the common clinical phenotype is supposed to be due to a lesion of one or more defined brain areas. Here, averaging of results for different subjects is necessary. Each individual brain has to be transferred into stereotaxic space and, in a second step, the arithmetic averaging of the results at a voxel by voxel level is possible. The averaging of the individual results can be performed in two different ways:

- For each single subject's data in MNI space, the FA map is calculated separately and arithmetic averaging of the FA maps is performed afterwards. This produces a group specific FA map FAM1.
- Each DTI data set is normalized and before FA mapping the whole DTI-data sets are averaged. FA-parameterization of these group-averaged DTI data leads to a group specific FA map FAM2.

Single subject FA maps can be averaged arithmetically taking the necessary corrections during MNI normalization (see below) into account. Figure 5 (lower panel) shows the comparison of an FA map created as the arithmetic average of 10 control FA maps (FAM1) and an FA map calculated from a DTI data set which is the arithmetic average from 10 control data sets (cf. also (Mueller et al., 2007a))

5. Fiber Tracking

Assuming that the direction of the major Eigenvector corresponds to the main fiber orientation inside the voxel, fiber tracts can be reconstructed step by step (Mori et al., 1999, Basser et al., 2000, Conturo et al., 1999). In the calculation of the diffusion spheroid, the Eigenvector corresponding to the largest Eigenvalue represents the direction of fastest diffusion and indicates the fiber direction in white matter regions. Based on this directional information, different methods and algorithms had been proposed to estimate white matter connectivity. Here, the conservative streamline tracking technique (STT) is described in more detail. STT models the propagation in the major Eigenvector field (Basser and Jones, 2002, Mori et al., 1999). Generally, the FT positions result from float numbers. The corresponding Eigenvector direction (which can be used to obtain the consecutive FT position) is the interpolation of the directions of the neighboured voxels weighted by the proportionate position (linear nearest neighbour interpolation):

$$\vec{v}_{new}(i,j,k) = \sum_{w=1}^{8} a_w \vec{v}(l_w, m_w, n_w) \tag{10}$$

is the resulting vector at the new position *i,j,k* (float numbers), and *l,m,n* are the voxel coordinates (integer numbers) of the 8 neighboured voxels. The factors are the respective 8 weighting factors for the interpolation.

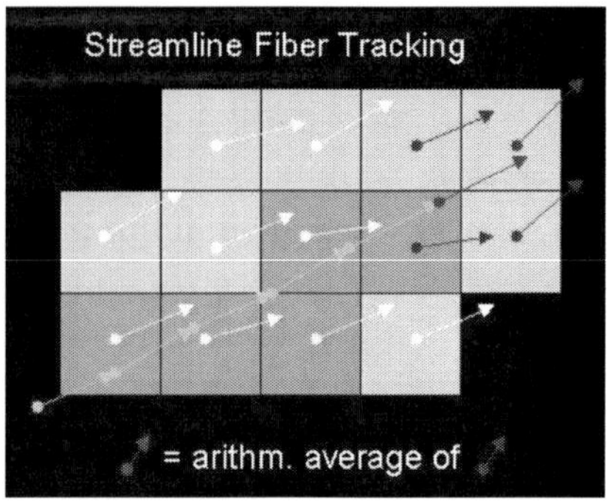

Figure 6. Schematical display of a streamline fiber tracking algorithm: following the green arrows (which are the arithmetic average of the neighbored major Eigenvectors - blue), voxels that belong to the fiber tract (amber) were detected.

The following set of parameters is used for FT:

- The threshold for the scalar product of the major Eigenvectors (angle between directions of two consecutive FT positions and the major Eigenvector directions), common used values are about 0.9.
- The distance between two FT positions, i.e. the stepwidth, common used values are about 0.5 mm, or about 0.5 voxels.

A scheme is provided in Figure 6.

6. The TIFT Software

The analyses shown in the following are performed by the software package *Tensor Imaging and Fiber Tracking* (TIFT) (Mueller et al., 2007a). The software package TIFT provides analysis tools for the following respects: analysis in terms of FA and MD maps, MNI normalization, group comparison in terms of FA and MD, several FT possibilities, FT on group averaged DTI data and the corresponding statistical analysis, allowing to perform a variety of analysis in one software environment (Mueller et al., 2007a, Mueller et al., 2007b, Mueller et al., 2008, Mueller et al., 2009). The structure of the software aims at a minimization of operator-dependency providing analysis in a fast and reproducible way. The TIFT software is steadily under development and therefore new tasks in MR data analysis with main topic DTI are always under construction.

E. Group Comparison and Fiber Tracking on Group-Averaged Data

1. Whole Brain Based Spatial Statistics

The concept of whole brain-based statistical analysis (WBSA) comprises a voxelwise comparison of normalized FA or MD values between two subject groups in stereotaxic space.

The scheme for this kind of analysis is visualized in Figure 7. For each voxel position, Student's t-test is performed voxelwise, i.e. the FA-values of the patients´ FA-maps are compared with the corresponding FA-values of the healthy volunteers´ FA-maps for each voxel separately. Prior to calculation, the FA-maps are masked by a WM mask derived by a thresholding procedure by which FA-values below 0.2 are not considered for calculation since cortical gray matter shows FA-values up to 0.2 (Kunimatsu et al., 2004).

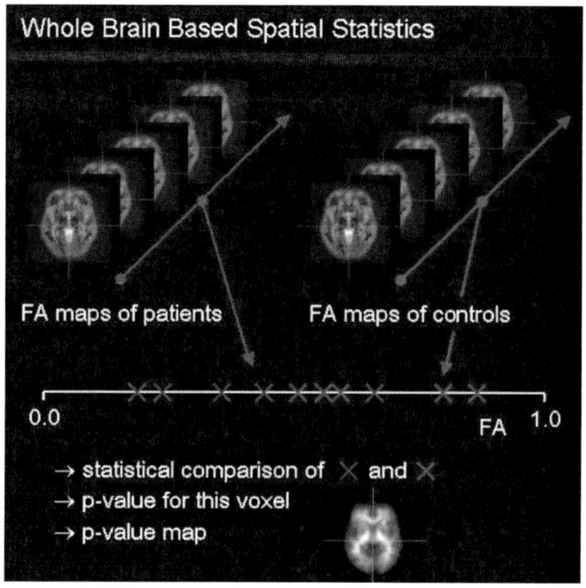

Figure 7. Schematical display of whole brain-based statistical analysis (WBSA). A voxelwise statistical comparison of FA maps of patients with the same pathology and FA maps of controls leads to a p-value map.

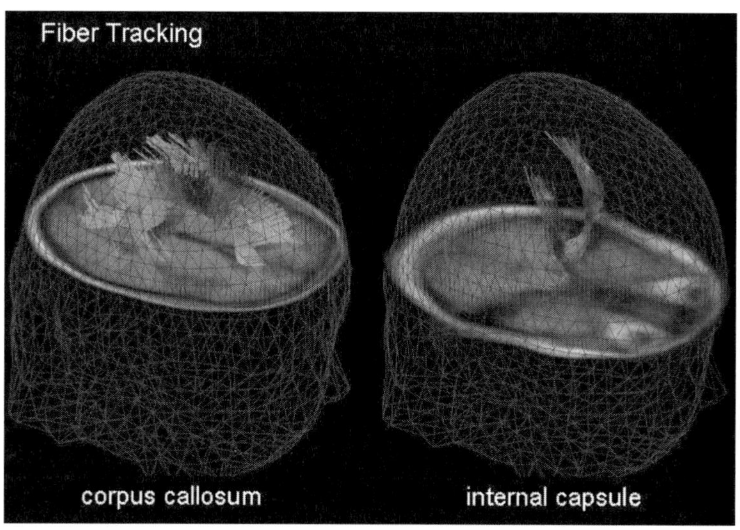

Figure 8. Fiber tracking with starting points in the corpus callosum (left) and in the posterior limb of the internal capsule (right) for an averaged DTI data set (arithmetically averaged from the DTI data of 13 normalized controls). Display background is a customized averaged T1-weighted data set.

The statistical results should be corrected for multiple comparisons using e.g. the false-discovery-rate (FDR) algorithm (Genovese et al., 2002). Further reduction of the alpha error could be performed by a spatial correlation algorithm that eliminates isolated voxels or small isolated groups of voxels (cluster threshold size).

2. Fiber Tracking on Group Averaged Data

In order to apply fiber tracking (FT) algorithms, additional data processing steps were performed. According to the methods previously described (Mueller et al., 2007b, Mueller et al., 2009), averaged DTI data sets were generated. The averaging requires a careful treatment of the orientational information (which has been preserved during the normalization process – for details refer to (Mueller et al., 2009) that used techniques described in (Alexander et al., 2001)). Tractography was performed by application of a FT algorithm (STT) (Figure 6). The group averaged DTI data sets were the basis for the consecutive FT which was performed using manually defined seed points adjacent to the local maxima detected by the whole-brain based FA analysis within the CC and the posterior limb of the internal capsule. Figure 8 shows examples for FT with starting points in the corpus callosum and in the posterior limb of the internal capsule for an averaged DTI data set.

In order to quantify the tractography results, a technique named tractwise fractional anisotropy statistics (TFAS) (Mueller et al., 2007b) could be applied. Here, the fiber tracts that had been created on the averaged DTI data set were used for the selection of the voxels that contribute to a statistical comparison between FA maps from the averaged DTI data of the patients and the FA maps from the averaged DTI data of the controls. All resulting voxels showing a FA above 0.2 were considered for statistical analysis by use of a Student's t-test.

Alternative techniques were reported by Corouge and co-workers (Corouge et al. 2006) or Behrens and co-workers (Behrens et al. 2007).

There is an ongoing discussion about the choice of the optimum FT algorithm (Ni et al., 2006). The minimum number of recorded directions is six. Nevertheless, for an improved signal-to-noise ratio in consecutive FT, at least 32 or even better 64 gradient directions should be recorded. In order to improve the SNR in recordings with a low number of gradient directions, the TBSS algorithm offers solutions. TBSS aims at the improvement of sensitivity, objectivity and interpretability of multi-subject diffusion imaging studies. Therefore, no exact spatial alignment and no smoothing are needed in TBSS. The main difference to the TFAS approach is that averaged DTI data sets have been created before they undergo the statistical analysis. The advantage of using averaged DTI data sets (including a correct realignment of all diffusion properties) is that due to the arithmetical averaging an improvement in the signal-to-noise ratio is achieved.

Unfortunately, the STT approach of reconstructing the fiber tracts step by step can often not resolve crossing fibers which appear in a one recorded element due to restricted voxel resolution of some millimeters. The problem is partially overcome with either preprocessing of tensor field (Basser et al., 2000, Coulon et al., 2004) or more sophisticated methods of tractography involving either regularization of the bundle trajectories (Poupon et al., 2000, Lazar et al., 2003) or probabilistic strategies based on Monte Carlo sampling and models of uncertainty about fiber orientations (Behrens et al., 2003b). The fiber orientation distribution function (ODF) inferred in each voxel from DTI, however, is not sufficient to map successfully the large-scale connectivity of the cortex because of the amount of crossings involved (Mangin et al., 2002). A method beyond DTI to carry further the idea that the fiber directions can be inferred from the local maxima of the amplitude of water molecule radial displacements is the so-called `q-ball imaging` (Tuch et al., 2003) - the interested reader is referred to (Perrin et al., 2008, Tuch, 2004).

3. Complementary Data Analyses

FA or MD analysis alone is sensitive to many factors and is very heterogeneous throughout the brain; in particular, FA may appear low in regions of crossing fibers or where there is no preferred fiber bundle direction dominating the region (Kolind et al., 2008). In order to detect regional alterations throughout the whole brain, i.e. also in areas where one MRI modality alone may fail due to its inherent limits of sensitivity, comprehensive information obtained from different analysis techniques could be helpful to get a widespread impression of WM differences between patients and controls. Therefore, several studies over the last years have reported the multiparametric approach (Reich et al., 2007) and the complementary approach (Kolind et al., 2008). The combination of MRI parameters that have a different access to human brain structures is useful, such as DTI as a technique for microstructural mapping of directionality and integrity of the white matter fiber bundles and e.g. intensity-based analysis of 3-D T1-weighted MRI as a technique for mapping macrostructural alterations in the human brain. A future concept for MRI is an integrative concept as a synopsis for congruent and comprehensive analysis (Mueller et al., 2008, Mueller et al., 2009).

Software solutions such as TIFT are able to process DTI data analysis, 3-D T1-weighted MRI analysis and other analysis techniques, e.g. functional MRI and/or magnetization transfer MRI analyses, allowing a variety of different analyses in one software environment. However, specialized analysis software exists for each analysis technique and the concept of TIFT also includes the import of results from other analysis software to perform complementary data display.

F. Applications of DTI in WM Pathology

1. DTI Parameters in WM Pathology Studies

DTI provides quantitative metrics of the diffusion process, the FA and MD. These metrics have already been shown to be sensitive markers for studying a wide range of WM pathologies, such as stroke (Pierpaoli et al., 2001, Werring et al., 2001), multiple sclerosis (Reich et al., 2008), schizophrenia (Kyriakopoulos et al., 2008, Seok et al., 2007), amyotrophic lateral sclerosis (Abe et al., 2004, Agosta et al., 2007, Ciccarelli et al., 2006, Ciccarelli et al., 2008, Sage et al., 2007, Toosy et al., 2003, Wang et al., 2006), idiopathic restless leg syndrome (Unrath et al., 2008) and several other neurological disorders (Borroni et al., 2007, Thomas et al., 2005, Wieshmann et al., 1999),

Additionally, DTI with FT was used to identify white matter tracts (Reich et al., 2008). This technique, while still in relative infancy and not in general clinical use, is emerging as a powerful tool for assessing pathway-specific abnormalities in neurological disease. Within the identified tracts, various quantitative MRI indices derived from DTI and additional acquisitions (e.g. T2 and/or magnetization transfer) that are anatomically coregistered to the DTI data could be measured. Thus, each index could be calculated as a function of position within the tract, referring to plots depicting their spatial variation as tract profiles.

Within the complementary or multiparametric approach, DTI is combined with other MRI modalities, such as T2- or T1-weighted imaging or MRS (Magnetic Resonance Spectroscopy) in order to obtain comprehensive and complementary information. Kolind et al. (Kolind et al., 2008) and independently Reich et al. (Reich et al., 2008) correlated DTI-FA results with T2 relaxation time in multiple sclerosis. Verma et al. (Verma et al., 2008) performed tissue characterization by the combination of DTI and 3-D T1-weighted imaging.

2. Special Example: Thinning of the Corpus Callosum

As an example for a groupwise comparison, differences between patients with atrophy of the CC and age-matched healthy controls were mapped. As a model of CC alteration, subjects with the neurodegenerative disease of complicated hereditary spastic paraparesis were investigated as a prototype of morphological alterations (thinning) along the whole structure of the corpus callosum (CC) (Kassubek et al., 2007, Sullivan et al., 2006). The CC is the most appropriate structure in the brain to be analyzed by FT since it is one of the white matter structures with the most strongly directed fibers (Fink, 2006).

All DTI data were acquired on the same 1.5 T scanner (Symphony, Siemens Medical, Erlangen, Germany). Thirteen patients with thinned CC (tCC) (male/female ration 8/5, 30.4 ± 10.0 years) and age- and gender-matched 13 healthy controls (male/female ration 8/5, average age 33.2 ± 4.5 years) underwent MRI. DTI data were acquired as 13 volumes (45 slices, 128 x 128 voxels, slice thickness 2.2 mm, in-plane voxel size 1.5 mm x 1.5 mm, in-plane field of view (FOV) 192 mm x 192 mm), representing 12 gradient directions and one scan with b=0, i.e. gradients were not performed,. Echo time (TE) and repetition time (TR) were 93 ms and 8000 ms, respectively. b was 800 s/mm^2, five scans were k-space averaged online by the Siemens SYNGO® operating software. Acquisition scheme was interleaved.

In order to obtain a morphological background, a high-resolution T1-weighted magnetization-prepared rapid-acquisition gradient echo sequence was used (MPRAGE, TR = 9.7 ms, TE = 3.93 ms, flip angle 15°, matrix size 256 x 256 mm^2, voxel size 1.0 x 0.96 x 0.96 mm^3, in-plane FOV 246 mm x 246 mm, 192 slices).

Figure 9, upper panel, shows differences in FA maps (left) as well as in MD maps (right) of 13 patients with a pathological thinning of the corpus callosum and 13 controls. Highly significant differences appear throughout the whole white matter. Figure 9, lower panel, shows the complementary display of statistical comparison of FA maps and of MD maps.

For the exemplary data shown here, highly significant differences ($p < 0.001$) in TFAS analysis for FT in the CC (Figure 8, left) as well as for FT in the posterior limb of the internal capsule (Figure 8, right) were found.

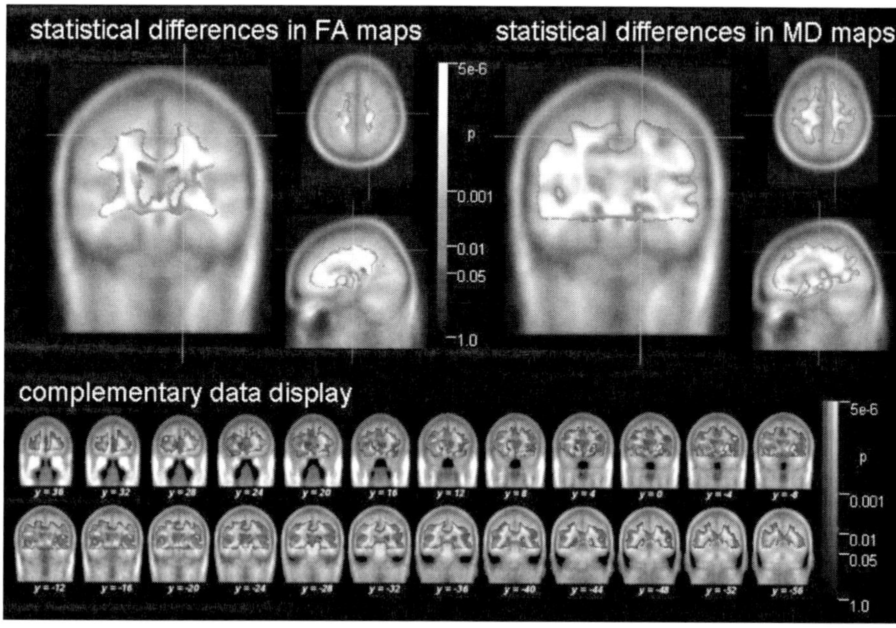

Figure 9. upper panel: statistical differences in FA maps (left) and in MD maps (right) of 13 patients with a pathological thinning of the corpus callosum and 13 controls for $p < 0.05$, corrected for multiple comparisons. Display background is an averaged T1-weighted data set (from the 26 subjects participating in the study). Lower panel: complementary display of statistical comparison of FA maps and of MD maps ($p < 0.05$, corrected) of 13 patients with a pathological thinning of the corpus callosum and 13 controls. Differences in FA are in red color, green color shows the differences in MD and the overlap is marked in yellow.

G. Discussion and Summary

In this chapter, an overview of the DTI technique and the analysis and quantification of diffusion properties was presented. After the transformation into a stereotaxic standard space, averaging of single subject diffusion properties becomes feasible, thus allowing for subject averaging to increase the SNR and also for the comparison of subject groups. Supplementary results could be provided by combined analysis of advanced MRI-based techniques of the computational neuroanatomy, i.e. the combination of DTI with e.g. intensity-based analysis of 3-D T1-weighted MRI in standard space (Mueller et al., 2011).

MNI normalization of DTI data is an ongoing topic of research (Lee et al., 2009). Recent works suggest viscous fluid models or mutual information for the non-rigid, i.e. non-affine, image registration (D'Agostino et al., 2003, D'Agostino et al., 2006, D'Agostino et al., 2007, VanHecke et al., 2007) resulting in the construction of a stereotaxic DTI atlas of the healthy human brain containing full diffusion tensor information (VanHecke et al., 2008, Verhoeven et al., 2010).

Several DTI parameterization and tractography techniques were discussed that are presently being used for white matter characterization and tract tracing. Beyond the differentiation of patient and control groups in many fields of white matter pathologies by the FA and the MD metrics, even the simple FT methodologies are already able to visualize major white matter connections *in vivo* in humans (Yamada et al., 2009). Upon using these data to investigate specific neuroanatomical questions, it is very important to keep in mind the limitations of the DTI method used to acquire them. First of all, this technique can be only used for analysis of white matter architecture, but not to address connectivity questions at the cellular level.

One particular limit related to this macroscopic character of DTI is the mixing of axonal tracts with different orientations within a voxel. DTI may be able to locate where these problematic voxels are, but it is difficult to decipher axonal information in such voxels (Mori and van Zijl, 2002). Furthermore, there are some approaches that circumwent this issue under favorable conditions, such as the use of reference ROI placement based on prior knowledge. The most important conclusion that can be drawn in this initial phase of the field is that DTI tractography can indeed delineate the core of large white matter tracts as judged from the encouraging results from initial validation studies (Mori and van Zijl, 2002).

At present, there are no other non-invasive techniques that can provide equivalent information on white matter structures and, as a consequence, DTI tractography is expected to be a powerful technique to investigate white matter anatomy and disease *in vivo* in humans (Mori and van Zijl, 2002).

References

Abe O, Yamada H, Masutani Y, Aoki S, Kunimatsu A, Yamasue H, Fukuda R, Kasai K, Hayashi N, Masumoto T, Mori H, Soma T, Ohtomo K. Amyotrophic lateral sclerosis: Diffusion tensor tractography and voxel-based analysis. *NMR in Biomedicine* 2004; *17*:411–416.

Agosta F, Pagani E, Rocca MA, Caputo D, Perini M, Salvi F, Prelle A, Filippi M. Voxel-based morphometry study of brain volumetry and diffusivity in amyotrophic lateral sclerosis patients with mild disability. *Human Brain Mapping* 2007; *28*:1430–1438.

Andersson JLR, Skare S, Ashburner J. How to correct susceptibility distortions in spin-echo echo-planar images: application to diffusion tensor imaging. *Neuroimage* 2003, *20*: 870-888.

Alexander DC, Pierpaoli C, Basser PJ, Gee JC. Spatial transformations of diffusion tensor magnetic resonance images. *IEEE Transactions on Medical Imaging* 2001, *20*: 1131-1139.

Ashburner J, Friston KJ. Nonlinear Spatial Normalization Using Basis Functions. *Human Brain Mapping* 1999, *7*: 254-266.

Basser PJ, Mattiello J, LeBihan D. Estimation of the effective self-diffusion tensor from the NMR spin echo. *Journal of Magnetic Resonance B* 1994, *103*: 247-254.

Basser PJ, Mattiello J, LeBihan D. MR Diffusion Tensor Spectroscopy and Imaging. *Biophysical Journal* 1994, *66*: 259-267.

Basser PJ, Pajevic S, Pierpaoli C, Duda J, Aldroubi A. In vivo tractography using DT-MRI data. *Magnetic Resonance in Medicine* 2000, *44*: 625-632.

Basser PJ, Jones DK. Diffusion-tensor MRI: theory, experimental design and data analysis – a technical review. *NMR in Biomedicine* 2002, *15*: 456-467.

Beaulieu C. The basis of anisotropic water diffusion in the nervous system. *NMR in Biomedicine* 2002; *15*: 435–455.

[Behrens2003a] Behrens TE, Johansen-Berg H, Woolrich MW, Smith SM, Wheeler-Kingshott CA, Boulby PA, Barker GJ, Sillery EL, Sheehan K, Ciccarelli O, Thompson AJ, Brady JM, Matthews PM. Non-invasive mapping of connections between human thalamus and cortex using diffusion imaging. *Nature Neuroscience* 2003, *6*: 750-757.

Behrens TE, Woolrich MW, Jenkinson M, Johansen-Berg H, Nunes RG, Clare S, Matthews PM, Brady JM, Smith SM. Characterization and propagation of uncertainty in diffusion-weighted MR imaging. *Magnetic Resonance in Medicine* 2003; *50*: 1077-1088.

Behrens TE, Berg HJ, Jbabdi S, Rushworth MF, Woolrich MW. Probabilistic diffusion tractography with multiple fibre orientations: What can we gain? *Neuroimage* 2007, *34*: 144-155.

Borroni B, Brambati SM, Agosti C, Gipponi S, Bellelli G, Gasparotti R, Garibotto V, Di Luca M, Scifo P, Perani D, Padovani A. Evidence of white matter changes on diffusion tensor imaging in frontotemporal dementia. *Archives of Neurology* 2007; *64*: 246–251.

Boss A, Martirosian P, Klose U, Nägele T, Claussen CD, Schick F. FAIR-TrueFISP imaging of cerebral perfusion in areas of high magnetic susceptibility differences at 1.5 and 3 Tesla. *Journal of Magnetic Resonance Imaging* 2007; *25*: 924-931.

Brett M, Johnsrude IS, Owen AM. The problem of functional localization in the human brain. *Nature Reviews Neuroscience* 2002, *3*: 243-249.

Bueltmann E, Erb G, Kirchin MA, Klose U, Naegele T. Intra-individual crossover comparison of gadobenate dimeglumine and gadopentetate dimeglumine for contrast-enhanced magnetic resonance angiography of the supraaortic vessels at 3 Tesla. *Investigative Radiology* 2008; *43*: 695-702.

Chenevert TL, Brunberg JA, Pipe JG. Anisotropic diffusion in human white matter: demonstration with MR techniques in vivo. *Radiology* 1990, *177*: 401-405.

Ciccarelli O, Behrens TE, Altmann DR, Orrell RW, Howard RS, Johansen-Berg H, Miller DH, Matthews PM, Thompson AJ. Probabilistic diffusion tractography: A potential tool to assess the rate of disease progression in amyotrophic lateral sclerosis. *Brain* 2006; *129*:1859–1871.

Ciccarelli O, Behrens TE, Johansen-Berg H, Talbot K, Orrell RW, Howard RS, Nunes RG, Miller DH, Matthews PM, Thompson AJ, Smith SM. Investigation of white matter pathology in ALS and PLS using tract-based spatial statistics. *Human Brain Mapping* 2009; *30*: 615–624.

Collins DL, Neelin P, Peters TM, Evans AC. Automatic 3-D intersubject registration of MR volumetric data in standardized Talairach space. *Journal of Computer Assisted Tomography* 1994, *18*: 192-205.

Conturo TE, Lori NF, Cull TS, Akbudak E, Snyder AZ, Shimony JS, McKinstry RC, Burton H, Raichle ME. Tracking neuronal fibre pathways in the living human brain. *Proceedings of the National Academy of Sciences USA*. 1999, *96*: 10422-10427.

Corouge I, Fletcher PT, Joshi S, Gouttard S, Gerig G. Fibre tract-oriented statistics for quantitative diffusion tensor MRI analysis. *Medical Image Analysis* 2006, *10*: 786-798.

Coulon O, Alexander DC, Arridge S. Diffusion tensor magnetic resonance image regularization. *Medical Image Analysis* 2004, *8*: 47–67.

D'Agostino E, Maes F, Vandermeulen D, Suetens P. A viscous fluid model for multimodal non-rigid image registration using mutual information. *Medical Image Analysis* 2003; *7*: 565-575.

D'Agostino E, Maes F, Vandermeulen D, Suetens P. An information theoretic approach for non-rigid image registration using voxel class probabilities. *Medical Image Analysis* 2006; *10*: 413-431.

D'Agostino E, Maes F, Vandermeulen D, Suetens P. Atlas-to-image non-rigid registration by minimization of conditional local entropy. *Information Processing in Medical Imaging.* 2007; *20*: 320-332.

Douek P, Turner R, Pekar J, Patronas N, Le Bihan D. MR color mapping of myelin fiber orientation. *Journal of Computer Assisted Tomography* 1991, *15*: 923-929.

Fink JK. Hereditary spastic paraplegia. *Current Neurology and Neuroscience Report* 2006, *6*: 65-76.

Friston KJ, Ashburner J, Frith CD, Poline J-B, Heather JD, Frackowiak RSJ. Spatial registration and normalization of images. *Human Brain Mapping* 1995, *2*: 165-189.

Genovese CR, Lazar NA, Nichols T. Thresholding of statistical maps in functional neuroimaging using the false discovery rate. *Neuroimage* 2002, *15*: 870-878.

Johansen-Berg H, Behrens TE. Just pretty pictures? What diffusion tractography can add in clinical neuroscience. *Current Opinion in Neurology* 2006, *19*: 379-385.

Jones DK, Simmons A, Williams SCR, Horsfield MA. Non-invasive assessment of axonal fiber connectivity in the human brain via diffusion tensor MRI. *Magnetic Resonance in Medicine* 1999, *42*: 37-41.

Jones DK, Symms MR, Cercignani M, Howard RJ. The effect of filter size on VBM analyses of DT-MRI data. *Neuroimage* 2005; *26*: 546-554.

Kanaan RA, Shergill SS, Barker GJ, Catani M, NG VW, Howard R, McGuire PK, Jones DK. *Psychiatry Research* 2006, *146*: 73-82.

Kassubek J, Juengling FD, Baumgartner A, Ludolph AC, Sperfeld A-D. Different regional brain volume loss in pure and complicated hereditary spastic paraparesis: a voxel-based morphometry study. *Amyotrophic Lateral Sclerosis* 2007, *8*: 328-336.

Kochunov P, Thompson PM, Lancaster JL, Bartzokis G, Smith S, Coyle T, Royall DR, Laird A, Fox PT. Relationship between white matter fractional anisotropy and other indices of cerebral health in normal aging: tract-based spatial statistics study of aging. *Neuroimage* 2007, *35*: 478-87.

Kolind SH, Laule C, Vavasour IM, Li DK, Traboulsee AL, Mädler B, Moore GR, Mackay AL. Complementary information from multi-exponential T2 relaxation and diffusion tensor imaging reveals differences between multiple sclerosis lesions. *Neuroimage* 2008; *40*: 77-85.

Kunimatsu A, Aoki S, Masutani Y, Abe O, Hayashi N, Mori H, Masumoto T, Ohtomo K. The optimal trackability threshold of fractional anisotropy for diffusion tensor tractography of the corticospinal tract. *Magnetic Resonance in Medical Sciences* 2004; *3*:11-17.

Kyriakopoulos M, Vyas NS, Barker GJ, Chitnis XA, Frangou S. A diffusion tensor imaging study of white matter in early-onset schizophrenia. *Biological Psychiatry* 2008; *63*: 519–523.

Lazar M, Weinstein DM, Tsuruda JS, Hasan KM, Arfanakis K, Meyerand ME, Badie B, Rowley HA, Haughton V, Field A, Alexander AL. White matter tractography using diffusion tensor deflection. *Human Brain Mapping* 2003, *18*: 306-321.

Le Bihan D, Breton F, Lallemand D, Aubin ML, Vignaud J, Laval-Jeantet M. Separation of diffusion and perfusionin intravoxel incoherent motion MR imaging. *Radiology* 1988, *168*: 497-505.

Le Bihan D, Mangin JF, Poupon C, Clark CA, Pappata S, Molko N, Chabriat H, Diffusion tensor imaging : concepts and applications. *Journal of Magnetic Resonance Imaging* 2001, *13*: 534-546.

Lee JE, Chung MK, Lazar M, DuBray MB, Kim J, Bigler ED, Lainhart JE, Alexander AL. A study of diffusion tensor imaging by tissue-specific, smoothing-compensated voxel-based analysis. *Neuroimage* 2009; *44*: 870-883.

Lehmann TM, Gonner C, Spitzer K. Survey: interpolation methods in medical image processing. *IEEE Transactions on Medical Imaging* 1999, *18*: 1049-1075.

Lori NF, Akbudak E, Shimony JS, Cull TS, Snyder AZ, Guillory RK, Conturo TE. Diffusion tensor fibre tracking of human brain connectivity: aquisition methods, reliability analysis and biological results. *NMR in Biomedicine* 2002, *15*: 494-515.

Mangin JF, Poupon C, Cointepas Y, Rivière D, Papadopoulos-Orfanos D, Clark CA, Régis J, Le Bihan D. A framework based on spin glass models for the inference of anatomical connectivity from diffusion-weighted MR data - a technical review. *NMR in Biomedicine* 2002; *15*: 481-492.

Matiello J, Basser JP, Le Bihan D. Analytical expression for the b-matrix in NMR diffusion imaging and spectroscopy. *Journal of Magnetic Resonance A* 1994; *108*: 131-141.

Matiello J, Basser JP, Le Bihan D. The b matrix in diffusion tensor echo-planar imaging. *Magnetic Resonance in Medicine* 1997; *37*: 292-300.

Mishra A, Lu Y, Meng J, Anderson AW, Ding Z. Unified framework for anisotropic interpolation and smoothing of diffusion tensor images. *Neuroimage* 2006; *31*: 1525-1535.

Mishra A, Lu Y, Choe AS, Aldroubi A, Gore JC, Anderson AW, Ding Z. An image-processing toolset for diffusion tensor tractography. *Magnetic Resonance Imaging* 2007, *25*:365-376.

Morgan PS, Bowtell RW, McIntyre DJ, WorthingtonBS. Correction of spatial distortion in EPI due to inhomogeneous static magnetic fields using the reversed gradient method. *Journal of Magnetic Resonance Imaging* 2004; *19*: 499-507.

Mori S, Cain B, Chacko VP, van Zijl PCM. Three dimensional tracking of axonal projections in the brain by magnetic resonance imaging. *Annals of Neurology* 1999; *45*: 265-269.

Mori S, van Zijl PCM. Fiber tracking: principles and strategies - a technical review. *NMR in Biomedicine* 2002; *15*: 468 – 480.

Moseley ME, Cohen Y, Kucharczyk J, Mintorovitch J, Asgari HS, Wendland MF, Tsuruda J, Norman D. Diffusion-weighted MR imaging of anisotropic water diffusion in cat central nervous system. *Radiology* 1990; *176*: 439–445.

Müller H-P, Unrath A, Ludolph AC, Kassubek J. Preservation of Diffusion Tensor Properties during Spatial Normalization by use of Tensor imaging and Fiber Tracking on a Normal Brain Database. *Physics in Medicine and Biology* 2007; *52*: N99-109.

Müller H-P, Unrath A, Sperfeld AD, Ludolph AC, Riecker A, Kassubek J. Diffusion tensor imaging and tractwise fractional anisotropy statistics: quantitative analysis in white matter pathology. *BioMedical Engineering OnLine* 2007, *6*: 42.

Müller H-P, Kassubek J. Multimodal imaging in neurology – special focus on MRI applications and MEG. *Lectures in Biomedical Engineering*, BME brochure, Morgan and Claypool Publishers (San Rafael, CA, USA), 2008; ISBN: 1598295500 (paperback), 1598295519 (ebook): 1-75.

Müller H-P, Unrath A, Riecker A, Pinkhardt EH, Ludolph AC, Kassubek J. Inter-subject variability in the analysis of diffusion tensor imaging at the group level: fractional anisotropy mapping and fiber tracking techniques. *Magnetic Resonance Imaging* 2009; *27*: 324-334.

Müller HP, Lulé D, Unrath A, Ludolph AC, Riecker A, Kassubek J. Complementary Image Analysis of Diffusion Tensor Imaging and 3-Dimensional T1-Weighted Imaging: White Matter Analysis in Amyotrophic Lateral Sclerosis. *Journal of Neuroimaging* 2011; *21:* 24-33.

Neeman M, Freyer JP, Sillerud LO. Pulsed-gradient spin-echo studies in NMR imaging. Effects of the imaging gradients on the determination of diffusion coefficients. *Journal of Magnetic Resonance* 1990; *90*: 303-312.

Ni H, Kavcic V, Zhu T, Ekholm S, Zhong J. Effects of number of diffusion gradient directions on derived diffusion tensor imaging indices in human brain. *American Journal of Neuroradiology* 2006; *27*: 1776-1781.

Pagani E, Filippi M, Rocca MA, Horsfield MA. A method for obtaining tract-specific diffusion tensor MRI measurements in the presence of disease: Application to patients with clinically isolated syndromes suggestive of multiple sclerosis. *Neuroimage* 2005; *26*: 258-265.

Pajevic S, Pierpaoli C. Color schemes to represent the orientation of anisotropic tissues from diffusion tensor data: application to white matter fiber tract mapping in the human brain. *Magnetic Resonance in Medicine* 1999; *42*: 526-540.

Perrin M, Cointepas Y, Cachia A, Poupon C, Thirion B, Rivière D, Cathier P, El Kouby V, Constantinesco A, Le Bihan D, Mangin JF. Connectivity-Based Parcellation of the Cortical Mantle Using q-Ball Diffusion Imaging. *International Journal of Biomedical Imaging* 2008; *2008*: 368406.

Pierpaoli C, Basser PJ. Toward a quantitative assessment of diffusion anisotropy. *Magnetic Resonance in Medicine* 1996; *36*: 893-906.

Pierpaoli C, Barnett A, Pajevic S, Chen R, Penix L, Virta A, Basser P. Water diffusion changes in Wallerian degeneration and their dependence on white matter architecture. *Neuroimage* 2001; *13*:1174–1185.

Poupon C, Clark CA, Frouin V, Régis J, Bloch I, Le Bihan D, Mangin J. Regularization of diffusion-based direction maps for the tracking of brain white matter fascicles. *Neuroimage*. 2000, *12*: 184-195.

Press WH, Flannery BP, Teukolsky SA, Vetterling WT. Numerical Recipes in C: The Art of Scientific Computing. *Cambridge University Press*, New York, 1992.

Reich DS, Zackowski KM, Gordon-Lipkin EM, Smith SA, Chodkowski BA, Cutter GR, Calabresi PA. Corticospinal tract abnormalities are associated with weakness in multiple sclerosis. *American Journal of Neuroradiology* 2008; *29*: 333–339.

Rosenfeld, Kak AC. Digital Picture Processing, 2nd ed. Vol. I and II, *Academic Press*, Orlando, 1982.

Rossi C, Boss A, Steidle G, Martirosian P, Klose U, Capuani S, Maraviglia B, Claussen CD, Schick F. Water diffusion anisotropy in white and gray matter of the human spinal cord. *Journal of Magnetic Resonance Imaging*. 2008; *27*: 476-482.

Rossi C, Boss A, Lindig TM, Martirosian P, Steidle G, Maetzler W, Claussen CD, Klose U, Schick F. Diffusion tensor imaging of the spinal cord at 1.5 and 3.0 Tesla. *Fortschritte auf dem Gebiete der Röntgenstrahlen und der Nuklearmedizin* 2007; *179*: 219-224.

Sage CA, Peeters RR, Gorner A, Robberecht W, Sunaert S. Quantitative diffusion tensor imaging in amyotrophic lateral sclerosis. *Neuroimage* 2007; *34*: 486–499.

Sage CA, Van Hecke W, Peeters R, Sijbers J, Robberecht W, Parizel P, Marchal G, Leemans A, Sunaert S. Quantitative diffusion tensor imaging in amyotrophic lateral sclerosis: Revisited. *Human Brain Mapping* 2009; *30*: 3657-3675.

Seok JH, Park HJ, Chun HW, Lee SK, Cho HS, Kwon JS, Kim JJ. White matter abnormalities associated with auditory hallucinations in schizophrenia: A combined study of voxel-based analyses of diffusion tensor imaging and structural magnetic resonance imaging. Psychiatry Res: *Neuroimage* 2007; *156*: 93–104.

Shen Y, Larkman DJ, Counsell S, Pu IM, Edwards D, Hajnal JV. Correction of High-Order Eddy Current Induced Geometric Distortion in Diffusion-Weighted Echo-Planar images. *Magnetic Resonance in Medicine* 2004; *52*: 1184-1189.

Smith SM, Jenkinson M, Johansen-Berg H, Rueckert D, Nichols TE, Mackay CE, Watkins KE, Ciccarelli O, Cader MZ, Matthews PM, Behrens TE. Tract-based spatial statistics: voxelwise analysis of multi-subject diffusion data. *Neuroimage* 2006; *31*: 1487-1505.

Smith SM, Johansen-Berg H, Jenkinson M, Rueckert D, Nichols TE, Miller KL, Robson MD, Jones DK, Klein JC, Bartsch AJ, Behrens TE. Acquisition and voxelwise analysis of multi-subject diffusion data with tract-based spatial statistics. *Nature Protocols* 2007; *2*: 499-503.

Sullivan EV, Adalsteinsson E, Pfefferbaum A. Selective age-related degradation of anterior callosal fibre bundles quantified in vivo with fibre tracking. *Cerebral Cortex* 2006, *16*:1030-1039.

Talairach J, Tournoux P. Coplanar stereotaxic atlas of the human brain. Thieme Medical, New York, 1988.

Thomas B, Eyssen M, Peeters R, Molenaers G, Van Hecke P, De Cock P, Sunaert S. Quantitative diffusion tensor imaging in cerebral palsy due to periventricular white matter injury. *Brain* 2005; *128*: 2562–2577.

Toosy AT, Werring DJ, Orrell RW, Howard RS, King MD, Barker GJ, Miller DH, Thompson AJ. Diffusion tensor imaging detects corticospinal tract involvement at multiple levels in amyotrophic lateral sclerosis. *Journal of neurology, neurosurgery, and psychiatry* 2003; *74*: 1250–1257.

Tuch DS, Reese TG, Wiegell MR, Wedeen VJ. Diffusion MRI of complex neural architecture. *Neuron* 2003; *40*: 885–895.

Tuch DS. Q-ball imaging. *Magnetic Resonance in Medicine* 2004; *52*: 1358–1372.

Turner R, Le Bihan D, Maier J, Vavrek R, Hedges LK, Pekar J. Echo-planar imaging of intravoxel incoherent motion. *Radiology* 1990; *177*: 407-414.

Uluğ AM, van Zijl PC. Orientation-independent diffusion imaging without tensor diagonalization: anisotropy definitions based on physical attributes of the diffusion ellipsoid. *Journal of Magnetic Resonance Imaging* 1999; *9*: 804-813.

Unrath A, Müller H-P, Ludolph AC, Riecker A, Kassubek J. Cerebral white matter alterations in idiopathic restless legs syndrome, as measured by diffusion tensor imaging. *Movement Disorders* 2008; *23*: 1250-1255.

Van Hecke W, Sijbers J, D'Agostino E, Maes F, De Backer S, Vandervliet E, Parizel PM, Leemans A. On the construction of an inter-subject diffusion tensor magnetic resonance atlas of the healthy human brain. *Neuroimage*. 2008 *15*; 43: 69-80.

Van Hecke W, Leemans A, D'Agostino E, De Backer S, Vandervliet E, Parizel PM, Sijbers J. Nonrigid coregistration of diffusion tensor images using a viscous fluid model and mutual information. *IEEE Transactions on Medical Imaging* 2007; *26*: 1598-1612.

Verhoeven JS, Sage CA, Leemans A, Van Hecke W, Callaert D, Peeters R, De Cock P, Lagae L, Sunaert S. Construction of a stereotaxic DTI atlas with full diffusion tensor information for studying white matter maturation from childhood to adolescence using tractography-based segmentations. *Human Brain Mapping* 2010; *31*: 470-486.

Verma R, Zacharaki EI, Ou Y, Cai H, Chawla S, Lee SK, Melhem ER, Wolf R, Davatzikos C. Multiparametric tissue characterization of brain neoplasms and their recurrence using pattern classification of MR images. *Academic Radiology* 2008;*15*: 966-977.

Wang S, Poptani H, Bilello M, Wu X, Woo JH, Elman LB, McCluskey LF, Krejza J, Melhem ER. Diffusion tensor imaging in amyotrophic lateral sclerosis: Volumetric analysis of the corticospinal tract. *American Journal of Neuroradiology* 2006; *27*: 1234–1238.

Werring DJ, Clark CA, Droogan AG, Barker GJ, Miller DH, Thompson AJ. Water diffusion is elevated in widespread regions of normal-appearing white matter in multiple sclerosis and correlates with diffusion in focal lesions. *Multiple Sclerosis* 2001; *7*: 83–89.

Wilson M, Tench CR, Morgan PS, Blumhardt LD. Pyramidal tract mapping by diffusion tensor magnetic resonance imaging in multiple sclerosis: improving correlations with disability. *Journal of neurology, neurosurgery, and psychiatry* 2003; *74*: 203-207.

Wieshmann UC, Clark CA, Symms MR, Franconi F, Barker GJ, Shorvon SD. Reduced anisotropy of water diffusion in structural cerebral abnormalities demonstrated with diffusion tensor imaging. *Magnetic Resonance Imaging* 1999; *17*:1269–1274.

Wu YC, Field AS, Chung MK, Badie B, Alexander AL. Quantitative analysis of diffusion tensor orientation: theoretical framework. *Magnetic Resonance in Medicine* 2004; *52*: 1146-1155.

Yamada K, Sakai K, Akazawa K, Yuen S, Nishimura T. MR tractography: a review of its clinical applications. *Magnetic Resonance in Medical Sciences* 2009; *8*:165-174.

Section IV. Advances in Treatment

In: Handbook of Technology in Psychology ... ISBN: 978-1-62100-004-4
Editor: Luciano L'Abate and David A. Kasier

Chapter 14

USING VIRTUAL REALITY FOR CLINICAL ASSESSMENT AND INTERVENTION

Albert "Skip" Rizzo[1], Thomas D. Parsons, Patrick Kenny and J. Galen Buckwalter
[1]University of Southern California, US

Virtual Reality (VR) technology offers new opportunities for the development of innovative assessment and intervention tools. VR-based testing, training, and treatment approaches that would be difficult, if not impossible, to deliver using traditional methods are now being developed that take advantage of the assets available with VR technology. If empirical studies continue to demonstrate effectiveness, VR applications could provide new options for targeting the cognitive, psychological, motor and functional impairments that result from various psychological and physical disorders and conditions. VR allows for the precise presentation and control of stimuli within dynamic multi-sensory 3D computer generated environments, as well as providing advanced methods for capturing and quantifying behavioral responses. These characteristics serve as the basis for the rationale for VR applications in the clinical assessment, intervention and training domains. This chapter will begin with a brief review of the history and rationale for the use of VR with clinical populations followed by a description of the technology for creating and using VR clinically. The chapter will then focus on reviewing three fundamental areas where Clinical VR has shown significant potential to enhance clinical practice and research (Exposure Therapy, Neuropsychological Assessment and Clinical Training with Virtual Patient agents). At the end of each of these sections, a detailed use-case will be presented. As in all areas of new technology design and development, it is easy for one to get caught up in excitement that surrounds the potential clinical possibilities, while casting a blind eye to the pragmatic challenges that exist for building useful and usable applications.

The goal of this chapter is to present a clear rationale for VR use across diverse areas of clinical practice and present examples of how this has been done successfully. While significant work has been done in other areas of Clinical VR (e.g. pain distraction, eating disorders, motor rehabilitation, etc.), a full treatment of such a broad literature is beyond the scope of this chapter. Thus, we have opted to provide more depth on specific clinical areas where VR has been applied to address Anxiety Disorders with exposure therapy, neuropsychological assessment and new visionary work developing virtual humans to serve the role of virtual standardized patients for clinical training.

THE HISTORY AND RATIONALE FOR CLINICAL VIRTUAL REALITY

Virtual reality (VR) has undergone a transition in the past few years that has taken it out of the realm of expensive toy and into that of functional technology. Over the last 15 years, a virtual revolution has taken place in the use of VR simulation technology for clinical purposes. Although media hype may have oversold VR's potential during the early stages of the technology's development, a uniquely suited match exists between the assets available with VR technology and applications in the clinical sciences. The capacity of VR technology to create controllable, multisensory, interactive 3D stimulus environments, within which human behavior can be motivated and measured, offers clinical assessment and intervention options that were not possible using previously available approaches. The unique match between Virtual Reality technology assets and the needs of various clinical application areas has been recognized by a determined and expanding cadre of researchers and clinicians who have not only recognized the potential impact of VR technology, but have now generated a significant research literature that documents the many clinical and research targets where VR can add value over traditional assessment and intervention methods (Glantz et al., 2003; Holden, 2005; Parsons and Rizzo, 2008a; Parsons, Rizzo, Rogers, and York, 2009; Powers and Emmelkamp, 2008; Rizzo et al., 2004; Rizzo and Kim, 2005; Rizzo et al., 2011abc) Rose, Brooks and Rizzo, 2005; Riva, 2011). Based on this, VR has now emerged as a promising tool in many domains of clinical care and research.

Virtual environments (VEs) have been developed that are now demonstrating effectiveness in a number of areas in clinical psychology, neuropsychology and in both cognitive and motor rehabilitation. A short list of areas where Clinical VR has been usefully applied includes fear reduction in persons with simple phobias (Parsons and Rizzo, 2008a; Powers and Emmelkamp, 2008), treatment for PTSD (Rothbaum et al., 2001; Difede et al., 2002, 2007; Rizzo et al., 2010ab, 2011b), stress management in cancer patients (Schneider et al., 2010), acute pain reduction during wound care and physical therapy with burn patients (Hoffman et al., 2011) and in other painful procedures (Gold et al., 2006), body image disturbances in patients with eating disorders (Riva, 2011), navigation and spatial training in children and adults with motor impairments (Stanton et al., 1998; Rizzo et al., 2004), functional skill training and motor rehabilitation with patients having central nervous system dysfunction (e.g., stroke, TBI, SCI, cerebral palsy, multiple sclerosis, etc.) (Holden, 2005; Merians et al., 2010), and for the assessment and rehabilitation of attention, memory, spatial skills and other cognitive functions in both clinical and unimpaired populations (Brooks et al., 1999; Brown et al., 1998; Matheis et al., 2005; Pugnetti et al., 1995; Rose et al., 2005; Rizzo et al., 2006, Parsons, Rizzo, Rogers, and York, 2009). To do this, VR scientists have constructed virtual airplanes, skyscrapers, spiders, battlefields, social settings, beaches, fantasy worlds and the mundane (but highly relevant) functional environments of the schoolroom, office, home, street and supermarket. Emerging research and development is also producing artificially intelligent virtual human patients that are being used to train clinical skills to health professionals (Kenny et al., 2010; Parsons et al., 2008b; Rizzo et al., in press; Lok et al., 2007).

In essence, clinicians can now create simulated environments that mimic the outside world and use them in clinical settings to immerse patients in simulations that support the aims and mechanics of a specific assessment or therapeutic approach. And this state of affairs now stands to transform the vision of future clinical practice and research in the disciplines of psychology, medicine, neuroscience, physical and occupational therapy, and in the many allied health fields that address the therapeutic needs of children and adults with healthcare issues and clinical disorders. As well, the clinical and research targets chosen for these applications reflect an informed appreciation for the assets that are available with VR technology (Rizzo et al., 2004) by clinicians/developers initially designing and using systems in this area. These initiatives give hope that in the 21st century, new and useful tools will be developed that will advance clinical areas that have long been mired in the methods of the past.

By its nature, VR simulation technology is well suited to simulate the challenges that people face in naturalistic environments, and consequently can provide objective simulations that can be useful for clinical assessment and intervention purposes. Within these environments, researchers and clinicians can present ecologically relevant stimuli embedded in a meaningful and familiar context. From this, VR offers the potential to create systematic human testing, training and treatment environments that allow for the precise control of complex, immersive, dynamic 3D stimulus presentations, within which sophisticated interaction, behavioral tracking and performance recording is possible. Much like an aircraft simulator serves to test and train piloting ability under a variety of controlled conditions, VR can be used to create relevant simulated environments where assessment and treatment of cognitive, emotional and motor problems can take place under a range of stimulus conditions that are not easily deliverable and controllable in the real world.

When combining these assets within the context of functionally relevant, ecologically enhanced VEs, a fundamental advancement could emerge in how human assessment and intervention can be addressed in many clinical and research disciplines. For example, instead of relying solely on unverifiable imagery processes in clients with anxiety disorders to produce the therapeutic effects of habituation, graduated exposure to feared or trauma-relevant stimuli can be delivered systematically in VR. As well, rather than try to predict real world functional performance from a decontextualized measure of attention when assessing children suspected of having ADHD, one can look at the effects of systematically increasing ecologically relevant attentional demands in a virtual environment, such as a classroom, social setting or home. These examples illustrate how VR technology can be used to provide exquisite timing and control over context-relevant imagery and stimulus load/complexity, all of which can be manipulated in a dynamic fashion contingent on the needs and responses of the client or research participant. Within such VEs, human performance can be digitally captured in real time in support of a precise and detailed analysis of relevant responses in relation to systematic stimulus presentations. In this regard, VR can be seen as capable of producing the "ultimate Skinner Box" for conducting human research, assessment and intervention.

Revolutionary advances in the underlying VR enabling technologies (i.e., computation speed and power, graphics and image rendering software, display systems, interface devices, immersive audio, haptics tools, wireless tracking, voice recognition, intelligent agents, and authoring software) have supported the creation of low-cost and usable VR systems capable of running on a commodity level personal computer. Such advances in technological "prowess" and accessibility have provided the hardware platforms needed for the conduct of human research and clinical intervention within more usable and useful VR scenarios. This convergence of the exponential advances in underlying VR enabling technologies with a growing body of clinical research and experience has fueled the evolution of the discipline of *Clinical Virtual Reality*. This has now supported the emergence of accessible VR systems that can uniquely target a wide range of psychological, cognitive and physical clinical targets and research questions.

This is in sharp contrast to what was possible in the mid-1990s when discussion of the potential for VR applications in the clinical assessment and intervention domains first emerged (Pugnetti, Mendozzi, Motta, Cattaneo, Barbieri, and Brancotti, 1995; Rizzo, 1994; Rose, Attree and Johnson, 1996). At that point in time, the technology to deliver on the anticipated VR "vision" was not in place. Consequently, during these early years, VR suffered from a somewhat imbalanced "expectation-to-delivery" ratio, as most users who eagerly lined up to try such systems during that time will attest. The "real" thing never quite measured up to expectations generated by some of the initial media hype, as delivered for example in the films "*The Lawnmower Man*" and "*Disclosure*"! Yet the idea of producing simulated virtual environments that allowed for the systematic delivery of ecologically relevant challenges was compelling and made intuitive sense. As well, the long and rich history of encouraging findings from the predecessor literature in aviation simulation (Hays et al., 1992) lent support to the concept that testing, training and treatment in highly proceduralized VR simulation environments would be a useful direction for clinical disciplines to explore (Johnston, 1995; Rizzo, 1994). Within this context, a small group of innovative clinicians and researchers also began the initial work of exploring the use of VR technology for applications designed to treat simple phobias (Hodges et al., 1995; Lamson, 1994; Rothbaum et al., 1995), while others addressed cognitive/functional performance in populations with central nervous system dysfunction (Brown et al., 1998; Pugnetti et al., 1995; Rizzo, 1994; Rose et al., 1996). While a good deal of this early work employed the costly, cumbersome, low resolution VR head mounted displays (HMDs) that were available at the time or simply used flatscreen monitors or stereoscopic projection approaches, these systems began to produce encouraging results (Cromby et al., 1996; Rizzo et al., 1998; Rose et al., 2000; Stanton et al., 1998). From these nascent efforts, findings emerged that began to demonstrate the unique value of the technology, served to inform ideas for future applications and created a grassroots level of enthusiasm for using VR that has continued grow and be supported into the present day. However, now instead of having to rely on $200,000 graphic workstations (and other expensive peripheral technologies) that were required back in 1990's, clinicians and researchers in the 21st century can now create and deliver compelling virtual worlds using a standard laptop and a $1500 HMD or stereo television. The technology has now caught up with the vision and such exponential advances are expected to continue to advance the science and practice in the discipline of Clinical VR.

Virtual Reality Definitions and Technology

Virtual Reality has been very generally defined as "...a way for humans to visualize, manipulate, and interact with computers and extremely complex data." (Aukstakalnis and Blatner, 1992). From this baseline perspective, VR can be seen as an advanced form of human-computer interface (Rizzo, Buckwalter and Neumann, 1997) that allows the user to "interact" with computers and digital content in a more natural or sophisticated fashion relative to what is afforded by standard mouse and keyboard input devices. And in some cases, with the aid of specialized VR display devices, users can become "immersed" *within* a computer generated simulated environment that changes in a natural/intuitive way with user interaction. VR sensory stimuli can be delivered by using various forms of visual display technology that can present real-time computer graphics and/or photographic images/video along with a variety of other sensory display devices that can present audio, "force-feedback" haptic/touch sensations and even olfactory content to the user.

However, VR is not defined or limited by any one technological approach or hardware set up. The creation of an engaged virtual reality *user experience* can be accomplished using combinations of a wide variety of interaction devices, sensory display systems, and in the design of content presented in a computer-generated graphic world. For example, *Immersive VR* can be produced by combining computers, head mounted displays (HMDs), body tracking sensors, specialized interface devices and real-time graphics to immerse a participant in a computer-generated simulated world that changes in a natural way with head and body motion. Thus, an engaged immersive virtual experience can be supported by employing specialized tracking technology that senses the user's position and movement and uses that information to update the sensory stimuli presented to the user to create the illusion of being immersed "in" a virtual space in which they can interact. One common configuration employs a combination of a HMD and head tracking system that allows delivery of real-time computer-generated images and sounds of a simulated virtual scene rendered in relation to user movements that corresponds to what the individual would see, hear and feel if the scene were real. Another method uses stereoscopic projection screens arrayed in various configurations, including six-walled rooms known as CAVES that allow users to interact in a less encumbered, wide field of view simulation environment. However, such CAVE systems are more costly and complex and are typically beyond the practical resources of a clinical service provider or basic researcher. In these immersive systems, one of the key aims is to perceptually replace the outside world with that of the simulated environment to create a specific user experience. Immersive HMD VR has been most commonly employed in applications where a controlled stimulus environment is desirable for constraining a user's perceptual experience within a specific synthetic world. This format has been often used in Clinical VR applications for anxiety disorder exposure therapy, analgesic distraction for patients suffering from acutely painful medical procedures and in the cognitive assessment of users with CNS dysfunction to measure performance under a range of systematically delivered task challenges and distractions.

By contrast, *Non-Immersive VR* is commonly experienced using modern computer and console games systems (as well as in non-game research lab generated systems). This format presents a three-dimensional (3D) graphic environment on a flatscreen monitor, projection system or television (no real world occlusion) within which the user can navigate and interact.

Albeit delivered on a less immersive display, such graphic worlds are still essentially a virtual reality *environment*. VEs presented on these widely available commodity display systems have the capacity to provide the user with significant options for interaction with dynamic digital content using traditional computer and game interface devices (e.g., keyboard, mouse, game pads, joysticks, etc.) in addition to more complex interaction devices that can track more natural user activity (e.g., data gloves, 3D mice, treadmills and some high-end "force feedback" exoskeleton devices).

The use of such ubiquitous display and interface devices has promoted widespread access to this form of non-immersive interactive media, primarily in the domain of entertainment. Moreover, researchers have investigated the value and usability of commercially available interaction devices and methods that can be used with flatscreen-delivered VEs that can allow users to interact with digital content using more naturalistic body actions beyond what is possible with traditional game interfaces (e.g. *Konami Dance Dance Revolution, Sony Eyetoy, Nintendo Wii, Novint Falcon, Microsoft Kinect*, etc.) (Lange et al., 2009, 2010). Regardless of the hardware and display format, the capacity of VR technology to create controllable, multisensory, interactive 3D stimulus environments, within which human performance can be motivated, captured and measured, offers clinical and research options that are not possible using traditional methods.

The following sections of this chapter will detail the history, rationales and key research for Clinical VR application in three areas: 1) Exposure therapy for Anxiety Disorders; 2) Neuropsychological Assessment for Central Nervous System (CNS) dysfunction; and 3) Virtual Patients for Clinical Training. The VR Exposure therapy and Neuropsychological areas were selected based on the consistent evolution of the research and the growing clinical adoption of applications in these areas. The Virtual Patients area, although in a nascent stage of development, was selected based on its estimated potential for future growth and clinical impact by the authors. At the end of each section, use-cases are presented of applications from our lab that illustrate the process for design, development and evaluation of a system in each of the outlined areas. While significant VR research and development activity is ongoing in the areas of substance abuse, eating disorders, pain distraction, social skills training, game-based cognitive and motor rehabilitation, etc., it was necessary to constrain the scope of what could presented with sufficient detail within the available page limitations for this chapter.

EXPOSURE THERAPY

The use of VR to address psychological disorders began in the mid-nineties with its use as a tool to deliver prolonged exposure (PE) therapy targeting anxiety disorders, primarily for specific phobias (e.g., heights, flying, spiders, enclosed spaces). PE is a form of individual psychotherapy based on the Foa and Kozak (1986) emotional processing theory, which posits that phobic disorders and PTSD involve pathological fear structures that are activated when information represented in the structures is encountered. Emotional processing theory purports that fear memories include information about stimuli, responses, and meaning (Foa and Kozak, 1986; Foa, Skeketee, and Rothbaum, 1989) and that fear structures are composed of harmless stimuli that have been associated with danger and are reflected in the belief that the world is a dangerous place. This belief then manifests itself in cognitive and behavioral avoidance strategies that limit exposure to potentially corrective information that could be incorporated into and alter the fear structure. As escape and avoidance from feared situations are intrinsically (albeit, temporarily) rewarding, phobic disorders can perpetuate without treatment. Consequently, several theorists have proposed that conditioning processes are involved in the etiology and maintenance of anxiety disorders. These theorists invoke Mowrer's (1960) two-factor theory, which posits that both Pavlovian and instrumental conditioning are involved in the acquisition of fear and avoidance behavior. Successful treatment requires emotional processing of the fear structures in order to modify their pathological elements so that the stimuli no longer invoke fear, and any method capable of activating the fear structure and modifying it would be predicted to improve symptoms of anxiety.

Imaginal PE entails engaging mentally with the fear structure through repeatedly revisiting the feared or traumatic event in a safe environment. The proposed mechanisms for symptom reduction involves activation and emotional processing, extinction/habituation of the anxiety, cognitive reprocessing of pathogenic meanings, the learning of new responses to previously feared stimuli, and ultimately an integration of corrective nonpathological information into the fear structure (Foa et al., 1996; Bryant et al., 2003). Thus, VR was seen early on to be a potential tool for the treatment of anxiety disorders; if an individual can become immersed in a feared virtual environment, activation and modification of the fear structure was possible. From this, the use of VR to deliver PE was the first psychological treatment area to gain traction clinically, perhaps in part due to the intuitive match between what the technology could deliver and the theoretical requirement of PE to systematically expose/engage users to progressively more challenging stimuli needed to activate the fear structure.

Moreover, even during the early days of VR, this was not so technically challenging to achieve. VEs could be created that required little complex user interaction beyond simple navigation within a simulation that presented users with scenarios that represented key elements of the targeted fear structure that could be made progressively more provocative (views from tall buildings, aircraft interiors, spiders in kitchens, etc.). And even with the limited graphic realism available at the time, phobic patients were observed to be "primed" to suspend disbelief and react emotionally to virtual content that represented what they feared. In general, the phenomenon that users of VR could become immersed in VE's provided a potentially powerful tool for activating relevant fears in the PE treatment of specific phobias in the service of therapeutic exposure.

From this starting point, a body of literature evolved that suggested that the use of virtual reality exposure therapy (VRET) was effective. Case studies in the 1990's initially documented the successful use of VR in the treatment of fear of flying (Rothbaum, Hodges, Watson, Kessler, and Opdyke, 1996; Smith, Rothbaum, and Hodges, 1999), claustrophobia (Botella et al., 1998), acrophobia (Rothbaum et al., 1995), and spider phobia (Carlin, Hoffman, and Weghorst, 1997). For example, in an early wait list controlled study, VRET was used to treat the fear of heights, exposing patients to virtual footbridges, virtual balconies, and a virtual elevator (Rothbaum et al., 1995). Patients were encouraged to spend as much time in each situation as needed for their anxiety to decrease and were allowed to progress at their own pace. The therapist saw on a computer monitor what the participant saw in the virtual environment and therefore was able to comment appropriately.

Results showed that anxiety, avoidance, and distress decreased significantly from pre- to post-treatment for the VRE group but not for the wait list control group. Examination of attitude ratings on a semantic differential scale revealed positive attitudes toward heights for the VRE group and negative attitudes toward heights for the wait list group. The average anxiety ratings decreased steadily across sessions, indicating habituation for those participants in treatment. Furthermore, 7 of the 10 VRE treatment completers exposed themselves to height situations in real life during treatment although they were not specifically instructed to do. These exposures appeared to be meaningful, including riding 72 floors in a glass elevator and intentionally parking at the edge of the top floor of a parking deck.

This research group then compared VRET to both an in vivo PE therapy condition and to a wait list (WL) control in the treatment of the fear of flying (Rothbaum et. al., 2000). Treatment consisted of eight individual therapy sessions conducted over six weeks, with four sessions of anxiety management training followed either by exposure to a virtual airplane (VRET) or exposure to an *actual airplane* at the airport (PE). For participants in the VRE group, exposure in the virtual airplane included sitting in the virtual airplane, taxi, take off, landing, and flying in both calm and turbulent weather according to a treatment manual (Rothbaum et. al., 1999). For PE sessions, in vivo exposure was conducted at the airport during Sessions 5 - 8. Immediately following the treatment or wait list period, all patients were asked to participate in a behavioral avoidance test consisting of a commercial round-trip flight.

The results indicated that each active treatment was superior to WL and that there were no differences between VRET and in vivo PE. For WL participants, there were no differences between pre and post self-report measures of anxiety and avoidance, and only one of the 15 wait-list participants completed the graduation flight. In contrast, participants receiving VRET or in vivo PE showed substantial improvement, as measured by self-report questionnaires, willingness to participate in the graduation flight, self-report levels of anxiety on the flight, and self-ratings of improvement. There were no differences between the two treatments on any measures of improvement. Comparison of post-treatment to the 6-month follow-up data for the primary outcome measures for the two treatment groups indicated no significant differences, indicating that treated participants maintained their treatment gains. By the 6-month follow-up, 93% of treated participants had flown since completing treatment. Since that time, an evolved body of literature of controlled studies has emerged and two recent meta-analyses of the available literature (Parsons and Rizzo, 2008a; Powers and Emmelkamp, 2008) concurred with the finding that VR is an efficacious approach for delivering PE, that it outperformed imaginal PE and was as effective as in vivo exposure.

VR has also been applied as a method for delivering for PE for posttraumatic stress disorder (PTSD). Among the many approaches that have been used to treat PTSD, exposure therapy appears to have the best-documented therapeutic efficacy (NAS, 2007). Such treatment typically involves the graded and repeated imaginal reliving of the traumatic event within the therapeutic setting. Similar to PE for specific phobias, this approach is believed to provide a low-threat context where the patient can begin to therapeutically process the emotions that are relevant to the traumatic event as well as de-condition the learning cycle of the disorder via a habituation/extinction process. However, while the efficacy of imaginal exposure has been established in multiple studies with diverse trauma populations (Bryant, 2005; Rothbaum and Schwartz, 2002; Van Etten and Taylor, 1998), many patients are unwilling or unable to effectively visualize the traumatic event. This is a crucial concern since avoidance of cues and reminders of the trauma is one of the cardinal symptoms of the DSM diagnosis of PTSD. In fact, research on this aspect of PTSD treatment suggests that the inability to emotionally engage (*in imagination*) is a predictor for negative treatment outcomes (Jaycox, Foa and Morral, 1998). To address this problem, researchers have recently turned to the use of VR to deliver exposure therapy by immersing clients in simulations of trauma-relevant environments that allow for precise control of stimulus conditions.

The first effort to apply VRET began in 1997 when researchers at Georgia Tech and Emory University began testing the *Virtual Vietnam* VR scenario with Vietnam veterans diagnosed with PTSD (Rothbaum et al., 2001). This occurred over 20 years after the end of the Vietnam War. During those intervening years, in spite of valiant efforts to develop and apply traditional psychotherapeutic and pharmacological treatment approaches to PTSD, the progression of the disorder in some veterans significantly impacted their psychological well-being, functional abilities and quality of life, as well as that of their families and friends. This initial effort yielded encouraging results in a case study of a 50-year-old, male Vietnam veteran meeting *DSM* criteria for PTSD (Rothbaum et al., 1999).

Results indicated post-treatment improvement on all measures of PTSD and maintenance of these gains at a 6-month follow-up, with a 34% decrease in clinician-rated symptoms of PTSD and a 45% decrease on self-reported symptoms of PTSD. This case study was followed by an open clinical trial with Vietnam veterans (Rothbaum et al., 2001). In this study, 16 male veterans with PTSD were exposed to two HMD-delivered virtual environments, a virtual clearing surrounded by jungle scenery and a virtual Huey helicopter, in which the therapist controlled various visual and auditory effects (e.g. rockets, explosions, day/night, shouting). After an average of 13 exposure therapy sessions over 5-7 weeks, there was a significant reduction in PTSD and related symptoms. *For more information, see the 9-minute Virtual Vietnam Documentary video at*: http://www.youtube.com/watch?v=C_2ZkvAMih8.

Similar positive results were reported by Difede et al. (2002) for PTSD that resulted from the attack on the World Trade Center in a case study using VRET with a patient who had failed to improve with traditional imaginal exposure therapy. This group later reported positive results from a wait-list controlled study using the same World Trade Center VR application (Difede et al., 2007). The VR group demonstrated statistically and clinically significant decreases on the "gold standard" Clinician Administered PTSD Scale (CAPS) relative to both pre-treatment and to the wait-list control group with a between-groups post treatment effect size of 1.54. Seven of 10 people in the VR group no longer carried the diagnosis of PTSD, while all of the wait-list controls retained the diagnosis following the waiting period and treatment gains were maintained at 6-month follow-up. Also noteworthy was the finding that five of the 10 VR patients had previously participated in imaginal exposure treatment with no clinical benefit. Such initial results are encouraging and suggest that VR may be a useful component within a comprehensive treatment approach for persons with combat/terrorist attack-related PTSD. *For more information, see the Virtual World Trade Center video at*: http://www.youtube.com/watch?v=XAR9QDwBILc

Use Case: The Virtual Iraq/Afghanistan PTSD Exposure Therapy Project

With this history in mind, the University of Southern California (USC) Institute for Creative Technologies (ICT) created an immersive VRET system for combat-related PTSD.

The treatment environment was initially based on recycling virtual assets that were built for the commercially successful X-Box game and tactical training simulation scenario, *Full Spectrum Warrior.* Over the years other existing and newly created assets developed at the ICT have been integrated into this continually evolving application. The *Virtual Iraq/Afghanistan* application consists of a series of virtual scenarios designed to represent relevant contexts for VR exposure therapy, including middle-eastern themed city and desert road environments.

The *Virtual Iraq/Afghanistan* PTSD Exposure Therapy System consists of Middle Eastern themed city and desert road environments (see Figure 1) and was designed to resemble the general contexts that most Service Members (SMs) experience during deployment to Iraq. The 24 square block "City" setting has a variety of elements including a marketplace, desolate streets, checkpoints, ramshackle buildings, warehouses, mosques, shops and dirt lots strewn with junk. Access to building interiors and rooftops is available and the backdrop surrounding the navigable exposure zone creates the illusion of being embedded within a section of a sprawling densely populated desert city.

Figure 15.1. Scene from *Virtual Iraq/Afghanistan* City and Desert Road HUMVEE scenarios.

Vehicles are active in streets and animated virtual pedestrians (civilian and military) can be added or eliminated from the scenes. The software has been designed such that users can be "teleported" to specific locations within the city, based on a determination as to which components of the environment most closely match the patient's needs, relevant to their individual trauma-related experiences. The "Desert Road" scenario consists of a roadway through an expansive desert area with sand dunes, occasional areas of vegetation, intact and broken down structures, bridges, battle wreckage, a checkpoint, debris and virtual human figures. The user is positioned inside of a HUMVEE that supports the perception of travel within a convoy or as a lone vehicle with selectable positions as a driver, passenger or from the more exposed turret position above the roof of the vehicle. The number of soldiers in the cab of the HUMVEE can also be varied as well as their capacity to become wounded during certain attack scenarios (e.g., IEDs, rooftop/bridge attacks).

Both the city and desert road HUMVEE scenarios are adjustable for time of day or night, weather conditions, illumination, night vision and ambient sound (wind, motors, city noise, prayer call, etc.). Users can navigate in both scenarios via the use of a standard gamepad controller, although the option for use of a replica M4 weapon with a "thumb-mouse" controller that supports movement during the city foot patrol is also available. This was based on repeated requests from experienced SMs who provided frank feedback indicating that to walk within such a setting without a weapon in-hand was completely unnatural and distracting! However, there is no option for firing a weapon within the VR scenarios. It is our firm belief that the principles of exposure therapy are incompatible with the cathartic acting out of a revenge fantasy that a responsive weapon might encourage.

In addition to the visual stimuli presented in the VR Head-Mounted Display (HMD), directional 3D audio, vibrotactile and olfactory stimuli can be delivered into the *Virtual Iraq* scenarios in real-time by the clinician. The presentation of additive, combat-relevant stimuli into the VR scenarios can be controlled in real time via a separate "Wizard of Oz" clinician's interface, while the clinician is in full audio contact with the patient. The clinician's interface is a key feature that provides a clinician with the capacity to customize the therapy experience to the individual needs of the patient. This interface allows a clinician to place the patient in VR scenario locations that resemble the setting in which the trauma-relevant events occurred and ambient light and sound conditions can be modified to match the patients description of their experience. The clinician can then gradually introduce and control real time trigger stimuli (visual, auditory, olfactory and tactile), via the clinician's interface, as required to foster the anxiety modulation needed for therapeutic habituation and emotional processing in a customized fashion according to the patient's past experience and treatment progress. The clinician's interface options have been designed with the aid of feedback from clinicians with the goal to provide a usable and flexible control panel system for conducting thoughtfully administered exposure therapy that can be readily customized to address the individual needs of the patient. Such options for real time stimulus delivery flexibility and user experience customization are key elements for these types of VR exposure therapy applications.

The specification, creation and addition of trigger stimulus options into the *Virtual Iraq* system has been an evolving process throughout the development of the application based on continually solicited patient and clinician feedback. This part of the design process began by including options that have been reported to be relevant by returning soldiers and military subject matter experts. For example, Hoge et al., (2004) presented a listing of emotionally challenging combat-related events that were commonly reported by their Iraq/Afghanistan

SM sample. These events provided a useful starting point for conceptualizing how relevant trigger stimuli could be presented in a VR environment. Such commonly reported events included: "*Being attacked or ambushed...receiving incoming artillery, rocket, or mortar fire... being shot at or receiving small-arms fire...seeing dead bodies or human remains...*" (p. 18). From this and other sources, we began our initial effort to conceptualize what was both functionally relevant and technically possible to include as trigger stimuli.

Currently the system offers a variety of auditory trigger stimuli (e.g., incoming mortars, weapons fire, voices, wind, etc.) that are actuated by the clinician via mouse clicks on the clinician's interface. Clinicians can also similarly trigger dynamic audiovisual events such as helicopter flyovers, bridge attacks, exploding vehicles and IEDs. The creation of more complex events that can be intuitively delivered in *Virtual Iraq/Afghanistan* from the clinician's interface while providing a patient with options to interact or respond in a meaningful manner is one of the ongoing focuses in this project. However, such trigger options require not only interface design expertise, but also clinical wisdom as to how much and what type of exposure is needed to produce a positive clinical effect. These issues have been keenly attended to in initial non-clinical user-centered tests with Iraq-experienced SMs and in the current clinical trials with patients. This expert feedback is essential for informed VR combat scenario design and goes beyond what is possible to imagine from the "Ivory Tower" of the academic world.

Whenever possible, *Virtual Iraq/Afghanistan* was designed to use off the shelf equipment in order to minimize costs and maximize the access and availability of the finished system. The minimum computing requirements for the current application is a Pentium 4 computer with 1 GB RAM, and a 128 MB DirectX 9-compatible 3D graphics card. Two computer monitors are required, one to display the clinician's interface and the second one displays the actual simulation scenes that is presented to the user in their head-mounted display (HMD) as they navigate using an interface (gamepad or gun controller). The HMD that was chosen was the *eMagin z800*, with displays capable of 800x600 resolution within a 40-degree diagonal field of view (http://www.emagin.com/). The major selling point for using this HMD was the presence of a built-in head tracking system. At under $1500 per unit with built-in head tracking, this integrated display/tracking solution was viewed as the best option to minimize costs and maximize the access to this system. The simulation's real-time 3D scenes are presented using Emergent's *Gamebryo* rendering engine. Pre-existing art assets were integrated using *Alias' Maya 6* and *AutoDesk 3D Studio Max 7* with new art created primarily in *Maya*.

Olfactory and tactile stimuli can also be delivered into the simulation to further augment the experience of the environment. Olfactory stimuli are produced by the *Enviroscent, Inc. Scent Palette*. This is a USB driven device that contains eight pressurized chambers, within which individual smell cartridges can be inserted, a series of fans and a small air compressor to propel the customized scents to participants. The scent delivery is controlled by mouse clicks on the clinician's interface. Scents may be employed as direct stimuli (e.g., scent of smoke as a user walks by a burning vehicle) or as cues to help immerse users in the world (e.g., ethnic food cooking). The scents selected for this application include burning rubber, cordite, garbage, body odor, smoke, diesel fuel, Iraqi food spices, and gunpowder. Vibration is also used as an additional user sensory input. Vibration is generated through the use of a *Logitech* force-feedback game control pad and through low cost (<$120) audio-tactile sound transducers from *Aura Sound Inc*. located beneath the patient's floor platform and seat. Audio files are customized to provide vibration consistent with relevant visual and audio stimuli in the scenario.

For example, in the HUMVEE desert road scenario, the user experiences engine vibrations as the vehicle moves across the virtual terrain and a shaking floor can accompany explosions. This package of controllable multisensory stimulus options was included in the design of *Virtual Iraq/Afghanistan* to allow a clinician the flexibility to engage users across a wide range of unique and highly customizable levels of exposure intensity. As well, these same features have broadened the applicability of *Virtual Iraq* as a research tool for studies that require systematic control of stimulus presentation within combat relevant environments (Rizzo et al., 2012). A direct link to a *YouTube* channel with videos that illustrate features of this system and of former patients discussing their experience with the VRET approach can be found at: http://www.youtube.com/user/AlbertSkipRizzo

The *Virtual Iraq/Afghanistan* system was designed and built from a user-centered design process that involved feedback from active duty SMs and veterans that began with solicited responses to the initial prototype. User-centered design feedback needed to iteratively evolve the system was gathered from a system deployed in Iraq with an Army Combat Stress Control Team and from returning OIF/OEF Veterans and patients in the US. Thus, leading up to the first clinical group test of treatment effectiveness, initial usability studies and case reports were published with positive findings in terms of SMs acceptance, interest in the treatment, and clinical successes (Gerardi et al., 2008; Reger et al., 2008, 2009, 2011; Wilson et al., 2008).

The Office of Naval Research, who had funded the initial system development of *Virtual Iraq/Afghanistan*, also supported an initial open clinical trial to evaluate the feasibility of using VRET with active duty participants. The study participants were recently redeployed from Iraq/Afghanistan at the Naval Medical Center San Diego and at Camp Pendleton and had engaged in previous PTSD treatments (e.g., group counseling, EMDR, medication, etc.) without benefit. The standard treatment protocol consisted of 2X weekly, 90-120 minute sessions over five weeks. The VRET exposure exercises followed the principles of prolonged exposure therapy (Foa et al., 1999) and the pace was individualized and patient-driven. Physiological monitoring (heart rate, galvanic skin response and respiration) was used to provide additional user state information to the clinician to help inform their pacing of the VRET.

The first VRET session consisted of a clinical interview that identified the index trauma, provided psychoeducation on trauma and PTSD, and instruction on a deep breathing technique for general stress management purposes. The second session provided instruction on the use of Subjective Units of Distress (SUDS), the rationale for PE, including imaginal exposure and in-vivo exposure. The participants also engaged in their first experience of imaginal exposure of the index trauma and an in-vivo hierarchical exposure list was constructed, with the first item assigned as homework. Session 3 introduced the rationale for VRET and the participant experienced the VR environment without recounting the index trauma narrative for approximately 25 minutes with no provocative trigger stimuli introduced. Sessions 4-10 focused on the participant engaging in the VR while recounting the trauma narrative.

Generally, participants were instructed that they would be asked to recount their trauma in the first person, as if it were happening again with as much attention to sensory detail as they could provide. Using clinical judgment, the therapist might prompt the patient with questions about their experience or provide encouraging remarks as deemed necessary to facilitate the recounting of the trauma narrative. The treatment included homework, such as requesting the participant to listen to the audiotape of their exposure narrative from the most recent session as a form of continual exposure for processing the index trauma to further enhance the probability for habituation to occur. Self-report measures were obtained at baseline and prior to sessions 3,5,7,9,10 and one week and three months post treatment to assess in-treatment and follow-up symptom status. The measures used were the PTSD Checklist-Military Version (PCL-M) (Blanchard et al., 1996), Beck Anxiety Inventory (BAI) (Beck et al., 1988) and Patient Health Questionnaire-Depression (PHQ-9) (Kroenke and Spitzer, 2002).

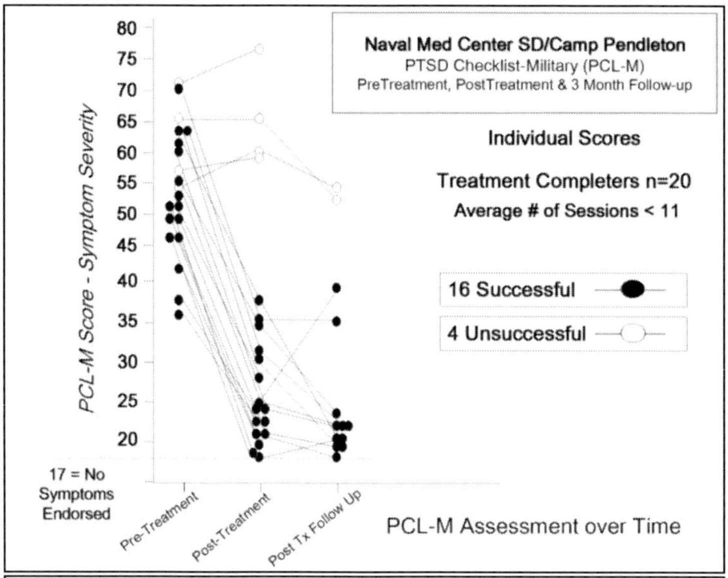

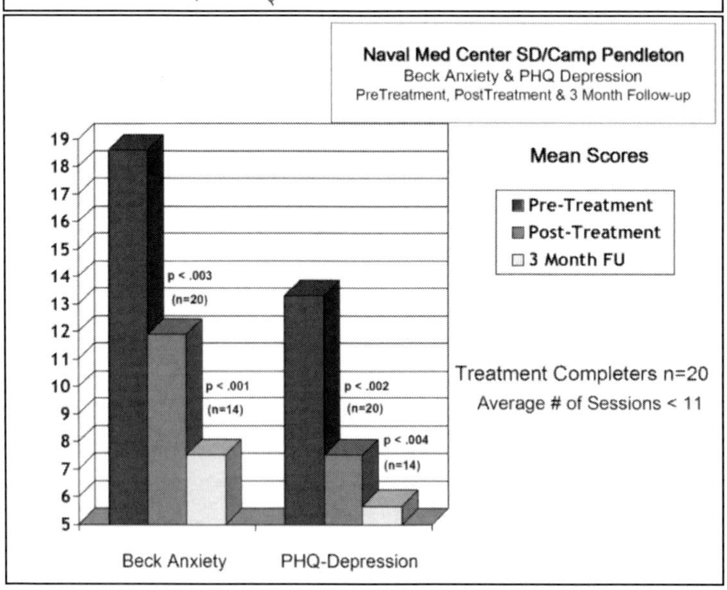

Figure 15.2. PCL-M scores across treatment Figure 15.3. BAI and PHQ-Depression scores.

Analyses of the first 20 active duty service members to complete treatment (19 male, 1 female, Mean Age=28, Age Range: 21-51) produced positive clinical outcomes. For this sample, mean pre/post PCL-M scores decreased in a statistical and clinically meaningful fashion; 54.4 (SD =9.7) to 35.6 (SD = 17.4). Paired pre/post t-test analysis showed these differences to be significant (t=5.99, df=19, p < .001). Correcting for the PCL-M no-symptom baseline of 17 indicated a greater than 50% decrease in symptoms and 16 of the 20 completers no longer met DSM criteria for PTSD at post treatment. Five participants in this group with PTSD diagnoses had pre-treatment baseline scores below the conservative cutoff value of 50 (pre-scores = 49, 46, 42, 36, 38) and reported decreased values at post treatment (post-scores = 23, 19, 22, 22, 24, respectively). Individual participant PCL-M scores at baseline, post treatment and 3-month follow-up are in Figure 2. Mean Beck Anxiety Inventory scores significantly decreased 33% from 18.6 (SD = 9.5) to 11.9 (SD = 13.6), (t=3.37, df=19, p < .003) and mean PHQ-9 (depression) scores decreased 49% from 13.3 (SD= 5.4) to 7.1 (SD = 6.7), (t=3.68, df=19, p < .002) (see Figure 3). The average number of sessions for this sample was just under 11. Also, two of the successful treatment completers had documented mild and moderate TBIs, which provide an early indication that this form of exposure therapy can be useful (and beneficial) for this population. Results from uncontrolled open trials are difficult to generalize from and we are cautious not to make excessive claims based on these early results. However, using an accepted military-relevant diagnostic screening measure (PCL-M), 80% of the treatment completers in the initial VRET sample showed both statistically and clinically meaningful reductions in PTSD, anxiety and depression symptoms, and anecdotal evidence from patient reports suggested that they saw improvements in their everyday life. These improvements were also maintained at three-month post-treatment follow-up.

Other studies have also reported positive outcomes. Two early case studies have been published that reported positive results using this system (Gerardi et al., 2008; Reger and Gahm, 2008). Following those, an open clinical trial with active duty soldiers (n=24) produced significant pre/post reductions in PCL-M scores and a large treatment effect size (Cohen's *d* = 1.17) (Reger et al., 2011). After an average of 7 sessions, 45% of those treated no longer screened positive for PTSD and 62% had reliably improved. These VRET results also outperformed a treatment-as-usual (TAU) Cognitive Behavioral Group approach (G. Reger, personal communication, January 5, 2009). Interesting mixed results have been reported from an ongoing study that used a combined sample of active duty soldiers (n=15) who had undergone either VR or imaginal PE therapy (Roy et al., 2010). While this combined sample revealed only modest pre/post treatment gains on the self-report Clinician Administered PTSD Scale (CAPS) (Blake et al., 1990), fMRI scans conducted at pre/post treatment with eight treatment completers produced an interesting desynchrony of response systems; activation changes in the amygdala and key frontal regions of interest for PTSD indicated a significantly normalized brain response following treatment. Such conflicting results bring up the thorny issue of the reliability of self-report PTSD measures when there may be incentives to not report improvement in symptoms and this will likely be an area of interest for some time to come.

Three randomized controlled trials (RCT) are ongoing with the *Virtual Iraq/Afghanistan* system with active duty and Veteran populations. Two RCTs are focusing on comparisons of treatment efficacy between VRET and imaginal PE, while the third RCT investigates the additive value of supplementing VRET and imaginal PE with a cognitive enhancer called D-

Cycloserine (DCS). DCS, a N-methyl-d-aspartate partial agonist, has been shown to facilitate extinction learning in laboratory animals when infused bilaterally within the amygdala prior to extinction training (Walker, Ressler, Lu, and Davis, 2002). The first clinical test in humans that combined DCS with VRET was performed by Ressler et al. (2004) with participants diagnosed with acrophobia (n=28). Participants who received DCS + VRET had significantly enhanced decreases in fear within the virtual environment 1 week and 3 months post-treatment, and reported significantly more improvement than the placebo group in their overall acrophobic symptoms at 3 month follow-up and on a psychophysiological measure of anxiety. The current multi-site PTSD RCT will test the effect of DCS vs. Placebo when added to VRET and PE with active duty and veteran samples (n=300).

This research has been supported by the relatively quick adoption of the VRET approach by approximately 55 Military, VA and University clinic sites over the last three years. Based on the outcomes from our initial open clinical trial and similar positive results from other research groups, we are encouraged by these early successes and continue to gather feedback from patients regarding the therapy and the *Virtual Iraq/Afghanistan* treatment environments. Patient feedback is particularly relevant now that the *Virtual Iraq/Afghanistan* project is undergoing a full rebuild using advanced software tools (*Unity 3D* Software) to provide more diversity of content and added functionality. In this regard, the new system has its design "roots" from feedback acquired from non-diagnosed SMs as well as the clinicians and PTSD patients who have used the VRET system thus far. The new system is also being designed to facilitate the development, exploration and testing of hypotheses relevant to improving PTSD treatment, and also for the use of the simulation for other purposes including PTSD and neurocognitive assessment, and for the creation of a stress resilience training system (Rizzo et al., 2011b; 2012).

Other research groups are also currently in the early stages of applying VR to treat PTSD in survivors of war and terrorist attacks (Gamito et al., 2007; Josman et al., 2006; Wood et al., 2008). One research group has reported "partial remission" of PTSD symptoms in four of six Iraq war SMs using VR combined with meditation and attentional refocusing (Wood et al., 2008). However, it is difficult to draw any conclusions from this work as the authors did not provide a clear specification or statistical analysis of the pre/post PTSD Checklist-Military version (PCL-M) scores other than to report a baseline PCL-M mean of 47.3, which is actually below the PTSD cutoff of fifty specified in Weathers et al. (1993). Another research group in Portugal, where there are an estimated 25,000 survivors with PTSD from their 1961-1974 colonial wars in Mozambique, Angola and Guiné, has constructed a VR exposure scenario by modifying a common PC-based combat game (Gamito et al., 2007). This group has reported an initial case study where the patient did not complete treatment due to experiencing a distressing flashback following the seventh session. While this is rarely reported in the exposure literature, this report highlights the need for well-trained clinicians with expertise in the delivery of exposure therapy at a rate that the patient can effectively handle and process and in the sensitive monitoring of patient status. Finally, in Israel, Josman et al. (2006) are implementing a VR terrorist "bus bombing" PTSD treatment scenario in which participants are positioned in an urban cafe across the street from a site where a civilian bus may explode.

Interest in VR technology to create tools for enhancing exposure therapy practice and research has grown in recent years as initial positive outcomes have been reported with its implementation. The enthusiasm that is common among proponents of the use of VR for exposure-based treatment partly derives from the view that VR technology provides the capacity for clinicians to deliver specific, consistent and controllable trauma-relevant stimulus environments that do not rely exclusively on the hidden world and variable nature of a patient's imagination. Moreover, the technology required to produce and use VR systems has advanced concomitantly as system costs have decreased. Additionally, if one reviews the history of the impact of war on advances in clinical care it could be suggested that VR is an idea whose time has come. For example, during WW I, the Army Alpha/Beta test emerged from the need for better cognitive ability assessment and that development later set the stage for the civilian intelligence testing movement during the mid-20th Century.

As well, the birth of clinical psychology as a treatment-oriented profession was borne out of the need to provide care to the many Veterans returning from WW II with "shellshock". Based on these examples, one of the outcomes of the OIF/OEF conflicts could be the military's support for research and development in the area of Clinical VR that could potentially drive increase recognition and use in the civilian sector. However, this will only occur if positive cost-effective outcomes are produced with military VRET applications. It should also be noted that any rush to adopt VRET should not disregard principles of evidence-based and ethical clinical practice. While novel VR systems can extend the skills of a well-trained clinician, they are not intended to be automated treatment protocols that are administered in a "self-help" format. The presentation of such emotionally evocative VR combat-related scenarios, while providing treatment options not possible until recently, will most likely produce therapeutic benefits when administered within the context of appropriate care via a thoughtful professional appreciation of the complexity and impact of these behavioral health challenges.

NEUROPSYCHOLOGICAL VR APPLICATIONS

In the broadest sense, neuropsychology is an applied science that evaluates how specific activities in the brain are expressed in observable behaviors (Lezak, 1995). Effective neuropsychological assessment (NA) is a prerequisite for both the scientific analysis and treatment of CNS-based cognitive/functional impairments as well as for research investigating normal functioning. The NA of persons with CNS disorders using psychometric evaluation tools serves a number of functions. These include the determination of a diagnosis, the provision of normative data on the status of impaired cognitive and functional abilities, the production of information for the design of rehabilitative strategies, and the measurement of treatment efficacy. NA also serves to create data for the scientific understanding of brain functioning through the examination of measurable sequelae that occur following brain damage or dysfunction.

Thus, treatment and rehabilitation of the cognitive, psychological and motoric sequelae of CNS dysfunction often relies on assessment devices to inform diagnosis and to track changes in clinical status. Typically, these assessments employ chapter and pencil psychometrics, hands-on analog tests, computer-delivered continuous performance tests, and observation/rating of behavior in real world functional environments (or within the context of physical mock-ups). On one end of the spectrum, traditional neuropsychological methods applied to impairment assessment and rehabilitation commonly use paper and pencil-based psychometric tests. Although these approaches provide highly systematic control and delivery of constrained performance challenges, they have also been criticized as limited in the area of *ecological* validity, that is, the degree of relevance or similarity that a test or training system has relative to the "real" world, and in its value for predicting or improving "everyday" functioning (Neisser, 1978). Adherents of this view challenge the usefulness of constrained chapter and pencil tests and analog tasks for addressing the complex integrated functioning that is required for successful performance in the real world. On the other end of the spectrum, a common method applied in the occupational sciences discipline to assess and rehabilitate functional abilities employs behavioral observation and ratings of human performance in the "real-world" or via physical mock-ups of functional environments (Weiss and Jessel, 1998). Mock-ups of daily living environments (i.e., kitchens, bathrooms, etc.) and workspaces (i.e., offices, factory settings, etc.) are typically built, within which persons with cognitive impairments are observed while their performance is evaluated.

Aside from the real economic costs to physically build the environments and to provide human resources to conduct such evaluations, this approach is limited in the systematic control of real-world stimulus challenges and in its capacity to provide detailed performance data capture. As well, many functional environments in everyday life do not easily lend themselves to mock-ups, as is readily apparent in the domain of driving skill assessment and training. In this regard, "Behind-the-Wheel" driving assessments, considered to be the gold standard in this area, are often conducted in only the safest possible conditions (i.e., good weather, low traffic roadways, etc.), and actually provide a limited "window" into how driving performance would fare under more realistic (and often unpredictable) conditions.

Neuropsychology is where VR stands to have significant impact. While the last 15 years have seen the majority of VR research and clinical application occur in the area of exposure therapy, the considerable potential of VR for the study, assessment and rehabilitation of human cognitive and functional processes been also been recognized (Pugnetti, Mendozzi, Motta, et.al., 1995; Rizzo, 1994; Rizzo, Buckwalter and van der Zaag, 2002; Rizzo et al., 2004; Rose, 1996; Rose et al., 2005). Indeed, in a U.S. National Institute of Health report of the National Advisory Mental Health Council (1995), the impact of virtual reality environments on cognition was specifically cited with the recommendation that: "Research is needed to understand both the positive and negative effects of such participation on children's and adults' perceptual and cognitive skills..." (p.51). One area where the potential for both "positive and negative effects" exists, is in the application of VR for neuropsychological assessment, rehabilitation and research. In this regard, VR could serve to advance the study of brain/behavior relationships as well as produce innovative evaluation and intervention options that are unavailable with traditional methods.

What makes VR application development in this area so distinctively important is that it represents the potential for more than a simple linear extension of existing computer technology for human use. This was recognized early on in a visionary article ("The Experience Society") by VR pioneer, Myron Kruegar (1993), in his prophetic statement that, "...Virtual Reality arrives at a moment when computer technology in general is moving from automating the paradigms of the past, to creating new ones for the future". (p. 163). In this comment Krueger encapsulated what had also been so limited in neuropsychology's approach to using computer and information technology at that time and opened a conceptual door to VR's potential to advance neuropsychological research and practice. Indeed, neuropsychology's use of technology up to that time could be characterized as mainly translating existing traditional chapter and pencil tools directly into computer delivered formats. In its defense, neuropsychology has been increasingly integrating advanced neural imaging technology tools (i.e., fMRI, DTI, SPECT, QUEEG, CT, etc.) in its quest for a better accounting of the structure and process underlying brain/behavior relationships. However, while these advances in *response* measurement have led to new findings and conceptualizations, the *stimulus* delivery end of the equation has lagged behind. By way of VRs capacity to place a person within an immersive, interactive computer generated simulation environment, new possibilities exist that go well beyond simply automating the delivery of existing chapter and pencil testing and training tools on a personal computer. VR offers the potential to fundamentally advance this area with innovative applications that leverage the immersive, involving and interactive assets available in VEs to deliver quantifiable analog-like stimulus protocols within the context of functionally relevant (and controllable) environments. Until now, these features have not been pragmatically available with existing methods in neuropsychology and thus VR may have plenty to offer in these vital and challenging areas.

A primary strength that VR offers assessment is in the creation of simulated "everyday" environments in which performance can be tested in systematic fashion. By designing VEs that not only *look like* the real world, but actually incorporate challenges that require integrated functional behaviors, the ecological validity of assessment methods could be enhanced. As well, within a VE, the experimental control required for rigorous scientific analysis and replication can still be maintained within simulated contexts that embody the complex challenges found in naturalistic settings. Thus, on a theoretical level, VR-derived results could have greater predictive validity and clinical relevance for the challenges that patients face in everyday life compared to chapter and pencil testing. On a more pragmatic level, rather than relying on costly physical mock-ups of functional assessment and rehabilitation environments, VR offers the option to produce and distribute identical "standard" simulation environments. Within such digital assessment (and rehabilitation) scenarios, normative data can be accumulated for performance comparisons in an automated fashion, needed for assessment/diagnosis and for treatment/rehabilitation purposes.

As well, a challenge to the conceptual growth of the field of psychometric testing methods was leveled in a 1997 *American Psychologist* article by intelligence theorist, Robert Sternberg (1997) in which he compared currently used intelligence and ability tests to black and white TV, rotary-dial phones, and the UNIVAC computer. His argument started by observing that the first edition of the most widely used intelligence test, the *Wechsler Adult Intelligence Scale* appeared in 1939, well before the UNIVAC. However, while computer and other information technology and telecommunication tools (i.e., TV, telephones, and sound recording) had undergone a revolution since then, with the exception of essentially cosmetic changes, tests of cognitive ability have remained essentially unchanged. Sternberg posited that "dynamic" interactive testing would be needed to provide a new option that could supplement traditional "static" tests. The "dynamic" assessment approach requires the provision of guided performance feedback as a component in tests that measure learning. This method appears well suited to the assets available with VR technology. In fact, VR might be the most efficient vehicle for conducting dynamic testing in an "ecologically valid" manner while still maintaining an acceptable level of experimental control.

Thus, it is possible that the use of VR technology could revolutionize our approach to NA. However, while encouraging on a theoretical level, the value of this technology for neuropsychology still needs to be substantiated via systematic empirical research with normal and clinical populations that can be replicated by others. To accomplish this first requires specification as to the real assets that VR offers that add value over existing assessment methodologies, as well as further exploration of its current limitations. The current status of VR technology applied to clinical populations, while provocative, is still limited by the small (but growing) number of controlled studies in this area. This is to be expected, considering the technology's relatively recent development and the lack of familiarity with VR technology by established researchers employing the traditional tools and tactics of their fields. In spite of this, a nascent body of work has emerged which can provide knowledge for guiding future research efforts.

In the mid-90's, using graphic imagery that would be considered primitive by today's standards, Pugnetti et al. (1995,1998) developed a HMD-delivered VR scenario that embodied the cognitive challenges that characterize the Wisconsin Card Sorting Test (WCST). The scenario consisted of a virtual building within which users were required to use environmental clues to aid in the correct selection of appropriate doorways needed to pass from room to room through the structure. The doorway choices varied according to the categories of shape, color and number of portholes. Similar to the WCST, the correct choice criteria were changed after a fixed number of successful trials, and the user was then required to shift cognitive set, look for clues and devise a new choice strategy in order to successfully pass into the next room. In one study, Pugnetti et al. (1998) compared a mixed group of neurological patients (multiple sclerosis, stroke, and traumatic brain injury) with normals' performance on both the WCST and on this head mounted display executive function system. Results indicated that the VR results mirrored previous anecdotal observations by family members of everyday performance deficits in the patient populations. Though the psychometric properties of the VE task were comparable to the WCST in terms of gross differentiation of patients and controls, weak correlations between the two methods suggested that the methods measured different aspects of these functions. A detailed analysis of the VR task data indicated that specific preservative errors appeared earlier in the test sequence compared to the WCST. The authors suggested that "...this finding depends on the more complex (and complete) cognitive demands of the VE setting at the beginning of the test when perceptuomotor, visuospatial (orientation), memory, and conceptual aspects of the task need to be fully integrated into an efficient routine" (p.160). The detection of these early "integrative" difficulties for this complex cognitive function may be particularly relevant for the task of predicting real world capabilities from test results. This was further evidenced in a detailed single subject case study of a stroke patient using this system. In this report (Mendozzi, Motta, Barbieri, Alpini and Pugnetti, 1998), results indicated that the VR system was more accurate in identifying executive function deficits in a highly educated patient two years post-stroke, who had a normal WCST performance. The VR system, although using graphic imagery that would never be mistaken for the real world, was successful in detecting deficits that had been reported to be limiting the patient's everyday performance, yet were missed using existing NP tests. These results are in line with the observation that patients with executive disorders often perform relatively well on traditional NP tests of "frontal lobe function", yet show marked impairment in controlling and monitoring behavior in real-life situations (Shallice and Burgess, 1991).

Similar findings were reported by McGeorge et al. (2001) in a study comparing real world and virtual world "errand running" performance in five traumatic brain injury patients and five matched normal controls. The selection of the patient sample for this study was based on staff ratings that indicated poor planning skills. However, the patient and control groups did not differ significantly from normative values on the Behavioural Assessment of the Dysexecutive Syndrome (BADS) battery (Wilson et al., 1996). Videotaped performance of subjects was coded and compared while performing a series of errands in the University of Aberdeen psychology department (real world) and within a flatscreen VR scenario modeled after this environment. Performance in both the real and virtual environment, as defined as the number of errands completed in a 20-minute period, was highly correlated ($r = .79$; $p<.01$). Interestingly, while the groups did not differ on age-corrected standardized scores on the BADS, significant differences *were* found between the groups in both the real world and virtual testing. This finding suggests several things. First, performance in the real and virtual world was functionally similar, second, patient and control groups could be discriminated equally using real and virtual tests while this discrimination was not picked up by standardized testing with the BADS, and third, that both measures of real and virtual world performance showed concordance with staff observations of planning skills. That these results support the view that VR testing may possess higher ecological value is in line with the observation by Shallice and Burgess (1991) that traditional neuropsychological tests do not demand the planning of behavior over more than a few minutes, or the prioritization of competing subtasks and may result in less effective prediction of real world performance.

Other research has also demonstrated VR's usefulness for neuropsychological assessment, particularly for visuospatial processes (Kaufmann et al., 2005; Koenig et al., 2009, 2010; McGee et al., 2000; Parsons, et al., 2005; Rizzo et al., 2001). For example, VR systems have been used to assess mental rotation, a cognitive function whereby a person needs to visualize the movement and organization of objects in a 3-dimensional space (Shepard and Metzler, 1971). Mental rotation is important for everyday tasks such as driving, organizing items in a limited space and any activity that relies on dynamic imagery for prediction of object movement. In the normal population, men outperform women on the mental rotation task. However, "hands-on" performance on a VR spatial rotation task that mimicked the task structure of the original mental rotation task, showed no gender difference and mental rotation was shown to be dramatically improved in both women and low performing men after a brief 10-15 minute period of interaction with the hands-on virtual reality interaction task (Rizzo et al., 2001). Along similar lines, the Morris Water Maze test of spatial navigation and place learning, commonly used with rodents, has also been simulated in a virtual environment as a test for humans (Astur et al., 2002, 2004). In this application, the person being tested must use visual cues in the surrounding environment to help guide navigation to a hidden platform. Used in conjunction with fMRI, the test can be applied to determine whether a person has decreased hippocampal activity that might be indicative of Alzheimer's disease (Shipman, and Astur, 2008) and with Schizophrenia (Folley et al., 2010). Such integration of VR as a complex stimulus delivery tool with advanced brain imaging and psychophysiological techniques may allow neuropsychology to reach its stated purpose, that of determining unequivocal brain-behavior relationships.

Along with this seminal work with adults, other researchers have followed the aircraft simulation metaphor with the creation of VR worlds designed to assess cognitive and functional performance in children with a range of CNS-related disease or injury conditions. For example, researchers have created virtual homes, classrooms, public spaces and traffic-filled streets to test and train children with autistic spectrum disorder (ASD) and other developmental disorders on activities relevant to fire safety (Rizzo, Strickland and Bouchard, 2004), social skills (Parsons et al., 2000), functional attention (Rizzo et al., 2006), street-crossing (Strickland et al., 1996; Bart et al., 2008) and earthquake safety (Raloff, 2006). Even a virtual obstacle course was created for determining if a child is capable of using a motorized wheelchair safely and effectively, and to support training if they're not quite ready yet (Inman et al. 1997). For those interested in more detail on Clinical VR research in children, see Parsons et al., (2009) and Rizzo et al., (2011a).

USE CASE: THE VIRTUAL CLASSROOM ATTENTION ASSESSMENT PROJECT

The original Virtual Classroom project began in 1999 as part of a basic research application program at the University of Southern California aimed at developing VR technology applications to improve our capacity to understand, measure, and treat the cognitive/functional impairments commonly found in clinical populations with CNS dysfunction as well as advance the scientific study of normal processes and function. The Virtual Classroom was designed as a HMD VR system for the assessment of attention processes in children. Efforts to target this cognitive process were supported by the widespread occurrence and relative significance of attention impairments seen in a variety of clinical conditions that effect children. Notable examples of childhood clinical conditions where attention difficulties are seen include Attention Deficit Hyperactivity Disorders (ADHD), Traumatic Brain Injury and Fetal Alcohol Syndrome. With these clinical conditions, VR technology provides specific assets for assessing attention that are not available using existing methods. For example, HMDs that serve to occlude the distractions of the outside world are well suited for these types of cognitive assessment applications. Within a HMD, researchers and clinicians can provide a controlled stimulus environment where attention (and other cognitive) challenges can be presented along with the precise delivery and control of "distracting" auditory and visual stimuli within the virtual environment. This level of experimental control allows for the development of attention assessment/rehabilitation tasks that are more similar to what is found in the real world and when delivered in the context of a relevant functional virtual environment, could improve on the ecological validity of measurement and treatment in this area. The first project with the Virtual Classroom focused on attention assessment in children with ADHD. The heterogeneous features of ADHD, a behavioral disorder marked by inattention, impulsivity, and/or hyperactivity, have made consensus regarding its diagnosis difficult. Furthermore, traditional methods for assessing ADHD in children have been questioned regarding issues of reliability and validity. Popular behavioral checklists have been criticized as biased and not a consistent predictor of ADHD, and correlations between concordant measures of ADHD, such as parent and teacher ratings of hyperactivity, have been repeatedly shown to be modest at best and frequently low or absent (Barkley, 1990; Colegrove et al., 1997). Due to the complexity of the disorder and the limitations of traditional assessment techniques, diagnostic information is required from multiple types of ADHD measures and a variety of sources in order for the diagnosis to be given (American Psychiatric Association, 2000; Greenhill, 1998). Thus, in the area of ADHD assessment where traditional diagnostic techniques have been plagued by subjectivities and inconsistencies, it was believed that an objective and reliable VR strategy might add value over existing approaches and methods.

Figure 15.4. Scenes from the Virtual Classroom.

The initial research version of the system was run on a standard Pentium 3 processor with the nVIDIA G2 graphics card. The HMD used in this study was the V8 model from Virtual Research. Tracking of the head, arm and leg used three 6-Degree of Freedom magnetic "Flock of Birds" trackers from Ascension Technology Corp. In addition to tracking head movement in real time to update the graphics display in the HMD, the tracking system also served to capture body movement metrics from the tracked locations. This provided concurrent data on the hyperactivity component that is a commonly observed feature of ADHD. The research version of the Virtual Classroom scenario consisted of a standard rectangular classroom environment containing desks, a female teacher, a blackboard across the front wall, a side wall with a large window looking out onto a playground and street with moving vehicles, and on each end of the opposite wall, a pair of doorways through which activity occurred (see Figure 4). Within this scenario, children's attention performance was assessed while a series of common classroom distracters (i.e., ambient classroom noise, activity occurring outside the window, etc.) were systematically controlled and manipulated within the virtual environment. The child sat at a virtual desk within the virtual classroom and on-task attention was measured in terms of reaction time performance and error profiles on a variety of attention challenge tasks that were delivered visually using the blackboard or auditorily via a virtual teacher's voice.

Prior to any clinical tests with the Virtual Classroom, we applied a user-centered design methodology to insure that the application was usable and safe for children. User-centered approaches generally require the involvement of the targeted user group in the early design and development phase of scenario development. This involves a series of tight, short heuristic and formative evaluation cycles conducted on basic components of the system. Consideration of user characteristics in this fashion has increasingly become standard practice in VR development (Hix and Gabbard, 2002). In the Virtual Classroom's user-centered design evaluation phase, twenty non-diagnosed children (ages 6-12) tried various evolving forms of the system over the first year of development and their performance was observed while trying out a variety of basic selective and alternating attention tasks. One such task involved having users recite the letters that appeared on the blackboard, while naming the color of a virtual chapter airplane that passed across the classroom at random intervals. We also solicited feedback pertaining to aesthetics and usability of the VE and incorporated some of this feedback into the iterative design-evaluate-redesign cycle. Overall, these initial results indicated little difficulty in adapting to use of the HMD, no self-reported occurrence of side effects as determined by post-test interviews using the Simulator Sickness Questionnaire (SSQ) (Kennedy et al., 1993) and excellent performance on the stimulus tracking challenges.

Following this user centered design phase, we conducted a clinical trial that compared eight physician-referred ADHD males (age 6-12) with ten non-diagnosed male children. The groups did not significantly differ in mean age, grade level, ethnicity, or handedness and all ADHD diagnosed children were currently taking stimulant medication as treatment for their condition. However, ADHD participants were off medication during the testing period with all testing occurring between 9am and 11am prior to normal medication ingestion. ADHD participants were excluded from the study if they presented with comorbid autism, mental retardation, Full Scale IQ score < 85, or head injury with loss of consciousness greater than 30 minutes. These same exclusion criteria were applied to the normal control group. Research participants were instructed to view a series of letters presented on the blackboard and to hit the response button only after he viewed the letter "X" preceded by an "A" (successive discrimination task). This AX continuous performance task consisted of the letters A, B, C, D, E, F, G, H, J, L, and X. The letters were white on a gray background (the virtual blackboard) and presented in a fixed position directly in front of the participant. The stimuli remained on the screen for 150 msec, with a fixed interstimulus interval of 1350 msec. The target letter X (correct hit stimuli) and the letter X without the A (incorrect hit stimuli) each appeared with equal probability of 10%. The letters A and H both appeared with a frequency of 20%. The remaining eight letters occurred with 5% probability.

Participants were instructed to press the mouse button as quickly and accurately as possible (with their dominant hand) upon detection of an X after an A (correct hit stimuli) and withhold their response to any other sequence of letters. Four hundred stimuli were presented during each of two 10-minute conditions. The two 10-minute conditions consisted of: one without distraction and one with distractions (pure audio-classroom noises, pure visual-chapter airplane flying across the visual field and mixed audiovisual-a car "rumbling" by the window and a person walking into the classroom with hall sounds occurring when the door to the room was opened). Distracters were each displayed for 5 sec, and presented in randomly assigned equally appearing intervals of 10 sec, 15 sec, or 25 sec. and 36 distracters (9 of each) were included in the 10-minute condition. As well, six degree of freedom tracking from the head, arm and leg was used to produce movement metrics needed to analyze the motor hyperactivity component in conjunction with the cognitive performance. VR performance was also compared with results from standard neuropsychological testing. The following results summarize the outcomes of this study:

- No significant side effects were observed in either group based on pre- and post-VR Simulator Sickness Questionnaire testing.
- Children with ADHD had *slower* correct hit reaction time compared with normal controls on the distraction condition (760ms vs. 610ms; $t\ (1,16) = -2.76$, $p < .03$).
- Children with ADHD had higher reaction time *variability* on correct hits compared with normal controls on both the no-distraction (SD= 220ms vs. 160ms; $t\ (1,16) = -2.22$, $p < .05$) and distraction conditions (SD= 250ms vs. 170ms; $t\ (1,16) = -2.52$, $p < .03$).
- Children with ADHD made more Omission errors (missed targets) compared with normal controls on both the no-distraction (14 vs. 4.4; $t\ (1,16) = -4.37$, $p < .01$) and distraction conditions (21 vs. 7.2; $t\ (1,16) = -4.15$, $p < .01$).
- Children with ADHD made more Commission errors (impulsive responding in the absence of a target) compared with normal controls on both the no-distraction (16 vs. 3.7; $t\ (1,16) = -3.15$, $p < .01$) and distraction conditions (12.1 vs. 4.2; $t\ (1,16) = -3.22$, $p < .01$)
- Children with ADHD made more Omission errors in the distraction condition compared to the non-distraction condition (21 vs. 14; $t\ (1,16) = -3.50$, $p < .01$). No such differences on Omission and Commission errors were found with the non-diagnosed children across no-distraction and distraction conditions.
- Exploratory analysis of motor movement in children with ADHD (tracked from head, arm and leg) indicated higher activity levels on all metrics compared to non-diagnosed children across both conditions.

- Exploratory analysis of motor movement in children with ADHD also indicated higher activity levels on all metrics in the distraction condition compared to the non-distraction condition. This difference was not found with the normal control children.
- Exploratory analysis using a neural net algorithm (support vector machine analysis) trained to recognize a stereotypic leg movement on the first five participants in each group was able to accurately discriminate the remaining subjects to groups at 100%.

These data suggested that the Virtual Classroom had good potential as an efficient, cost-effective and scalable tool for conducting attention performance measurement beyond what existed using traditional methodologies. The system allowed for controlled performance assessment within an ecologically valid environment and appeared to parse out significant effects due to the presence of distraction stimuli. Additionally, the capacity to integrate measures of movement via the tracking technology further added value to this form of assessment when compared to traditional analog tests and rating scales. In this regard, a HMD appeared to be the optimal display format. Although one of the common criticisms of HMD technology concerns field of view (FOV) limitations, in this application the limited FOV fostered head movement to supplant eye movement as the primary method for scanning the Virtual Classroom. This type of "poor-man's" tracking of behavioral attention within the controlled stimulus environment obtained in the HMD allowed for ongoing documentation as to where the user was "looking" during test stimulus content delivery.

For example, a child missing a target while directly looking at the blackboard is illustrating an attentional error that is fundamentally different from the occurrence of a missed target due to the child looking out the window at a distraction. The documentation provided by head tracking in a HMD can be used to produce metrics of percent time on task during stimulus "hit" trials as well as allowing for a re-creation of a naturalistic behavioral performance record for later review. In the current research, ADHD children were found to miss targets due to looking away from the blackboard during 25% of the "hit" trials as opposed to normal subjects who were documented to be looking away at less than 1% of the time. This form of integrated cognitive/behavioral performance record of attention performance during delivery of systematic distraction is simply not obtainable using other methods. More detailed information on the rationale, methodology, other studies and long-term vision for this project can be found in (Rizzo et al., 2006; Parsons et al., 2007; Pollack et al., 2010; Gilboa et al., 2011). *For more information, see a Virtual Classroom Video at*: http://www.youtube.com/watch?v=JBIhey7sjzg and http://www.youtube.com/watch?v= daUu3iXyWWY.

VIRTUAL PATIENTS FOR CLINICAL TRAINING

Clinical interest in artificially intelligent (AI) agents designed for interaction with humans can trace its roots to the work of MIT AI researcher, Joe Weizenbaum. In 1966, he wrote a language analysis program called ELIZA that was designed to imitate a Rogerian therapist. The system allowed a computer user to interact with a virtual therapist by typing simple sentence responses to the computerized therapist's questions. Weizenbaum reasoned that simulating a non-directional psychotherapist was one of the easiest ways of simulating human verbal interactions and it was a compelling simulation that worked well on teletype computers (and is even instantiated on the internet today; http://www-ai.ijs.si/eliza-cgi-bin/eliza_script). In spite of the fact that the illusion of Eliza's intelligence soon disappears due to its inability to handle complexity or nuance, Weizenbaum was reportedly shocked upon learning how seriously people took the ELIZA program (Howell and Muller, 2000). And this led him to conclude that it would be immoral to substitute a computer for human functions that "...involves interpersonal respect, understanding, and love." (Weizenbaum, 1976).

More recently, seminal research and development has appeared in the creation of highly interactive, artificially intelligent and natural language capable virtual human (VH) agents. No longer at the level of a prop to add context or minimal faux interaction in a virtual world, these agents are designed to perceive and act in a 3D virtual world, engage in face-to-face spoken dialogues with real users (and other VHs) and in some cases, they are capable of exhibiting human-like emotional reactions. Previous classic work on VHs in the computer graphics community focused on perception and action in 3D worlds, but largely ignored dialogue and emotions. This has now changed. Artificially intelligent VH agents can now be created that control computer generated bodies and can interact with users through speech and gesture in virtual environments (Gratch et al., 2002). Advanced virtual humans can engage in rich conversations (Traum et al., 2008), recognize nonverbal cues (Morency et al., 2008), reason about social and emotional factors (Gratch and Marsella, 2004) and synthesize human communication and nonverbal expressions (Thiebaux et al., 2008). Such fully embodied conversational characters have been around since the early 90's (Bickmore and Cassell, 2005) and there has been much work on full systems that have been designed and used for training (Evans et al., 1989; Kenny et al., 2007; Prendinger and Ishizuka, 2004; Rickel et al., 2001), intelligent kiosks (McCauley and D'Mello, 2006), and virtual receptionists (Babu et al., 2006). Both in appearance and behavior, VHs have now passed through "infancy" and are ready for service in a variety of clinical and research applications.

An integral part of medical and psychological clinical education involves training in interviewing skills, symptom/ability assessment, diagnosis and interpersonal communication. In the medical field, students initially learn these skills through a mixture of classroom lectures, observation, and role-playing practice with *standardized patients*–persons recruited and trained to take on the characteristics of a real patient, thereby affording medical students a realistic opportunity to practice and be evaluated in a simulated clinical environment.

Although a valuable tool, there are several limitations with the use of standardized patients that can be mitigated through VR simulation technology. First, standardized patients are expensive. For example, although there are 130 medical schools in the U.S., few sites provide standardized patient assessments as part of the U.S. Medical Licensing Examination at a cost of several thousand dollars per student (ECFMG, 2011). Second, there is the issue of standardization. Despite the expense of standardized patient programs, the standardized patients themselves are typically low skilled actors making about $15/hr and administrators face constant turnover resulting in considerable challenges to the consistency of patient portrayals. This limits the value of this approach for producing reliable and valid interactions needed for the psychometric evaluation of clinicians in training. Finally, the diversity of the conditions that standardized patients can characterize is limited by availability of human actors and their skills. This is even a greater problem when the actor needs to be a child, adolescent, elder, or in the mimicking of nuanced or complex symptom presentations.

The situation is even more challenging in the training of clinical psychology and social work students. Rarely are live standardized patients used in such clinical training. Most direct patient interaction skills are acquired via role-playing with supervising clinicians and fellow graduate students, with closely supervised "on-the-job" training providing the brunt of experiential training. While one-way mirrors provide a window for the direct observation of trainees, audio and video recordings are a more common method of providing supervisors with information on the clinical skills of trainees. As well, the imposition of recording has been reported to have demonstrable effects on the therapeutic process that may confound the end goal of clinical training (Bogolub, 1986).

In this regard, Virtual Patients can fulfill the role of standardized patients by simulating diverse varieties of clinical presentations with a high degree of consistency, and sufficient realism (Stevens et al., 2005), as well as being always available for *anytime-anywhere* training. Similar to the compelling case made over the years for Clinical VR generally, VP applications can likewise enable the precise stimulus presentation and control (dynamic behavior, conversational dialog and interaction) needed for rigorous laboratory research, yet embedded within the context of an ecologically relevant simulated environment. Toward this end, there is a growing literature on the use of VPs in the testing and training of bioethics, basic patient communication, interactive conversations, history taking, clinical assessment, and clinical decision-making (Bickmore and Giorgino, 2006; Bickmore et al., 2007; Lok et al., 2007, 2011; Kenny et al., 2007, 2011; Parsons et al., 2008b; Rizzo et al., in press; Wendling et al, 2011). There has also been a significant amount of work creating physical manikins and virtual patients that represent physical problems (Kotranza et al, 2008), such as how to attend to a wound, or interviewing a patient with a stomach problem to assess the condition, and initial results suggest that VPs can provide valid and reliable representations of live patients (Triola et al., 2006).

However, research into the use of VPs in psychology and related psychosocial clinical training has been limited and can best be considered "fledgling". For example, Beutler and Harwood (2005) describe the development of a VR system for training in psychotherapy (characters primarily with psychological disorders or medical conditions in which a psychological condition may complicate a straightforward medical diagnosis) and summarize training-relevant research findings. Other than the above cited work, by Beutler et al., the Lok lab, and the Kenny, Parsons and Rizzo lab, we could find no other references on the use and evaluation of VPs in clinical psychology training to date, despite online searches through MEDLINE, Ovid, and the psychotherapy literature. From this, there appears to be a gap in research into the design and evaluation of intelligent VPs that have realistic interaction and communication capabilities for training clinical interviewing, diagnostic assessment and therapy skills in novice mental health clinicians.

USE CASE: VIRTUAL PATIENTS FOR CLINICAL TRAINING

The remainder of this chapter will detail some of our work developing and evaluating the use of artificially intelligent VHs designed to serve the role of virtual patients for training clinical interaction skills in novice clinicians. While we believe that the use of VHs to serve the role of virtual therapists is still fraught with both technical and ethical concerns, we have had success in the initial creation of VHs that can mimic the content and interaction of a patient with a clinical disorder for training purposes. We will also briefly discuss our related emerging work developing an online VH presence (SimCoach) for providing assistance to military personnel and significant others in the access of relevant psychological health and TBI care information. This project aims to break down barriers to care (e.g. unawareness, stigma, complexity of the military psychological healthcare system, etc.) and assist users in the process of initiating a first contact with a live human healthcare provider.

The art and science of evaluating interviewing skills using VPs is still a young discipline with many challenges. One formative approach is to compare performances obtained during interviews with both live standardized patients and with VPs, and then to conduct correlational analyses of metrics of interest. This information can then be evaluated relative to an Objective Structured Clinical Examination (OSCE) (Walters et al, 2005). Such tests typically take from 20-30 minutes and require a faculty member to watch the student perform a clinical interview while being videotaped. The evaluation consists of a self-assessment rating along with faculty assessment and a review of the videotape. This practice is common, although is applied variably, based on the actors, available faculty members and space and time constraints at the training site. A general complication involved in teaching general interviewing skills is that there are multiple theoretical orientations and techniques to choose from and the challenge will be to determine what commonality exists across these methods for the creation of usable and believable VPs that are adaptable to all clinical orientations. To minimize this problem in our initial efforts, we have concentrated on assessing the skills required to diagnose very specific mental disorders (i.e., conduct disorder, PTSD, depression, etc.). We also use the setting of an initial intake interview to constrain the test setting for acquiring comprehensible data to drive future research. In our test protocols, clinicians are typically provided some knowledge as to why the virtual patient is there (i.e., a referral question), but need to ask the patient strategic questions to obtain a detailed history useful for specifying a clinical condition in support of coming to a differential diagnosis and for formulating a treatment plan. In this manner, the system is designed to allow novice clinicians the opportunity to practice asking interview questions that eventually lead to the narrowing down of the alternative diagnostic options, leading to the arrival of a working diagnosis based on the VP meeting the criteria for a specific DSM diagnosis.

Our initial project in this area involved the creation of a virtual patient, named "Justin". Justin portrays a 16-year old male with a conduct disorder who is being forced to participate in therapy by his family. The system was designed to allow novice clinicians to practice asking interview questions, to attempt to create a positive therapeutic alliance and to gather clinical information from this very challenging VP. Justin was designed as a first step in our research. At the time, the project was unfunded and thus required our lab to take the economically inspired route of recycling a virtual character from a military negotiation-training scenario to play the part of Justin. The research group agreed that this sort of patient was one that could be convincingly created within the limits of the technology (and funding) available to us at the time.

For example, such resistant patients typically respond slowly to therapist questions and often use a limited and highly stereotyped vocabulary. This allowed us to create a believable VP within limited resources for dialog development. As well, novice clinicians have been typically observed to have a difficult time learning the value of "waiting out" periods of silence and non-participation with these patients. We initially collected user interaction and dialog data from a small sample of psychiatric residents and psychology graduate students as part of our iterative design process to evolve this application area. The project produced a successful proof of concept demonstrator, which then led to the acquisition of funding that currently supports our new projects in this area.

Following our successful Justin proof of concept, our 2nd VP project involved the creation of an adolescent female sexual assault victim, "Justina". The aim of this work was two fold: 1. Explore the potential for creating a system for use as a clinical interview trainer for promoting sensitive and effective clinical interviewing skills with a VP that had experienced significant personal trauma; and 2. Create a system whereby the dialog content could be manipulated to create multiple versions of Justina to provide a test of whether novice clinicians would ask the appropriate questions to assess whether Justina met the criteria for the DSM-4r diagnosis of PTSD based on symptoms reported during the clinical interview.

For the PTSD content domain, 459 questions were created that mapped roughly 4 to 1 to a set of 116 responses. The aim was to build an initial language domain corpus generated from subject matter experts and then capture novel questions from a pilot group of users (psychiatry residents) during interviews with Justina. The novel questions that were generated could then be fed into the system in order to iteratively build the language corpus. We also focused on how well subjects asked questions that covered the six major symptom clusters that can characterize PTSD following a traumatic event. While this approach did not give the Justina character a lot of depth, it did provide more breadth for PTSD-related responses, which for initial testing seemed prudent for generating a wide variety of questions for the next Justina iteration. (This 2nd iteration is currently in progress.)

In the initial test, a total of 15 Psychiatry residents (6 females, 9 males; mean age = 29.80, SD 3.67) participated in the study and were asked to perform a 15-minute interaction with the VP to take an initial history and determine a preliminary diagnosis based on this brief interaction with the character.

The participants were asked to talk normally, as they would to a standardized patient, but were informed that the system was a research prototype that uses an experimental speech recognition system that would sometimes not understand them. They were instructed that they were free to ask any kind of question and the system would try to respond appropriately, but if it didn't, they could ask the same question in a different way.

From post questionnaire ratings on a 7-point likert scale, the average subject rating for believability of the system was 4.5. Subjects reported their ability to understand the patient at an average of 5.1, but rated the system at 5.3 as frustrating to talk to, due to speech recognition problems, out of domain answers or inappropriate responses. However, most of the participants left favorable comments that they thought this technology would be useful in the future, and that they enjoyed the experience of trying different ways to talk to the character in order to elicit a relevant response to a complex question. When the patient responded back appropriately to a question, test subjects informally reported that the experience was very satisfying.

Analysis of concordance between user questions and VP response pairs indicated moderate effects sizes for Trauma inquiries ($r = 0.45$), Re-experiencing symptoms ($r = 0.55$), Avoidance ($r = 0.35$), and in the non-PTSD general communication category ($r = 0.56$), but only small effects were found for Arousal/Hypervigilance ($r = 0.13$) and Life impact ($r = 0.13$). These relationships between questions asked by a novice clinician and concordant replies from the VP suggest that a fluid interaction was sometimes present in terms of rapport, discussion of the traumatic event, the experience of intrusive recollections and discussion related to the issue of avoidance.

Low concordance rates on the arousal and life impact criteria, indicated that a larger domain of possible questions and answers for these areas was not adequately modeled in this pilot effort and this is now being addressed in the 2nd iteration of Justina. We have also commenced two projects that present military VPs for the training of both MDs and Social Workers for better performance in their interactions with both Active Duty and Veteran patients (see Figure 5).

If this exploratory work continues to show promise, we intend to address a longer-term vision—that of creating a comprehensive DSM diagnostic trainer that has a diverse library of VPs modeled after each diagnostic category. The VPs would be created to represent a wide range of age, gender and ethnic backgrounds and could be interchangeably loaded with the language and emotional models defined by the criteria specified in any of the DSM disorders. We believe this vision will also afford many research opportunities for investigating the functional and ethical issues involved in the process of creating and interacting with virtual humans and patients. While ethical challenges may be more intuitively appreciated in cases where the target user is a patient with a clinical condition seeking a virtual clinician, the training of clinicians with VPs will also require a full appreciation of how this form of training impacts clinical performance with *real* patients. These are not trivial concerns and will require careful ethical and scientific consideration.

The systematic use of artificially intelligent virtual humans in clinical virtual reality applications is still clearly in its infancy. But the days of limited use of VH's as simple props or static elements to add realism or context to a clinical VR application are clearly in the past. In this section we have presented an example of the creation and use of VH characters to serve the role of digital "standardized patients" for training clinical skills, in both psychological and medical care domains. This initial effort and other VP project we have in progress (Rizzo et al., in press) have also lead to new opportunities for exploring the use of VHs to serve as online mental healthcare guides or coaches (Rizzo et al., 2011c).

This work is focused on breaking down barriers to care (stigma, unawareness, complexity, etc.) by providing military Service Members, Veterans, and their families with confidential help in exploring and accessing psychological healthcare content and for promoting the initiation of care with a live provider if needed. These projects are nascent efforts in this area, yet in spite of the current limits of the technology, it is our view that the clinical targets selected can still be usefully addressed.

The capacity to conduct clinical training within simulations that provide access to credible virtual human patients where novice clinicians can gain exposure to the presentation of a variety of clinical conditions will soon provide a safe and effective means for learning skills before actual training with real patients and for supplementing continuing education throughout the professional lifespan. And as the underlying enabling technologies continue to advance, significant opportunities will emerge that will reshape the clinical training landscape. The birth of this field has already happened, the next step is to insure that it has a healthy upbringing. *For more information, see a Virtual Patient and SimCoach Video at*: http://www.youtube.com/watch?v=PPbcl8Z-8Ec and http://www.youtube.com/watch?v=JiHlTioZktc

Figure 15.5. Military Male Virtual Patient Characters (Versions 1 and 2).

CONCLUSION

The use of computer-based VR simulation technology will play an increasing role in how clinical care (and training) and scientific research is conducted in the future. Advances in the underlying enabling technologies and continuing cost reductions in system hardware are expected to make it possible for VR to shortly become a mainstream tool in this area. With this view in mind, this chapter has aimed to provide a detailed specification of rationales, research and development across three major areas of Clinical VR. However, VR is not a panacea for all the challenges that occur in clinical care.

By presenting the history, rationale and brief reviews of key literature, along with use-cases for the application of VR in each section, it is hoped that clinicians and researchers may use this information to understand the basis for when VR is well matched to the needs of clinical care in certain areas and when it is technological overkill in others. As well, the task of building really good VR systems that are both usable and useful is a challenging endeavor that requires a interdisciplinary mix of domain-specific knowledge. This sort of collaboration can be supported by providing clinicians with an informed view of what is possible with the technology and why, with what makes the most sense from a clinical perspective. Armed with this information, clinical experts can begin to partner with scientists in other more technical disciplines for the thoughtful development of Clinical VR applications that meet the needs of clinical populations of interest.

Finally, one of the guiding principles in the Clinical VR area concerns how novel and innovative VR approaches can extend the skills of a well-trained clinician. VR approaches are not intended to be automated assessment or treatment protocols that are administered in a "self-help" format without the need for informed clinical expertise. As long as VR technology is viewed in the context of its value as a *tool* for extending clinical skill rather than a replacement for it, we believe that effective VR system development and application will bring the benefits of the information age to clinical populations in ways that go beyond the limits of the past.

REFERENCES

American Psychiatric Association. (2000). *Diagnostic and statistical manual of mental disorders* (Revised 4th ed.). Washington, DC: American Psychiatric Association.

Astur, R.S., Tropp, J., Sava, S., Constable, R.T., and Markus, E.T. (2004). Sex differences in a virtual Morris water task and a virtual eight-arm maze. *Behavioural Brain Research.* 151(1-2):103-15.

Astur, R.S., Taylor, L.B., Mamelak, A.N., Philpott, L. and Sutherland, R.J. (2002). Humans with hippocampus damage display severe spatial memory impairments in a virtual Morris water task. *Behavioural Brain Research*, 132, 77-84.

Aukstakalnis, S., and Blatner, D. (1992). *Silicon mirage: The art and science of virtual reality*. Berkeley, CA: Peachpit Press.

Babu, S., Schmugge, S., Barnes, T., and Hodges, L. (2006). What Would You Like to Talk About? An Evaluation of Social Conversations with a Virtual Receptionist. In J. Gratch et al. (Eds.): *IVA 2006*, LNAI 4133, Springer-Verlag, Berlin, Germany. pp. 169-180.

Barkley, R.A. (1990). Attention Deficit Hyperactivity Disorder: A Handbook for Diagnosis and Treatment. New York, NY: Guilford Press.

Beck, A.T., Epstein, N., Brown, G. and Steer, R.A. (1988). An inventory for measuring clinical anxiety: psychometric properties *Journal of Consulting and Clinical Psychology, 56,* 893-897.

Bickmore, T. and Cassell, J. (2005). Social Dialogue with Embodied Conversational Agents. In J. van Kuppevelt, L. Dybkjaer, and N. Bernsen (eds.), *Advances in Natural, Multimodal Dialogue Systems*. New York: Kluwer Academic.

Bickmore, T. and Giorgino, T. (2006). Health Dialog Systems for Patients and Consumers. *Journal of Biomedical Informatics. 39*(5), 556-571.

Bickmore, T., Pfeifer, L., and Paasche-Orlow, M. (2007). Health Document Explanation by Virtual Agents. *The Proc. of the Intelligent Virtual Agents Conference*. Paris, Springer.

Blanchard, E.B., Jones-Alexander, J., Buckley, T.C. and Forneris, C.A. (1996). Psychometric properties of the PTSD Checklist (PCL). *Behaviour Research and Therapy, 34*(8), 669-673.

Botella, C., Banos, R. M., Perpina, C., Villa, H., Alcaniz, M., and Rey, A. (1998). Virtual reality treatment of claustrophobia: a case report. *Behaviour Research and Therapy*. 36. 239-246.

Brooks, B.M., McNeil, J.E., Rose, F.D., Greenwood, R.J., Attree, E.A. and Leadbetter, A.G. (1999). Route learning in a case of amnesia: A preliminary investigation into the efficacy of training in a virtual environment. *Neuropsychological Rehabilitation*, *9*, 63-76.

Brown, D. J., Kerr, S. J., and Bayon, V. (1998). The development of the Virtual City: A user centred approach. In Sharkey, P., Rose, D. and Lindstrom, J. (Eds.), Proceedings of the 2nd European Conference on Disability, Virtual Reality and Associated Technologies (ECDVRAT). Reading UK: University of Reading. 11-16.

Bart, O., Katz, N., Weiss, P.T. and Josman, N. (2008). Street Crossing by Typically Developed Children in Real and Virtual Environments. *OTJR: Occupation, Participation and Health*. 28(2). 89-96.

Beutler, L.E. and Harwood, T.M. (2004). Virtual reality in psychotherapy training. Journal of Clinical Psych., *60*, 317–330.

Bogolub, E.B. (1986). Tape Recorders in Clinical Sessions: Deliberate and Fortuitous Effects. *Clinical Social Work Journal.* 14(4), 349-360.

Bryant, R. A., Moulds, M. L., Guthrie, R. M., Dang, S. T. and Nixon, R. D. (2003). Imaginal exposure alone and imaginal exposure with cognitive restructuring in treatment of posttraumatic stress disorder. *Journal of Consulting and Clinical Psychology. 71*(4). 706-712.

Bryant, R.A. (2005) Psychosocial Approaches of Acute Stress Reactions. *CNS Spectrums 10*(2), 116-122.

Carlin, A. S., Hoffman, H. G., and Weghorst, S. (1997). Virtual reality and tactile augmentation in the treatment of spider phobia: A case report. Behavior Research and Therapy, 35, 153-158.

Colegrove, R., Homayounjam, H., Williams, J., Hanken, J. and Horton, N.L. (1999). The problem of overreliance upon behavioral checklists in the diagnosis of ADHD. Chapter presented at: The 107th Annual Conference of the American Psychological Association; August, 1999, Washington, D.C.

Cromby, J., Standen, P., Newman, J., Tasker, H. (1996). Successful transfer to the real world of skills practiced in a virtual environment by student with severe learning disabilities. In: Sharkey, P.M. (Ed) *Proceedings of the 1st European Conference on Disability, Virtual Reality and Associated Technologies*. Reading, UK: University of Reading, 305-313.

Difede, J. and Hoffman, H. G. (2002). Virtual reality exposure therapy for World Trade Center Post-traumatic Stress Disorder: a case report. C*yberpsychology and Behavior,* 5, 529-535.

Difede, J., Cukor, J., Jayasinghe, N., Patt, I., Jedel, S., Spielman, L., et al. (2007). Virtual Reality exposure therapy for the treatment of posttraumatic stress disorder following September 11, 2001. *Journal of Clinical Psychiatry, 68,* 1639-1647.

ECFMG, 2011. Website accessed on April 15, 2011 at: http://www.ecfmg.org/ Evans, D., Hern, M., Uhlemann, M. and Lvey, A. (1989). *Essential Interviewing: A Programmed Approach to Effective Communication* (3rd Ed): Brooks/Cole Publishing Company

Foa, E. B. and Kozak, M. J. (1986). Emotional processing of fear: Exposure to corrective information. *Psychological Bulletin, 99,* 20-35.

Foa, E. B., Steketee, G., and Rothbaum, B. (1989). Behavioral/cognitive conceptualizations of post-traumatic stress disorder. *Behavior Therapy*. 20. 155-176.

Foa, E. B. and Hearst-Ikeda, D. (1996). Emotional Dissociation in Response to Trauma: An Information-Processing Approach. In L. K. Michelson and W. J. Ray (Eds.), *Handbook of Dissociation:Theoretical and Clinical Perspectives* (pp. 207-222). New York: Plenum Press.

Foa, E. B., Davidson, R. T. and Frances, A. (1999). Expert Consensus Guideline Series: Treatment of posttraumatic stress disorder. *American Journal of Clinical Psychiatry, 60,* 5-76.

Folley. B.S., Astur, R.S., Jagannathan, K., Calhoun, V.D., Pearlson, G.D. (2010). Anomalous neural circuit function in schizophrenia during a virtual Morris water task. *Neuroimage*, 49(4), 3373-84.

Gamito, P., Oliveira, J., Morais,, D., Saraiva, T., Rosa, P. et al. (2007). War PTSD: A VR pre-trial case study. *Annual Review of Cybertherapy and Telemedicine*, 191-198.

Gilboa, Y., Rosenblum, S., Fattal-Valevski, A., Toledano-Alhadef, H., Rizzo, A., Josman, N. (in press). Describing the Attention Deficit profile of Children with Neurofibromatosis

Type 1 Using a Virtual Classroom Environment. *Proceedings of the 2011 International Conference on Virtual Rehabilitation.*

Glantz, K., Rizzo, A.A. and Graap, K. (2003). Virtual reality for psychotherapy: Current reality and future possibilities. *Psychotherapy: Theory, Research, Practice, Training.* 40(1), 55-67.

Gold, J.I., Kim, S.H., Kant, A.J., Joseph, M.H. and Rizzo, A.A. (2006). Effectiveness of Virtual Reality for Pediatric Pain Distraction during IV Placement. *CyberPsychology and Behavior.* 9(2), 207-213.

Gratch, J., Rickel, J., Andre, E., Cassell, J., Petajan, E., Badler, N. (2002). Creating Interactive Virtual Humans: Some Assembly Required. *IEEE Intelligent Systems.* July/August: 54-61.

Gratch, J. and Marsella, S. (2004). A domain independent framework for modeling emotion. *Journal of Cognitive Systems Research. 5*(4): 269-306.

Greenhill, L.L. (1998). Diagnosing Attention-Deficit/Hyperactivity Disorder in Children. Journal of Clinical Psychiatry. 59/Suppl. 7, 31–41.

Hays, R. T., Jacobs, J. W., Prince, C., and Salas, E. (1992). Requirements for future research in flight simulation training: Guidance based on a meta-analytic review. *International Journal of Aviation Psychology, 2*, 143-158.

Hix, D. and Gabbard, J.L. (2002). Usability engineering of virtual environments. In Stanney K, ed. Handbook of Virtual Environments. New York, NY: L.A. Earlbaum; 681-700.

Hodges, L.F., Kooper, R., Meyer, T., Rothbaum, B.O., Opdyke, D. de Graaff, J.J., Williford, J.S. and North, M. (1995). Virtual Environments for treating the fear of heights.*Computer*. 28(7). 27-34.

Hoffman, H.G., Chambers, G.T., Meyer, W.J., Araceneaux, L.L., Russell, W.J., Seibel, E.J., Richards, T.L. and Sharar, S.R. (2011). Virtual Reality as an Adjunctive Non-pharmacologic Analgesic for Acute Burn Pain During Medical Procedures, *Annals of Behavioral Medicine*. 41(2), 183-191.

Hoge, C.W., Castro, C.A., Messer, S.C., McGurk, D., Cotting, D.I. and Koffman, R.L. (2004) Combat Duty in Iraq and Afghanistan, Mental Health Problems, and Barriers to Care. *New England Jour of Medicine. 351* (1). 13-22.

Holden, M.K. (2005). Virtual Environments for Motor Rehabilitation: Review. *CyberPsych. and Behav. 8*(3).187-211.

Howell, S.R. and Muller, R. (2000). Computers in Psychotherapy: A New Prescription. Chapter downloaded on April 23, 2011 at: http://www.psychology.mcmaster.ca/beckerlab/showell/ComputerTherapy.PDF

Inman, D., Loge, K. and Leavens, J. (1997). VR education and rehabilitation. *Commun.of the ACM*. 40,8. 53-58.

Jaycox, L.H., Foa, E.B., and Morral, A.R. (1998). Influence of emotional engagement and habituation on exposure therapy for PTSD. *Journal of Consulting and Clinical Psychology, 66,* 186–192.

Josman, N., Somer, E., Reisberg, A., Weiss, P.L., Garcia-Palacios, A. and Hoffman, H. (2006). BusWorld: designing a virtual environment for PTSD in Israel: a protocol. *Cyberpsychology and Behavior. 9*(2), 241-244.

Johnston, R. (1995). Is it live or is it memorized? *Virtual Reality Special Report. 2*, 53-56.

Kaufmann, H., Steinbügl, K., Dünser, A. and Glück, J. (2005). General training of spatial abilities by geometry education in augmented reality. Annual Review of CyberTherapy and Telemedicine: A Decade of VR. 3: 65-77.

Kenny, P., Hartholt, A., Gratch, J., Swartout, W., Traum, Marsella, S., Piepol, D. (2007). Building Interactive Virtual Humans for Training Environments. *Proceedings of the Interservice/Industry Training, Simulation and Education Conference (I/ITSEC).*

Kenny, P., Rizzo, A.A., Parsons, T., Gratch, J. and Swartout W. (2007). A Virtual Human Agent for Training Clinical Interviewing Skills to Novice Therapists. *Annual Review of Cybertherapy and Telemedicine 2007. 5,* 81-89.

Kenny, P., Parsons, T. and Garrity, P. (2010). Emerging Concepts and Innovative Technologies: Virtual Patients for Virtual Sick Call Medical Training. *Proceedings of the Interservice/Industry Training, Simulation, and Education Conference 2010.*

Kennedy, R.S., Lane, N.E., Berbaum, K.S. and Lilienthal, M.G. (1993). Simulator sickness questionnaire: An enhanced method for quantifying simulator sickness. *Inter. Jour. of Aviation Psy.* 3(3). 203-220.

Koenig, S.T., Curcian, G.P., Dalrymple-Alford, J.C. and Dünser, A. (2009). Virtual Reality Rehabilitation of Spatial Abilities after Brain Damage. Annual Review of CyberTherapy and Telemedicine. 7: 105-107.

Koenig, S.T., Crucian, G.P., Dalrymple-Alford, J.C. and Dünser, A. (2010). Assessing navigation in real and virtual environments: a validation study, Proc. 8th Intl Conf. on Disability, Virtual Reality and Assoc. Technologies, P M Sharkey, J Sánchez (Eds), Viña del Mar/Valparaíso, Chile. 7–16.

Kotranza, A., Lind, D., Pugh, C. and Lok, B. (2008). Virtual Human + Tangible Interface = Mixed Reality Human. An Initial Exploration with a Virtual Breast Exam Patient. *Proceedings of the IEEE Virtual Reality 2008 Conference*, pp. 99-106.

Kroenke, K., and Spitzer RL. (2002). The PHQ-9: A new depression and diagnostic severity measure. *Psychiatric Annals, 32*, 509-521.

Krueger, M.W. (1993). The Experience Society. *Presence: Teleoperators and Virtual Environments.* 2(2). 162-168.

Lamson, R. (1994). Virtual therapy of anxiety disorders. *CyberEdge Journal.* 4(2). 1,6-8.

Lange, B., Flynn, B. and Rizzo, A.A. (2009) Initial usability assessment of off-the-shelf video game consoles for clinical game-based motor rehabilitation. *Physical Therapy Reviews.*14(5). 355-63.

Lange, B.S., Flynn, S.M., Proffitt, R., Chang, C.Y. and Rizzo, A.A. (2010). Development of an interactive game-based rehabilitation tool for dynamic balance training. *Topics in Stroke Rehabilitation.*17(5). 345–52.

Lezak, M. D. (1995). *Neuropsychological Assessment.* Oxford University Press, New York.

Lok, B., Ferdig, R.E., Raij, A., Johnson, K., Dickerson R., Coutts, J., Lind, D.S. (2007). Applying Virtual Reality in Medical Communication Education: Current Findings and Potential Teaching and Learning Benefits of Immersive Virtual Patients. *Jour. of Virtual Reality. 10*(3-4), 185-195.

Matheis, R., Schultheis, M.T., Tiersky, L.A., DeLuca, J., Mills, S.R. and Rizzo, A.A. (2007). Is learning and memory different in a virtual environment? *The Clinical Neuropsychologist.* 21, 146-161.

McCauley, L. and D'Mello, S. (2006). A Speech Enabled Intelligent Kiosk. In J. Gratch et al. (Eds.): *IVA 2006*, LNAI 4133, Springer-Verlag, Berlin, Germany. 132-144.

McGee, J. S., van der Zaag, C., Rizzo, A.A., Buckwalter, J. G., Neumann, U., and Thiebaux, M. (2000). Issues for the Assessment of Visuospatial Skills in Older Adults using Virtual Environment Technology, *CyberPsychology and Behavior,* 3(3), 469-482.

McGeorge, P., Phillips, L.H., Crawford, J.R., Garden, S.E., Della Sala, S., Milne, A.B., Hamilton, S. and Callander, J. (2001). Using virtual environments in the assessment of executive dysfunction. *Presence-Teleoperators and Virtual Environments. 10.* 375-383.

Mendozzi, L., Motta, A., Barbieri, E., Alpini, D. and Pugnetti, L. (1998). The application of virtual reality to document coping deficits after a stroke: Report of a case. *Cyberpsychology and Behavior, 1.* 79-91.

Merians, A.S., Fluet, G.G., Qiu, Q., Saleh, S., Lafond, I. and Adamovich, S.V. (2010). Integrated arm and hand training using adaptive robotics and virtual reality simulations. *Proceedings of the 2010 International Conference on Disability, Virtual Reality and Associated Technology.* 213-222.

Morency, L.-P., de Kok, I. Gratch, J. (2008). Context-based Recognition during Human Interactions: Automatic Feature Selection and Encoding Dictionary. *10th International Conference on Multimodal Interfaces*, Chania, Greece, IEEE.

Mowrer, O.A. (1960) *Learning and behavior.* New York: Wiley.

National Academies of Science Institute of Medicine Committee on Treatment of Posttraumatic Stress Disorder (2007). Treatment of Posttraumatic Stress Disorder: An Assessment of the Evidence. ISBN: 0-309-10925-6, 200 pages. Downloaded on 10/24/2007 from: http://www.nap.edu/catalog/ 11955.html

National Institutes of Health. (1995). *Basic behavioral science research for mental health: A national investment.* A report of the U.S. National Advisory Mental health Council, (NIH Publication No. 95-3682). Rockville, MD: Author.

Neisser, U. (1978). Memory: what are the important questions? In: Gruneberg, M.M., Morris, P.E., Sykes, R.N., eds. *Practical Aspects of Memory.* London, UK: Academic Press. 3-24.

North, M.M., North, S.M. and Coble, J.R. (1997). Virtual Reality Therapy: An Effective Treatment for the Fear of Public Speaking. *Proceedings of the Medicine Meets Virtual Reality Conference.* San Diego, CA.

Parsons, S., Beardon, L., Neale, H.R., Reynard, G., Eastgate, R., Wilson, J.R. et al. (2000). Development of social skills amongst adults with Asperger's Syndrome using virtual environments: The 'AS Interactive' project. In: Sharkey PM, Cesarani A, Pugnetti L, Rizzo S, editors) *Proc. 3rd Interl Conf. Disability, Virtual Reality and Assoc. Tech.*; 2000 Sep 23-25; Alghero, Sardinia, Italy: UK: Univ. of Reading. p.163-170.

Parsons, TD, Rizzo, AA, van der Zaag, C, McGee, JS, and Buckwalter, JG. (2005). Gender Differences and Cognition Among Older Adults. *Aging, Neuropsychology, and Cognition.*12(1), 78-88.

Parsons, T., Bowerly, T., Buckwalter, J.G. and Rizzo, A.A. (2007). A controlled clinical comparison of attention performance in children with ADHD in a virtual reality classroom compared to standard neuropsychological methods. *Child Neuropsychology.13,* 363-381.

Parsons, T. D. and Rizzo, A. A. (2008a). Affective Outcomes of Virtual Reality Exposure Therapy for Anxiety and Specific Phobias: A Meta-Analysis. *Journal of Behavior Therapy and Experimental Psychiatry*, 39, 250-261.

Parsons, T. D., Kenny, P., Ntuen, C.A., Pataki, C.S., Pato, M.T., Rizzo, A. Sugar, J. (2008b). Objective Structured Clinical Interview Training using a Virtual Human Patient. *Stud. in Health Tech. and Informatics. 132*, 357-362.

Parsons, T.D., Rizzo, A.A., Rogers, S. and York, P. (2009). Virtual reality in paediatric rehabilitation: A review. *Developmental Neurorehabilitation*, 12:4 224-238.

Pollak, Y., Barhoum-Shomaly, H., Weiss, P. L., Rizzo, A.A., Gross-Tsur, V. (2010). Methylphenidate effect in children with ADHD can be measured by an ecologically valid continuous performance test embedded in virtual reality. *CNS Spectrums*. 15.125-130.

Powers, M. and Emmelkamp, P. M. G. (2008). Virtual reality exposure therapy for anxiety disorders: A meta-analysis. *Journal of Anxiety Disorders. 22*, 561-569.

Prendinger H. and Ishizuka, M. (2004). *Life-Like Characters – Tools, Affective Functions, and Applications*, Springer.

Pugnetti, L., Mendozzi, L., Motta, A., Cattaneo, A., Barbieri, E., and Brancotti, S. (1995). Evaluation and retraining of adults' cognitive impairments: Which role for virtual reality technology? *Computers in Biology and Medicine*, *25*, 213-227.

Pugnetti, L., Mendozzi, L., Attree, E., Barbieri, E., Brooks, B.M., Cazzullo, C.L., Motta, A. and Rose, F.D. (1998). Probing memory and executive functions with virtual reality: Past and present studies. *CyberPsychology and Behavior,1,* 151-162.

Raloff J. (2006). Virtual reality for earthquake fears. *Science News* [Internet]. Aug 5 [cited 30 Oct 2010]. Available from: http://www.thefreelibrary.com/Virtual+reality+for+earthquake +fears-a0151188408.

Reger, G. and Gahm, G. (2008). Virtual Reality Exposure Therapy for Active Duty Soldiers. *Journal of Clinical Psychology*. *64,* 940-946.

Reger, G. M., Gahm, G. A., Rizzo, A. A., Swanson, R.A. and Duma, S. (2009). Soldier evaluation of the virtual reality Iraq. *Telemedicine and e-Health Journal, 15,* 100-103.

Reger, G. M., Holloway, K. M., Rothbaum, B.O, Difede, J., Rizzo, A. A., and Gahm, G. A. (2011). Effectiveness of virtual reality exposure therapy for active duty soldiers in a military mental health clinic. *Journal of Traumatic Stress.* 24(1), 93–96.

Ressler, K.J., Rothbaum, B.O., Tannenbaum, L., Anderson, P., Zimand, E., Hodges, L. and Davis, M. (2004). Facilitation of Psychotherapy with D-Cycloserine, a Putative Cognitive Enhancer. *Archives of General Psychiatry, 61,* 1136-1144.

Rickel, J., Gratch, J., Hill, R., Marsella, S. and Swartout, W. (2001). Steve Goes to Bosnia: Towards a New Generation of Virtual Humans for Interactive Experiences. *The Proceedings of the AAAI Spring Symposium on AI and Interactive Entertainment*, Stanford University, CA.

Rizzo, A.A. (1994). Virtual Reality applications for the cognitive rehabilitation of persons with traumatic head injuries. In Murphy, H.J. (ed.), *Proceedings of the 2nd International Conference on Virtual Reality And Persons With Disabilities.* CSUN: Northridge.

Rizzo, A.A., Buckwalter, J.G., and Neumann, U. (1997). Virtual reality and cognitive rehabilitation: A brief review of the future. *The Journal of Head Trauma Rehabilitation,*12(6),1-15.

Rizzo, A.A., Buckwalter, J.G., Neumann, U., Kesselman, C., Thiebaux, M., Larson, P., and Van Rooyan, A. (1998). The virtual reality mental rotation/spatial skills project: Preliminary findings. *CyberPsychology and Behavior*, 1(2), 107-113.

Rizzo, A.A., Buckwalter, J.G., Bowerly, T., McGee, J., van Rooyen, A., van der Zaag, C., Neumann, U., Thiebaux, M., Kim, L., Pair, J. and Chua, C. (2001). Virtual Environments for Assessing and Rehabilitating Cognitive/Functional Performance: A Review of Project's at the USC Integrated Media Systems Center. *Presence: Teleoperators and Virtual Environments.*Vol. 10 (4), 359-374.

Rizzo, A.A., Buckwalter, J.G. and van der Zaag, C. (2002). Virtual Environment Applications for Neuropsychological Assessment and Rehabilitation. In Stanney, K. (Ed.) *Handbook of Virtual Environments.* L.A. Earlbaum: New York. pp. 1027-1064.

Rizzo, A.A., Strickland, D. and Bouchard, S. (2004). Issues and challenges for using virtual environments in telerehabilitation. *Telemedicine Journal and e-Health.* 10(2).184-95.

Rizzo, A.A., Schultheis, M.T., Kerns, K. and Mateer, C. (2004). Analysis of Assets for Virtual Reality Applications in Neuropsychology. *Neuropsychological Rehabilitation.*14(1/2) 207-239.

Rizzo, A.A. and Kim, G. (2005). A SWOT analysis of the field of Virtual Rehabilitation and Therapy. *Presence: Teleoperators and Virtual Environments.* 14(2), 1-28.

Rizzo, A.A., Bowerly, T., Buckwater, J.G., Klimchuk, D., Mitura, R. and Parsons, R.D. (2006). A virtual reality scenario for all seasons: the virtual classroom. *CNS Spectums.* 11(1). 35-44.

Rizzo, A., Difede, J., Rothbaum, B.O., and Reger, G. (2010a). Virtual Iraq/Afghanistan: Development and early evaluation of a Virtual Reality Exposure Therapy System for combat-related PTSD. *Annals of the New York Academy of Sciences (NYAS)*, 1208, 114-125.

Rizzo, A. (2010b). Virtual Iraq/Afghanistan and how it is helping some troops and vets with PTSD. Available at *Veterans Today*: http://www.veteranstoday.com/2010/07/29/virtual-iraqafghanistan-and-how-it-is-helping-some-troops-and-vets-with-ptsd/

Rizzo, A., Lange, B., Suma, E., and Bolas, M. (2011a). Virtual Reality and Interactive Digital Game Technology: New Tools to Address Childhood Obesity and Diabetes. *The Journal of Diabetes Science and Technology. 5(2),* 256-264.

Rizzo, A., Parsons, T.D., Lange, B., Kenny, P., Buckwalter, J.G., Rothbaum, B.O., Difede, J., Frazier, J., Newman, B., Williams, J. and Reger, G. (2011b). Virtual Reality Goes to War: A Brief Review of the Future of Military Behavioral Healthcare, *Journal of Clinical Psychology in Medical Settings.*

Rizzo, A., Lange, B., Buckwalter, J.G., Forbell, E., Kim, J., Sagae, K., Williams, J., Rothbaum, B.O., Difede, J., Reger, G., Parsons, T. and Kenny, P. (2011c). An Intelligent Virtual Human System for Providing Healthcare Information and Support. In J.D. Westwood et al. (Eds.), *Technology and Informatics.* Amsterdam, NL: IOS Press. 503-509.

Rizzo, A.A., Buckwalter, J.G., John, B., Newman, B., Parsons, T., Kenny, P. & Williams, J. (2012). STRIVE: Stress Resilience in Virtual Environments: A Pre-Deployment VR System for Training Emotional Coping Skills and Assessing Chronic and Acute Stress Responses. In J.D. Westwood et al. (Eds.), *Technology and Informatics.* Amsterdam, NL: IOS Press.

Rizzo, A.A., Parsons, T., Buckwalter, J.G. and Kenny, P. (in press). The birth of intelligent virtual patients in clinical training. *American Behavioral Scientist.*

Riva, G. (2011). The Key to Unlocking the Virtual Body: Virtual Reality in the Treatment of Obesity and Eating Disorders. Journal of Diabetes Science and Technology. 5(2), 283-292.

Rose, F.D., Attree, E.A., and Johnson, D. A. (1996). Virtual reality: an assistive technology in neurological rehabilitation. *Curr Opin Neurol., 9*, 461-467.

Rose, D., Brooks, B.M. and Attree, E.A. (2000). Virtual reality in vocational training of people with learning.

disabilities. In Sharkey, P., Cesarani, A., Pugnetti, L. and Rizzo, A. (Eds.), *Proceedings of the 3rd International Conference on Disability, Virtual Reality, and Associated Technology*. Reading, UK: University of Reading. 129-136.

Rose, F.D., Brooks, B.M. and Rizzo, A.A. (2005). Virtual Reality in Brain Damage Rehabilitation: Review. *CyberPsychology and Behavior*. 8(3), 241-262.

Rothbaum, B.O., Hodges, L.F. and Kooper, R. (1995). Effectiveness of Virtual Reality Graded Exposure in the Treatment of Acrophobia. *Behavior Therapy*. 26, 547-554.

Rothbaum, B.O., Hodges, L., Watson, B.A., Kessler, G.D. and Opdyke, D. (1996). Virtual reality exposure therapy in the treatment of fear of flying: A case report. *Behaviour Research and Therapy*. 34. 477-481.

Rothbaum, B. O. and Hodges, L. F. (1999). The use of virtual reality exposure in the treatment of anxiety disorders. *Behavior Modification, 23,* 507-525.

Rothbaum, B. O., Hodges, L. F., and Smith, S. (1999). Virtual reality exposure therapy abbreviated treatment manual: Fear of flying application. *Cognitive and Behavioral Practice*. 6. 234-244.

Rothbaum, B. O., Hodges, L. F., Smith, S., Lee, J. H., and Price, L. (2000). A controlled study of virtual reality exposure therapy for the fear of flying. *Journal of Consulting and Clinical Psychology*. 68. 1020-1026.

Rothbaum, B.O., Hodges, L., Ready, D., Graap, K. and Alarcon, R. (2001). Virtual reality exposure therapy for Vietnam veterans with posttraumatic stress disorder. *Journal of Clinical Psychiatry, 62,* 617-622.

Rothbaum, B.O. and Schwartz, A. (2002). Exposure therapy for posttraumatic stress disorder. *American Journal of Psychotherapy, 56*, 59-75.

Roy, M.J., Francis, J., Friedlander, J., Banks-Williams, L., Lande, R.G., Taylor' P., et al. (2010). Improvement in cerebral function with treatment of posttraumatic stress disorder. *Annals of the New York Academy of Sciences (NYAS)*. 1208, 142-149.

Schneider, S.M., Kisby, C.K. and Flint, E.P. (2010, December 10, 2010). Effect of virtual reality on time perception in patients receiving chemotherapy. *Supportive Care in Cancer*, Retrieved from: http://www.springerlink.com/content/?k=(au%3a(Susan+Schneider)+OR+ed%3a(Susan+Schneider))+pub%3a(Supportive+Cancer+Care)

Shallice, T. and Burgess, P. (1991). Higher order cognitive impairments and frontal lobe lesions in man. In: Levin H.S., Eisenberg, H.M. and Benton, A.L., Eds. *Frontal Lobe Function and Dysfunction*. Oxford University Press. 125-138.

Shepard, R.N. and Metzler, J. (1971). Mental rotation of three-dimensional objects. Science. 171:701-703.

Shipman, S., and Astur, R. (2008). Factors affecting the hippocampal BOLD response during spatial memory. *Behavioural Brain Research*. 187(2). 433-441.

Shepard, R.N. and Metzler, J. (1971). Mental rotation of three-dimensional objects. Science. 171. 701-703.

Smith, S., Rothbaum, B. O., and Hodges, L. F. (1999). Treatment of fear of flying using virtual reality exposure therapy: A single case study. *The Behavior Therapist*. 22. 154-158.

Stanton, D., Foreman, N. and Wilson, P. (1998). Uses of virtual reality in clinical training: Developing the spatial skills of children with mobility impairments. In Riva, G., Wiederhold, B. and Molinari, E. (Eds.), *Virtual reality in clinical psychology and neuroscience*. Amsterdam: IOS Press. 219-232.

Sternberg, R.J. (1997). Intelligence and lifelong learning: What's new and how can we use it? *American Psychologist. 52*. 1134-1139.

Stevens, A., Hernandez, J., Johnsen, K., Dickerson, R., Raij, A., Harrison, C.,...Lind, D.S. (2005). The use of virtual patients to teach medical students communication skills. *The Association for Surgical Education Annual Meeting*. New York.

Strickland, D., Marcus, L.M., Mesibov, G.B. and Hogan, K. (1996). Brief report: two case studies using virtual reality as a learning tool for autistic children. *J Autism Dev Disord*. 26(6). 651-9.

Thiebaux, M., Marshall, A., Marsella, S., Fast, E., Hill, A., Kallmann, M.,...Lee, J. (2008). SmartBody: Behavior Realization for Embodied Conversational Agents. *Proceedings of the International Conference on Autonomous Agents and Multi-Agent Systems (AAMAS)*. Portugal.

Traum, D., Marsella, S., Gratch, J., Lee, J., Hartholt, A. (2008). Multi-party, Multi-issue, Multi-strategy Negotiation for Multi-modal Virtual Agents. *8th International Conference on Intelligent Virtual Agents*. Tokyo, Japan, Springer.

Triola, M., Feldman, H., Kalet, A.L., Zabar, S. Kachur, E.K., Gillespie, C.,...Lipkin, M. (2006). A randomized trial of teaching clinical skills using virtual and live standardized patients. *Journal of General Internal Medicine. 21*, 424-429.

Van Etten, M.L. and Taylor, S. (1998). Comparative efficacy of treatments of posttraumatic stress disorder: An empirical review. *Journal of the American Medical Association. 268*. 633-638.

Walker, D.L., Ressler, K.J., Lu, K.T. and Davis, M. (2002). Facilitation of conditioned fear extinction by systemic administration or intra-amygdala infusions of D-cycloserine as assessed with fear-potentiated startle in rats. *Journal of Neuroscience, 22*, 2343-2351.

Walters, K., Osborn, D. and Raven, P. (2005). The development, validity and reliability of a multimodality objective structure clinical examination in psychiatry. *Medical Education. 39*, 292–298.

Weathers, F.W., Litz, B., Herman, D.S., Huska, J.A. and Keane, T.M. (1993) The PTSD Checklist (PCL): Reliability, Validity, and Diagnostic Utility. Chapter presented at the *Annual Meeting of International Society for Traumatic Stress Studies*. San Antonio, TX, October, 1993.

Wendling, A., Halan, S., Tighe, P., Le, L., Euliano, T., and Lok, B. (2011). Virtual humans versus standardized patients in presenting abnormal physical findings. *Academic Medicine*. 86(3). 384-388.

Weiss, P. and Jessel, A.S. (1998). Virtual reality applications to work. *Work*.11. 277-293.

Weizenbaum, J. (1976). *Computer Power and Human Reason*. San Francisco: W. H Freeman.

Wilson, B.A., Alderman, N., Burgess, P.W., Emslie, H. and Evans, J.J. (1996). *BADS:Behavioural Assessment of the Dysexecutive Syndrome*. Bury St Edmunds, UK: Thames Valley Test Company.

Wilson, J., Onorati, K., Mishkind, M., Reger, M. and Gahm, G.A. (2008). Soldier attitudes about technology-based approaches to mental healthcare. *Cyberpsychology and Behavior, 11*, 767-769.

Wood, D.P., Murphy, J.A., Center, K., Russ, C. McLay, R.N. et al. (2008). Combat Related Post Traumatic Stress Disorder: A multiple case report using virtual reality graded exposure therapy with physiological monitoring. In *Medicine Meets Virtual Reality 16,* Amsterdam: IOS Press, 556-561.

In: Handbook of Technology in Psychology ...
Editor: Luciano L'Abate and David A. Kasier
ISBN: 978-1-62100-004-4

Chapter 15

TRANSCRANIAL MAGNETIC AND DEEP BRAIN STIMULATION

Dafne C. Andrade[1], Igor C. Borges, Rodrigo F. Cury and Felipe Fregni
Universidade Federal da Bahia, UFBA, Brasil

The use of electric currents to modulate brain function has a long history, although the contemporary techniques have only been recently developed. The first reports of its effects on the nervous system date from 43–48 AD, when Scribonius Largus used a live torpedo fish to relieve a patient's headache. Since then, many empirical clinical attempts have been made to develop electrical brain therapy. It is worthy to note that, in 1804, Giovanni Aldini was successful in treating depressed patients with galvanic currents administered over the skull. In 1930, Bini and Cerletti developed the method of Electric Convulsive Therapy (ECT), which has been extensively used since that time for the treatment of neuropsychiatric disorders (Priori et al., 2003).

In the 1960s, researchers began the systematic investigation of noninvasive brain stimulation techniques, mainly through animal studies. In 1980, Merton and Morton showed that it was possible to electrically stimulate the brain through an intact scalp, giving rise to Transcranial Electric Stimulation (TES). Transcranial Magnetic Stimulation (TMS) was introduced some decades after, by Barker et al. in 1985, originally as a neurophysiologic diagnostic tool to study neural excitability, but then evolving into a versatile therapeutic option in neurology, psychiatry, and rehabilitation (Priori et al., 2003). In 1998, it was published the first safety guidelines for the use of single pulse and repetitive TMS in clinical research and, in 2008, these guidelines were reviewed during a consensus conference. In the same year, the United States Food and Drug Administration (FDA) approved the use of a specific protocol of repetitive TMS as a valid therapy for treatment-resistant major depression.

In parallel to the development of non-invasive brain stimulation techniques, invasive brain stimulation was initially used in 1947 as an intraoperative tool – a stereotactic apparatus that could play a role in ablative procedures. At that time, it was mainly used to explore targeted brain areas prior to surgical lesioning. In 1952, Delgado et al. proposed a technique that consisted of the insertion of intracerebral electrodes in psychotic patients for chronic recording and stimulation. This technique was then shown to have a therapeutic effect compared to that of electroconvulsive therapy, evidenced by a decrease in psychotic symptoms. The modern form of Deep Brain Stimulation (DBS) has only arisen recently in 1987, when Benabid et al. reported the effects of high frequency stimulation on the thalamus. It was then shown that DBS could have the same effects as a functional lesional surgery, therefore demonstrating its use as a less aggressive treatment option. Since then, the therapeutic nature of DBS, combined with its significant research potentials, have been attracting the interest of neurosurgeons, clinicians and other professionals worldwide, and thus ensuing significant increase on its scientific basis.

Transcranial Magnetic Stimulation: Overview of the Technique

TMS applied through different protocols of delivery (single-, paired- and repetitive-pulse) has been widely used in the study of normal human neurophysiology and has both diagnostic and therapeutic roles in several neuropsychiatric disorders. The basic principle of this technique is based on Faraday's law of electromagnetic induction. In TMS, brief pulses of an electric current pass through the magnetic coil. This creates rapidly changing pulses of a magnetic field that penetrate the scalp and skull, reaching the brain with minimal attenuation. These magnetic pulses induce an electric field, that when falls in contact with a conductor, such as the brain tissue, generates current flow. It is this final electrical current that affects the functioning of neural tissues, directly interacting with nerve cells in superficial layers of the brain. However, the effects of TMS are not restricted to the area directly stimulated. Besides modifying local intracortical excitability, it activates distant cortical, subcortical, and spinal structures along specific connections, thus enabling the use of this technique to evaluate and intervene in many neural areas (Wagner et al., 2007).

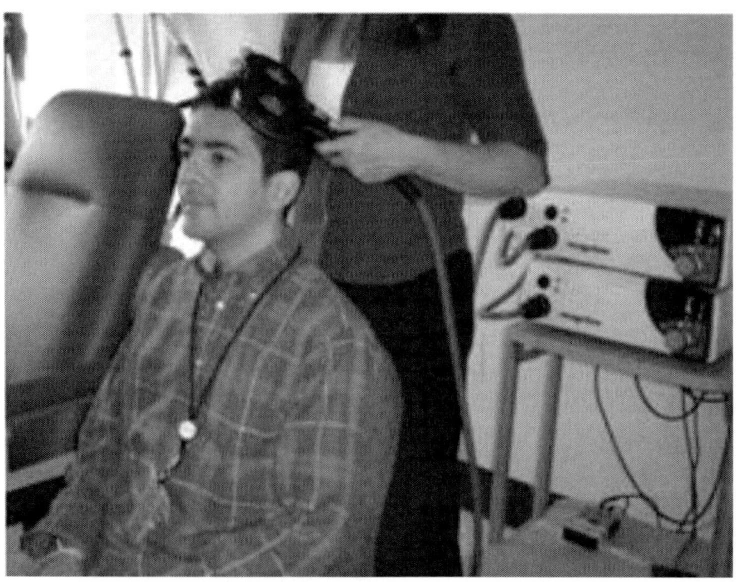

TMS applied over the dorsolateral prefrontal cortex with a figure of eight coil.

TMS can be delivered as single, paired or repetitive pulses. Single and paired pulse TMS is used to measure cortical excitability while repetitive TMS is used to change brain excitability and thus used as a potential treatment tool. When using TMS, one important parameter is the intensity of stimulation that is modulated by changing the intensity of current flowing in the coil. This intensity can be measured by applying single-pulse TMS to the primary motor cortex (M1) and using electromyography (EMG) on the targeted muscle to measure response. One used measure is the motor threshold that is measured as the stimulus with the intensity for eliciting a small response of at least 5 out of 10 trials of 50 μV peak-to-peak amplitude in a resting muscle. This is commonly used as a reference for stimulation intensity, given as a percentage of rMT. From the rMT, it is possible to calculate other parameters of cortical excitability such as motor evoked potential (MEP), silent period, intracortical facilitation and inhibition (Kobayashi et al., 2003).

The intensity of the stimulus is also related to the size of the area stimulated, specifically the depth of the activated neural layers, and the size of the effect. For example, when using intensities below the rMT on motor cortical areas, the stimulation will not elicit significant MEPs. In addition to the intensity of the pulse, the area stimulated depends on the focus of the pulse. The focal area of the magnetic field depends on the shape of the coil used. Between the two most commonly used coils, the figure-of-eight shape allows a more focal stimulation of superficial cortical areas as it induces a more focal peak of the current, while the circular shape results in less focal area but permits direct stimulation of deeper neuronal populations using higher intensities (Kobayashi et al., 2003).

REPETITIVE REGULAR AND PATTERNED TMS

Repetitive-pulse Transcranial Magnetic Stimulation (rTMS) allows cortical areas to be stimulated by trains of magnetic pulses at a given frequency and intensity. It is the form of TMS used for therapeutic purposes because of its neuromodulatory properties. Although it disrupts cortical function during trains of stimulation, it can also modulate the cortical excitability by inhibiting or facilitating the depolarization of the neuronal membranes (Bestmann, 2008). There are several variations of rTMS protocols that are currently in use: (1) - stimulus spaced by regular interstimulus intervals (conventional rTMS-protocol); (2) - variable interstimulus intervals, such as theta-burst stimulation (TBS), which generally consists of brief bursts of rTMS (3 pulses at 50Hz) repeated at intervals of 200 ms (Di Lazzaro et al., 2005; Huang et al., 2005); and (3) - Paired Associative Stimulation (PAS), consisting of low-frequency repetitive peripheral nerve stimulation combined with timed TMS over the contralateral motor cortex (Stefan et al., 2000; Wolters et al., 2003). The excitatory or inhibitory effect of each protocol depends on: stimulus frequency, stimulus intensity, total number of stimuli, and interstimulus interval. For the conventional rTMS-protocol, low-frequency rTMS (rates ≤ 1 Hz) is generally associated with inhibition while high-frequency rTMS (≥ 5 Hz, generally 10-20 Hz) is generally associated with increase of cortical excitability. The duration of the effects of brief rTMS stimulation is usually short-lasting, but interestingly, longer-lasting after-effects can be achieved by using protocols that include longer periods of stimulation or even multiple sessions of rTMS, as demonstrated by several clinical studies (Hallett, 2007).

SAFETY

Concerning the safety of TMS, the currently published protocols are associated with mostly minor, and transient side effects (Rossi et al., 2009). The most common side effects reported are headache or local pain/discomfort, especially during active rTMS. Transient hearing changes may be observed because of an acoustic artifact produced by the TMS stimulating coil. This side effect can be minimized with the use of auditory protection during the stimulation. The most serious adverse event related to TMS is seizure induction during or shortly after stimulation, which may result very rarely from single-pulse TMS, usually at high stimulus intensities, and from high-frequency rTMS. However, this side effect is extremely rare in the general population, and regarding rTMS protocols, the risk in epileptic patients was estimated at 1.4% (Bae et al., 2007) and less than 1% in patients without this condition. Excluding the studies involving patients with pre-existing neurological disorders known to increase the risk of seizures (e.g. epilepsy, stroke), there have been only 12 case reports of seizures during rTMS to date, although most were related to the use of parameters outside of the recommended guidelines or to the use of pro-epileptogenic medications (Loo et al., 2007). Syncope is another adverse event that may also occur, but not related to direct brain effects from the electric current. Instead, it seems to be from neurocardiogenic origin, as a common reaction to anxiety or psycho-physical discomfort. Other adverse effects that may occur do not seem to represent much concern and structural brain changes or histotoxicity were found to be inconsistent up to now.

There are also contraindications for TMS that are clearly outlined in the literature. Importantly, TMS should be avoided in patients with metal implants or electronic devices close to the stimulated area. This is considered an absolute contraindication, in order to avoid interaction of the magnetic field generated by the coil with ferromagnetic components or electrical networks in those devices. Shifting or malfunctioning of implanted devices can result in serious injury. In addition, due to the risk of seizure, those with epilepsy or seizure disorders should not participate in TMS protocols, as a relative contraindication. In pregnant women, there are no specific side effects that have been documented that affect the fetus long-term. However, due to the risk of seizures, pregnant women should also be excluded from TMS protocols (Rossi et al., 2009).

In summary, TMS is considered a relatively safe technique if the current guidelines are followed, both in healthy subjects and those who may have a neurological illness. However, safe parameters still needed to be established for most recent patterns of rTMS, such as TBS, and for prolonged protocols of stimulations, with continued or intermittent interventions.

MECHANISMS OF ACTION

Although its mechanisms of action are not completely understood, some studies have tried to explain how the underlying neural circuits are altered by application of electrical fields and what correlations these changes may have with clinical outcomes.

Different TMS protocols, applied at appropriate stimulation frequency and intensity, can transiently activate or disrupt focal brain activity. The former refers to the ability of TMS to directly induce activating currents in cortical neuronal networks and evoke artificial activity in brain areas while the latter means that the perturbation caused by the stimulation may interrupt the ongoing processing of an area, causing a functional deficit, also known as a "virtual lesion" (Bestmann, 2008). Nonetheless, one must keep in mind that TMS is not that focal or specific. In addition the term "virtual lesion" needs to be carefully considered as rTMS effects are not strong enough to induce a block in local activity. The relatively low spatial resolution of the technique does not assure that the surrounding tissue of the targeted cortical area is not concurrently stimulated. In addition, the nonspecific nature of neural activation may not predict if excitatory or inhibitory networks will be activated. Another limitation on the interpretation and prediction of TMS interventions is that compensatory regulatory mechanisms will be evoked as a homeostatic response for the maintenance of normal brain function. This phenomenon, also known as metaplasticity, is a plasticity-regulating mechanism that tends to maintain the outputs of the nervous system at a balance point, as it has been proposed (Rich and Wenner, 2007). The neural effect of TMS is also state-dependent: variations in pre-existing activity level contribute to the variability of TMS responses. Typically, higher pre-TMS activity predicts greater post-TMS activity (Hamada et al., 2008, 2010). These factors may partially account for the explanation of variability in response to electrical stimulation found in previous studies.

The possibility of rTMS to induce changes in brain excitability that outlast the stimulation period favors the viability of its use for therapeutic purposes. The after-effects of rTMS in humans are often described as LTD- or LTP-like mainly because the duration of the neurophysiological and behavioral alterations seems to implicate neuroplastic changes. Although there are several studies strongly suggesting that rTMS can lead to alterations in synaptic plasticity, currently there is not enough direct evidence linking rTMS to synaptic changes in the human brain.

Pharmacological evidence using drugs that block the action of N-Methyl-D-Aspartate Receptor (NMDAR) in humans have already demonstrated that the after-effects of different rTMS protocols depend on this receptor. Huang et al (2008) demonstrated that the use of memantine, a NMDAR antagonist, blocks the suppressive effect of continuous TBS and the facilitatory effect of intermittent TBS in the motor cortex of healthy volunteers. Stefan et al (2002) attested these results antagonizing NMDA receptors with dextromethorphan in a PAS protocol. This drug suppressed the after-effects of the repetitive application of single afferent electric stimuli delivered to the right median nerve, paired with single pulse TMS over the optimal site for activation of the abductor pollicis brevis muscle, in healthy volunteers. According to the classical model of induction of LTP and LTD, postsynaptic NMDA receptors induce Ca^{2+} influx which triggers a series of reactions that lead to long-term changes in synaptic strength, thus supporting the notion that NMDA receptors are essential for the occurrence of this process (Thickbroom, 2007). The results of these two studies give some evidence toward the linkage between rTMS and synaptic plasticity.

In addition to the relationship between NMDA receptors and TMS effects, this brain stimulation technique seems also involved with modulation of inhibitory GABAergic systems and monoamine-mediated neurotransmission (Post et al., 2001). The most studied neuromodulatory substance in this pool is dopamine, because of its possible relationship to pathophysiologic mechanisms of psychiatric disorders. Furthermore, there is some evidence showing an increase of the release of dopamine following TMS in rats and humans (Padberg et al., 2009).

There is also evidence demonstrating that rTMS may exert its effects through induction of gene expression. Animal studies have already shown that acute rTMS can cause transient increase in glial fibrillary acidic protein (GFAP; an indicator of reactive astrogliosis) mRNA in the hippocampal dentate gyrus in mice (Fujiki and Steward, 1997). In another study, long-term rTMS treatment performed under conditions comparable to those in clinical use led to an increase in brain-derived neurotrophic factor (BDNF) mRNA and cholecystokinin (CCK) in several brain areas in rats (Muller et al., 2000). Accordingly, it was also demonstrated that after chronic rTMS c-fos mRNA was markedly increased in neurons in layers I-IV and VI of the parietal cortex of rats (Hausmann et al., 2000) and that rTMS led to immediate early gene expression of c-fos mRNA in the paraventricular nucleus of the thalamus of rats (Ji et al., 1998).

Although the mechanisms of rTMS are still elusive, the putative mechanisms of this technique support its therapeutic exploration for neuropsychiatric disorders. As the pathophysiology of these disorders is conceived to comprise dysfunction in neuronal circuits, neuromodulatory techniques have the potential to intervene in this process and possibly restore the normal activity of the affected areas. Two of the most studied conditions in which rTMS may have a therapeutic effect are major depression and Parkinson's disease. Current evidence on rTMS in these clinical settings will be discussed as a means to highlight the clinical potential of rTMS.

DEPRESSION AND RTMS

The treatment of major depression with rTMS has been one of the most studied areas in this field. Most of the published studies showed results favoring active rTMS as compared to sham rTMS and there are already large clinical trials, some of them multicenter randomized controlled trials, and meta-analyses corroborating this evidence. In October 2008, the FDA approved the use of a specific TMS device, NeuroStar (Neuronetics Inc., Malvern, PA, USA), for the treatment of a medication-resistant depression condition. However, there are still concerning points regarding safety issues and optimal parameters of stimulation.

The specific mechanisms by which rTMS would manifest antidepressant properties are not clearly understood. The DLPFC was initially targeted for rTMS based on imaging findings linking dysfunction in this region activity with the pathophysiology of depression and antidepressant effects of therapeutic interventions (Steele et al., 2007; Bench et al., 1995). It has been demonstrated that major depression is associated with a decreased level of activity in the left DLPFC, along with other specificities of activity in interconnected brain areas, and that these abnormalities would be causally associated with depressive symptoms. Thus, the main purpose of stimulation would be to increase activity in the DLPFC to revert this condition. This is possible by directly stimulating this area with high-frequency rTMS or by modulating interhemispheric interactions through the delivery of low-frequency rTMS to the right DLPFC (Speer et al., 2000, 2008; Kito et al., 2008). In the last case, the right prefrontal cortex is considered to have inhibitory influence over the left one, so by decreasing its activity, the left-right imbalance would be reverted and activity restored on the left side.

Another possible mechanism is related to the evidence that rTMS is able to increase the expression and release of BDNF, a neurotrophic factor associated with neuronal survival and differentiation, also being involved with the occurrence of neuroplastic processes, such as LTP and LTD. Several studies have also shown that there is a decrease in BDNF expression in patients with major depression and that there is normalization of the levels of this substance after different antidepressant treatments. Hence, one of the hypothesized mechanisms of generation of depressive symptoms is a decreased plasticity in areas associated with positive emotional processing. Thus, rTMS could interfere with this process by enhancing the expression of BDNF (Post et al., 2001).

Additionally, the utility of rTMS in the treatment of depression is also associated with a theory widely used to justify the efficacy of pharmacotherapy. This theory states that this disorder is linked to a decrease in dopamine neurotransmission and that rTMS, with the capability to enhance dopaminergic pathways, could reverse this imbalance (Keck et al. 2002; Zangen and Hyodo, 2002; Ohnishi et al., 2009). To complement the myriad of possibilities, rTMS could also interact with the pathophysiologic mechanisms of major depression by attenuating the activity in the hypothalamic-pituitary-adrenocortical (HPA) system. Disinhibition of this system is a known feature of major depression and, just as hypothesized for most of the pharmacologic antidepressant therapies with similar effect above the HPA system, the regulation of this neuroendocrine system could account for the therapeutic effect of the transcranial stimulation (Post et al., 2001).

Currently, the most accepted parameter of stimulation for the treatment of major depression consists of suprathreshold rTMS applied to the left DLPFC at 10 Hz using an elevated number of pulses (i.e. 3000 pulses) per session, usually delivered daily for 5 sessions per week. The number of stimulation sessions is another important feature, as it has been found by some studies that rTMS has dose-dependent antidepressant effects, in which more sessions are associated with better response and longer lasting effects (Avery et al., 2006; Gross et al., 2007). Most of the earlier studies used 5-20 sessions (1-4 weeks) of stimulation, while the most recent ones have used up to 20-30 sessions (4-6 weeks), what in part may explain the difference in the clinical significance of the outcomes of these group of studies (Gross et al., 2007). The two most important studies to date, the industry-sponsored one that supported the FDA approval for the use of rTMS in the treatment of unipolar major depressive disorder (O`Reardon et al., 2007) and the NIH-sponsored study that replicated its results (George et al., 2010), demonstrated that the parameter of stimulation above mentioned, with 20-30 stimulation sessions and 3000 pulses, was effective in treating major depression with a good tolerability profile. However, the results of the study of O'Reardon and colleagues did only manage to obtain the FDA approval to the population of adults who have failed to achieve satisfactory improvement from one adequate antidepressant medication trial and using these specific parameters of rTMS. This specification was due to the observation that across the various degrees of refractoriness of the patients from the study, only this subgroup showed significant positive results after post-hoc analysis. There is also increasing evidence that other parameters of stimulation, such as right DLPFC low-frequency rTMS (Pallanti et al., 2010; Schutter, 2010) and bilateral DLPFC rTMS (high-frequency on the left and low-frequency on the right) (Fitzgerald et al., 2006a), might also be useful strategies for the treatment of major depression.

Although these data suggest that rTMS has significant antidepressant effects, there are studies showing that these clinical benefits may relapse in less than one year (at a mean time of 5-10 months, depending on the study) (Fitzgerald et al., 2006b; Demirtas-Tatlidede et al., 2008). This implicates the need for rTMS maintenance in order to maintain clinical improvements; however, there is limited data on this approach. Sustained clinical response to retreatment and also prevention of relapses with intermittent rTMS (1 or 2 times a week) were already obtained in a few naturalistic studies (O'Reardon et al., 2005; Fitzgerald et al., 2006b; Demirtas-Tatlidede et al., 2008). Nonetheless, the safety of this procedure has not yet been assured and more trials are needed to confirm these results.

Another important aspect to be highlighted is the low number of remitters and responders following rTMS. Remission rates are estimated at 17% and response rate at 25% while the number needed to treat (NNT) calculated for clinical remission is 7 and response is 6 in short-duration treatment (1-4 weeks), based on results of studies involving patients with treatment-resistant depressive disorder (Lam et al., 2008). At first glance, these numbers seem to be disappointing, but the comparison of remission rates between rTMS and conventional pharmacological or psychotherapy treatment show an overall equality or advantage of rTMS in this scenario, because prior resistance to antidepressants treatment significantly diminishes the likelihood of responding to subsequent interventions, according to studies with conventional treatments (Rush et al., 2006). This is also true for rTMS, as demonstrated that less treatment-resistant depression is associated with better outcomes after stimulation, compared to more treatment-resistant patients (Fregni et al., 2006; Brakemeier et al., 2007; Lisanby et al., 2009). Thus, there is still to be defined the range of refractoriness in which rTMS would be superior to conventional treatment. Although rTMS was originally suggested as a potential substitute for ECT for refractory patients, it is still not clear whether rTMS effects would be reach the same magnitude as those of ECT (Grunhaus et al., 2000; O'Connor et al., 2003; Eranti et al., 2007). Nevertheless, rTMS is still an attractive option, since it is related to a better profile of adverse effects.

Concerning the response and remission rates, predictors of better response to rTMS are desired, as it would improve the potential benefits of this procedure. Besides the above mentioned level of treatment resistance, younger age is also related to better outcomes following rTMS in the treatment of depression (Fregni et al., 2006; Aguirre et al., 2010), but these two variables are still too limited and with restricted applicability in the therapeutic setting of this pathology. Of interest, there is also evidence that genetic factors may predict some rTMS effects (Kleim et al., 2006; Cheeran et al., 2008), giving rise to the question that if combination of rTMS with genetic testing would improve the rates of responders and allow a more efficient treatment.

In summary, there is significant evidence of the clinical utility of rTMS in the therapeutic of treatment-resistant depression and future studies need to elucidate optimal parameters of stimulation for this purpose, the efficacy and safety of prolonged protocols of stimulation, identify reliable and useful predictors of response to the treatment and discuss the ethical considerations of the introduction of this new alternative in the practice of therapeutic medicine, especially after FDA approval.

Parkinson Disease and rTMS

Parkinson's Disease (PD) is a neurodegenerative disorder caused by the progressive loss of dopaminergic neurons in the substantia nigra of the nigrostriatal pathway. The current treatment of PD relies mainly on pharmacological approaches, specially the administration of

L-Dopa to restore dopamine levels. However, L-Dopa is suspected to accelerate the loss of dopaminergic neurons through a neurotoxic effect (Arias-Carrion et al., 2008). In addition, it is also associated to the development of dyskinesias and its therapeutic efficacy diminishes over the time of utilization. Hence, new therapies for PD are necessary, which have to be effective, have long-lasting effects, and cannot present serious adverse reactions (Fregni et al., 2005).

The effects of rTMS on Parkinson's disease (PD) have been intensively studied on the past two decades. The rationale for the application of rTMS for PD is the modulation of brain areas which are involved in the dopaminergic system and are dysfunctional in this condition, i.e. the basal ganglia and related cortical regions including the primary motor cortex (M1), the premotor cortex and the prefrontal cortex. (Arias-Carrion et al., 2008; Shimamoto et al., 2003). For instance, animal studies have already demonstrated that the descending pathways originated in the frontal cortex control dopamine release in subcortical regions such as the striatum. The regulation of these pathways, either the mesostriatal dopaminergic pathway or the direct corticostriatal pathway, is believed to be the mechanism through which rTMS might act on PD.

There are several reports from studies both in animals and humans showing the influence of rTMS on dopaminergic pathways. Stimulation of the DLPFC by rTMS was demonstrated to increase dopamine release in the caudate nucleus in healthy humans (Straella et al., 2001). Similarly, the stimulation of M1 was also related to an increase in dopamine release in the ventral striatum and in the putamen, which is the principal input nucleus for somatic motor control in the basal ganglia and receives corticostriatal projections from the frontal cortex (Strafella et al., 2003; Ohnishi et al., 2004). These results support the hypothesis that stimulation of a specific cortical site by rTMS results in increased dopamine release in its known projection area. This effect promoted by rTMS has already been demonstrated to have a similar magnitude as the one promoted by the administration of D-amphetamine, a substance known to increase synaptic dopamine (Pogarell et al., 2007). A recent trial correlated the motor improvement of PD patients after rTMS with the levels of dopamine on plasma samples, raising the possibility that the beneficial effects of rTMS may be due to an increase in cerebral dopamine as well (Khedr et al., 2007).

Furthermore, there is also evidence from an animal model of nigrostriatal degeneration that rTMS prevented neuronal death following the toxic lesion on injured dopaminergic neurons (Funamizu et al., 2005). It is also suggested that rTMS may induce an in situ differentiation of subventricular zone precursor cells into dopaminergic neurons. These neuroprotective and neurogenic effects might also account for a possible long-term effect of rTMS in PD, by preventing and compensating for the neuronal loss (Arias-Carrion et al., 2008); though direct evidence is lacking. There is some clinical evidence that the use of rTMS can actually slow the progression of PD (Mally et al., 2004); nevertheless, further studies are still needed in order to confirm this effect of rTMS.

Another pathophysiological feature of PD that may be affected by rTMS is the excitability and neuroplastic changes in the basal-ganglia-related cortical areas. TMS and imaging studies have demonstrated that M1 in PD patients shows an excessive tonic activity at "rest", reduced intracortical inhibition and plasticity and a relative failure to recruit neurons progressively and correctly when active (Cantello et al., 2002). These maladaptive compensatory changes are important for the decreased performance of the motor system, and may be corrected by rTMS.

Through the modulation of cortical excitability and intracortical inhibition, rTMS would compensate the lack of basal ganglia drive to the motor cortex (Siebner et al 2000; Filipovic et al., 2010). Lastly, there is evidence that a change in the activity of thalamic neurons in the motor circuits occurs in PD (Halliday., 2009). It is believed that the pathologic alterations in the thalamus may also account for abnormalities present in PD; hence modulation of thalamic activity is one additional mechanism by which rTMS could promote motor improvement in PD patients.

The first evidence of the efficacy of rTMS as a therapeutic tool for PD was provided in 1994, through the stimulation of 6 patients on M1 at a frequency of 5 Hz (Pascual-Leone et al., 1994). Several studies have then been conducted to demonstrate the effects of rTMS on PD, which are summarized in Table 1. As it may be noted, the methodology of these studies is highly heterogeneous, with great variability of rTMS parameters and patients' characteristics.

Table 1. Summary of the studies on rTMS and motor function in PD patients

Author	Year	Patients	Frequency	Pulses	Intensity	Coil	Sessions	Site	Results
Pascual-Leone	1994	6	5	NA	NA	NA	NA	M1	+
Mally	1999	10	1	30*	0.2MT	C	10	Vertex	+
Mally	1999	12	1	30	0.34T	CC	10	Vertex	o
		12	1	30	0.34T	C	14	Vertex	+
		12	1	30	0.57T	C	10	Vertex	+
		12	1	30	0.80T		10	Vertex	+
Tergau	1999	7	1,5,10,20	1000	0.90MT	C	4	NA	o
Siebner	1999	12	5	2250	0.90MT	F8	1	M1	+
Siebner	2000	10	10	2000	0.90MT	F8	1	M1	+
Boylan	2001	10	10	2000	1.1MT	F8	1	SMA	–
De Groot	2001	9	5	2250	0.90MT	F8	1	M1	+
Shimamoto	2001	9	0.2	60	700V	C	8	Frontal	+
Dragasevic	2002	10	0.5	200	1.1MT	C	10	Prefrontal	+
Ikeguchi	2003	1	0.2	60	0.7%	C	6	Prefrontal	+
Khedr	2003	19	5	2000	1.2MT	F8	10	M1	+
Okabe	2003	85	0.2	100	1.1MT	C	8	Vertex	o
Fregni	2004	21	15	3000	1.1MT	F8	10	Prefrontal	+
Lefaucheur	2004	12	0.5	600	0.80MT	F8	1	M1	+
		12	10	2000	0.80MT	F8	1	M1	+
Boggio	2005	13	15	200	1.1MT	F8	10	Left DLPFC	o
Lomarev	2006	9	25	1200	1MT	F8	1	DLPFC/M1	+
Khedr	2006	35	25	3000	1MT	F8	6	M1	+
		10	10	3000	1MT	F8	6	M1	+
del Olmo	2007	8	10	450	0.90MT	F8	10	DLPFC	o
Khedr	2007	20	25	3000	1MT	F8	6	M1	+
Hamada	2008	55	5	1000	1.1AMT	F8	8	SMA	+
Sedlácková	2009	10	10	1350	1MT	F8	1	PMd/DLPFC	o
Bäumer	2009	15	1	1200	0.80AMT	F8		PMd	o
Arias	2010	9	1	100	0.90MT	C	10	Vertex	o
Filipović	2010	10	1	1800	0.90MT	F8	4	M1	o

*=twice a day; MT: Motor threshold; NA: Not available; M1 Primary motor cortex; SMA: Supplementary motor cortex; DLPFC: Dorsolateral prefrontal cortex; PMd: Premotor dorsal cortex; C: Circular; F8: Figure of eight; +: Improvement of motor function; o: no effects on motor function; –: worsening in motor function. NA: Not Available

A meta-analysis has evaluated the effects of 12 clinical trials and its results supported the efficacy of rTMS in the improvement of motor function on PD patients. The clinical effect of rTMS was significant even on the subset of studies that compared it to the effect of placebo, which was reported to be present in 17% of PD patients submitted to novel therapies. It was also demonstrated that the studies with larger clinical effects where the ones in which several sessions of stimulation were provided. Furthermore, the reported side effects of rTMS on the trials with PD patients were benign, mainly a mild headache, neck pain, light burning sensation on the scalp and increase in salivation (Fregni et al., 2005).

However, the clinical effect of rTMS demonstrated on PD patients is modest, and not all of the trials demonstrated positive results. The small size of the samples, the confounding effect of the use of antiparkinsonian drugs and other methodological issues might account for these results. The site of stimulation seems to be important for the motor improvement and the coil type may alter the precision of the stimulation (Fregni et al., 2005).

A more recent meta-analysis studied separately the effects of high- and low- frequency rTMS on the motor symptoms of PD. It demonstrated that high frequency rTMS promoted a significant improvement of motor function and is a promising treatment for PD. This result corroborated those of the aforementioned meta-analysis. However, low frequency rTMS was not shown to produce any significant overall improvement in motor function (Elahi et al., 2009).

RTMS may also be used on the treatment of other non-motor aspects of PD. There is evidence that rTMS improved treatment-resistant depression in PD patients in multiple functional domains, such as anxiety and mood (Epstein et al., 2007; Pal et al., 2010). Actually, the efficacy of rTMS as a therapy for depression in PD patients has been shown to have the same antidepressant effects as fluoxetine and last at least eight weeks after treatment (Fregni et al., 2004). Improvements in cognition and time perception have also been reported (Boggio et al., 2005; Koch et al., 2004). Characteristic sleep disturbances in PD could also be partly reversed by rTMS (van Dijk et al., 2009). These effects may significantly improve the quality of life of PD patients, even if unrelated to motor improvement. RTMS has also been shown to ameliorate L-DOPA-induced-dyskinesia (Rektorova et al., 2008; Brusa et al., 2006). Future studies with rTMS should explore not only changes in motor function but also other symptoms that might also respond well to rTMS.

Deep Brain Stimulation: Overview of the Technique

Deep Brain Stimulation (DBS) refers to a functional neurosurgical procedure, based on the delivery of high frequency (usually higher than 100 Hz) electric pulse trains applied through a single or a pair of electrodes stereotactically implanted, which are connected to a subcutaneously internalized programmable pacemaker. Although available for several decades, the technique, as we know it today, re-emerged only in the early 1990's for the treatment of parkinsonian and essential tremor, by means of stimulating the Ventral intermedius nucleus (Vim) of the thalamus (Benabid et al., 1991). It was then established that either ablation or electrical stimulation of the subthalamic nucleus (STN) in nonhuman primates treated with 1-methyl-4-phenyl-1,2,3,6-tetrahydropyridine (MPTP) resulted in significant improvement of their iatrogenic parkinsonian motor disturbances. After that, the first three severe Parkinson's Disease (PD) patients were successfully treated with DBS (Limousin et al., 1995). Since this early developing period, thousands of patients have been receiving DBS yearly, and the surgical approach consists on locating the target nuclei and/or tracts by determination of their anatomic landmarks – indentified on MRI scans – and then assigning stereotactic coordinates to them; it is followed by placement of appropriate burr holes in the skull and insertion of the electrodes; and finally, and if necessary, the targets can still be reassessed intraoperatively by using neurophysiologic techniques, as the different subcortical structures have different electrographic patterns (Dormont et al., 2010). The advantage of DBS over the ablative surgeries relies on the fact that it drives brain's activity in a controlled manner, and thus its effects may be tailored for one person's individual needs. Further, DBS also allows the conduction of blinded studies.

Mechanisms of Action

The mechanisms by which DBS exerts its biological effects remains under intense dispute; it depends on the exact location of the electrodes, on the geometric disposition and distance of the surrounding neural elements in relation to the stimulation site, and obviously, on the stimulation parameters. However, regardless of the specific technique, some principles can be extrapolated to all cases. Most importantly, the pulse duration and the pulse strength that are required to elicit a discrete response in the neural membrane are negatively correlated in a nonlinear fashion; insofar, a decrease in pulse strength requires an increase in pulse duration to ensue an effect of the same magnitude, and vice-versa. This strength-duration curve differs from that of a typical capacitor in the sense that it flattens out with long stimulus durations, reaching an asymptote that is called the rheobase. In this regard, the rheobase parameter refers to the minimal pulse strength required to stimulate a membrane; any stimuli bellow the rheobase is ineffective.

The chronaxie time is the minimal pulse duration – required for stimulation – that uses twice the strength dictated by the rheobase (Ranck, 1975). The chronaxie is thus a mathematical constant that can be used to describe the membrane excitability properties of the different nervous tissue elements. As the chronaxie of cell bodies and dendrites (1–10 ms) were described to be substantially larger than the chronaxie of myelinated axons (30–200 μs), and considering the usual DBS therapeutic parameters (e.g., pulse width in the order of 100 μs), one can expect that postsynaptic responses from DBS result mainly from the activation of axons rather than from cell bodies (Kringelbach et al., 2007).

As DBS seemingly parallels the clinical outcomes observed in the surgical ablation of subcortical structures, it was once hypothesized that the main effect of the stimulation was local inhibition or disruption of neural activity. Indeed, a PET study compared the effects of STN stimulation and STN lesioning in PD patients, disclosing that both procedures shared several metabolic features, including declines in the local metabolism of the midbrain, internal globus pallidus (GPi), and cingulate motor region, with increases in the parietal cortex (Trost et al., 2006). However, another study (Haslinger et al., 2005) demonstrated a tight negative correlation between increasing STN-DBS stimulation frequency and cortical resting regional Cerebral Blood Flow (rCBF) changes in the motor cortex, while a positive correlation between rCBF and the stimulation frequency was found at the STN; these findings argue against the concept of a local decrement of neurocircuitry activity at the electrode vicinities.

Accordingly to the aforementioned view, in vitro studies have found that high frequency DBS produce depolarization of rat thalamic neurons, and it was postulated that mechanisms of synaptic depression and axonal filtering may prevent concomitant efferent cortical excitation. Conversely, low frequency stimulation (< 50 Hz) should enable axons and synapses to convey signals with high fidelity (Kiss, 2009). In fact, a computer-based model of the electric field generated by a DBS electrode was used to examine the effects of high-frequency extracellular stimulation on a thalamocortical relay neuron (McIntyre et al., 2004). The main finding of this study was that the activity in the cell body may be decoupled from the activity in the axon, so that a single neuron may simultaneously exhibit suppression of intrinsic activity in the soma (due to stimulation-induced trans-synaptic inputs) and excitation in the axon.

There is also strong evidence that, in untreated parkinsonism, GPi and the dorsal STN neurons exhibit an anomalous tendency to synchronize their activity in the beta frequency range around 20 Hz. This characteristic may interfere with adequate movement execution, and can be reversed, after dopaminergic therapy, to local field potentials that oscillate within a much higher gamma frequency band (Weinberger et al., 2006). It was also shown that, in PD patients, DBS stimulation of the subthalamic area at 25 Hz is associated with an increase in the troublesome beta activity in GPi, a nucleus that receives its primary source of excitatory input from STN. Conversely, stimulation at a therapeutic frequency (> 70 Hz) had the opposite effect, suppressing the spontaneous tendency of GPi to get synchronized around 20 Hz (Brown et al., 2004).

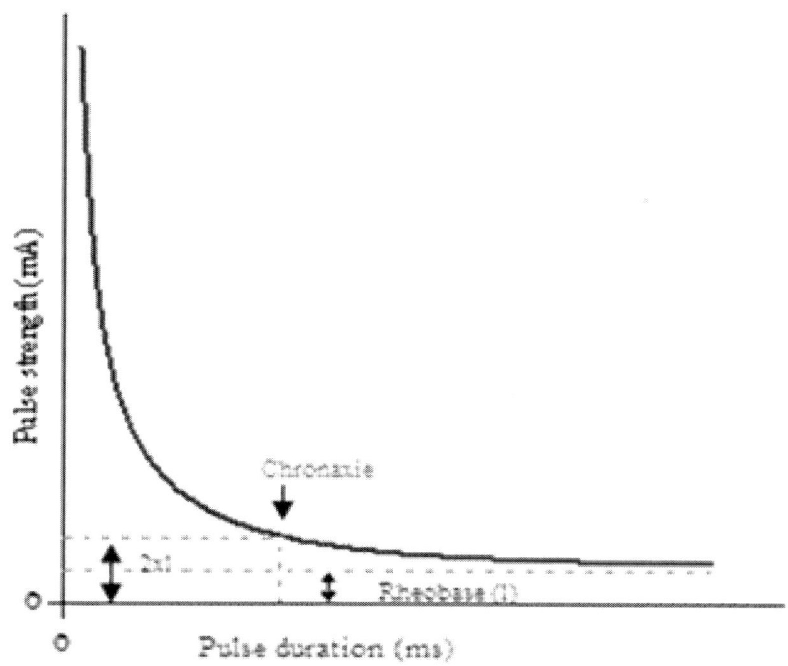

The strength-duration curve for DBS (adapted from ref. 64).

Although the uttermost importance of the localization issue, DBS probably impacts not only local neurocircuitry, but also projecting fibers from adjacent subcortical nuclei and even direct cortical afferents, which can be excited antidromically; and it was recently proposed that the optimal DBS frequencies may be resonant frequencies of the targeted cortico-basal-ganglia-thalamo-cortical oscillator loop. In conclusion, the most likely mechanism through which DBS exerts its action should be a stimulation-induced entrainment of the brain's intrinsic networks, resulting in "jamming" of the pathological inputs and reinforcement of the physiological ones. These clinical and experimental data are summarized by excellent reviews elsewhere (McIntyre and Hahn, 2010; Kringelbach et al., 2010; Kringelbach et al., 2007).

Notwithstanding the aforementioned mechanisms are more suited to explain rapid-onset effects of DBS (e.g., tremor and bradykinesia amelioration), the same basis may not hold true for some specific scenarios (e.g., dystonia and psychiatric disorders), in which there may be a latency of days to months until a clinical response is evident. This notion only highlights the role of DBS as a powerful neuromodulatory tool, which may lead to long lasting plastic changes in the brain; insofar, judicious use of the technique is advocated, as its effects may not be completely known neither completely reversible.

DBS in the Clinical Setting

As the FDA requires separate approvals for each different DBS indication, allowance for clinical use has historically lagged the related medical research. In this sense, unilateral thalamic stimulation was approved for the suppression of medically refractory essential tremor and parkinsonian tremor in July 1997; the stimulation of the GPi and STN gained approval as adjunctive therapies for PD in January 2002; and the stimulation of GPi and STN received a FDA Humanitarian Device Exemption (HDE) for the treatment of medically refractory primary dystonia in April 2003 (Collins et al., 2010). DBS has also received the first HDE approval for a psychiatric indication, namely chronic severe Obsessive-Compulsive Disorder (OCD), in February 2009. More recently, the FDA Neurological Devices Panel recommended approval with conditions of DBS as an adjunctive treatment for partial-onset seizures in adults with medically refractory epilepsy, in March 2010.

In terms of PD, the ideal PD patient candidate for DBS implantation should have a well-established story of a disabling disease plus motor fluctuations, but still be respondent to L-dopa therapy. While Vim thalamic stimulation should be reserved only for selected patients in whom tremor is a major disabling feature, both GPi-DBS and STN-DBS were shown to significantly improve all the three major PD motor symptoms (i.e., tremor, rigidity and bradykinesia). A randomized controlled clinical trial (RCCT) assessed the outcomes in the 24-month follow-up of the two most common functional PD treatments, disclosing that, while the extent of motor function improvement was the same for both groups, those patients undergoing STN-DBS required a lower dose of dopaminergic agents than did those undergoing GPi-DBS (Follett et al., 2010). Nonetheless, STN-DBS cases were more prone to a cognitive decline (visuomotor processing speed) and to worsening of previous depression, while GPi-DBS improved this last condition. Serious adverse events, most of which were completely reversible, occurred in 51% of patients undergoing pallidal stimulation and in

56% of those undergoing subthalamic stimulation, and this difference was not statistically significant.

Two RCCTs have compared the functional neurosurgical procedure alone (i.e., without pharmacological add-on) with the best medical therapy alone for the treatment of advanced PD at 6-month follow-up. One of these studies (Deuschl et al., 2006) has found that STN-DBS was even better than the best medical therapy for the amelioration of the overall PD symptoms, while serious adverse effects were more common (13% vs. 4%) within the stimulation arm. The other RCCT (Weaver et al., 2009) has found that either GPi-DBS or STN-DBS were associated with a mean gain of 4.6 hours/day of "on time" without disabling motor fluctuations, compared with 0 hour/day for patients who received the best medical therapy alone; on the other hand, at least one serious adverse event occurred in 40% of DBS cases, versus 11% of best medical therapy patients. Finally, a recent RCCT trial enrolled 366 patients in two arms: 183 subjects received the functional neurosurgery approach plus best medical therapy, while 183 subjects received the best medical therapy alone. At one-year follow-up, the self-reported overall quality of life of the DBS group still excelled that of the medical therapy alone, and serious surgery-related adverse effects accounted for 19% of the implanted patients (Williams et al., 2010). DBS is also successfully used in other refractory Movement Disorders settings. The stimulation of the pedunculopontine nucleus (PPN), for instance, may be a safe target for the amelioration of primary progressive freezing of gait, a neurodegenerative disorder that can be associated to parkinsonian symptoms and that causes gait freezing, postural instability and eventually gait akinesia (Wilcox et al., 2010). Regarding medically-intractable essential tremor, Vim-DBS seems to have overcome the major drawbacks related to classic thalamotomy, a procedure associated with increased morbidity and mortality. While unilateral Vim-DBS is highly effective for dominant extremity tremor, axial tremor and bilateral extremity tremor are better relieved by bilateral Vim-DBS (Flora et al., 2010). The posterioventral lateral portion of the GPi is a typical target for DBS in the treatment of disabling dystonia; stimulation of the GPi for dystonia induces slow, progressive improvement over days to months (Collins et al., 2010). Meige Syndrome is another disabling condition that may have a good response rate to GPi-DBS, but requires an extended period of observation as well (Lyons et al, 2010). DBS has also been used for the hodiernal treatment of psychiatric diseases, resulting from the serendipitous observation of the effects of the classic ablative psychosurgeries and the effects of DBS devices implanted on Movement Disorders' patients. This adds much to the view that psychiatric and neurological conditions are part of a same biological continuum. The modern DBS-based functional psychosurgery approach has its roots in the much less amenable procedures performed between the 1940's and 50's, such as the prefrontal leukotomy and transorbital lobotomy, which were associated with some improvement in psychiatric symptoms, but at the expense of irreversible serious side effects. Along with the concomitant evolution of the stereotactic techniques, brain targets became more circumscribed to some specific cortical-limbic projections areas. However, with the popularization of the neuroleptics by the 1960's, there was a dramatic decline in the number of such procedures (Ward et al., 2010). The renewed interest in stereotactic ablative neurosurgery for the treatment of Movement Disorders, specially PD patients, by the early 1990's, culminated with the use of DBS initially for these neurological conditions, but soon after it was evident that the effects of electrical therapy was not restricted to alleviating parkinsonian motor symptoms. For instance, some of the adverse effects of STN-DBS may include pathological laughter and/or crying spells, acute mania/hypomania, aggressive behavior and dysphoria (Krack et al., 2010). In fact, stimulating different targets along the cortico-basal-ganglia-thalamo-cortical loops may lead to a significant improvement in the mood and/or affect, and this approach can so be a valid option for those medically refractory depressive patients who fail or who refuse the well-established electroconvulsive therapy (ECT) method. Much of the recent success of electric stimulation for depression stemmed from studies with medically resistant OCD patients, as an improvement in mood was noted in many cases. To date, a collaborative study have assessed 52 of these patients over 8 years; the target included the ventral anterior limb of the internal capsule and adjacent ventral striatum (VC/VS), which is thought to represent a node within the cortico-basal-ganglia-thalamo-cortical loop. VC/VS-DBS was associated to improvement in over 60% of that OCD cases, and it was overall a well-tolerated treatment, despite the incidence of 42% of serious, but reversible, adverse effects (Greenberg et al., 2010).

Chronic stimulation of the white-matter tracts adjacent to the subgenual cingulate gyrus was reported to be associated with a striking and sustained remission of symptoms in four out of six severely depressed patients (Mayberg et al., 2005). Stimulation of the subcallosal cingulate gyrus (SCG) is also associated with improvement in the mood. A survey of 20 patients who underwent SCG-DBS, up to 12-month follow-up, has disclosed that the clinical response rate was 60%, and 35% of patients met criteria for remission at 6 months after the surgery; none of them sustained permanent side effects (Lozano et al., 2008). A recent review on 17 severe refractory depression patients who received bilateral VC/VS-DBS with a stimulation frequency of 100 to 130 Hz and amplitude of 2.5 to 8 V has shown that a positive clinical response was achieved by 53% of the patients at 3 months, and the response rate increased to 71% by the most recent follow-up; adverse effects (paresthesias, anxiety, mood changes and autonomic effects) were all reversible with adjustment of stimulation parameters (Malone, 2010). Bilateral DBS to the nucleus accumbens (Nacc) may have immediate effects on depression scales of medically refractory patients, and the mood change is corroborated by concomitant metabolic activations (VC, dorsolateral and dorsomedial prefrontal cortex, cingulate cortex and amygdala) and deactivations (ventromedial and ventrolateral prefrontal cortex, dorsal caudate nucleus, and thalamus) of brain's networks associated with the depressive state (Schlaepfer et al., 2008). Although encouraging preliminary results, it can never be overstated that DBS in the psychiatric clinical setting remains mainly experimental. Lack of control groups in most of DBS studies is an important limitation; thus results need to be interpreted with caution. Thus emphasis should be put on engaging patients in RCCTs, as well as avoiding the former abuses of the Pychosurgery Era.

Emerging DBS Indications

More recently, DBS has been used in the management of several medically-intractable neuropsychiatric conditions, with generally good outcomes when considering highly selected cases. Stimulation of the posterior hypothalamic gray matter has been used in medically-intractable chronic cluster headache for up to ten years now (Leone et al., 2001; Grover et al., 2009; Leone et al., 2010). Targeting the feeding center in lateral hypothalamus, the satiety center in the ventromedial hypothalamus, or more promisingly, aiming the food reward sensations and palatability in the Nacc may be helpful for controlling morbid obesity

(Halpern et al., 2008; Pisapia et al., 2010); electric manipulation of the Nacc may also have a role for controlling addiction, but with the risk of provoking anhedonia (Ward et al., 2010). The centromedian thalamic nucleus (CMN) may be targeted for aleviating Tourette Syndrome and the GPi for treating primary dystonia (Lipsman et al., 2010). Finally, minimally conscious state patients may have awareness improved by stimulation of the central lateral nucleus of the intralaminar thalamic nuclei, but the procedure will eventually raise ethical considerations (Schiff et al., 2007; Sen et al., 2010).

Another emerging and promising indication for DBS is the adjunctive treatment of refractory epilepsy, being the CMN and the anterior nucleus of the thalamus (AN) the main targets by now, in addition to the possible direct stimulation of the hippocampus or amygdala (in mesial temporal lobe epilepsy), the STN or even the posterior hypothalamus (Rahman et al., 2010). A recent RCCT assessed the clinical utility of AN-DBS for patients with intractable epilepsy (SANTE Trial; Fisher et al., 2010), enrolling 110 participants in two study arms; one month after the surgical procedure, the participants were randomized to stimulation with the following parameters: 5 V and 90 μs pulse width at 145 Hz pulse frequency ("ON" state for 1 min, and "OFF" state for 5 min) versus no stimulation at 0 V (control). Four months after the implantation of the electrodes, it was noted a 14.5% reduction of seizures in the control group versus a 40.4% reduction in the stimulated group, and this difference was statistically significant. At completion of this first blinded phase, 108 participants entered the unblinded phase of the trial and all of them received stimulation; follow-up by 2 years revealed that there was a 56% median reduction in seizure frequency and 54% of patients had a seizure reduction of at least 50%, with 14 patients seizure-free for at least 6 months. However, only the patients with seizures of temporal lobe origin had a significant decrease in seizure frequency. Five deaths occurred but none were device-related and asymptomatic hemorrhage occurred in 4.5% of patients.

Conclusion

Although rTMS and DBS are conceptually different neuromodulatory approaches, they both act by injecting electrical currents into the brain, cortically or subcortically, through the generation of repetitive electric pulses. As a function of the expansion of their therapeutic scopes, along with the recent and future FDA approvals, and coupled with the exponential evolution of biomedical engineering, they might be promising therapeutic tools in the near future.

Acknowledgments

We are grateful to Kayleen Weaver and Laura Sherman for proofreading this chapter.

References

Aguirre I, Carretero B, Ibarra O, Kuhalainen J, Martínez J, Ferrer A, Salva J, Roca M, Gili M, Montoya P, Garcia-Toro M. (2010). Age predicts low-frequency transcranial magnetic stimulation efficacy in major depression. *J. Affect Disord.* Nov 17 [Epub ahead of print].

Arias-Carrión O. (2008). Basic mechanisms of rTMS: Implications in Parkinson's disease; *Int. Arch. Med.* Apr 15;1(1):2.

Arias P, Vivas J, Grieve KL, Cudeiro J. (2010). Controlled trial on the effect of 10 days low-frequency repetitive transcranial magnetic stimulation (rTMS) on motor signs in Parkinson's disease; *Mov. Disord.* Sep 15;25(12):1830-8.

Avery DH, Holtzheimer PE 3rd, Fawaz W, Russo J, Neumaier J, Dunner DL, Haynor DR, Claypoole KH, Wajdik C, Roy-Byrne P. (2006). A controlled study of repetitive transcranial magnetic stimulation in medication-resistant major depression. *Biol. Psychiatry.* Jan 15;59(2):187-94.

Bae EH, Schrader LM, Machii K, Alonso-Alonso M, Riviello JJ Jr, Pascual-Leone A, Rotenberg A. (2007). Safety and tolerability of repetitive transcranial magnetic stimulation in patients with epilepsy: a review of the literature. *Epilepsy Behav.* Jun;10(4):521-8.

Bäumer T, Hidding U, Hamel W, Buhmann C, Moll CK, Gerloff C, Orth M, Siebner HR, Münchau A. (2009). Effects of DBS, premotor rTMS, and levodopa on motor function and silent period in advanced Parkinson's disease. *Mov. Disord.* Apr 15;24(5):672-6.

Benabid AL, Pollak P, Gervason C, Hoffmann D, Gao DM, Hommel M, Perret JE, de Rougemont J. (1991). Long-term suppression of tremor by chronic stimulation of the ventral intermediate thalamic nucleus. *Lancet* 337, 403–406.

Bench CJ, Frackowiak RS, Dolan RJ. (1995). Changes in regional cerebral blood flow on recovery from depression. *Psychol. Med.* Mar;25(2):247-61.

Bestmann S. (2008). The physiological basis of transcranial magnetic stimulation. *Trends Cogn. Sci.* Mar;12(3):81-3.

Boylan LS, Pullman SL, Lisanby SH, Spicknall KE, Sackeim HA. (2001). Repetitive transcranial magnetic stimulation to SMA worsens complex movements in Parkinson's disease; *Clin. Neurophysiol.* Feb;112(2):259-64.

Brakemeier EL, Luborzewski A, Danker-Hopfe H, Kathmann N, Bajbouj M. (2007). Positive predictors for antidepressive response to prefrontal repetitive transcranial magnetic stimulation (rTMS). *J. Psychiatr. Res.* Aug;41(5):395-403.

Brown P, Mazzone P, Oliviero A, Altibrandi MG, Pilato F, Tonali PA, Di Lazzaro V. (2004). Effects of stimulation of the subthalamic area on oscillatory pallidal activity in Parkinson's disease. *Exp. Neurol.* 188, 480–490.

Brusa L, Versace V, Koch G, Iani C, Stanzione P, Bernardi G, Centonze D. (2006). Low frequency rTMS of the SMA transiently ameliorates peak-dose LID in Parkinson's disease. *Clin. Neurophysiol.* Sep;117(9):1917-21.

Cantello R. (2002). Applications of transcranial magnetic stimulation in movement disorders; *J. Clin. Neurophysiol.* Aug;19(4):272-93.

Cheeran B, Talelli P, Mori F, Koch G, Suppa A, Edwards M, Houlden H, Bhatia K, Greenwood R, Rothwell JC. (2008). A common polymorphism in the brain-derived neurotrophic factor gene (BDNF) modulates human cortical plasticity and the response to rTMS. *J. Physiol.* Dec 1;586(Pt 23):5717-25.

Collins KL, Lehmann EM, Patil PG. (2010). Deep brain stimulation for movement disorders. *Neurobiol. Dis.* Jun;38(3):338-45.

del Olmo MF, Bello O, Cudeiro J. (2007). Transcranial magnetic stimulation over dorsolateral prefrontal cortex in Parkinson's disease. *Clin. Neurophysiol.* Jan;118(1):131-9.

Demirtas-Tatlidede A, Mechanic-Hamilton D, Press DZ, Pearlman C, Stern WM, Thall M, Pascual-Leone A. (2008). An open-label, prospective study of repetitive transcranial magnetic stimulation (rTMS) in the long-term treatment of refractory depression: reproducibility and duration of the antidepressant effect in medication-free patients. *J. Clin. Psychiatry.* Jun;69(6):930-4.

Deuschl G, Schade-Brittinger C, Krack P, Volkmann J, Schäfer H, Bötzel K, Daniels C, Deutschländer A, Dillmann U, Eisner W, Gruber D, Hamel W, Herzog J, Hilker R, Klebe S, Kloss M, Koy J, Krause M, Kupsch A, Lorenz D, Lorenzl S, Mehdorn HM, Moringlane JR, Oertel W, Pinsker MO, Reichmann H, Reuss A, Schneider GH, Schnitzler A, Steude U, Sturm V, Timmermann L, Tronnier V, Trottenberg T, Wojtecki L, Wolf E, Poewe W, Voges J; German Parkinson Study Group, Neurostimulation Section. (2006). A randomized trial of deep-brain stimulation for Parkinson's disease. *N. Engl. J. Med.* 355, 896–908.

Dormont D, Seidenwurm D, Galanaud D, Cornu P, Yelnik J, Bardinet E. Neuroimaging and deep brain stimulation. (2010). *AJNR Am. J. Neuroradiol.* Jan;31(1):15-23.

Dragasevic N, Potrebić A, Damjanović A, Stefanova E, Kostić VS. (2002). Therapeutic efficacy of bilateral prefrontal slow repetitive transcranial magnetic stimulation in depressed patients with Parkinson's disease: an open study. *Mov. Disord.* May;17(3):528-32.

Epstein CM, Evatt ML, Funk A, Girard-Siqueira L, Lupei N, Slaughter L, Athar S, Green J, McDonald W, DeLong MR. (2007). An open study of repetitive transcranial magnetic stimulation in treatment-resistant depression with Parkinson's disease. *Clin. Neurophysiol.* Oct;118(10):2189-94.

Eranti S, Mogg A, Pluck G, Landau S, Purvis R, Brown RG, Howard R, Knapp M, Philpot M, Rabe-Hesketh S, Romeo R, Rothwell J, Edwards D, McLoughlin DM. (2007). A randomized, controlled trial with 6-month follow-up of repetitive transcranial magnetic stimulation and electroconvulsive therapy for severe depression. *Am. J. Psychiatry.* Jan;164(1):73-81.

Filipović SR, Rothwell JC, Bhatia K. (2010). Slow (1 Hz) repetitive transcranial magnetic stimulation (rTMS) induces a sustained change in cortical excitability in patients with Parkinson's disease. *Clin. Neurophysiol.* Jul;121(7):1129-37.

Filipović SR, Rothwell JC, Bhatia K. (2010). Low-frequency repetitive transcranial magnetic stimulation and off-phase motor symptoms in Parkinson's disease. *J. Neurol. Sci. Apr.* 15;291(1-2):1-4.

Fisher R, Salanova V, Witt T, Worth R, Henry T, Gross R, Oommen K, Osorio I, Nazzaro J, Labar D, Kaplitt M, Sperling M, Sandok E, Neal J, Handforth A, Stern J, DeSalles A, Chung S, Shetter A, Bergen D, Bakay R, Henderson J, French J, Baltuch G, Rosenfeld W, Youkilis A, Marks W, Garcia P, Barbaro N, Fountain N, Bazil C, Goodman R, McKhann G, Babu Krishnamurthy K, Papavassiliou S, Epstein C, Pollard J, Tonder L,

Grebin J, Coffey R, Graves N; SANTE Study Group. (2010). Electrical stimulation of the anterior nucleus of thalamus for treatment of refractory epilepsy. *Epilepsia*, May;51(5):899-908.

Fitzgerald PB, Benitez J, de Castella A, Daskalakis ZJ, Brown TL, Kulkarni J. (2006). A randomized, controlled trial of sequential bilateral repetitive transcranial magnetic stimulation for treatment-resistant depression. *Am. J. Psychiatry*. Jan;163(1):88-94.

Fitzgerald PB, Benitez J, de Castella AR, Brown TL, Daskalakis ZJ, Kulkarni J. (2006). Naturalistic study of the use of transcranial magnetic stimulation in the treatment of depressive relapse. *Aust. N Z J. Psychiatry*. Sep;40(9):764-8.

Flora ED, Perera CL, Cameron AL, Maddern GJ. (2010). Deep Brain Stimulation for Essential Tremor: A Systematic Review. *Mov. Disord.* Aug 15;25(11):1550-9.

Follett KA, Weaver FM, Stern M, Hur K, Harris CL, Luo P, Marks WJ Jr, Rothlind J, Sagher O, Moy C, Pahwa R, Burchiel K, Hogarth P, Lai EC, Duda JE, Holloway K, Samii A, Horn S, Bronstein JM, Stoner G, Starr PA, Simpson R, Baltuch G, De Salles A, Huang GD, Reda DJ; CSP 468 Study Group. (2010). Pallidal versus subthalamic deep-brain stimulation for Parkinson's disease. *N. Engl. J. Med.* Jun 3;362(22):2077-91.

Fregni F, Marcolin MA, Myczkowski M, Amiaz R, Hasey G, Rumi DO, Rosa M, Rigonatti SP, Camprodon J, Walpoth M, Heaslip J, Grunhaus L, Hausmann A, Pascual-Leone A. (2006). Predictors of antidepressant response in clinical trials of transcranial magnetic stimulation. *Int. J. Neuropsychopharmacol.* Dec;9(6):641-54.

Fregni F, Santos CM, Myczkowski ML, Rigolino R, Gallucci-Neto J, Barbosa ER, Valente KD, Pascual-Leone A, Marcolin MA. (2004). Repetitive transcranial magnetic stimulation is as effective as fluoxetine in the treatment of depression in patients with Parkinson's disease. *J. Neurol. Neurosurg. Psychiatry*. Aug;75(8):1171-4.

Fregni F, Simon DK, Wu A, Pascual-Leone A. (2005). Non-invasive brain stimulation for Parkinson's disease: a systematic review and meta-analysis of the literature. *J. Neurol. Neurosurg. Psychiatry,* Dec;76(12):1614-23.

Fujiki M, Steward O. (1997). High frequency transcranial magnetic stimulation mimics the effects of ECS in upregulation astroglia gene expression in the murine CNS. *Molecular Brain Research,* 44: 301–8.

Funamizu H, Ogiue-Ikeda M, Mukai H, Kawato S, Ueno S. (2005). Acute repetitive transcranial magnetic stimulation reactivates dopaminergic system in lesion rats. *Neurosci. Lett.* Jul 22-29;383(1-2):77-81.

George MS, Lisanby SH, Avery D, McDonald WM, Durkalski V, Pavlicova M, Anderson B, Nahas Z, Bulow P, Zarkowski P, Holtzheimer PE 3rd, Schwartz T, Sackeim HA. (2010). Daily left prefrontal transcranial magnetic stimulation therapy for major depressive disorder: a sham-controlled randomized trial. *Arch. Gen. Psychiatry,* May;67(5):507-16.

Greenberg BD, Gabriels LA, Malone DA Jr, et al. (2010). Deep brain stimulation of the ventral internal capsule/ventral striatum for obsessive-compulsive disorder: worldwide experience. *Molecular Psychiatry*, 15:64–79.

Gross M, Nakamura L, Pascual-Leone A, Fregni F. (2007). Has repetitive transcranial magnetic stimulation (rTMS) treatment for depression improved? A systematic review and meta-analysis comparing the recent vs. the earlier rTMS studies. *Acta Psychiatr. Scand.* Sep;116(3):165-73.

Grover PJ, Pereira EA, Green AL, Brittain JS, Owen SL, Schweder P, Kringelbach ML, Davies PT, Aziz TZ. (2009). Deep brain stimulation for cluster headache. *J. Clin. Neurosci.* Jul;16(7):861-6.

Grunhaus L, Dannon PN, Schreiber S, Dolberg OH, Amiaz R, Ziv R, Lefkifker E. (2000). Repetitive transcranial magnetic stimulation is as effective as electroconvulsive therapy in the treatment of nondelusional major depressive disorder: an open study. *Biol. Psychiatry.* Feb 15;47(4):314-24.

Hallett M. (2007). Transcranial magnetic stimulation: a primer. *Neuron. Jul.* 19;55(2):187-99.

Halliday GM. (2009). Thalamic changes in Parkinson's disease; Parkinsonism Relat Disord. Dec;15 Suppl 3:S152-5.

Halpern CH, Wolf JA, Bale TL, Stunkard AJ, Danish SF, Grossman M, Jaggi JL, Grady MS, Baltuch GH. (2008). Deep brain stimulation in the treatment of obesity. *J. Neurosurg.* Oct;109(4):625-34.

Hamada M, Terao Y, Hanajima R, Shirota Y, Nakatani-Enomoto S, Furubayashi T, Matsumoto H, Ugawa Y. (2008). Bidirectional long-term motor cortical plasticity and metaplasticity induced by quadripulse transcranial magnetic stimulation. *J. Physiol.* Aug 15;586(16):3927-47.

Hamada M, Ugawa Y. (2010). Quadripulse stimulation - a new patterned rTMS. *Restor. Neurol. Neurosci.* 28(4):419-24.

Hamada M, Ugawa Y, Tsuji S. (2008). Effectiveness of rTMS on Parkinson's Disease Study Group, Japan; High-frequency rTMS over the supplementary motor area for treatment of Parkinson's disease. *Mov. Disord.* Aug 15;23(11):1524-31.

Hariz MI, Blomstedt P, Zrinzo L. (2010). Deep brain stimulation between 1947 and 1987: the untold story; *Neurosurg. Focus,* Aug;29(2):E1.

Haslinger B, Kalteis K, Boecker H, Alesch F, Ceballos-Baumann AO. (2005). Frequency-correlated decreases of motor cortex activity associated with subthalamic nucleus stimulation in Parkinson's disease. *Neuroimage*, 28:598 – 606.

Hausmann A, Weis C, Marksteiner J, Hinterhuber H, Humpel C. (2000). Chronic repetitive transcranial magnetic stimulation enhances c-fos in the parietal cortex and hippocampus. *Brain Res. Mol. Brain Res.* Mar 29;76(2):355-62.

Hilker R, Voges J, Ghaemi M, Lehrke R, Rudolf J, Koulousakis A, Herholz K, Wienhard K, Sturm V, Heiss WD. (2003). Deep brain stimulation of the subthalamic nucleus does not increase the striatal dopamine concentration in parkinsonian humans. *Mov. Disord.* 18(1):41-8.

Huang YZ, Rothwell JC, Edwards MJ, Chen RS. (2008). Effect of physiological activity on an NMDA-dependent form of cortical plasticity in human. *Cereb. Cortex.* Mar;18(3):563-70.

Ikeguchi M, Touge T, Nishiyama Y, Takeuchi H, Kuriyama S, Ohkawa M. (2003). Effects of successive repetitive transcranial magnetic stimulation on motor performances and brain perfusion in idiopathic Parkinson's disease. *J. Neurol. Sci.* May 15;209(1-2):41-6.

Ji RR, Schlaepfer TE, Aizenman CD, Epstein CM, Qiu D, Huang JC, Rupp F. (1998). Repetitive transcranial magnetic stimulation activates specific regions in rat brain. *Proc. Natl. Acad. Sci. USA.* Dec 22;95(26):15635-40.

Keck ME, Welt T, Müller MB, Erhardt A, Ohl F, Toschi N, Holsboer F, Sillaber I. (2002). Repetitive transcranial magnetic stimulation increases the release of dopamine in the mesolimbic and mesostriatal system. *Neuropharmacology.* Jul;43(1):101-9.

Khedr EM, Farweez HM, Islam H. (2003). Therapeutic effect of repetitive transcranial magnetic stimulation on motor function in Parkinson's disease patients. *Eur. J. Neurol.* Sep;10(5):567-72.

Khedr EM, Rothwell JC, Shawky OA, Ahmed MA, Foly N, Hamdy A. (2007). Dopamine levels after repetitive transcranial magnetic stimulation of motor cortex in patients with Parkinson's disease: preliminary results. *Mov. Disord.* May 15;22(7):1046-50.

Khedr EM, Rothwell JC, Shawky OA, Ahmed MA, Hamdy A. (2006). Effect of daily repetitive transcranial magnetic stimulation on motor performance in Parkinson's disease; *Mov. Disord.* Dec;21(12):2201-5.

Kiss ZH. (2009). Clinical and experimental aspects of deep brain stimulation. *Conf. Proc. IEEE Eng. Med. Biol. Soc.* 2009:2371-4.

Kito S, Fujita K, Koga Y. (2008). Regional cerebral blood flow changes after low-frequency transcranial magnetic stimulation of the right dorsolateral prefrontal cortex in treatment-resistant depression. *Neuropsychobiology*, 58(1):29-36.

Kleim JA, Chan S, Pringle E, Schallert K, Procaccio V, Jimenez R., Cramer SC. (2006). BDNF val66met polymorphism is associated with modified experience-dependent plasticity in human motor cortex. *Nat. Neurosci.* Jun;9(6):735-7.

Kobayashi M, Pascual-Leone A. (2003) Transcranial magnetic stimulation in neurology. *Lancet Neurol.* Mar;2(3):145-56.

Koch G, Oliveri M, Brusa L, Stanzione P, Torriero S, Caltagirone C. (2004). High-frequency rTMS improves time perception in Parkinson disease. *Neurology.* Dec 28;63(12):2405-6.

Krack P, Hariz MI, Baunez C, Guridi J, Obeso JA. (2010). Deep brain stimulation: from neurology to psychiatry? *Trends Neurosci.* Oct;33(10):474-84.

Kringelbach ML, Jenkinson N, Owen SL, Aziz TZ. (2007). Translational principles of deep brain stimulation. *Nat. Rev. Neurosci.* Aug;8(8):623-35.

Kringelbach ML, Green AL, Owen SL, Schweder PM, Aziz TZ. (2010). Sing the mind electric – principles of deep brain stimulation. *Eur. J. Neurosci.* Oct;32(7):1070-9.

Lam RW, Chan P, Wilkins-Ho M, Yatham LN. (2008). Repetitive transcranial magnetic stimulation for treatment-resistant depression: a systematic review and meta-analysis. *Can. J. Psychiatry,* Sep;53(9):621-31.

Lefaucheur JP, Drouot X, Von Raison F, Ménard-Lefaucheur I, Cesaro P, Nguyen JP. (2004). Improvement of motor performance and modulation of cortical excitability by repetitive transcranial magnetic stimulation of the motor cortex in Parkinson's disease. *Clin. Neurophysiol.* Nov;115(11):2530-41.

Leone M, Franzini A, Bussone G. (2001). Stereotactic stimulation of posterior hypothalamic gray matter in a patient with intractable cluster headache. *N. Engl. J. Med.* 345:1428–9.

Leone M, Franzini A, Cecchini AP, Broggi G, Bussone G. (2010). Hypothalamic deep brain stimulation in the treatment of chronic cluster headache. *Ther. Adv. Neurol. Disord.* May;3(3):187-95.

Lipsman N, Ellis M, Lozano AM. (2010). Current and future indications for deep brain stimulation in pediatric populations. *Neurosurg. Focus.* Aug;29(2):E2.

Limousin P, Pollak P, Benazzouz A, Hoffmann D, Le Bas JF, Broussolle E, Perret JE, Benabid AL. (1995). Effect of parkinsonian signs and symptoms of bilateral subthalamic nucleus stimulation. *Lancet,* 345, 91–95.

Lisanby SH, Husain MM, Rosenquist PB, Maixner D, Gutierrez R, Krystal A, Gilmer W, Marangell LB, Aaronson S, Daskalakis ZJ, Canterbury R, Richelson E, Sackeim HA,

George MS. (2009). Daily left prefrontal repetitive transcranial magnetic stimulation in the acute treatment of major depression: clinical predictors of outcome in a multisite, randomized controlled clinical trial. *Neuropsychopharmacology*, Jan;34(2):522-34.

Lomarev MP, Kanchana S, Bara-Jimenez W, Iyer M, Wassermann EM, Hallett M. (2006). Placebo-controlled study of rTMS for the treatment of Parkinson's disease; *Mov. Disord.* Mar;21(3):325-31.

Loo CK, McFarquhar TF, Mitchell PB. (2008). A review of the safety of repetitive transcranial magnetic stimulation as a clinical treatment for depression. *Int. J. Neuropsychopharmacol.* Feb;11(1):131-47.

Lozano AM, Mayberg HS, Giacobbe P, Hamani C, Craddock RC, Kennedy SH. (2008). Subcallosal cingulate gyrus deep brain stimulation for treatment-resistant depression. *Biol. Psychiatry*, 64 (6), 461–467 Sep 15.

Lyons MK, Birch BD, Hillman RA, Boucher OK, Evidente VG. (2010). Long-term follow-up of deep brain stimulation for Meige syndrome, *Neurosurg. Focus.* Aug;29(2):E5.

Mally J, Stone TW. (1999). Improvement in Parkinsonian symptoms after repetitive transcranial magnetic stimulation; *J. Neurol. Sci.* Jan 15;162(2):179-84.

Mally J, Stone TW. (1999). Therapeutic and "dose-dependent" effect of repetitive microelectroshock induced by transcranial magnetic stimulation in Parkinson's disease; *J. Neurosci. Res.* Sep 15;57(6):935-40.

Mally J, Farkas R, Tóthfalusi L, Stone TW. (2004). Long-term follow-up study with repetitive transcranial magnetic stimulation (rTMS) in Parkinson's disease; *Brain Res. Bull.* Sep 30;64(3):259-63.

Malone DA Jr. (2010). Use of deep brain stimulation in treatment-resistant depression. *Cleve. Clin. J. Med.* Jul;77 Suppl 3:S77-80.

Mayberg HS, Lozano AM, Voon V, McNeely HE, Seminowicz D, Hamani C, Schwalb JM, Kennedy SH. (2005). Deep brain stimulation for treatment-resistant depression. *Neuron*, 45, 651-660.

McIntyre CC, Grill WM, Sherman DL, Thakor NV. (2004). Cellular effects of deep brain stimulation: Model-based analysis of activation and inhibition. *J. Neurophysiol.* 91, 1457-1469.30.

McIntyre CC, Hahn PJ. (2010). Network perspectives on the mechanisms of deep brain stimulation. *Neurobiol. Dis.* Jun;38(3):329-37.

Müller MB, Toschi N, Kresse AE, Post A, Keck ME. (2000). Long-term repetitive transcranial magnetic stimulation increases the expression of brain-derived neurotrophic factor and cholecystokinin mRNA, but not neuropeptide tyrosine mRNA in specific areas of rat brain. *Neuropsychopharmacology,* Aug;23(2):205-15.

Naskar S, Sood SK, Goyal V, Dhara M. (2010). Mechanism(s) of deep brain stimulation and insights into cognitive outcomes in Parkinson's disease. *Brain Res. Rev.* Oct 5;65(1):1-13.

O'Connor M, Brenninkmeyer C, Morgan A, Bloomingdale K, Thall M, Vasile R, Leone AP. (2003). Relative effects of repetitive transcranial magnetic stimulation and electroconvulsive therapy on mood and memory: a neurocognitive risk-benefit analysis. *Cogn. Behav. Neurol.* Jun;16(2):118-27.

Okabe S, Ugawa Y, Kanazawa I. (2003). Effectiveness of rTMS on Parkinson's Disease Study Group; 0.2-Hz repetitive transcranial magnetic stimulation has no add-on effects as compared to a realistic sham stimulation in Parkinson's disease. *Mov. Disord.* Apr;18(4):382-8.

Ohnishi T, Hayashi T, Okabe S, Nonaka I, Matsuda H, Iida H, Imabayashi E, Watabe H, Miyake Y, Ogawa M, Teramoto N, Ohta Y, Ejima N, Sawada T, Ugawa Y. (2004). Endogenous dopamine release induced by repetitive transcranial magnetic stimulation over the primary motor cortex: an [11C]raclopride positron emission tomography study in anesthetized macaque monkeys; *Biol. Psychiatry*. Mar 1;55(5):484-9.

O'Reardon JP, Blumner KH, Peshek AD, Pradilla RR, Pimiento PC. (2005). Long-term maintenance therapy for major depressive disorder with rTMS. *J. Clin. Psychiatry*. Dec;66(12):1524-8

O'Reardon JP, Solvason HB, Janicak PG, Sampson S, Isenberg KE, Nahas Z, McDonald WM, Avery D, Fitzgerald PB, Loo C, Demitrack MA, George MS, Sackeim HA. (2007). Efficacy and safety of transcranial magnetic stimulation in the acute treatment of major depression: a multisite randomized controlled trial. *Biol. Psychiatry*. Dec 1;62(11):1208-16.

Padberg F, George MS. (2009). Repetitive transcranial magnetic stimulation of the prefrontal cortex in depression. *Exp. Neurol*. Sep;219(1):2-13.

Pal E, Nagy F, Aschermann Z, Balazs E, Kovacs N. (2010). The impact of left prefrontal repetitive transcranial magnetic stimulation on depression in Parkinson's disease: a randomized, double-blind, placebo-controlled study. *Mov. Disord*. Oct 30;25(14):2311-7.

Pallanti S, Bernardi S, Di Rollo A, Antonini S, Quercioli L. (2010). Unilateral low frequency versus sequential bilateral repetitive transcranial magnetic stimulation: is simpler better for treatment of resistant depression? *Neuroscience,* May 5;167(2):323-8.

Pascual-Leone A, Valls-Solé J, Brasil-Neto JP, Cammarota A, Grafman J, Hallett M. (1994). Akinesia in Parkinson's disease. II. Effects of subthreshold repetitive transcranial motor cortex stimulation. *Neurology*, May;44(5):892-8.

Pisapia JM, Halpern CH, Williams NN, Wadden TA, Baltuch GH, Stein SC. (2010). Deep brain stimulation compared with bariatric surgery for the treatment of morbid obesity: a decision analysis study. *Neurosurg. Focus,* Aug;29(2):E15.

Pogarell O, Koch W, Pöpperl G, Tatsch K, Jakob F, Mulert C, Grossheinrich N, Rupprecht R, Möller HJ, Hegerl U, Padberg F. (2007). Acute prefrontal rTMS increases striatal dopamine to a similar degree as D-amphetamine. *Psychiatry Res*. Dec 15;156(3):251-5.

Post A, Keck ME. (2001).Transcranial magnetic stimulation as a therapeutic tool in psychiatry: what do we know about the neurobiological mechanisms? *J. Psychiatr. Res*. Jul-Aug;35(4):193-215.

Priori A. (2003). Brain polarization in humans: a reappraisal of an old tool for prolonged non-invasive modulation of brain excitability. *Clin. Neurophysiol*. Apr;114(4):589-95.

Rahman M, Abd-El-Barr MM, Vedam-Mai V, Foote KD, Murad GJ, Okun MS, Roper SN. (2010). Disrupting abnormal electrical activity with deep brain stimulation: is epilepsy the next frontier? *Neurosurg. Focus,* Aug;29(2):E7.

Ranck, J. B. Jr. (1975). Which elements are excited in electrical stimulation of mammalian central nervous system: a review. *Brain Res*. 98, 417–440.

Rektorova I, Sedlackova S, Telecka S, Hlubocky A, Rektor I. (2008). Dorsolateral prefrontal cortex: a possible target for modulating dyskinesias in Parkinson's disease by repetitive transcranial magnetic stimulation. *Int. J. Biomed. Imaging,* 2008:372125.

Rich MM, Wenner P. (2007). Sensing and expressing homeostatic synaptic plasticity. *Trends Neurosci.* Mar;30(3):119-25.

Rossi S, Hallett M, Rossini PM, Pascual-Leone A. (2009). Safety of TMS Consensus Group. Safety, ethical considerations, and application guidelines for the use of transcranial magnetic stimulation in clinical practice and research. *Clin. Neurophysiol.* Dec;120(12):2008-39.

Rush AJ, Trivedi MH, Wisniewski SR, Nierenberg AA, Stewart JW, Warden D, Niederehe G, Thase ME, Lavori PW, Lebowitz BD, McGrath PJ, Rosenbaum JF, Sackeim HA, Kupfer DJ, Luther J, Fava M. (2006). Acute and longer-term outcomes in depressed outpatients requiring one or several treatment steps: a STAR*D report. *Am. J. Psychiatry.* Nov;163(11):1905-17.

Schiff ND, Giacino JT, Kalmar K et al. (2007). Behavioural improvements with thalamic stimulation after severe traumatic brain injury. *Nature* 448, 600-603.

Schlaepfer, T.E., Cohen, M.X., Frick, C., Kosel, M., Brodesser, D., Axmacher, N., Joe, A.Y., Kreft, M., Lenartz, D., Sturm, V. (2008). Deep brain stimulation to reward circuitry alleviates anhedonia in refractory major depression. *Neuropsychopharmacology*, 33, 368–377.

Schutter DJ. (2010). Quantitative review of the efficacy of slow-frequency magnetic brain stimulation in major depressive disorder. *Psychol. Med.* Nov;40(11):1789-95.

Sedlácková S, Rektorová I, Srovnalová H, Rektor I. (2009). Effect of high frequency repetitive transcranial magnetic stimulation on reaction time, clinical features and cognitive functions in patients with Parkinson's disease. *J. Neural. Transm.* Sep;116(9):1093-101.

Sen AN, Campbell PG, Yadla S, Jallo J, Sharan AD. (2010). Deep brain stimulation in the management of disorders of consciousness: a review of physiology, previous reports, and ethical considerations. *Neurosurg. Focus*, Aug;29(2):E14.

Shimamoto H, Takasaki K, Shigemori M, Imaizumi T, Ayabe M, Shoji H. (2001). Therapeutic effect and mechanism of repetitive transcranial magnetic stimulation in Parkinson's disease; *J. Neurol.* Sep;248 Suppl 3:III48-52.

Siebner HR, Mentschel C, Auer C, Conrad B. (1999). Repetitive transcranial magnetic stimulation has a beneficial effect on bradykinesia in Parkinson's disease. *Neuroreport.* Feb 25;10(3):589-94.

Siebner HR, Mentschel C, Auer C, Lehner C, Conrad B. (2000). Repetitive transcranial magnetic stimulation causes a short-term increase in the duration of the cortical silent period in patients with Parkinson's disease. *Neurosci. Lett.* Apr 28;284(3):147-50.

Siebner HR, Rossmeier C, Mentschel C, Peinemann A, Conrad B. (2000). Short-term motor improvement after sub-threshold 5-Hz repetitive transcranial magnetic stimulation of the primary motor hand area in Parkinson's disease. *J. Neurol. Sci.* Sep 15;178(2):91-4.

Speer AM, Benson BE, Kimbrell TK, Wassermann EM, Willis MW, Herscovitch P, Post RM. (2009). Opposite effects of high and low frequency rTMS on mood in depressed patients: relationship to baseline cerebral activity on PET. *J. Affect Disord.* Jun;115(3):386-94.

Speer AM, Kimbrell TA, Wassermann EM, D Repella J, Willis MW, Herscovitch P, Post RM. (2000). Opposite effects of high and low frequency rTMS on regional brain activity in depressed patients. *Biol. Psychiatry,* Dec 15;48(12):1133-41.

Steele JD, Currie J, Lawrie SM, Reid I. (2007) Prefrontal cortical functional abnormality in major depressive disorder: a stereotactic meta-analysis. *J. Affect Disord.* Aug;101(1-3):1-11.

Stefan, K., Kunesch, E., Cohen, L.G., Benecke, R., and Classen, J. (2000). Induction of plasticity in the human motor cortex by paired associative stimulation. *Brain,* 123, 572–584.

Stefan K, Kunesch E, Benecke R, Cohen LG, Classen J. (2002). Mechanisms of enhancement of human motor cortex excitability induced by interventional paired associative stimulation. *J. Physiol.* Sep 1;543(Pt 2):699-708.

Strafella AP, Paus T, Barrett J, Dagher A. (2001). Repetitive transcranial magnetic stimulation of the human prefrontal cortex induces dopamine release in the caudate nucleus. *J. Neurosci.* Aug 1;21(15):RC157.

Strafella AP, Paus T, Fraraccio M, Dagher A. (2003). Striatal dopamine release induced by repetitive transcranial magnetic stimulation of the human motor cortex. *Brain,* Dec;126(Pt 12):2609-15.

Tergau F, Wassermann EM, Paulus W, Ziemann U. (1999). Lack of clinical improvement in patients with Parkinson's disease after low and high frequency repetitive transcranial magnetic stimulation. *Electroencephalogr. Clin. Neurophysiol. Suppl.*51:281-8.

Thickbroom GW. (2007). Transcranial magnetic stimulation and synaptic plasticity: experimental framework and human models. *Exp. Brain Res.* Jul;180(4):583-93.

Trost M, Su S, Su P, et al. (2006). Network modulation by the subthalamic nucleus in the treatment of Parkinson's disease. *Neuroimage*, 2006;31:301– 07.

Tsuji S, Akamatsu N. (2003). Does transcranial magnetic stimulation improve the motor symptoms of Parkinson disease? *J. Neurol.* Oct;250 Suppl 3:III47-50.

van Dijk KD, Møst EI, Van Someren EJ, Berendse HW, van der Werf YD. (2009). Beneficial effect of transcranial magnetic stimulation on sleep in Parkinson's disease. *Mov. Disord.* Apr 30;24(6):878-84.

Wagner T, Valero-Cabre A, Pascual-Leone A. (2007). Noninvasive human brain stimulation. *Annu. Rev. Biomed. Eng.* 2007;9:527-65.

Ward HE, Hwynn N, Okun MS. (2010). Update on deep brain stimulation for neuropsychiatric disorders. *Neurobiol. Dis.* Jun;38(3):346-53.

Weaver FM, Follett K, Stern M, Hur K, Harris C, Marks WJ Jr, Rothlind J, Sagher O, Reda D, Moy CS, Pahwa R, Burchiel K, Hogarth P, Lai EC, Duda JE, Holloway K, Samii A, Horn S, Bronstein J, Stoner G, Heemskerk J, Huang GD; CSP 468 Study Group. (2009). Bilateral deep brain stimulation vs best medical therapy for patients with advanced parkinson disease: a randomized controlled trial. *JAMA*, 301, 63–73.

Weinberger M, Mahant N, Hutchison WD, Lozano AM, Moro E, Hodaie M, Lang AE, Dostrovsky JO. (2006). Beta oscillatory activity in the subthalamic nucleus and its relation to dopaminergic response in Parkinson's disease. *J. Neurophysiol.* 96, 3248–3256.

Wilcox RA, Cole MH, Wong D, Coyne T, Silburn P, Kerr G. (2010). Pedunculopontine nucleus deep brain stimulation produces sustained improvement in primary progressive freezing of gait. *J. Neurol. Neurosurg. Psychiatry,* Oct 22.

Williams A, Gill S, Varma T, Jenkinson C, Quinn N, Mitchell R, Scott R, Ives N, Rick C, Daniels J, Patel S, Wheatley K; PD SURG Collaborative Group. (2010). Deep brain stimulation plus best medical therapy versus best medical therapy alone for advanced Parkinson's disease (PD SURG trial): a randomised, open-label trial. *Lancet Neurol.* Jun;9(6):581-91.

Wolters A, Sandbrink F, Schlottmann A, Kunesch E, Stefan K, Cohen LG, Benecke R, and Classen J. (2003). A temporally asymmetric Hebbian rule governing plasticity in the human motor cortex. *J. Neurophysiol.* 89, 2339–2345.

Zangen A, Hyodo K. (2002). Transcranial magnetic stimulation induces increases in extracellular levels of dopamine and glutamate in the nucleus accumbens. *Neuroreport,* Dec 20;13(18):2401-5.

In: Handbook of Technology in Psychology …
Editor: Luciano L'Abate and David A. Kasier
ISBN: 978-1-62100-004-4

Chapter 16

An Introduction to Neurotherapy

David A. Kaiser and Andrea Meckley
WaveAtlas, Online Case Study

Neurotherapy is as much about mammalian phylogeny and technology as it is about volition

In the 1950s neuroscientists developed equipment that allowed rats to stimulate the pleasure centers of the brain (e.g., Olds, 1958). A lever was connected to a weak electrical generator which itself was connected by a wire to the deepest part of the rat's brain. Whenever a rat tapped the lever, a miniscule amount of electricity flowed from the generator into the rat's brain, into the pleasure center of the brain. Rats wired in this fashion quickly caught on to the nature of the task and after little training would tap the lever thousands of times an hour to receive the impulse. The rodent wireheads would pass up sleep, comfort, food, and even a receptive mate to work the bar. It was found to be the strongest form of learning ever to occur on the planet, and extremely resistant to extinction, to use the terminology of the day. Most rats would indulge in day-long binges of self-stimulation without a moment's rest and even when an electrified grid was placed between them and the lever, the rats would risk a gauntlet of shocks to gain access to their beloved.

The implications for studying addiction were obvious; the implications for learning and brain normalization training less so. But rats are rats. Would humans fare any better if allowed to reward themselves similarly, that is, intra-cranially without restraint? What do you think? A few epileptic patients stimulated themselves into convulsions before the experiments were called off.

EEG operant conditioning does not stimulate the brain directly – rewards come in through the eyes and ears and mind of the individual – but by rewarding EEG changes with visual and auditory rewards a therapist is able to help shape an individual's brainwave activity toward normal patterns of activity. The advent of powerful personal computers and advances in miniaturization and amplification has allowed anyone who doesn't mind getting gel in their hair the ability to visualize the electrical activity of their own brain and once you can visualize your behavior, you can change it.

Brainwaves are minute electrical voltages generated by the cortex which can be detected by modern electrical equipment. Sensors placed on the scalp record these tiny voltage changes across the scalp and analyze the signals looking for specific rhythms. When brain rhythms are normal, an individual is rewarded, usually by means of a sound (bell or chime) or light or video event. But when his or her brainwave activity deviates from normal, this positive feedback stops and a negative response may be provided such as a red flashing light or even a buzzer. In this fashion good brain-behaviors are exercised and undesirable brain-behaviors are not reinforced, with a goal being the accumulation of good brain-behaviors. A reasonably large repertoire of healthy behaviors is the basis for cognitive flexibility and self-regulation. An important point to realize is that no electrical current is put into the brain. The brain's electrical activity are merely registered passively at the scalp and these brain energies are relayed to a computer.

Neurotherapists attempt to restore impaired collaborations within and between brain areas by momentarily adjusting brain activity measured by the EEG toward a preconceived ideal based on a client's age or maturational state. Mental fitness is conceived as a healthy blend of EEG rhythms and synchronies. By analyzing EEG synchronization between electrode sites, we evaluate the likelihood that cortical networks are functional enough to manage age-appropriate challenges, especially self-regulation. Does a specific challenge such as reading or rest have anticipated psychophysiology? To what extent is the dominant frequency present? Is there age-appropriate hyperpolarization? Are sufficient fast frequencies present or do slow rhythms dominate? Are faster frequencies synchronized with slower ones or acting independently?

Human EEG consists of random events and rhythms. By means of operant conditioning – rewarding the presence of certain brain activity patterns and not others – healthy brain habits can be learned and unhealthy brain habits unlearned. Therapists typically focus on brain rhythms which have been studied for decades, which are called alpha (8-12 Hertz or cycles per second), beta (15-40 Hz), gamma (40+ Hz), delta (0.1-4 Hz) – the first four letters of the Greek alphabet – plus theta (4-8 Hz), the 8th letter, and the more English-sounding SMR (12-15 Hz) which stands for sensorimotor rhythm. Because these rhythms encompass a variety of physiological processes, however, each rhythm is often also dissected into smaller frequency ranges and such "narrow bands" are identified by their numerical range (8-10 Hz, 10-11 Hz).

When someone closes their eyes, alpha activity occurs across most or all of the brain. When he or she opens his eyes in a well-lighted room, alpha rhythms are replaced by beta rhythms, which are fast and low-amplitude waves. The amount of replacement and brain locations where these replacements occur varies depending upon the complexity, novelty, and meaningfulness of the environment, among other factors. Alpha rhythm replacement may involve all of the electrode positions or be selective and only occur at a few sites. Drugs, drowsiness, drive, and time of day generally influence every part of the brain whereas sensory and cognitive demands activate only a selected few brain areas.

Electrodes are positioned on the scalp according to a 50-year standard known as the International 10-20 system which divides the head into proportional distances –10% or 20 % of the way between the dent of the nose (nasion), protrusion in the back of the head (inion), and preauricular points directly in front of each ear. Labels reflect underlying brain areas: FP for frontal pole, F for frontal, P for parietal, C for central, T for temporal, and O for occipital. Sites are numbered with zero or "z" in the middle of the head (midline), followed by larger numbers as electrodes are positioned farther out to either side, with odd numbers alternating with even numbers between the left and right hemispheres (i.e., odd on the left, even on the right). Electrodes are spaced 6 or 7 cm apart on most heads. If more coverage is needed, additional electrodes may be placed halfway between any pair of electrodes. This system owes its endurance to its simplicity and fortuitous division of the scalp into brain regions that remain useful for cognitive and psychiatric research. Finally an EEG signal is always the difference in electrical potential between two electrodes on the scalp. Each electrode may be compared to its neighbor (e.g., C3 to Cz, P3 to Pz), or every electrode can be compared to the same electrode (C3 to Cz, P3 to Cz, O1 to Cz, etc).

EEG training allows an individual to monitor his or her own brain behavior, making visible and discrete what is normally hidden and continuous. This transformation of the invisible to the visible allows anyone to alter the behavior of his or her own brain. Without technological assistance, brain behaviors would be simply too subtle or ambiguous for proper detection and training (i.e., operational conditioning). This is the strength of neurofeedback – operant conditioning of psychophysiological responses beyond the level of normal (unassisted) observation. EEG biofeedback acts like a telescope to the mental sky. But with this strength comes some ambiguity as brain behaviors (neurophysiological responses) are not so readily classified as good or bad as motor actions can be. That is why children are often trained towards a database norm. The rationale is that if most children on average show a specific brain behavior, a certain incidence of this or that brain rhythm, this behavior ought to be generally healthy and positive. Many neurotherapists guide training by performing a normative EEG assessment of an individual before training and even at regular intervals during training.

Behavioral and mental states such as mathematical processing, reading, or relaxation are believed to consist of unique and distinct perceptual and cognitive operations and every mental operation has its own unique EEG profile – that is, a unique pattern of rhythmic activity in various parts of the brain. This concept is the foundation of functional neuroimaging including functional magnetic resonance imaging (fMRI), a popular method of investigating cerebral blood flow. This concept also provides the rationale for EEG normalization training: One's brain activity is trained toward a population norm because any deficit or excess of rhythmic activities is likely a result of abnormal neurophysiology and mental irregularity.

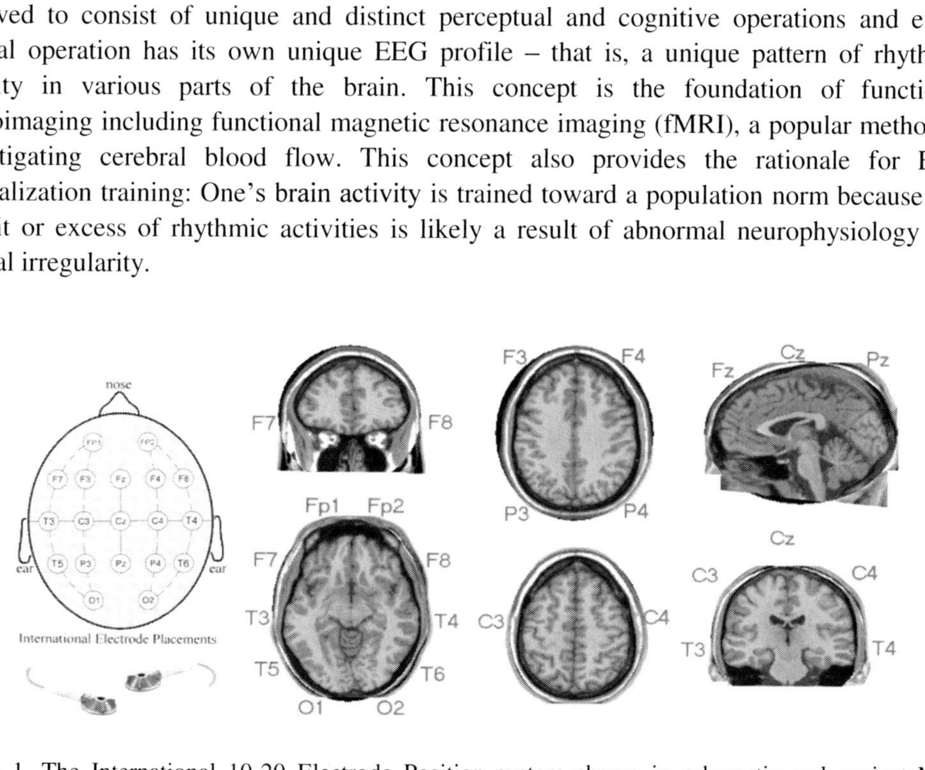

Figure 1. The International 10-20 Electrode Position system shown in schematic and against MRI slices.

Running on a treadmill helps a physician determine how well a patient's heart handles work or stress. Running through a test battery of reading, math, and problem-solving during the acquisition of EEG signals helps determine how well an individual's brain handles work or stress. Eyes closed relaxation or simply opening the eyes may reveal mental shortcomings for some individuals while others require challenges such a general test battery to reveal suspected or known deficits. Continuous attention tasks are often used to reveal processing deficits in attention deficit hyperactivity disorder (ADHD) children and executive control and inhibition tasks for identifying disturbances in the frontal lobe.

A minute of EEG contains a vast amount of information. Frequency analysis reduces EEG to a manageable handful of numbers. Spectral information, as it is called because we examine the entire frequency spectrum for rhythmic patterns, can be presented in tables, histograms, line graphs or the popular brain maps. Brain maps convert numbers into colors, which allows the human eye to quickly detect important patterns. In addition to unusual amounts of energy in various parts of the brain, we can also depict abnormal network activity. Too little or too much reciprocity between brain areas is quantified by means of comodulation and coherence analysis. Comodulation captures the reciprocity of energies between brain areas and coherence captures the timing relationships.

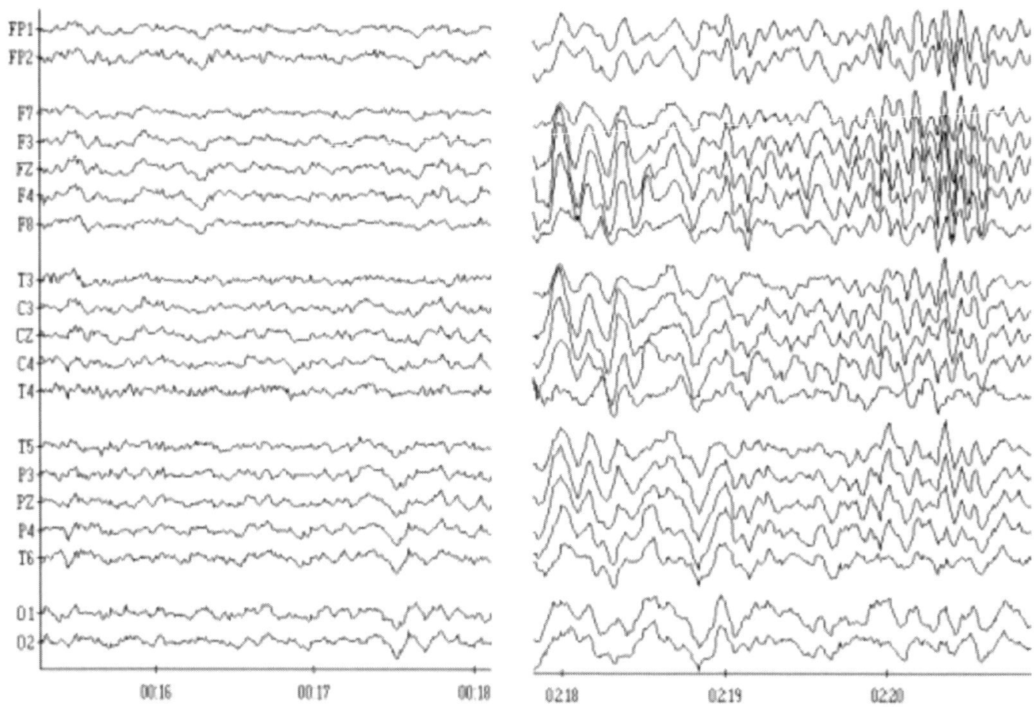

Figure 2. Six seconds of EEG data recorded at 19 electrode sites from the same child recorded minutes apart. One might mistakenly characterize the intial 3-s segment as inactive and the latter as alert and active. The fast anterior sinusoidal rhythm in the latter segment is a sleep spindle. The child was alert with eyes open during the first part of the record but in stage two sleep a few minutes later.

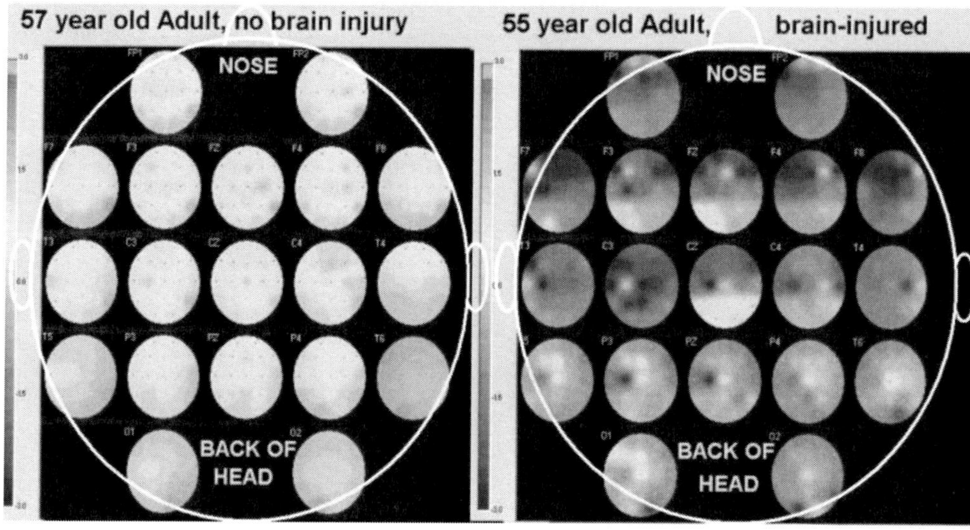

Figure 3. An example of a brain-network map of a healthy adult compared to a brain-injured adult of similar age. We are looking down from above on the head of each person, with their nose on top and ears to either side. Each circle is positioned at one of the 19 electrode sites sitting atop the head and depicts within itself all of the connections between electrodes relative to its position. The color yellow indicates normal function (i.e., database average), green and orange mostly normal, with blue and red indicating too little and too much connectivity, respectively.

As you can see, the healthy adult (left) shows relatively normal amounts of shared energy between electrodes for the rhythm under investigation while the brain-injured individual (right) shows too little communication or reciprocity between areas, notably between the frontal areas of the brain.

When we evaluate children we have to take into account a degree of neurological immaturity. Most of the energy of an infant's brain resides in the slow-wave delta rhythm and more than a decade of development may pass before an adult brainwave pattern emerges. Theta rhythms are prominent in many children diagnosed with attention deficit hyperactivity disorder (ADHD) because this reflects the immaturity of the ADHD brain. Theta rhythms in an adult or non-ADHD teenager often indicate brain-injury or neurological disease. The brainwave activity of any child may be compared to a database of many normal children in order to identify what parts of the brain, if any, differ in activity and what frequency rhythms are present at these locations.

Many psychiatric and neurological conditions manifest themselves more as disturbances in brain connections than as local damage or disorder. Therapy can focus on restoring activity to isolated brain areas or focus on re-establishing brain networks. Through trial and error any individual gradually develops mental strategies that modify his or her brain rhythms so as to maximize reward and in so doing alter these rhythms for the better. Neurofeedback works at the interface between volition and neurophysiology. An individual explores what is and what is not healthy willful behaviors, however indirectly, through the impact one's will has on one's brain rhythms. The brain is enormously plastic in terms of function as well as structure and is capable of altering neural pathways in response to reward. Much like our body, our brain also responds to exercise. Neurofeedback provides one of the best forms of exercise – regulatory practice, the brain practicing at regulating itself.

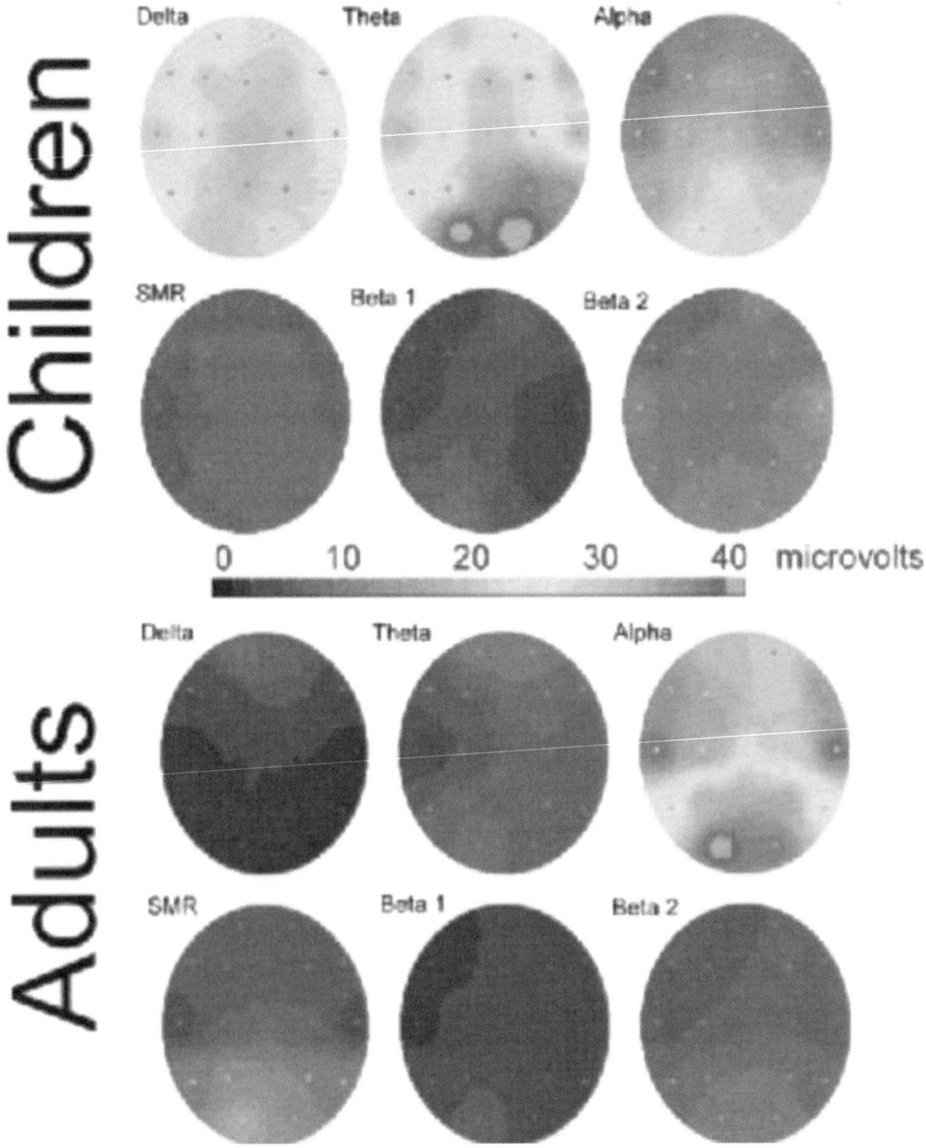

Figure 4. Typical EEG magnitude for children and adults. Each topograph is oriented so that the top is where the nose is, the sides where the ears are. The back of the head shows the heightened activity. The theta rhythm dominates a child's spectral energy while an adult brain contains most of its spectral energy in a faster rhythm, the alpha rhythm.

Lucas' first feature film from 1971, *THX-1138,* takes place in the 25th century. By this time the state has taken control of our mental health and everyone is drugged into a continual stupor. Loudspeakers in the home and at work constantly remind individuals to take the appropriate amount of sedatives. Avoiding one's "medication" is a crime. Near the end of the film our fate is tied to one of the men who has fought the system and failed. He's about to be arrested and "re-educated." With nowhere else to go or hide, he sits alone in a lobby, calmly awaiting the authorities. We see a line of children ride up an escalator near him. The children are noticeably subdued, passive, almost without curiousity. The camera focuses on an IV-tube strapped to each of their right arms. One of the children appears frightened and confused and approaches our man for help. His IV tube has come loose. Our man, drained of all his fight, reattaches the IV tube and with a sweet smile sends the child on his way... Perhaps this way is best for all of us, his weak smile conveys.

Neurotherapy relies on technology in a more benevolent manner, helping individuals learn their way out of poor behavioral and mental habits. Learning is a permanent change in behavior or behavioral potential that is produced as a result of prior experience. Most learning is adaptive, but not all. In fact mental health problems are often maladaptive learning such as depressed mental set, obsessions, or social tendencies that continue to thwart the will of individual. Changes in behavior due to learning also exclude those temporary or permanent changes associated with fatigue or injury and the change in behavior is not a direct product of maturation, disease, or instinct. There are two major laws of learning: Rewarded behavior is repeated, and punished behavior leads to avoidance of the punisher.

EEG rhythm learning began informally with Lord Adrian, who in 1934 watched his EEG in front of oscillograph and created alpha bursts at will. A quarter century later, Joe Kamiya at U Chicago developed alpha biofeedback or deep states training (1958). His first subject was remarkable, learning to discriminate his internal psychophysiological states. He began by hearing a tone and at that time he had to say whether the alpha activity was prominant in his EEG. He was correct half of the time, guessing. After a second block of trial he was accurate 65% of the time, and after another block of trials, 85%. During his 4th block, after initial mistakes, he correctly labelled his EEG state 400 times in a row! Had the first subject of Kamiya's experment been insensitive to his internal states, who knows what would have happened, but instead the work along with other psychophysiological researcher's work lead to the 1st congress in Aspen Colorado on this topic in 1968. Here the field was called "Biofeedback", as the competing term "autoregulation" was consider to associated with the teamster union.

A third of all patients who undergo surgery for intractable epilepsy continue to have seizures postoperatively (e.g., Kim et al., 2008). Brain areas are removed and the patient is no better off. That the supposed starting point for a seizure is removed and convulsions continue reflects how little we actually do know about epilepsy. It may be that the entire brain runs fast in such cases, making EEG training a viable option. The story of the treatment of epilepsy with neurofeedback begins in the 1960s, when Mercury astronauts claimed they saw Polynesian natives waving at them as they flew over the Pacific. The neurotoxicity of the capsule's rocket fuel was to blame, and in testing the mixture, Barry Sterman seredipitously discovered that SMR training countered its effect, a behavioral anticonvulsant. It was an incidental investigation of electroencephalographic correlates of behavioral inhibition in cats (Sterman et al., 1972). Neurofeedback now has a 40-year history for treating seizures (Sterman et al., 1969; Tan et al., 2009) and this technology and approach (operant conditioning) has since been adapted to treat an astounding array of human frailties including mood and sleep problems, attention and learning difficulties, as well as brain injury and drug abuse. The mysteries of brain function still remain, but with neurofeedback we've added an extremely powerful tool to resolving one of its chief disorders.

The first published report of human EEG was in 1929, from data collected in part as early as 1924. Berger (1929) compared his son's EEG against a steady signal marking time. Spectral analysis to determine the exact amount of each frequency was performed soon after (Dietsch, 1932; Berger, 1932). Since then the utility of EEG analysis for clinical and research purposes is without question – in staging sleep, detecting nonconvulsive seizures, and monitoring mental operations and their shortcomings, to name but a few applications for this sensitive technology. We have been analyzing EEG activity quantitatively in every decade, and there are more than 120,000 papers on EEG analysis indexed in Medline and other indices. That is a long time to be studying a single phenomenon, even if it is of our own creation, a correlate of human thought. The 17th century poet Alexander Pope once said "The proper study of mankind is Man" and much of modern science has followed his prescription, expanding our definition of humanity to include our thoughts, our behaviors, and the energy relating them, psychophysiology.

References

Adrian ED & Matthews BHC (1934). The interpretation of potential waves in the cortex. *Journal of Physiology, 81*, 440-471.

Berger H. (1929). Ueber das Elektroenkephalogramm des Menschen. *Archiv für Psychiatrie und Nervenkrankheiten, 87*, 527-570.

Berger, H. (1932). Über das Elektren-kephalogramm des Menschen. Fünfte Mitteilung. (Fifth Report) *Archiv für Psychiatrie und Nervenkrankheiten, 98*, 231-254.

Caton R (1875). The electric currents of the brain. *British Medical Journal, 2*, 278.

Dietsch, G. (1932). Fourier-analyse von Elektrenkephalogrammen des Menschen. *Pflüger's Arch. Ges. Physiol.*, 230, 106-112.

Kamiya, J. (1968, November). Conscious control of brain waves. *Psychology Today, 1*, 56-60.

Olds, J (1958). Self-stimulation of the brain. *Science 127*:315-24.

Kim SK, Wang KC, Hwang YS, Kim KJ, Chae JH, Kim IO, & Cho BK (2008). Epilepsy surgery in children: outcomes and complications. *Journal of Neurosurgery.Pediatrics, 1*, 277-83.

Meacham J (2009). A storm in the brain. *Newsweek*, Apr 11, 2009.

Sterman MB, Wyrwicka W, & Howe R (1969). Behavioral and neurophysiological studies of the sensorimotor rhythm in the cat. *Electroencephalography & Clinical Neurophysiology, 27*, 678-9.

Sterman MB, Fairchild MD, & McRae GL (1972). Effects of hydrazine on electrophysiology, behavior and runway performance in the cat. DTIC technical report AD0746014.

Sterman, M.B. & Kovalesky, R.A. (1979). Anticonvulsant effects of restraint and pyridoxine on hydrazine seizures in the monkey. *Experimental Neurology, 65*, 78-86.

Tan G, Thornby J, Hammond DC, Strehl U, Canady B, Arnemann K, & Kaiser DA (2009). Meta-analysis of EEG biofeedback in treating epilepsy. *Clinical and EEG Neuroscience, 40*, 173-9.

In: Handbook of Technology in Psychology ...
Editor: Luciano L'Abate and David A. Kasier
ISBN: 978-1-62100-004-4

Chapter 17

TELEHEALTH CONFERENCING IN THE NEUROSCIENCES

Carlos De las Cuevas*
Universidad de La Laguna, Spain

Telemedicine can be considered to be the use of telecommunications technology for the delivery of health services when the patient and health professional are at different locations. Videoconferencing constitutes the principal, although not the only, communication method in telemedicine applications' telepsychiatry and teleneurology. Advances in technology and changes in medical care delivery have enhanced the ability to develop effective telemedicine videoconferencing systems. At the present time, both telepsychiatry and teleneurology are in a slope of enlightenment phase where focused experimentation and solid hard work by an increasingly diverse range of organizations lead to a true understanding of the technology's applicability, risks and benefits. Telepsychiatry is nearer than teleneurology to a plateau of productivity phase where the real-world benefits of the technology will be demonstrated and accepted, tools and methodologies will be increasingly stable as they enter their second and third generations, and growing numbers of organizations will feel comfortable with the reduced level of risk; the rapid-growth phase of adoption begins. Nevertheless, it is essential to remember that both telepsychiatry and teleneurology are important enablers of care delivery processes, but they are merely enablers and must not be confused with the actual delivery of care.

Videoconferencing could be defined as a set of interactive telecommunications technologies which allow two or more locations interact via two-way, real-time transmission of data, audio and video signals between specialized devices or computers. Videoconferencing technologies can greatly expand the reach of healthcare practitioners by providing patients (particularly those in rural/remote areas) with unprecedented access to services, representing a fundamental shift in the way that healthcare professionals care for their patients. Thus, videoconferencing constitutes the principal, although not the only, communication method in telemedicine clinical applications..

*E-mail: cdelascuevas@gmail.com

Telemedicine can be considered to be the use of telecommunications technology for the delivery of health services when the patient and health professional are at different locations. Over the past 20 years, telemedicine has been promoted as an important change in the way that medicine is practiced, because of its potential to make medical care both more efficient and more equitable in terms of access. Clinical applications of telemedicine are marked by diversity. They differ in the medical problems addressed, the evidence base for decision making, the personnel and setting of care involved, the diagnostic and therapeutic strategies employed, and the organizational and cost implications of these strategies. Telepsychiatry is the branch of telemedicine that focuses on mental health applications and includes the practice of psychiatry and psychology at distance. Telepsychiatry is the "native application" for interactive-video telemedicine. Mental health interactions are straightforward, and demand little other than that the participants see and talk with one another. There is no need for sophisticated peripheral devices such as video oto-ophthalmoscopes or electronic stethoscopes. Perhaps for this reason, telepsychiatry was the first use of interactive video for medical purposes. In the other hand, neurology has been a late starter in realizing the benefits of telemedicine and most of the publications on teleneurology have been in the last decade, but interest has increased in the past few years. Neurology is a specialty that, because of the emphasis on accurate interpretation of a history, does lend itself to telemedicine.

VIDEOCONFERENCING HISTORY

Videoconferencing in its first forms has been around a lot longer than one might imagine. Videonference origins are closely linked to the development of telephone and television. Thus, first public demonstration of this technology took place on April 7, 1927, when Herbert Hoover - the Commerce Secretary of United States at that time - placed the first ever videoconference call from Washington DC to the president of Bell Labs, Walter Gifford, in New York. This first videoconference was unidirectional via telephone lines where live images and sound from Washington DC could be heard and seen in a television at New York. It was necessary to wait three years, till April 9, 1930, to the first public demonstration of two-way videoconference between two buildings separated by 2 miles, in New York City.

As a result of the economic recession of the mid-30s and the Second World War there were no major progresses in videonference until the development by Bell Labs of the first videophone prototype in late 50s. In 1964, Bell Labs demos the "Picturephone" at the World Fair of Flushing Meadows in New York. The phone featured a black and white screen which displayed the caller. Although Bell Labs launched the Picturephone in 1970, it did not gain widespread consumer acceptance because the equipment was bulky, difficult to use, and the picture was so tiny that it could hardly be seen. It was also expensive for that time since a three-minute Picturephone call from Washington to New York City cost $16. The most expensive connection, between New York City and Chicago, cost $27 for three minutes. In 1973, ATandT, the former Bell Labs, left the marketing of this service.

At this time, the idea of introducing videoconference at home or personal videonference was temporally abandoned and the efforts in the following years focused in group videoconferencing equipment.

The 1980s was crucial in the development of videoconferencing systems, since the introduction of technological breakthrough of solid-state memories that allowed the stored image processing, which directly influenced the development of cheaper codecs. It was only in the 1980s that digital telephony transmission networks became possible, such as ISDN, assuring a minimum bandwidth (usually 128 kilobits/sec) for compressed video and audio transmission. The digital compression and transmission technologies in the late 1980s, allowing point-to-point interactive videoconferencing to and from anywhere that had access to T1, fractional T1, or ISDN lines.

In 1982 Compression Labs launched the first videoconferencing system VTS 1.5 that required a T-1 line (24 channels of 64Kbps). This lowers the costs associated with the signal transmission, since it did not depend on the satellite to achieve the required bandwidth. It was also huge, and took up an enormous amount of resources, which could trip 15 amp circuit breakers. With a price tag of $250,000, it was definitely outside the reach of most businesses, let alone the fact that using a connection cost $1,000 per hour. Still, it was the only working videoconferencing system on the market.

It would be four more years before PictureTel, in Delaware, launched an opposition product. It was substantially cheaper, at only $80,000 for the system and $100 an hour for lines.

In 1991, PictureTel - a subsidiary of IBM - introduced the first PC-based videoconferencing system. This was a color video-based full-motion system. The big leap forward was that this system was PC based, and video conferencing was finally on track to be widely available. It was still pricey, at $20,000 for the system and $30 an hour. The other drawback was that you also still needed to use a conference room specially equipped for videoconferencing. This meant the participants had to relocate to the videoconferencing room, and also had to pay to use the facilities.

In 1992, ATandT introduced their new Videophone, which only cost $1,500. Still expensive for the general public, but business was definitely starting to get interested and finally videoconferencing began to be an an affordable option for the masses started to become reality. Thus, video teleconference systems throughout the 1990's rapidly evolved from highly expensive proprietary equipment, software and network requirements to standards based technology that is readily available to the general public at a reasonable cost.

Finally, in the 1990s, IP (Internet Protocol) based videoconferencing became possible, and more efficient video compression technologies were developed, permitting desktop, or personal computer (PC)-based videoconferencing.

CU-SeeMe was originally written by Tim Dorcey of the Information Technology department at Cornell University (Dorcey, 1995). It was first developed for the Macintosh in 1992 and later for the Windows platform in 1994. Originally it was video-only with audio added in 1994 for the Macintosh and 1995 for Windows, meeting the requirement for true videoconferencing with audio and video.

This was enormously important, because at this moment everyone could talk to another person anywhere in the world, and yet still be sitting in their own office or home. Since then, many high-tech companies joined the race and videoconferencing software and equipment options expanded.

In 1996, Microsoft introduced NetMeeting, a Voice over Internet Protocol and multi-point videoconferencing client in many versions of Microsoft Windows (from Windows 95) and now videoconferencing was basically available to anyone with a home computer. The same year, the software developer Vocaltec, launched its SurfandCall plug-in application which enables on-line users to call directly from Internet sites to company sales departments or call centers.

And by 2000, Samsung had developed the MPEG-4 streaming 3G video cell phone. By streaming the media, it can be read, heard and viewed the instant it's being delivered. It's now possible to videoconference anywhere in the world for as little as $12 a month and the cost of a good phone. Industry giants like Yahoo and MSN have gone one step further and made the service free.

VIDEOCONFERENCING TECHNOLOGY

In telemedicine, the level of technology quality of the specific equipment deployed is application dependent. The selection of the equipment for videoconferencing in telepsychiatry and teleneurology should be based on ease of use, image and sound quality, cost, and suitability to given applications. For instance, simple cognitive screening is not likely to require the resolution needed to detect extrapyramidal signs; an interview with an active youngster is likely to require more sophisticated equipment than one with a demented geriatric patient who moves and speaks little. In our experience high-quality enhanced video at 30 frames per second at 384 kilobits per second (Kbps) and full-duplex digital audio with noise suppression and echo cancellation assures good acceptance and satisfaction by telepsychiatry users. A voice tracking camera and track-to-preset function which automatically focuses on the current speaker is also important.

The major components of a videoconferencing system include monitors, cameras, CODEC, microphones and speakers, and a means of transmission.

Monitor

The monitor, or screen, shows the image of the people at the distant site(s); it may show a picture of the local site too. Monitor size only allows for a bigger picture. Clarity of picture and motion handling are primarily a function of bandwidth.

Camera

The camera captures images at the local site to send to other sites. There is a wide range of cameras available with little difference in the quality of the picture from camera to camera since the clarity of the image and the motion handling is primarily a function of the bandwidth and the algorithm used to compress the image prior to transmission to another site. Factors to weigh include lens quality, whether the lens is fixed or can "pan" about a room, and whether it can "zoom" in (for closeup) or out (for distant) views; a wide angle lens is useful if more than 1 or 2 people are to be viewed. It is really helpful for the provider to have control over remote camera zoom and scan. A document camera for transmission of graphic material is helpful but not essential.

Codec

In order to have a video conference, both the audio and video streams need to be coded and decoded. The CODEC (coder -decoder) is the heart of the teleconferencing device. It transforms the analog signal (the picture) picked up by a video camera to a digital signal and compresses it for transmission to the distant site; there, another CODEC transforms the digital signal back to an analog one for viewing on the video monitor. Compression is the reduction in size of data in order to save space or transmission time. Transmission of compressed video over a communications network requires sophisticated compression algorithms. The compression algorithms used are responsible for the quality of the signal ultimately received. There are industry standards for video and sound compression and for inter-network compatibility with other teleconferencing systems.

There are many different types of codec's available but it depends on what you need in order to figure out what codec is best for you. One of the bigger codec's, used for larger projects is the G.711 codec. This particular codec can sample the audio at 64,000 times per second. This will be enough for most of videoconferencing needs. Another common codec is the G.729A codec. This particular one samples at a rate of 8,000 times per second. This is a very commonly used codec and guarantees good sound and video concentration.

CONNECTING

The most significant distinction among videoconferencing systems is the method of transmission. Transmission is important because two systems cannot connect if they are using different transmission methods. Videoconferences can be transmitted over two protocols, H.320 - ISDN (phone) or H.323 - IP (Internet) lines. In the past, most videoconferences used ISDN lines; however, many people are now using IP connections due to cost savings. In order to connect two units using different transmission methods, a bridge must be used that will handle these mixed protocols. In an ISDN call, bandwidth is dedicated to only one videoconference, while in an IP call, bandwidth may be used to transmit for multiple uses. However, ISDN calls can be very costly since you may be making the call over a distance, in which case, long distance phone line charges apply, and ISDN lines take up 6 phone lines. Connections around the world average 384 kbps. Standards exist for health care in the form of legislative and regulatory pronouncements (e.g., Joint Accreditation of Healthcare Organizations), for telecommunication and information systems (e.g., International Multimedia Telecommunications Consortium, Inc. – ITU-T sector standards for multimedia teleconferencing, T.120, H.320, H.323, and H.324), and for general systems quality (e.g., International Standards Organization 9000:2000). Such standards provide a relevant foundation for quality requirements in the telemedicine field. Videoconferencing connections may be limited to a closed network (such as a LAN) or may use public networks (such as regular phone lines). Many connect via ISDN (Integrated Services Digital Network) because it is an economical solution for high-quality videoconferencing. ISDN works over regular copper telephone lines, transmits at a minimum of 128 Kbps per line, and provides dedicated bandwidth for smooth audio and video (15-30 frames per second). The preferred rate for events is at least 384 Kbps. Internet-based connections are increasing in popularity. IP-based video shares or competes for bandwidth with other Internet data which may cause audio clipping or delays that result in jerky video or slow networks. Increasing numbers of schools, libraries, and other institutions are developing high bandwidth networks and are experiencing good results with IP videoconferencing. Many believe that the Internet will eventually replace ISDN as the medium of choice for videoconferencing. However, for the highest quality, ISDN is still most widely recommended. The Moving Pictures Expert Group (MPEG) standards used in digital video applications have been adopted in videoconferencing systems. MPEG-1 was developed to compress video for playback on CD-ROM drives at 30 frames/second, with 352x240 pixel resolution and CD-quality audio. MPEG-2 was optimized for digital versatile disk (DVD) and cable and satellite television at 30 frames/second, with 720x480 pixel resolution and high-quality audio. The new MPEG-4 standard has been designed for streaming video applications using the internet.

Bandwidth

Bandwidth refers to the amount of data that can be transmitted electronically. An analog phone lines that operates at 56-64 Kbps is enough bandwidth to handle voice communication, but the transmission of video signals need more bandwidth to avoid motion and image distortion. For this reason digital ISDN lines (128 Kbps) are commonly used. Only a few years ago it was thought that high bandwidths (T1-384 Kbs) were essential for adequate resolution to assure clinical accuracy. With subsequent technologic advances it now appears that a bandwidth of 384 -128 Kbs is acceptable in most situations. The lower bandwidth (128 Kbps) appears suitable for many clinical applications and for administering questionnaires and rating scales. The higher bandwidth (384 Kbps) enhances recognition of negative symptoms of schizophrenia, recognition of manual tremors and pupillary reflexes, and it may enhance patient and provider satisfaction. Probably, future technologic advances may render bandwidth a non-issue.

Other Equipment

High quality microphones and speakers assure aural communication.

Videoconferencing in Telemedicine

The first use of videoconferencing in psychiatry took place in 1961 by Dr. Cecil Wittson (Wittson et al., 1961) at the University of Nebraska Medical Center in Omaha who began pilot testing telepsychiatry on the campus-wide tele-education system, which had been operating since the early 1950's. The trials, involving two psychiatrists and eight groups of four or five patients, went well; the researchers found that the televideo sessions were well accepted and appeared to be as effective as on-site sessions. In January, 1965 the first "production" telepsychiatry program was begun, a microwave-mediated link between Omaha and Norfolk State Mental Hospital, 112 miles away. It was supported by a 6-year grant from the National Institutes of Mental Health. During the period of the grant, over 300 hours of clinical telepsychiatry sessions were logged. The program didn't survive the end of the grant period, probably because of the formidable transmission costs between Omaha and Norfolk, which averaged about $48,000/ year (in 1960's dollars!).

Since the mid-1990s, telemedicine programs have become common throughout the world in nearly every specialty and area of healthcare. In recent years, growth in the use of telemedicine has been facilitated by several important factors: lower cost and more widely available communications; lower cost, higher performance computers; greater public confidence in the use of computer technology; greater acceptance of the technology by medical professionals, and emerging global standards in communications, video conferencing, and medical disciplines.

At the present time, telepsychiatry and teleneurology are clinical telemedicine applications with a different level of maturity. Telepsychiatry is a mature area of telemedicine more extended and characterized by: great quantity and high quality of research about technical feasibility, diagnostic accuracy, sensitivity, specificity and clinical outcome; high degree of acceptation and satisfaction by patients and mental health professionals; and by the availability of standards and protocols. In the other hand, teleneurology is a promising telemedicine application with a great potential to develop. In neurology, real-time telemedicine has the greater potential in the management of neurological emergencies such as status epilepticus and stroke both requiring urgent treatment within the window period for better outcome. Transporting these patients to a specialist centre might take several hours, resulting in high mortality and morbidity. Its uptake within the neurological community is low but increasing. The evidence to date is that teleneurology can narrow the gap between patients with neurological disease and the doctors who are trained to look.

The rationale for telemedicine in psychiatry and neurology is twofold: first, to provide services that cannot easily be provided face-to-face; and second, to improve the efficiency or effectiveness of existing services. The only reason to do telepsychiatry or teleneurology, or indeed any type of telemedicine, is to solve problems that cannot be solved easily by other means. It is emphatically not to use modern technology just because it is available.

Clinical applications using videoconferencing in telepsychiatry and teleneurology encompass diagnostic, therapeutic, and forensic modalities. Common applications include pre-hospitalization assessment and post-hospital follow-up care, medication management and consultation. Psychotherapy, including supportive, cognitive-behavioral, brief interpersonal, psychodynamically oriented, psychoanalytic, group and family therapy, is feasible. Commitment hearings, evaluation of competence and forensic evaluations are feasible. Other applications in telepsychiatry and teleneurology includes**: a** broad range of applications for distance learning, an effective method for administrative services and support that may achieve cost savings in large systems, and an effective and reliable means of gathering research data from certain clinical populations enabling multisite acquisition of information for large clinical databases.

Telemedicine videoconferencing has experienced rapid growth in the last decade primarily due to advances in technology (e.g., increased bandwidth; new cameras, monitors, and coder/decoders (CODECs)) and changes in the medical care environment. According to the literature, establishing systems for psychiatric and neurologic patient care using videoconference technology is feasible. However, there is little evidence from the literature of its economic benefit, especially regarding its cost-effectiveness when compared to face-to-face care. Tables 18.1 and 18.2 shows the most relevant papers published in the last decade evaluating the evidence of effectiveness and cost-effectiveness of remote consultation in psychiatry and neurology by videoconference in comparison with standard face-to-face consultation.

Table 18.1. Relevant papers in teleneurology evaluating effectiveness and cost-effectiveness of remote consultation in neurology by videoconference in comparison with standard face-to-face consultation

Authors/ Journal	Objetive	Design/Sample	Technical Aspects	Conclusions
Shafqat et al. Stroke, 1999	Aassess the validity and reliability of remote diagnosis of patient's neurological deficit after suffering a stroke	Prospective cohort study where twenty stroke patients underwent evaluations by remote and bedside examiners	VC VTEL SmartStation TC 1000 with an integrated PTZ Sony camera. Data were transmitted over 3 ISDN lines at speeds up to 384 kilobits per second.	The NIH Stroke Scale remains a swift and reliable clinical instrument when used over interactive video. Application of this technology can bring stroke expertise to the bedside, regardless of patient location.
Wiborg et al. Aktuelle Neurologie 2000	Assess the reliability of remote diagnosis of patient's neurological deficit after suffering a stroke	Prospective cohort study where forty-four stroke patients underwent evaluations by remote and bedside examiners	VC system Sony Contact. Data were transmitted over 3 ISDN lines at speeds up to 384 kilobits per second.	Interrater reliability in assessing both patients and CT scans by teleconsultation was shown to be comparable to the examination on the scene
Handschu et al. Stroke, 2003	Assess the reliability of remote diagnosis of patient's neurological deficit after suffering a stroke	Prospective cohort study where forty-one stroke patients underwent evaluations by remote and bedside examiners	EVITA: multimedia telesupport system for transfer of real-time video sequences and audio data. Within its modular component, a special Internet protocol–based communication protocol with a strong data security concept was used	Remote examination of acute stroke patients with a computer-based telesupport system is feasible and reliable when applied in the emergency room; interrater agreement was good to excellent in all items. For more widespread use, some problems that emerge from details like brightness, optimal camera position, and audio quality should be solved.
Wang et al. Stroke, 2003	Assess the reliability of remote diagnosis of patient's neurological deficit after suffering a stroke	Prospective cohort study where twenty stroke patients underwent evaluations by remote and bedside examiners	Axis 2130 Pan/Tilt/Zoom camera, a 1.5-GHz Dell PC workstation and LCD monitor running Microsoft Windows 2000 with Internet Explorer 5.5, a Linksys WET11 wireless bridge, Netgear 5 port ethernet switch.	The NIHSS can be reliably performed over the REACH system. This supports our endeavor to bring stroke expertise to rural community hospitals.
Meyer et al. Neurology 2005	Assess the validity and reliability of remote	Prospective cohort study Twenty-five stroke patients underwent	Operative system designed to care for stroke patients over the Internet with wireless	This Internet system is reliable and valid. Further studies should assess its use in acute stroke.

Authors/ Journal	Objetive	Design/Sample	Technical Aspects	Conclusions
	diagnosis of patient's neurological deficit after suffering a stroke	evaluations by remote and bedside examiners	connection (750 kbps) integrating videoconferencing system with online data recording and transmission of images	
Montani et al. J. Telemed and Telecare 1997	Assess the reliability of a videolinked diagnosis of hospitalized demented patients	Prospective cohort study where fifteen patients participated in a video-consultation and a conventional face-to-face consultation	Non described videoconferencing system	Although most patients (11 of the 15) expressed a preference for the face-to-face consultation, the video-consultations were judged to be acceptable by the patients and the psychologists.
Shores et al. Int J Geriatr Psychiatry 2004	Assess diagnostic reliability of telemedicine in examination for dementia	Prospective cohort study were sixteen patients participated in a video-consultation and a conventional face-to-face consultation	VC systemVCON Armada Cruiser at 384 kbps	Ttelemedicine was found as accurate as an in-person clinical examination in establishing the diagnosis of dementia.
Loh et al. Internal Medicine Journal, 2004	To assess the reliability of remote diagnosis of dementia and depression	Prospective cohort study where twenty patients participated in a video-consultation and a conventional face-to-face consultation	VC system VCON at 128, 256 and 384 kbps.	Remote assessments with SMMSE and GDS using Telehealth methods yielded similar results to direct assessments. However, there was a moderate difference between face-to-face and Telehealth assessments in some subjects, which could influence clinical decision-making.
Cullum et al. Assessment, 2006	Assess reliability of telemedicine in the neurocognitive assessment of patients with cognitive déficit or Alzheimer disease	Prospective cohort study Thirty-three patients participated in a video-consultation and a conventional face-to-face consultation	VC system Polycom iPower 680	Telecognitive assessment appears to be a valid means to conduct neuropsychological evaluation of older adults with cognitive impairment.
Lott et al. Alzheimer and Dementia, 2006	Study the feasibility of using telemedicine to evaluate symptoms of Alzheimer Disease in individuals with Down Syndrome	Prospective cohort study where 61 patients participated in a video-consultation and 29 conventional face-to-face consultation	VC system Polycom Model H323 at 384kbps	It is feasible to make a diagnosis of Alzheimer disease in Down Syndrome by Telemedicine

Table 18.1. (Continued)

Authors/ Journal	Objetive	Design/Sample	Technical Aspects	Conclusions
Theodoros et al. J Telemed and Telecare, 2003	Asses validity and reliability of telemedicine in motor speech disorders in adults with acquired neurological impairment	Prospective cohort study where ten patients participated in a video-consultation and a conventional face-to-face consultation	Standard Videoconferencing sistema at 128 kbps	Internet-based assessment has potential as a reliable method for assessing motor speech disorders
Craig et al. European J Neurol 2000	Asses the validity telemedicine in the diagnosis of neurological disorders	Prospective cohort study where twenty-five patients participated in a video-consultation and a conventional face-to-face consultation	VC system Sony with 3 ISDN lines at 384 kbps	It is practical to assess patients admitted with neurological symptoms to distant hospitals using interactive videoconsultation and this may result in more efficient use of in-patient resources
Craig et al. J Telemed and Telecare 2000	Asses the validity telemedicine in the diagnosis of neurological disorders	Prospective cohort study where twenty-five patients participated in a video-consultation and a conventional face-to-face consultation	VC system Sony with 3 ISDN lines at 384 kbps	The study showed that neurologists can deliver outpatient neurological care to distant patients using telemedicine
Authors/ Journal	Objetive	Design/Sample	Technical Aspects	Conclusions
Chua et al. J Neurol Neurosurg Psychaitry 2 001	Test the hypothesis that telemedicine for new patient referrals to neurological outpatients is as efficient and acceptable as conventional face to face consultation	A randomised controlled trial between two groups: face to face and telemedicine 141 patients: 76 in telemedicine and 85 in face to face modality	Commercial interactive video conferencing equipment with ISDN lines transmitting information at 384 kbps	Telemedicine for new neurological outpatients is possible and feasible but generates more investigations and is less well accepted than face to face examination
Ahmed et al. Epilepsi 2008	Cost analysis and patient satisfaction with telemedicine in epilepsy care	A randomised controlled trial between two groups: face to face and telemedicine 41 patients: 23 in telemedicine and 18 in face to face modality		Telemedicine can play a role in follow-up care of epilepsy patients, reduce patient costs, and improve patient satisfaction

Table 18.2. Relevant papers in telepsychiatry evaluating effectiveness and cost-effectiveness of remote consultation in neurology by videoconference in comparison with standard face-to-face consultation

Authors/ Journal	Objetive	Design/Sample	Teleconsultation technical aspects	Conclusions
Elford et al. J. Telemed and Telecare 2000	Assessment of the utility of a child telepsychiatry program.	23 patients with a mean age of 9 years were initially assessed by a psychiatrist using VC and then by a different psychiatrist FTF.	PC-based VC workstation at each site with H.320-compatible software. Three ISDN lines from the psychiatrist's room connected to the public utility switching centre and then returned to the patient's room along six SW-56 digital lines.	In 96% of cases, the diagnosis and treatment recommendations made via VC were the same as those made FTF. VC assessments were an adequate alternative to FTF and did not interfere with diagnosis.
Kirkwood et al. J Telemed and Telecare 2000	Neuropsychological assessment of individuals with a history of alcohol abuse comparing VC to FTF consultations.	27 subjects with a mean history of alcohol abuse of 15 years completed FTF and VC assessments on the same day, administered by the same individual. Half the subjects were given FTF assessments first and half were given the VC assessments first.	Two desktop VC units (VC7000, British Telecom) connected by ISDN at 128 kbps and a document imager.	For most of the outcome measures, cognitive assessment via VC produced similar results to FTF assessment. Most participants expressed high overall satisfaction with the teleconsultation
Mielonen et al. J Telemed and Telecare 2000	Assess the costs of psychiatric inpatient careplanning consultations to remote areas using VC, instead of hospital FTF consultations.	124 patients, relatives and healthcare personnel (14 VC sessions compared to 20 FTF).	VC equipment (VCS H.320 roll-about unit, Videra) consisted of a video codec and monitor, an adjustable camera and an audio unit with echo suppression. The connection was via three ISDN lines.	Of the respondents (90% response rate), 90% were satisfied with the quality of the VC communication.
Zaylor J Telemed and Telecare 1999	Compare clinical outcomes of patients seen by VC and those seen FTF to etermine whether VC reduced the quality of psychiatric care.	49 patients with either major depression or schizoaffective disorder. Retrospective (two-year) chart review. Biopsychological scale.	PC-based VC equipment running at 128 kbps.	No significant difference was found in the percentage change in clinical outcomes as measured by the Global Assessment of Functioning scores between the two groups. Patients seen by VC had a greater attendance rate and follow-up visits took less than half the time compared with FTF visits.

Table 18.2. (Continued)

Authors/ Journal	Objetive	Design/Sample	Teleconsultation technical aspects	Conclusions
Ball and Puffet J Telemed and Telecare 1998	Assess cognitive function in the elderly using VC.	8 subjects with a mean age of 73 years. Structured interview (the CAMCOG test) by VC and FTF, by an investigator blind to the results of the test in the other mode. Reassessments were carried out within one week of the initial assessment.	VC equipment was based on PC units (PictureTel) connected by ISDN 128 kbps.	11 subjects were initially involved in the study and eight completed both modes. The number of patients in this study is very small but the results suggest the feasibility of using the cognitive impairment test over a VC system without major modification.
Ruskin et al. Psychiatr Serv 1998	Examine the reliability of Psychiatric diagnoses.	Randomized study were 15 subjects randomly assigned to two FTF interviews, one by each physician; and 15 subjects were assigned to one FTF and one intervie via VC, one by each physician).	Two VC units (Omega Flex Plus, VSI) were used. The system was set up to use half T1 transmission speed (768 Kbps).	For each diagnosis, interrater reliability was identical or almost identical for the patients who had two FTF interviews and those who had an FTF interview and an interview via VC.
Trott and Blignault J Telemed and Telecare 1998	Cost comparison associated with delivering a mental health service by telepsychiatry and by FTF.	50 cases/month (40 cases for general adult psychiatry, four for child and adolescent mental health, four for psychology and two for forensic services).	VC system (PCS100, Picture Tel) with a superior camera and ISDN line connection.	The results of the study showed high savings from reduced travel by patients and health-care personnel.
De las Cuevas et al. Telemedicine and E-Health 2006	Evaluates the efficacy of telepsychiatry through VC in the treatment of mental disorders by comparing to F2FC treatment	Randomized clinical trial where 140 psychiatric outpatients were randomized to either F2FC treatment or VC telepsychiatry treatment	VC Polycom Viewstation 512	This study demonstrated that telepsychiatry treatment through VC has equivalent efficacy to F2FC psychiatric treatment. Telepsychiatry showed to be an effective mean of delivering mental health services to psychiatric outpatients living in remote areas with limited resources

Given the reported efficacy of telepsychiatry and teleneurology, the question that easily arises is why are them not more widely practiced? There are three main reasons. First, potential teleconsultants have to be able to deal with a technology, despite its wide adoption in society, is still relatively underused in healthcare, and many doctors are not familiar or comfortable with its use. Some health professionals are affected by "technophobia". Videoconferencing equipment also depends on the availability of digital lines, which can be expensive, although this situation is changing with the advent of different types of broadband connection, and videoconferencing might also become integrated with personal computers. The second problem is that both telepsychiatry and teleneurology require a change in the way in which health professional practice. New telecommunications technologies had profoundly changed the spatial and temporal relationship between health professional and patient and make necessary a challenge in the "behavioral inertia", the reluctance to change habits that are ingrained as a result of years of practice. The willingness of health professionals to change their practice has also been severely strained by the continual changes in health-care systems around the world, which might have made them reluctant to embark on yet another change unless it brings them obvious benefits. In telepsychiatry and teleneurology, the benefits accrue mostly to the patients rather than to the psychiatrists and neurologists, and this might be one reason why it has not gained more rapid acceptance. As the main beneficiaries are the patients, political forces might be expected to drive the wider introduction of these telemedicine applications. In our experience, they have so far failed to do so. Lastly, some doctors might be concerned about medico-legal issues if they use these healthcare technologies.

The literature shows that some health professionals considered telemedicine applications via videoconference to be impersonal and expressed a reluctance to use it, as well as their concern about confidentiality. It is clear that videoconferencing, like all tools, can be used poorly or well, and requires that the user learn a set of basic rules and etiquette to maximize the equipment's capabilities. Nevertheless, it is probably necessary to accomplish future studies to analyze what styles are the more adequate to use this type of technology. We cannot forget the possible relevance of economic factors, mainly the fear that telemedicine threatens "business-as-usual" and will generate "turf wars".

Apart from clinical applications, videoconferencing brings enormous opportunities for education, research, and administration (Hilty et al., 2004). Health care professionals need to continue their medical education, keep current on new medications and procedures, and refresh skills and knowledge. Videoconferencing is being used as an effective way to enhance and expand the learning environment (Loera et al., 2007; Al-Qirim, 2007). Eliminating the financial and geographical barriers of education delivery, can ensure all healthcare providers will be able to maintain competencies and stay current on advances occurring at larger institutions (Rossaro et al., 2007). Videoconferencing also permits those in distant sites to participate in meetings without the financial and time costs of traveling to the location where the meeting is being held. Videoconferencing is currently being used in proceedings for guardianship and advocacy tribunals, and Mental Health Review Tribunals. It is also being used in service delivery settings for staff support, debriefing, organizational management and clinical development.

If we consider the hype cycle of telepsychiatry and teleneurology, the maturity of these telemedicine applications are in the "slope of enlightenment" phase where focused experimentation and solid hard work by an increasingly diverse range of organizations lead to a true understanding of the technology's applicability, risks and benefits. At the present time, commercial, off-the-shelf methodologies and tools ease the development process. Telepsychiatry is nearer than teleneurology to reach the "plateau of productivity" phase where the benefits of it become widely demonstrated and accepted and tools and methodologies will be increasingly stable as they enter their second and third generations. In next future, growing numbers of organizations will feel comfortable with the reduced level of risk, and the rapid-growth phase of adoption will begin.

THE FUTURE

The evaluation of the videoconferencing encounter context in telepsychiatry and teleneurology has focused on its ability to simulate real-time experience in terms of image and interaction (Hilty et al., 2003). Telemedicine researchers have manifested their worries about the information cues present in the physical environment that could not be available in the virtual environment created when a videoconference system is used to bridge remote locations (Turner 2001; De las Cuevas and González de Rivera, 2004). Since this fact could have unknown effects on decisions made in these environments, the study and analysis of the concepts of presence and telepresence become necessary in an attempt to clarify the possible limitations of this welfare modality.

With the introduction of new 3D videoconferencing systems these worries will be partly overcome, since new 3D telepresence systems will give the ability to hold multi-site meetings where all the participants have the feeling that they are actually in the same room. The advantages over traditional videoconferencing and telepresence systems seems clear and include:

Table 18.3. Differences and advantages of software based high definition video collaboration solutions in comparison towards "traditional" hardware based Videoconferencing solutions

	LEGACY VIDEOCONFERENCING (small room equipment)	VIDEOCOLLABORATION SOFTWARE
FEATURES and FUNCTIONALITIES	SW presents a complete Integration of all set of services (Video, Voice and Data) allowing flexible media and conferencing switching. Includes a Presence Engine and Encrypted Instant Messenger system. Sessions can be easily recorded, labeled and storage which creates a very valuable data base of video-documented real cases archive.	
HQ Video	YES	YES
HQ Audio	YES	YES
Real time Sharing: Images and Documents	YES, but requires additional PC	YES
Integrated, Real Time Video files presentation	NO	YES
Files Transfer	NO	YES
Sessions Recording	NO, requires additional Hardware	YES
Session Recording Distribution	NO, requires additional Hardware	YES
Presence and Instant Messenger	NO, requires integration to 3rd party software	YES
TECHINICAL REQUIREMENTS	It is a myth that software based solutions cannot deliver a High Video Quality. Provided with the same dedicated network conditions, software solutions can provide as good quality as Legacy Videoconferencing; even better when bandwidth is scarce or functioning over the Internet. Software based solutions can also use the same peripherals (sound stations systems, or high definition cameras) as Legacy Videoconferencing or just basic headsets and webcams when the usage case allows it.	
Bandwidth (6 concurrent users)	5 MB in multipoint	2 MB in multipoint

	LEGACY VIDEOCONFERENCING (small room equipment)	VIDEOCOLLABORATION SOFTWARE
Bandwidth (1 to 1)	384 Kbps	256 Kbps
Hardware (HW)	Dedicated HW equipment Additional PC for data sharing	Standard PC
Devices	HQ video camera Sound Stations	Webcam or HQ video camera Head Set with Micro or Sound Stations
ESCALABILITY	Legacy videoconferencing requires intensive additional investments when scaling punctual experiences or pilots to a daily usage service for Care Delivery Organizations. It also requires strong re-dimensioning of network infrastructure. And deployment takes minimum 2 days per site. Software video collaboration works on the existing equipment and infrastructures (average standard PC´s and networks of any current Care Delivery Provider). Deployment is centralized and distribution is immediate with very few additional hardware requirements (PC, webcams and Headsets).	
Maximum number of connected concurrent points without additional HW	6 = three 1 to 1 sessions	100 = 50 one to one sessions
Additional deployment 5 users	Additional HW = MCU	Additional HW= Standard Server
Time to deploy	Min. 2 days per site	Max. 2 hours per site
PRICING (AVERAGE)	3.000 EUROS per fix point	800 Euros per user from any point
OTHER ADVANTAGES OF SOFTWARE BASED SOLUTIONS		
Interconnection:	Software solutions interconnect transparently with existing communication systems like PBX or even other Legacy Videoconferencing Systems.	
Flexibility	Software solutions work optimally on dedicated networks, but are also reliable over the Internet for many clinical applications. Their adapatative bandwidth management will always ensure the best available service quality. Besides that allows the connection of multiple devices like PC, PDA, Mobile Phones on the same session which adapts to any remote user	
Usabilityand Accessibility	Software solutions are user friendly and require very little expertise to become an advance user. Since they are web based accessibility is immediate at a click on the desktop or mobile device.	
INTEGRATION	Software solutions easily integrate with other important applications like Patient Data Records, Hospitals CRM and ERP.	

See eye-to-eye – with real eye-to-eye contact, rather than eye-to-camera, remote participants can engage with each other much more intimately and meaningfully than has previously been possible. This deeper level of interaction engenders deeper understanding, collaboration and motivation.

Pick up all the signs – the lifesize, three-dimensional 'projections' ensure that communication is far more than verbal. Just as in same-room meetings, body language is visible for all to see, interpret and react to. This leads to more open and transparent exchanges and, ultimately, more informed business decisions.

Get closer – by creating previously unimagined in-room presence, 3D TelePresence system provides an atmosphere in which personal interaction can achieve its full potential. Promoting stronger working relationships, the solution equips businesses with an exceptionally powerful tool for developing cross-department, cross-regional and cross-continental knowledge exchange and objective sharing. Telepsychiatry and teleneurology are here to stay and is likely to play an increasing role in future health care where every psychiatrist and neurologist may become “telecarers”, at least for part of their working time, delivering, responsive, high-quality information services, and supports remote patients or clients by using the most appropriate communication technology. The future of telemedicine holds innumerous opportunities in the healthcare field. As the population continues to grow and the paucity of qualified professionals widens, telemedicine clinical applications can assist in providing prompt and appropriate medical attention. The applications for telemedicine will dramatically grow as people become more familiar with and accepting of the technology. However, it is essential to remember that telemedicine does not represent a separate medical specialty. Telemedicine is a tool or a series of tools, a family of enabling technologies. It is simply a way of providing health care and is not an end it itself. Finally, a few words about the visionary present that involve software based high definition video collaboration solutions. A new world of possibilities has opened with the arising of specialized video collaboration software vendors and the enhancement of the IP networks and consequently the Internet. In general, a software solution brings full flexibility to enable any part of the medical processes with real time collaboration and communication services. That means increasing access to physicians and specialized knowledge, and facilitating Real Time and multipoint interactivity between patients and the medical care organization anytime that might be required. Furthermore, this can be achieved in a very simple manner, with no technological complexity for the end users.

Consequently, in any medical process where a software solution can be integrated, the main benefits for the Care Delivery Organization - when applied in a proper way- are: diminishing of processes latency, gain of efficiency, costs reduction and increase of frequency of interaction with patients or among physicians. We have studied the case of dialcom, one of the leading vendors, whose solution Spontania E-health “right care, right time, right place” is already being successfully used in Teleneurology (Teleictus), Telecare (dependent people), Teleoncology, Telenefrology, Teledermatology, as well as doctors Tele-training and emergencies group coordination.

It is also the chosen solution for some Military Agencies ought to its ability to deliver satisfactory video quality over satellites networks using very little bandwidth. Beside this, Dialcom- whose Telemedicine´s Division claim is “moving data, not patients”- is recognized as a Visionary vendor by Gartner, the leading IT consultancy firm. The main differences and advantages in comparison towards “traditional” hardware based Videoconferencing solutions that can be stated are described in the following table (Insert Table 18.3 about here).

References

Ahmed SN, Mann C, Sinclair DB, Heino A, Iskiw B, Quigley D, Ohinmaa A. Feasibility of epilepsy follow-up care through telemedicine: a pilot study on the patient's perspective. *Epilepsia* 2008, Apr;49(4): 573-85. Epub 2007 Dec 11.

Al-Qirim, N. Realizing telemedicine advantages at the national level: Cases from the United Arab Emirates. *Telemedicine Journal and E-Health* 2007, 13 (5): 545-555.

Ball C, Puffett A. The assessment of cognitive function in the elderly using videoconferencing. *Journal of Telemedicine and Telecare* 1998, 4 Suppl 1: 36-8.

Chua R, Craig J, Wootton R, Patterson V. Randomised controlled trial of telemedicine for new neurological outpatient referrals. Journal of Neurology, *Neurosurgery and Psychiatry* 2001, Jul; 71(1): 63-6.

Craig J, Chua R, Wootton R, Patterson V. A pilot study of telemedicine for new neurological outpatient referrals. *Journal of Telemedicine and Telecare* 2000, 6(4): 225-8.

Craig J, Patterson V, Rusell C, Wootton, R. Interactive viseoconsultation is a feasible methof for neurological in-patient assessment. *European Journal of Neurology* 2000, 7(6): 699-702.

Cullum CM, Weiner MF, Gehrmann HR, Hynan LS. Feasibility of telecognitive assessment in dementia. *Assessment*. 2006, Dec;13(4): 385-90.

De las Cuevas C, Artiles J, De la Fuente J, Serrano P: Telepsychiatry in the Canary Islands: user acceptance and satisfaction. *Journal of Telemedicine and Telecare* 2003, 9 (4): 221-224.

De las Cuevas C, Artiles J: Process Quality Analysis of Telepsychiatry. Contributions of Statistical Control Process and Critical Pathway Analysis. *Neuropsychiatrie* 2004, 18 S2: 100-105.

De las Cuevas C, González de Rivera JL: Some considerations about the concept of presence in telepsychiatry. *Neuropsychiatrie* 2004, 18 S2: 112-115.

De las Cuevas C, Arredondo MT, Cabrera MF, Sulzenbacher H, Meise U. Randomized Clinical Trial of Telepsychiatry through Videoconference versus Face-to-Face Conventional Psychiatric Treatment. *Telemedicine Journal and E-Health* 2006, 12(3): 341-350.

Dorcey T. CU-SeeMe Desktop VideoConferencing Software. Connexions, 9 (3), March 1995.

Doze S, Simpson J, Hailey D, Jacobs P. Evaluation of a telepsychiatry pilot project. *Journal of Telemedicine and Telecare* 1999, 5(1): 38-46.

Elford R, White H, Bowering R, Ghandi A, Maddiggan B, St John K, et al. A randomized, controlled trial of child psychiatric assessments conducted using videoconferencing. *Journal of Telemedicine and Telecare* 2000, 6(2): 73-82.

Frueh BC, Deitsch SE, Santos AB, Gold PB, Johnson MR, Meisler N, and others. Procedural and methodological issues in telepsychiatry research and program development. *Psychiatric Services* 2000, 51: 1522–7

Handschu R, Littmann R, Reulbach U, Gaul C, Heckmann JG, Neundörfer B, Scibor M. Telemedicine in emergency evaluation of acute stroke: interrater agreement in remote video examination with a novel multimedia system. *Stroke* 2003, Dec; 34(12): 2842-6.

Hilty DM, Marks SL, Urness D, Yellowlees PM, Nesbitt TS. MD5 Clinical and Educational Telepsychiatry Applications: A Review. *Canadian Journal of Psychiatry* 2004, 49(1): 12-23.

Jarvis-Selinger S, Chan E, Payne R, Plohman K, Ho K. Clinical telehealth across the disciplines: lessons learned. *Telemedicine Journal and E-Health* 2008, Sep; 14(7): 720-5.

Kennedy C, Yellowlees P. A community-based approach to evaluation of health outcomes and costs for telepsychiatry in a rural population: preliminary results. *Journal of Telemedicine and Telecare* 2000, 6 Suppl 1: S155-S157.

Kirkwood KT, Peck DF, Bennie L. The consistency of neuropsychological assessments performed via telecommunication and face to face. *Journal of Telemedicine and Telecare* 2000, 6(3): 147-51.

Loera, J.A., Kuo, Y.F., and Rahr, R.R. Telehealth distance mentoring of students. *Telemedicine Journal and E-Health* 2007, 13 (1): 45-50.

Loh PK, Ramesh P, Maher S, Saligari J, Flicker L, Goldswain P. Can patients with dementia be assessed at a distance? The use of Telehealth and standardised assessments. *Internal Medicine Journal* 2004, May; 34(5): 239-42.

Lott IT, Doran E, Walsh DM, Hill MA. Telemedicine, dementia and Down syndrome: Implications for Alazheimer disease. *Alzheimer´s and Dementia* 2006, 2(3): 179-184.

McLaren P, Ball CJ. Telemedicine: lessons remain unheeded. *British Medical Journal* 1995, 310:1390–1

Meyer BC, Lyden PD, Al-Khoury L, Cheng Y, Raman R, Fellman R, Beer J, Rao R, Zivin JA. Prospective reliability of the STRokE DOC wireless/site independent telemedicine system. *Neurology* 2005, Mar 22; 64(6): 1058-60.

Mielonen ML, Ohinmaa A, Moring J, Isohanni M. Psychiatric inpatient care planning via telemedicine. *Journal of Telemedicine and Telecare* 2000, 6(3): 152-7.

Montani C, Billaud N, Tyrrell J, Fluchaire I, Malterre C, Lauvernay N, Couturier P, Franco A. Psychological impact of a remote psychometric consultation with hospitalized elderly people. *Journal of Telemedicine and Telecare* 1997, 3(3): 140-5.

Noorani H, Picot J. Assessment of videoconferencing in telehealth in Canada. Ottawa: Canadian Coordinating Office for Health Technology Assessment; 2001. Technology report no. 14.

Patterson V. Teleneurology. *Journal of Telemedicine and Telecare* 2005, 11 (2): 55-59.

Patterson V, Wootton R. How can teleneurology improve patient care? *Nature Clinical Practice Neurology 2006*, 2 (7): 346-347

Rossaro, L., Tran, T.P., Ransibrahmanakul, K., Rainwater, J.A., Csik, G., Cole, S.L., et al. Hepatitis C videoconferencing: The impact on continuing medical education for rural healthcare providers. *Telemedicine Journal and E-Health* 2007, 13 (3): 269-277.

Ruskin PE, Reed S, Kumar R, Kling MA, Siegel E, Rosen M, et al. Reliability and acceptability of psychiatric diagnosis via telecommunication and audiovisual technology. *Psychiatric Services* 1998, 49(8): 1086-8.

Shafqat S, Kvedar JC, Guanci MM, Chang Y, Schwamm LH. Role for telemedicine in acute stroke. Feasibility and reliability of remote administration of the NIH stroke scale. *Stroke*, 1999, Oct; 30(10): 2141-5.

Shores MM, Ryan-Dykes P, Williams RM, Mamerto B, Sadak T, Pascualy M, Felker BL, Zweigle M, Nichol P, Peskind ER. Identifying undiagnosed dementia in residential care veterans: comparing telemedicine to in-person clinical examination. *International Journal of Geriatric Psychiatry* 2004, Feb; 19(2): 101-8.

Theodoros D, Rusell TG, Hill A, Cahill L, Clark K. Assessment of motor speech disorders online: a pilot study. *Journal of Telemedicine andTelecare* 2003, 9 suppl. 2: S66-S68.

Trott P, Blignault I. Cost evaluation of a telepsychiatry service in northern Queensland. *Journal of Telemedicine and Telecare* 1998, 4 Suppl 1: 66-8.

Wang S, Lee SB, Pardue C, Ramsingh D, Waller J, Gross H, Nichols FT 3rd, Hess DC, Adams RJ. Remote evaluation of acute ischemic stroke: reliability of National Institutes of Health Stroke Scale via telestroke. *Stroke* 2003, Oct; 34(10): e188-91. Epub 2003 Sep 18.

Wiborg A, Widder B, Riepe MW, Krauss M, Huber R, Schmitz B. Contribution of telemedicine to global stroke care in rural areas. *Aktuelle Neurologie* 2000, 27(3): 119-124.

Wittson CL, Affleck DC, Johnson V. Two-way television in group therapy. *Mental Hospitals* 1961, 2 :22-23.

Zaylor C. Clinical outcomes in telepsychiatry. *Journal of Telemedicine and Telecare* 1999, 5 Suppl 1: S59-S60.

In: Handbook of Technology in Psychology ...
Editor: Luciano L'Abate and David A. Kasier
ISBN: 978-1-62100-004-4

Chapter 18

WORKBOOKS: PROGRAMMED INTERACTIVE PRACTICE EXERCISES

Luciano L'Abate
Georgia State University, Atlanta, Georgia, US

Workbooks are a specialized technology of relatively late vintage that developed over the last two decades. Technology implies or denotes specialization in mechanical equipment, including computers and programs that offer systematic evaluation and treatment. By specialization is meant specificity in evaluation and treatment through a particular equipment or program produced to deal with a specific condition (L'Abate, 2010, in press-a). By systematic is meant a step-by-step process that allows replicability in standard operating procedures and practices. Both specificity and systematic, replicable procedures are the cornerstone of the scientific method. A procedure or program could be replicable and systematic but without specificity it may not produce the desired outcome (L'Abate, 2008b). Furthermore, science progresses with the accumulation of written records not just words (Beard, Myhill, Riley, and Nystrand, 2009; L'Abate, 2007b).

As long as talk remains the only medium of communication, as in face-to-face psychotherapy and preaching, it will be practically impossible to obtain cumulative records and progress on the basis of data that can be obtained only through writing not talk (L'Abate, 1999, 2011). Given these definitions, Workbooks defined as Programmed Interactive Practice Exercises (PIPEs) qualify eminently as a mental health technology that has emerged and surfaced during the last generation (L'Abate, 1986, 1990, 1991,1992, 1997, 2001, 2002, 2004a, 2004b, 2007c, 2008a, 2008c, 2011c; L'Abate and De Giacomo, 2003; L'Abate and Goldstein, 2007; L'Abate and Kern, 2002).

Together with recent advances in computer technology, the emergence of the Internet (Chapter 20 this volume), the growing importance of homework assignments (Kazantzis, Deane, Ronan, and L'Abate, 2005; Kazantzis and L'Abate, 2007), the growth of self-help (Harwood and L'Abate, 2010), and emergence of low-cost approaches to promote physical and mental health in their delivery (L'Abate, 2007d), there are now plenty of choices available to both professionals and participants in the mental health field, in promotion of health, and the prevention and psychotherapy of mental illness.

The issue remains in finding ways to evaluate participants to help them find which approach will be more helpful and less costly to them than other approaches (L'Abate, 2011a). These approaches would suggest targeted or step-by-step treatment, starting from the least expensive of professional help first and then moving on to the most expensive treatment, that is: face-to-face (f2f) talk-based (tb) contacts with professionals, especially if least expensive approaches are not working. For instance, depression in its least severe forms (determined by scores on the Beck Depression Inventory), could be dealt with first by using much less expensive activities such as running, physical therapy, yoga, or medi-tation. Somewhere in between least and most expensive, f2f tb approaches (Chapter 2, this volume), interactive practice exercises in Workbooks could be used either on their own, or in conjunction with least or with most expensive approaches.

Workbooks essentially perform as standardized *recipes*, furnishing professionals with specific ways of helping and intervening that can be replicated from one participant to another and from professional to another. There are no limits on how many times these Workbooks can be reproduced and administered to as many participants as possible (L'Abate, 2011c), just like recipes from a cookbook. If one particular Workbook does not work for a specific condition, let's say depression, one would have to: (1) question the initial reason for referral; (2) evaluate historical and developmental background for the referral; (3) evaluate results from a test battery administered even before seeing (or not seeing) the participant; (4) see whether another, different Workbook for depression may be more appropriate than the one originally administered; or (5) administer a completely different, not depression-related Workbook that may be more appropriate for the context of the referral, such as one related to over-all functioning rather than a specific Workbook related to a diagnosis or to the reason for the referral, as would be the case with Planned Parenting or Negotiation Workbooks (Figure 1).

The use of Workbooks in no way eliminates the important functions of professionals. On the contrary, a distance-writing approach requires that more attention be given to an objective evaluation for the referral reason, concern, or symptom. An objective evalua-tion is an aspect that may differentiate between professionals who chose to practice as artists from professionals who chose to practice as scientists, evaluating participants from the very outset of a professional relationship, after termination of that relationship, and on follow-up (L'Abate, 2011a). In addition to an evaluative function, the profess-sional/scientist needs to evaluate responses to each single Workbook and feedback appro-priate and specific information about the nature of those responses to participants. This function is implicit in the interactive nature of the process. Consequently, this conviction (L'Abate, 2011a) would not allow a professional to administer a entire series of Workbooks without breaking them down into one assignment after another, one assignment at one time, with corrective feedback on completion of each assignment, after an Informed Consent Form about the importance of pre-post- and follow-up evaluation is emphasized from the outset of the professional relationship (L'Abate, in press).

The Nature of Workbooks

Workbooks as programmed interactive practice exercises are composed by four characteristics: (1) administered in writing at a distance from professionals, with (2) participants in continuous interactions with professionals who (3) request consistent, reliable practice exercises, such as homework assignments, and who (4) tryto match a specific Workbook with the specific referral question, concern, diagnosis, or symptom.

Programmed

Workbooks are written according to systematic procedures outlined in Figures 19.1 and 19.2 to be administered at a distance between mental health professionals and participants, be participants composed of individuals, couples, or families. In the past, this distance may have been covered though regular mail. Nowadays, distance is covered through computers and the Internet (See Chapter 20 this volume). Being written means that these exercises are completely replicable and extremely specific for determined situations or disorders, as shown in self-explanatory Figures 19.1 and 19.2. There is no need to expand on these Figures here because they have been already expanded in previous and present publications cited above (L'Abate and Sweeney, 2011).

1. *Structured-Unstructured* can vary from *open*, as in diaries or *closed*, as in answering specific questions in writing on pre-established topic, either positive, happy or unhappy memories. Closed writing can be guided or programmed, as discussed in Figure19.2.
2. *Goals* could be *prescriptive* to produce specific outcomes or *cathartic* to produce discharge of and release from tensions or traumas.
3. *Content* can be painful, traumatic, neutral, banal, or joyous, emotional, rational, problem-solving, individual, or relational/
4. *Level of abstraction* ranging from very concrete, i.e., "Write what you ate for breakfast, to very abstract, i.e., "Write about the meaning of life."
5. *Specificity*, ranging from general, as in an autobiography or: "Write about the meaning of life," to extremely specific, concrete, and restricted, such as: "Write about all the clothes you have in your closet."

* Adapted from L'Abate (2004a).

Figure 19.1. Toward a Classification of Distance Writing Dimensions.*

By this qualification is meant that these exercises cannot and should not be administered without continuous corrective feedback between a professional and a participant. Inter-active means that Workbooks are or should be administered after a thorough, objective evaluation of the referring concern, question, or referral reason (L'Abate, 2011a). Furthermore, Workbooks should be administered after an Informed Consent Form (ICF) has been read and signed by all participants. Participants should know what possible consequences or unpredictable side effects may occur in many instances, especially when dealing with hurt feelings (L'Abate, 2011b) and how important evaluation is in finding which approach is more appropriate for which condition.

Therefore, the ICF should be administered to protect both professionals and participants, since one cannot predict whether: (1) the reason for referral is a real and accurate one; (2) whether the evaluation uncovered contextual aspects of the referral reason that may have not been considered through a subjective interview. For instance, a participant may want help to deal with self-diagnosed anxiety but without an evaluation the professional may accept uncritically the participant's opinion and administer an anxiety-related Workbook. When anxiety does not decrease on termination of the entire Workbook the patient may complain about the professional's incompetence and threatens legal action. On the basis of this reaction, the professional may discover that there may exist a whole delusional, paranoid system that could have been detected from the outset with a thorough objective evaluation. An ICF represents the contract between professionals and participants about the use of Workbooks. Copies of such ICFs are available in L'Abate (2011c), including also a control Workbook with seemingly neutral items to be used for research purposes (L'Abate, 2011c). Whoever professionals choose to administer Workbooks without a thorough, objective evaluation, a signed ICF, and continuous, exercise-by-exercise professional feedback, should be considered artistis/ charlatans and should receive appropriate retribution from professional organizations in which they belong (L'Abate, 2011a; in press).

A. *Automatic* was of questionable usefulness as a fad and in need of more controlled research before considering its use.
B. *Dictionary-aided* basic to many self-help practice exercises (L'Abate, 2007a).
C. *Expressive*, as in "Pennebaker's Paradigm" writing about hurts and traumas heretofore not shared with others for 15 minutes a day for four consecutive days. The literature on this approach is extensive and available in many old (Esterling, L'Abate, Murray, & Pennebaker, 1999) and new references (Kacewicz, Slatcher, & Pennebaker, 2007; Lepore & Smyth, 2002).
D. *Focused*, as in autobiographies to be mailed or sent online (L'Abate, 2007c)
E. *Guided*, as in answering written questions in writing, after completion, for instance, of an autobiography, journal, or other homework assignments.
F. *Open-ended*, as in personal information gathered through diaries or journals.
G. *Programmed* as in Workbooks for non-targeted conditions and life-long learning and at risk populations (Figure 19.4) targeted and psychiatrically diagnosed children and youth, single individuals, couples, and families (Figure 19.5), single- or multiple-score tests (Figure 19.6).

* Adapted from L'Abate (2004b; see also L'Abate and Sweeney, 2011).

Figure 19.2. A Classification of Distance Writing.*

Practice

Professionals should be careful to specify how, when, how often, and where Workbooks should take place (L'Abate, 1986), making sure that participants record place and time of occurrence and checking on complete homework before accepting it as given and administering the next assignment. For instance, in my experience with acting-out inmates, some of them tried to get away with as little work as possible. Consequently, they were required to repeat writing the same exercise until it showed that responsibility was taken to complete the assignment in the best possible manner. Incomplete or irrelevant answers are unacceptable and participants should not be allowed to get away with minimal performance. More instructions about this issue are available in L'Abate (2011c).

In a way Workbooks in and of themselves, allow professionals to evaluate whether there is motivation for change in participants or whether they expect the professional to produce change miraculously without their active participation (L'Abate, L'Abate, and Maino, 2005). In stepped treatment, one could make homework assignments and completion of Workbooks as a pre-condition for seeing the professional f2f through talk (Omer, 1985). The expensive presence of a professional should also be contingent on the completion of the least expensive treatment. This approach is especially useful to evaluate motivation to change in Custer B participants, who may use professional help as a way to avoid criminal or legal persecution (L'Abate, van Eigen, & Rigamonti, 2011).

Exercises

Figure 19.3 includes a self-explanatory classification of Workbooks. As in the previous paragraphs, the interested reader can find expanded explanations of this and the two previous Figures in references already cited. One important aspect of Workbooks contained in Figures 19.3 lies in their being produced from existing, validated test instruments, verified factor analyses, or symptom lists available in the DSM-IV (American Psychiatric Association, 1994). This is another way to indicate that PIPEs are isomorphic with their underlying composition based on validated measures. Hence, the gap between evaluation and intervention using PIPEs is greatly diminished, allowing professionals to match treatment with evaluation in ways that would be difficult, expensive, or even impossible to accomplish in f2f tb psychotherapy.

1. *Composition* of Participants: singles (adults, children, youth), couples, families, groups.
2. *Reason for Referral,* i. e., concern(s), diagnosis(es), single versus dual or multiple, problem(s), symptom(s),
3. *Level and Type of Functionality*: DSM-IV, Reason for Referral, or both
   a. Functional: No diagnosis
   b. Externalizations: Axis II. Cluster B (e. g., Aboujaoude & Koran, 2010).
   c. Internalizations : Axis II. Cluster C (e.,g. Millon, Blaney, & Davis, 1999)
   d. Borderline: Axis II. Cluster A (e. g., Millon, Blaney, & Davis, 1999)
   e. Severe: Axis I. (e. g., e.g., Millon, Blaney, & Davis, 1999)
4. Practice exercises for *specific symptoms* versus general conditions
5. *Symptom-free* versus symptom-related & diagnosis-linked
6. *Theory-derived*, theory-related, theory-independent (Cusinato & L'Abate, in press; L'Abate, 2005; L'Abate & Cusinato, 2007; L'Abate, Cusinato, Maino, Colesso, & Scilletta, 2010).
7. *Format*: (1) fixed (nomothetic); (2) flexible (idiographic); & (3) mixed (nomothetic and idiographic)
8. *Style*: Linear versus circular (paradoxical, i. e., Weeks & L'Abate (1982)
9. *Derivation*: Single versus multiple score tests, i. e., BDI vs. MMPI-2.
10. *Content*: clinical (addictions, affective disorders, Axis I and Axis II: Clusters A, B, & C etc. and non-clinical, life-long learning for individuals, couples, and families

*Adapted from L'Abate, 2004b.

Figure 19.3. Classification of Workbooks in Mental Health.*

Transforming inert, passive test or factor analyses items or psychiatric symptoms into dynamic Workbooks is a very simple process that apparently has not been thought of until recently (L'Abate, 2011c). This process occurs in three easy steps: (1) asking participants to define items of single score tests, or dimensions of multiple score tests, factor analyses, and DSM-IV symptoms lists with the help of a dictionary (L'Abate, 2007a), and give two, possibly concrete, examples of the item just defined, a nomothetic task; (2) rank-order items just defined according to how those items refer to oneself in order of importance and relevance, an idiographic task that individualizes immediately the sequence to follow in treatment, starting with the item that has been rank-ordered first, moving then to the item rank-ordered second, and so on, until most of the top items (no more than 5 or 6 in my experience) have been administered one by one; and (3) administer a standard practice exercise (L'Abate, 2011c) where the top item becomes the subject of an intense scrutiny, that is: asking participants to answer queries about the overall nature of that item, its developmental origins, duration, frequency, rate, and functional and dysfunctional outcomes to self and to loved ones.

Sometimes, if the behavior represented in a particular item is strongly ingrained in the individual's affective, cognitive, and behavioral repertoire, participants may be asked to produce and reproduce it at certain times and places to achieve greater control over it, following the principle that: "If we learn to approach and start a troublesome behavior, we can learn to stop it." (L'Abate, 1984), contrary to many behavioral dogmas.

Through this process, therefore, as shown in Figures to follow, it is possible to transform thousands of passively inert test instruments into active and interactive methods of treatment, linking evaluation with treatment in ways that would be difficult and expensive to achieve in self-help, preventive, promotional, or psychotherapeutic interventions using f2f tb approaches.

Workbooks for Preventive and Health Promotional Life-long Learning

Figure 19.4. includes Workbooks that can be used when no specific clinical concern is present or detected but there is interest and motivation to learn more about oneself in individuals, couples, or families.

Individuals

Character Strengths: This workbook was based on the work of Peterson and Seligman (2004).

Emotional Expression: This workbook was based on a list of feelings developed by Bonanno and Keltner (1997).

Development of Emotional Competence: The contents of this workbook were developed from a variety of sources (Averill, 1997; Cacioppo & Gardner, 1999; Gottman, Feinsilber, & Hooven. 1997; Kennedy-Moore & Watson, 1999; L'Abate, 2002, p. 116; Mascolo & Griffin, 1998).

Multiple Abilities: This workbook was developed from various theories of intelligence (Bodine & Crawford, 1999; Campbell, Campbell, & Dickinson, 1999; Fogarty & Bellanca, 1998; Gardner, 1999; Mayer & Salovey, 1995; Sternberg, 1996; Sternberg & Kaufman, 1998).

Normative Experiences: Form AB5C: This workbook was based on the work of L. E. Goldberg (1999) and published with his expressed permission.

Normative Experiences: Form IPIP: This workbook was based on the work of L. E. Goldberg (1999) and published with his expressed permission.

Normative Experiences: Form PSC: This workbook was based on the work of L. E. Goldberg (1999) and published with his expressed permission.

Priorities: This workbook was based on a relational model of priorities that was part of a contextual theory of personality socialization (Cusinato & L'Abate, 2012; L'Abate & Cusinato, 2007; L'Abate, Cusinato, Maino, Colesso, & Scilletta, 2010; L'Abate & De Giacomo, 2003).

Self-awareness: This workbook was based on a model of relationships developed by Piero De Giacomo translated into English and expanded by L'Abate (2002, pp. 117-118; L'Abate & De Giacomo, 2003).

Self- Others Importance: This workbook was developed from a Selfhood model that makes the attribution of importance to self and intimate others the cornerstone of intimate and non-intimate relationships (L'Abate & Cusinato, 2007; L'Abate, 2005; L'Abate et al., 2010; L'Abate & De Giacomo, 2003).

Social Skills: This workbook was developed from the Inventory by the same name developed by Lorr, Youniss, and Stefic (1991).

Speak UP For Yourself: This workbook was based on the need to help fearful or insecure partners express themselves assertively in their intimate relationships without anger or aggression.

Who Am I? Aspects of Identity Formation: This workbook was based on a relational model of identity-differentiation that is part of a contextual theory of personality socialization developed (Cusinato & L'Abate, in press; L'Abate, 2005; L'Abate & Cusinato, 2007; L'Abate et al., 2010; L'Abate & De Giacomo, 2003).

Couples

Premarital Preparation: This workbook was developed from clinical experience with couples, indicating the crucial importance of marital preparation for many to-be married or not-yet-married couples.

Relational Quality1: This workbook was developed from the research by Hassebrauck and Fehr (2002).

Relational Quality2: This workbook was developed from the research by Hassebrauck and Fehr (2002).

Relational Quality3: This workbook was developed from the research by Hassebrauck and Fehr (2002).

Relational Styles: This workbook was based on the Elementary Pragmatic Model of relationships developed by De Giacomo (L'Abate & De Giacomo, 2003).

Sexual Motivation: This workbook was developed from the research by Schachner and Shaver (2004).

Families

Foster/Adoptive Care: This workbook was developed by Mario Cusinato after an extensive review of the literature on adoptive parenting and translated by L'Abate.

Planned Parenting: This workbook was based on models that compose RCT (Cusinato & L'Abate, in press; L'Abate, 2005; L'Abate et al., 2010)

Figure 19.4. Self-help Workbooks for Individual, Couple, and Family life-long Learning*

These Workbooks, therefore, could be used in self-help (Harwood and L'Abate, 2010), prevention (L'Abate, 1990), and promotional (L'Abate, 2007d) activities with functional or semi-functional participants, especially those who are at risk for possible breakdown, such a adult children of alcoholic parents. These individuals, couples, and families may not be classified as yet according to a psychiatric label, but they may evidence prodromal characteristics that may foretell future disorders, especially in children.

Workbooks for Psychiatric Classification

Figure 19.5 includes Workbooks that were derived from list of symptoms in psychiatric diagnoses found in the DSM-IV (Author, 1994), Millon, Blaney, and Davis (1999) and other sources, showing that through programmed, distance writing it is possible to link evaluation with specific treatment in ways that would be difficult if not impossible to achieve through f2f, tb interventions.

Disorders of Internalization

Children and Youth

Anxiety: This workbook was based on the scale by the same name developed by Newcomer, Barenbaum, and Bryant (1994).

Anxiety, Depression, and Fears: This workbook was based on the work by Chorpita, Albano, and Barlow (1998).

Asperger Disorder: The list of symptoms used for this workbook was taken from the Diagnostic Statistical Manual f or Mental Disorders-IV (DSM-IV). Washington, DC: American Psychiatric Association.

Depression: This workbook was based on the Scale by the same name developed by Newcomer, Barenbaum, and Bryant (1994).

Post-Traumatic Stress: This workbook was based on the work of Anthony, Lonigan, and Hecht (1999).

Separation Anxiety: This workbook was based on the list of symptoms found in the *Diagnostic Statistical Manual for Mental Disorders-IV* (DSM-IV, 1994), Washington, DC: American Psychiatric Association.

Adults

Anxiety: The list of symptoms used for this workbook was taken from the *Diagnostic Statistical Manual f or Mental Disorders-IV (*DSM-IV, 1994). Washington, DC: American Psychiatric Association.

Depressive Personality: This workbook was based on the research by Hartlage, Arduino, and Alloy (1998) about depressive characteristics.

Loneliness: Although loneliness was not a psychiatric dimension or category, this condition affects a great many people who suffer also from other conditions. One could say that loneliness was a concomitant of many psychiatric and non-psychiatric conditions, especially in youth. These are the reasons for including this workbook in this Section. This workbook was developed from the work of Hurtug, Audy, and Cohen (1998) and of Peplau and Perlman (1982). A questionnaire found in Peplau and Perlman (1982) could be administered before and after completion of this workbook.

Phobias: This workbook was developed from the combined work of Levinson (1986) and Henley (1987).

Post-Traumatic Stress Disorder: This workbook was based on the factor analysis of research performed by King, Leskin, King, and Weathers (1998) as well as by Taylor, Koch, Koch, Crockett, & Passey (1998).

Procrastination: This workbook was derived from the work of Ferrari, Johnson, and McCown (1995).

Signs of Depression: This workbook was based on a variety of clinical and research sources.

Disruptive Developmental Disorders

Anger: This workbook was based on the factor analysis of Lahey, Frick, Loeber, Tannenbaum et al. (1990) and the work of Feindler (1995).

Conduct Disorder: The list of symptoms used for this workbook was taken from the *Diagnostic Statistical Manual f or Mental Disorders-IV* (DSM-IV, 1994). Washington, DC: American Psychiatric Association.

Hyperactive/Attention Deficit: The list of symptoms used for this workbook was taken from the *Diagnostic Statistical Manual f or Mental Disorders-IV* (DSM-IV, 1994). Washington, DC: American Psychiatric Association.

Juvenile Troublemaking: This workbook was based on the Child Troublemaking Scale developed by Lynam (1997).

Oppositional Defiant: The list of symptoms used for This workbook was taken from the *Diagnostic Statistical Manual for Mental Disorders-IV* (DSM-IV, 1994). Washington, DC: American Psychiatric Association.

Disorders of Externalization in Adults

Addendum to Relational Training: This workbook was an elaboration of the original workbook on Social Training (L'Abate, 1992).

Anger, Hostility, Aggression: This workbook was based on the work by Eckhardt and Deffenbacher (1995) and by Spielberger, Reheiser, and Sydeman (1995).

Troublemaking: This workbook was based on the research by Kosson, Steuerwald, Forth, and Kirkhart (1997).

Workbooks for Conflicting Couples and Families

Children and Families

Bing Eating: This workbook was developed in collaboration with Monique Gray and published with her kind permission.

Divorce Adjustment in Children: This workbook was developed by Karin B. Jordan, Ph. D., and published here with her kind permission.

Domestic Violence: This workbook was developed in collaboration with Fran Greenfield and published here with her kind permission.

Lying: This workbook was developed in collaboration with Maureen O'Toole and published here with her kind permission.

Shyness: This workbook was developed in collaboration with Sue Matthews and published here with her kind permission.

Stealing: This workbook was developed in collaboration with Ernestine Williams and published here with her kind permission.

Temper Tantrums: This workbook is a substantial revision of a previously published workbook by the same name (L'Abate, 1992).

Time Out: This workbook was based on the first author's clinical experience and one of the first Workbooks developed for children and their families.

Verbal Abuse: This workbookwas developed from the first author's clinical experiences and considered as part of the Abusive-Apathetic (AA) style in RCT (Cusinato & L'Abate, 2012; L'Abate, 2005; L'Abate et al., 2010).

Couples

A theory-derived workbook for intimate couples. This workbook was derived from a structured interview covering the 16 models of RCT (Cusinato & L'Abate, 2012; L'Abate, 2005; L'Abate et al., 2010).

Arguing or Fighting: This workbook was developed from clinical experience and was based on a model of paradoxical psychotherapy presented by Weeks and L'Abate (1982; L'Abate, 2002, pp.124-125).

Complaints: This workbook was based on the research by Coyne, Thompson, and Palmer (2003)

Depression: This workbook was based on relational model of depression developed by L'Abate (1986).

Difficulties: This workbook was developed jointly by Keith Sanford and Luciano L'Abate from Sanford's research (2004).

Intimacy: This workbook was based on a relational Model[15] of intimacy developed by Cusinato and L'Abate (1994).

Sexuality: The contents of this workbook were taken from the vast literature on sexuality, with no primary source being more relevant than others.

Violence: This workbook was based on the research by Eckhardt, Barbour, and Davison (1998).

Figure 19.5. (Continued)

Families

Hurt Feelings: This workbook is based on a model developed by L'Abate (2011b).

Intimacy: This workbook was based on a relational Model[15] of intimacy developed by L'Abate defined as the sharing of joys and hurts as well as of fears of being hurt (1986, 1997, 2005, 2011b; L'Abate et al., 2010).

Negotiation: This workbook was based on a relational model of negotiation developed by L'Abate (Cusinato & L'Abate, 2012; L'Abate, 1986, 2005; L'Abate et al., 2010).

When Parents Argue: From the Child's Eyes: This workbook was developed from the research by Weston, Boxer, and Heatherington (1998).

When Parents and Child Argue: From the Child's Eyes: This workbook was developed from the research by Weston, Boxer, and Heatherington (1998).

*Adapted from L'Abate, 2011c

Figure 19.5.

Workbooks from Research with Single- or Multiple-Score Tests

Another way to link evaluation with treatment is found in developing Workbooks directly from items found in factor analyses, single- and multiple-score tests as shown in Figure 19.6.

Single Score Tests

Children: Mixed Internalizing and Externalizing Disorders

School Conduct Problems: This workbook was developed from the work of Brown and Hammill (1978), Kelly (1990), Kelly and Vitali (1992), and more general sources, such as Hersen and Last (1990) and Mash and Barkley (1996).

School Social Skills: This workbook was developed from the work of Brown, Black, and Downs (1984).

Unusual or Troublesome Behavior: This workbook was developed from Aman and Singh's (undated) *Checklist for Aberrant Behavior.*

Adults

Butcher Treatment Planning: This workbook was developed from the scale developed by Butcher (1998; Perry & Butcher, 1999) to evaluate individuals who are resistant to psychotherapeutic interventions.

Axis II Cluster C Internalizing Personality Disorders

Beck Anxiety: This workbook was developed from research by Joiner, Steer, Beck, Schmidt, Rudd, and Catanzaro (1999).

Beck Depression: This workbook was developed from research by Dozois, Dobson, and Ahnberg (1998) and the review by Katz, Katz, and Shaw (1999).

Hamilton Anxiety: This workbook was developed from research of Moras, Di Nardo, and Barlow (1992).

Hamilton Depression: This workbook was developed from research of Moras, Di Nardo, and Barlow (1992).

Help-seeking (Dependent): This workbook was based on lists of attributes for personality disorders developed by Davis and Millon (1999, p. 517).

Self-suffering (Masochist): This workbook was based on lists of attributes for personality disorders developed by Davis and Millon (1999, p. 517).

Serious (Depressed): This workbook was based on lists of attributes for personality disorders developed by Davis and Millon (1999, p. 517).

What Are My Concerns? This workbook was developed from Blatt's et al., (1982) two factor model of depression: dependency and self-criticism.

Axis II Cluster B Selected Externalizing Personality Disorders

Anger Expression: This workbook was developed from the work of Forgays, Forgays, and Spielberger, (1997).

Compelling (Compulsive): This workbook was based on lists of attributes for personality disorders developed by Davis and Millon (1999, p. 517).

Distinct (Schizoid): This workbook was based on lists of attributes for personality disorders developed by Davis and Millon (1999, p. 517).

Exciting (Histrionic): This workbook was based on lists of attributes for personality disorders developed by Davis and Millon (1999, p. 517).

Non-conformist (Negativistic): This workbook was based on lists of attributes for personality disorders developed by Davis and Millon (1999, p. 517).

Private (Avoidant): This workbook was based on lists of attributes for personality disorders developed by Davis and Millon (1999, p. 517).

Self-interested (Narcissistic): This workbook was based on lists of attributes for personality disorders developed by Davis and Millon (1999, p. 517).

Unpopular (Antisocial): This workbook was based on lists of attributes for personality disorders developed by Davis and Millon (1999, p. 517).

Unusual (Sadistic): This workbook was based on lists of attributes for personality disorders developed by Davis and Millon (1999, p. 517).

Mixed Internalizing and Externalizing Disorders

Personality Disorders, Form 1989: This workbook was developed from research of Livesley, Jackson, and Schroeder (1989).

Personality Disorders, Form 1990: This workbook was derived from research of Clark (1990).

Personality Disorders, Form 1992: This workbook was developed from a factor analysis of Schroeder, Wormsworth, and Livesley (1992)

Multiple Score Tests

Personality Dimensions from the MMPI: This workbook was developed from the 15 Content Scales of the Minnesota Multiphasic Personality Inventory-2 (Greene & Anderson, 1999).

Personality Dimensions from Neuroticism Extraversion Openness Inventory (NEO): This workbook was developed from the work of McCrae (1994) and Costa (Costa & Widiger, 1994).

Personality Dimensions from Personality Assessment Inventory: This workbook was developed from the work of Morey (1999) and Morey and Henry (1994).

Severe Disorders

Individuals

Mood Swings: The list of symptoms used for this workbook was taken from the *Diagnostic Statistical Manual f or Mental Disorders-IV (*DSM-IV, 1994). Washington, DC: American Psychiatric Association.

Figure 19.6. (Continued)

Over-dependency: This workbook was based on a model of Codependency developed by L'Abate and Harrison (1992), which is part of a theory of relational competence and psychopathology in intimate relationships (Cusinato & L'Abate, in press; L'Abate, 2005; L'Abate et al., 2010).

Severe Concerns: The list of symptoms used for this workbook was taken from the *Diagnostic Statistical Manual for Mental Disorders-IV* (DSM-IV, 1994). Washington, DC: American Psychiatric Association.

Sexual Abuse: The contents of this workbook were taken from the vast literature on sexual abuse, with no single source being primary over others.

Couples

Improving Relationships: This workbook was developed by Brenda Whaley, Ph.D. (L'Abate, 2002, pp. 121-122) and published with her kind permission.

Relationship Conflict: This workbook was developed from Bodin's (1996) Relationship Conflict Inventory and was published with the kind permission of Dr. Bodin.

Couple Satisfaction: This workbook was developed from Snyder's Inventory by the same name (L'Abate, 2002, pp. 122-123; Snyder, 1981; Snyder & Aikman, 1999) and reprinted with the kind permission of Western Psychological Services.

Families

Profile Form: This workbook was developed from an instrument developed by L'Abate and Weinstein (1987) and validated by Kochalka and L'Abate (1997).

Functioning: This workbook was developed from the factor analysis of self-report measures reviewed by Bloom (1985).

*Adapted from L'Abate, 2011c.

Figure 19.6. Workbooks from Research with Single- and Multiple-score Tests.

RESEARCH RESULTS

Research about the effectiveness of Workbooks was conducted from the very beginning starting 30 or more years ago but reported more recently (L'Abate, Boyce, Fraizer, and Russ, 1992; L'Abate, 2004b, pp. 75-92) with unexpected paradoxical and sometimes inconsistent results. For instance, in "...few studies control groups obtained higher gain scores than experimental groups suggesting that this was an important area to investigate in the future (L'Abate, 2004b, pp. 91-92). To find a more definitive answer, Smyth and L'Abate (2001) performed a meta-analysis of 18 Workbooks, 12 in mental health and 6 in physical health. The overall effect size for mental health outcomes was $d = .44$ while the overall effect size for physical health outcomes was $d = .25$.

Results from this analysis as well as from the studies summarized in L'Abate (2004b, pp. 75-92) suggest that Workbooks may produce a medium effect size in mental health and a somewhat lower effect size for physical health, even though great variability was present for such outcomes. This effect size was quite similar to the one obtained from a meta-analysis of f2f tb psychological treatment in 90 studies conducted across a broad range of clinical settings (mean $d = .41$; Shadish, Matt, Navarro, and Phillips, 2000).

Here is where costs and cost-effectiveness come into being. If the same results are obtained by two different types of intervention, one would chose the less expensive intervention. This, however, is not a either-or but a both/and issue. Conceivably, f2f, tb interventions may needed with more severe cases and even in these cases, adding PIPEs may decrease the length of treatment or reduce the possibility of relapse.

However, this was not the case with outpatients who received Workbooks in addition to face-to-face, talk-based psychotherapy (L'Abate, L'Abate, and Maino, 2005). Those participants (individuals, couples, and families) who received Workbooks showed a significantly greater number of sessions than those who did not received them. In an inpatient setting, however, Goldstein (L'Abate and Goldstein, 2007), a problem-solving Workbooks reduced significantly the length of hospitalization in decompensating women who complete that Workbook versus a control group of women who did not receive it.

The inconsistency and contradictions found in the area of research with PIPEs cries out for more adequate and controlled research that could be conducted by just one investigator alone.

CONCLUSION

Either by themselves or in conjunction with medical, neurological, psychiatric, and psychological approaches, reviewed in this volume, Workbooks can perform self-help, preventive, promotional, psychotherapeutic, and rehabilitative functions that are either time-consuming and difficult to perform in f2f tb treatments. Only future research will determine who will be helped by which approach. Fortunately, the field of mental health in the last generation has been expanded by some many different approaches that will allow professionals to investigate which approach is more adequate, effective, and specific to deal with particular referral questions, concerns, and symptoms.

REFERENCES

Aboujaoude, E., and Koran, L. M. (2010). *Impulse control disorders*. New York: Cambridge University Press.

Aman, M. G., and Singh, N. N. (undated). *Aberrant behavior checklist (ABC): Residential and community version.* East Aurora, NY: Slosson Educational Publications.

Author (1994). *Diagnostic Statistical Manual for Mental Disorders-IV.* Washington, DC: American Psychiatric Association Press.

Anthony, J. L., Lonigan, C. J., and Hecht, S. A. (1999). Dimensionality of posttraumatic stress disorder concerns in children exposed to disaster: Results from confirmatory factor analysis. *Journal of Abnormal Psychology, 108*, 326-336.

Averill, J. R. (1997). The emotions: An integrative approach. In R. Hogan, J. Johnson, and S. Briggs (Eds.), *Sourcebook of personality psychology* (pp. 513-541). San Diego, CA: Academic Press.

Beard, R., Myhill, D., Riley, J., and Nystrand, M. (Eds.). (2009). *The SAGE handbook of writing development.* Thousand Oaks, CA: Sage.

Bearson, D. J., and Zimiles, H. Eds.).(1986). *Thought and emotion: Developmental perspectives.* Hillsdale, NJ: Erlbaum.

Beltman, B. D., and Nair, J. (Eds.). (2004). *Self-awareness deficits in psychiatric patients: Neurobiology, assessment, and treatment*. New York: Norton.

Blatt, S. J., Quinlan, D. M., Chevron, E. S., McDonald, C., and Zuroff, D. (1982). Dependency and criticism: Psychological dimensions of depression. *Journal of Consulting and Clinical Psychology, 50,* 113-124.

Bloom, B. L. (1985). A factor analysis of self-report measures of family functioning. *Family Process, 24*, 225-239.

Bodin, A. M. (1996). Relationship conflict – verbal and physical: Conceptualizing an inventory for assessing process and content. In F. W. Kaslow (Ed.), *Sourcebook of relational diagnoswas and dysfunctional family patterns* (pp. 371-393). New York: Wiley.

Bodine, R., J., and Crawford, D. K. (1999). *Developing emotional intelligence: A guide to behavior management and conflict resolution in schools*. Champlain, IL: Research Press.

Bonanno, G. A. and Keltner, D. (1997). Facial expression of emotion and the course of conjugal bereavement. *Journal of Abnormal Psychology, 106*, 126-137.

Brown, L., Black, D., and Downs, J. (1984). *School Social Skills (S3).* East Aurora, NY: Slosson Educational Publications.

Brown, L., and Hammill, D. D. (1978). *Behavior rating profile (BRP-2).* East Aurora, NY: Slosson Educational Publications.

Butcher, J. (1998). *Psychological Treatment Planning Inventory: Test manual and interpretative guide.* San Antonio, TX, Psychological Corporation.

Cacioppo, J. T., and Gardner, W. L. (1999). Emotion. *Annual Review of Psychology, 50*, 191-214.

Campbell, L., Campbell, b. and Dickinson, D. (1999). *Teaching and learning through multiple intelligence.* Boston, MA: Allyn and Bacon.

Chorpita, B. F., Albano, A. M., and Barlow, D. H. (1998). The structure of negative emotions in a clinical sample of children and adolescents. *Journal of Abnormal Psychology, 107,* 74-85.

Clark, L. A. (1990). Toward a consensual set of concern clusters for assessment of personality disorders. In J. H. Butcher and C. D. Spielberger (Eds.), *Advances in personality assessment* (pp. 243-266). Hillsdale, NJ: Erlbaum.

Costa, P. T. Jr., and Widiger, T. A. (1994). *Personality disorders and the five-factor model of personality.* Washington, D.C.: American Psychological Association.

Coyne, J. C., Thompson, R., and Palmer, S. C. (2003). Marital characteristic quality, coping with conflict, marital complaints, and affection in couples with a depressed wife. *Journal of Family Psychology, 16*, 26-37.

Cusinato, M., and L'Abate, L. (1994). A spiral model of intimacy. In S. M. Johnson and Greenberg, (Eds.), *The heart of the matter: Perspectives on emotion in marital therapy,* (pp. 108-123). New York: Brunner/Mazel.

Cusinato, M., and L'Abate, L. (Eds.). (2012). *Advances in relational competence theory: With special attention to alexithymia.* New York: Nova Science Publishers.

Davis, R. D., and Millon, T. (1999). Models of personality and its disorders. In T.Millon, T. H. Blaney, and R. D. Davis (Eds.), *Oxford textbook of psychopathology* (pp. 485-522). New York: Oxford.

Dozois, D. J. A., Dobson, K. S., and Ahnberg, J. L. (1998). A psychometric evaluation of the Beck Depression Inventory-II. *Psychological Assessment, 10*, 83-89.

Eckhardt, C. L., Barbour, K. A., and Davison, G. C. (1998). Articulated thoughts of marital violence and nonviolent men during anger arousal. *Journal of Consulting and Clinical Psychology, 66,* 259-269.

Eckhardt, C. I., and Deffenbacher, J. L. (1995) Diagnosis of anger disorders. In H. Kassinove (Ed.), *Anger disorders: Definition, diagnosis, and treatment* (pp. 27-47). Philadelphia, PA: Taylor and Francis.

Esterling, B. A., L'Abate, L., Murray, E., and Pennebaker, J. M. (1999). Empirical foundations for writing in prevention and psychotherapy: Mental and physical outcomes. *Clinical Psychology Review, 19*, 79-96.

Feindler, E. L. (1995). Ideal treatment package for children and adolescents with anger disorders. In H. Kassinove (Ed.). *Anger disorders: Definition, diagnosis, and treatment* (pp. 173-195). Philadelphia, PA: Taylor and Francis.

Ferrari, J. R., Johnson, J. L., and McCown, W. G. (1995). *Procrastination and task avoidance: Theory, research, and treatment.* New York: Plenum.

Fogarty, R., and Bellanca, J. (1998). *Multiple intelligences: A collection.* Boston, MA: Allyn and Bacon.

Forgays, D. G., Forgays, D. K., and Spielberger, C. D. (1997). Factor structure of the State-Trait Anger Expression Inventory. *Journal of Personality Assessment, 69*, 497-507.

Gardner, H. (1999). *Intelligence reframed: Multiple intelligences for the 21st century.* New York: Basic Books.

Goldberg, L. R. (1992). The development of markers for the Big-Five factor structure. *Psychological Assessment, 4*, 26-42.

Goldberg, L. E. (1999). The Curious Experiences Survey: A revised version of the Dissociative Experiences Scale: Factor structure, reliability, and relation to demographic and personality variables. *Psychological Assessment, 11*, 134-145.

Gottman, J. M., Fainsilber, L., and Hooven, C. (1997). *Meta-emotion: How families communicate emotionally.* Mahwah, NJ: Erlbaum.

Greene, R. L., and Anderson, E. R. (1999). Minnesota Multiphasic Personality Inventory (MMPI-2). In M. M., Maruish (Ed.). *The use of psychological testing for treament planning and outcomes assessment*_(pp. 1023-1050). Mahwah, NJ: Erlbaum.

Hartlage, S., Arduino, K., and Alloy, L. B. (1998). Depressive personality characteristics : State dependent Concomitants of depressive disorder and traits independent of current depression. *Journal of Abnormal Psychology, 107*, 349-354.

Harwood, T. M., and L'Abate, L. (2010). *Self-help in mental health: A critical evaluation.* New York: Springer-Science.

Hassebrauck , M., and Fehr, B. (2002). Dimensions of relationship characteristics quality. *Personal Relationships, 9*, 253-270.

Henley, A. R. (1987). *Phobias: The crippling fears.* New York: Avon Books.

Hersen, M., and Last, C. G. (Eds.).(1990). *Sourcebook of child and adult psychopathology: A longitudinal perspective.* New York: Pergamon.

Hurtug, J., Audy, J. R., and Cohen, Y. A. (1998). *The anatomy of loneliness.* New York: International Universities Press.

Joiner, T. E., Jr., Steer, R. A., Beck, A. T., Schmidt, N. D., Rudd, M. D., and Catanzaro, S. J. (1999). Physiological hyperarounsal: Construct vaidity of a central aspect of the tripartite model of depression and anxiety. *Journal of Abnomal Psychology, 108*, 290-298.

Kacewicz, E., Slatcher, R. B., and Pennebaker, J. W. (2007). Expressive writing: An alternative to traditional methods. In L. L'Abate (Ed.), *Low-cost approaches to promote physical and mental health: Theory, research, and practice* (pp. 271-284). New York: Springer.

Katz, R., Katz, J., and Shaw, B. F. (1999). Beck Depression Inventory and Helplessness Scale. In M. M., Maruish (Ed.), *The use of psychological testing for treament planning and outcomes assessment* (pp. 831-870). Mahwah, NJ: Erlbaum.

Kazantzis, N., Deane, F. P., Ronan, K. R., and L'Abate, L. (Eds.).(2005). *Using homework assignments in cognitive behavior therapy.* New York: Rutledge.

Kazantzis, N., and L'Abate, L. (Eds.). (2005). *Handbook of homework assignments in psychotherapy: Theory, research, and prevention.* New York: Springer.

Kelly, E. J. (1990). *Differential test of conduct and emotional problems.* East Aurora, NY: Slosson Educational Publications.

Kelly, E. J., and Vitali, G. J. (1992). *Conduct problem/Emotional problem interventions.* East Aurora, NY: Slosson Educational Publications.

Kennedy-Moore, E., and Watson, J. C. (1999). *Expressing emotion: Myths, realities, and therapeutic strategies.* New York: Gulford.

King, D. W., Leskin, G. A., King, L. A., and Weathers, F. W. (1998). Confirmatory factor analysis of the clinician-administered PTSD scale: Evidence for the dimentionality of posttraumatic stress disorder. *Psychological Assessment, 10,* 90-96.

Kochalka, J., and L'Abate, L. (1997). Linking evaluation with structured enrichment: The Family Profile Form. *American Journal of Family Therapy, 25*, 361-374.

Kosson, D. S., Steuerwald, B. L., Forth, A. E., and Kirkhart, K. J. (1997).A new method for assessing the interpersonal behavior of psychopathic individuals: Preliminary validation studies. *Psychological Assesent, 9*, 89-101.

L'Abate, L. (1984). Beyond paradox: Issues of control. *American Journal of Family Therapy, 12,* 12-20.

L'Abate, L. (1986). *Systematic family therapy.* New York: Brunner/Mazel.

L'Abate, L. (1990). *Building family competence: Primary and secondary prevention strategies.* Newbury Park, CA: Sage.

L'Abate, L. (1991). The use of writing in psychotherapy. *American Journal of Psychotherapy, 45*, 87-98.

L'Abate, L. (1992). *Programmed writing: A self-administered approach for interventions with individuals, couples and families.* Pacific Grove, CA: Brooks/Cole.

L'Abate, L. (1997). Distance writing and computer-assisted training. In S. R. Sauber (Ed.), *Managed mental health care: Major diagnostic and treatment approaches* (pp. 133-163). Bristol, PA: Brunner/Mazel.

L'Abate, L. (1999). Taking the bull by the horns: Beyond talk in psychological interventions. *The Family Journal: Counseling and Therapy for Couples and Families, 7*, 206-220.

L'Abate, L. (Ed.).(2001). *Distance writing and computer-assisted interventions in psychiatry and mental health.* Westport, CT: Ablex.

L'Abate, L. (2002). *Beyond psychotherapy: Programmed writing and structured computer-assisted interventions.* Westport, CT: Ablex.

L'Abate, L. (2004a). *A guide to self-help Workbooks for clinicians and researchers.* Binghamton, NY: Haworth.

L'Abate, L. (2004b). Systematically written homework assignment: The case for homework-based treatment. In L. L'Abate, *Using Workbooks in mental health: Resources in prevention, psychotherapy, and rehabilitation for clinicians and researchers* (pp. 65-102). Binghamton, NY: Haworth.

L'Abate, L. (2005). *Personality in intimate relationships: Socialization and psychopathology.* New York: Springer-Science.

L'Abate, L. (2007a). A completely preposterous proposal: The dictionary as an initial vehicle of change in the family. *The Family Psychologist, 23*, 39-42.

L'Abate, L (2007c). Introduction: Section IV. Writing. In L. L'Abate (Ed.), *Low-cost approaches to promote physical and mental health: Theory, research, and practice*_(pp. 219-266). New York: Springer-Science.

L'Abate, L. (Ed.). (2007d). *Low-cost approaches to promote physical and mental health: Theory, research, and practice.* New York: Springer-Science.

L'Abate, L. (2008a). A proposal for including distance writing in couple therapy. *Journal of Couple and Relationship Therapy, 7*, 337-362.

L'Abate, L. (Ed.). (2008b). *Toward a science of clinical psychology: Laboratory evaluations and interventions.* New York: Nova Science Publishers.

L'Abate, L. (2008c). Working at a distance from participants: Writing and nonverbal media. In L. L'Abate (Ed.), *Toward a science of clinical psychology: Laboratory evaluations and interventions* (pp. 355-383). Hauppauge, NY: Nova Science Publishers.

L'Abate, L. (2011a). Psychotherapy consists of homework assignments: A radical iconoclastic conviction. In H. Rosenthal (Ed.), *Favorite Counseling and Therapy Homework Techniques: Classic Anniversary Edition* (pp. 219-229). New York: Routledge.

L'Abate, L. (2011b). *Hurt feelings: Theory, Research, and Applications in Intimate Relationships.* New York: Cambridge University Press.

L'Abate, L. (2011c). *Sourcebook of Practice Exercises in Mental Health.* New York: Springer-Science.

L'Abate, L. (in press). *Clinical Psychology andPpsychotherapy as a Science: An iconoclastic Perspective.* New York: Springer-Science.

L'Abate, L. Boyce, J., Fraizer, I., and Russ, D. A. (1992). Programmed writing: Research in progress. *Comprehensive Mental Health Care, 2*, 45-62.

L'Abate, L., and Cusinato, M.. (2007). Linking theory with practice: Theory-derived interventions in prevention and family therapy. *The Family Journal: Counseling and Psychotherapy with Couples and Families., 15*, 318-327.

L'Abate, L., Cusinato, M., Maino, E., Colesso, W., and Scilletta, C. (2010). *Relational competence theory: Research and mental health applications.* New York: Springer-Science.

L'Abate, L., and De Giacomo, P. (2003). *Intimate relationships and how to improve them: Integrating theoretical models with preventive and psychotherapeutic approaches.* Westport, CT: Praeger.

L'Abate, L., and Goldstein D. (2007). Protocols to promote mental health and life-long learning. In L. L'Abate (Ed.), *Low-cost approaches to promote physical and mental health: Theory, research, and practice*_(pp. 285-302). New York: Springer-Science.

L'Abate, L., and Harrison, M. G. (1992). Treating codependency. In L. L'Abate, J. E. Farrar, and D. A. Serritella, (Eds.), *Handbook of differential treatments for addictions* (pp. 286-306). Boston, MA: Allyn and Bacon.

L'Abate, L., and Kern, R. (2002). Workbooks: Tools for the expressive writing paradigm. In S. J. Lepore, J. M. and Smyth (Eds.), *The writing cure: How expressive writing promotes health and emotional well-being* (pp. 239-255).Washington, DC: American Psychological Association.

L'Abate, L., L'Abate, B. L., and Maino, E. (2005). A review of 25 years of part-time professional practice: Workbooks and length of psychotherapy. *American Journal of Family Therapy, 33*, 19-31.

L'Abate, L., & Sweeney, L. G. (2011). *Research on writing approaches in mental health.* Bingley, UK: Emerald Group Publishing Limited.

L'Abate, L., van Eigen, A., and Rigamonti, S. (2011). Relational and cross-cultural perspectives on non-violent externalizing personality disordered women: Introduction to research. *American Journal of Family Therapy, 39*, 325-347.

L'Abate, L., and Weinstein, S. E. (1987). *Structured enrichment programs for couples and families.* New York: Brunner/Mazel.

Lahey, B. B., Frick, P. J., Loeber, R., Tannenbaum, D. A., et al. (1990). *Oppositional conduct disorder: A meta-analysis review.* Unpublished manuscript. University of Georgia, Athens.

Lepore, S. J., and Smyth, J. M. (Eds.). (2002). *The writing cure: How expressive writing promotes health and emotional well-being.* Washington, DC: American Psychological Association.

Levinson, H. N. (1986). *Phobia free.* New York: Evans.

ivesley, W. J., Jackson, D. N., and Schroeder, M. L. (1989). A study of the factorial structure of personality pathology. *Journal of Personality Disorders, 3*, 292-306.

Lorr, M., Younis, R. P., and Stefic, E. C. (1991). An inventory of social skills. *Journal of Personality Assessment, 57*, 506-520.

Lynam, D. R. (1997). Pursuing the psychopath: Capturing the fledging psychopath in a nomological net. *Journal of Abnormal Psychology, 106*, 425-438.

Mascolo, M. F., and Griffin, S. (Eds.).(1998). *What develops in emotional development?* New York: Plenum.

Mash, L. J., and Barkley, R. A. (1996). *Child psychopathology.* New York: Guilford.

Mayer, J. D., and Salovey, P. (1995). Emotional intelligence and the construction and regulation of feelings. *Applied and Preventive Psychology, 4,* 197-208.

McCrae, R. R. (1994). A reformulation of Axis II: Personality and personality-related problems. In P. T. Costa and T. A. Widiger (Eds.), *Personality disorders and the five-factor model of personality* (pp. 303-309). Wshington, DC: American Psychological Association.

Millon T., Blaney, P. H, and Davis, R. D. (Eds.). (1999). *Oxford textbook of psychopathology.* New York: Oxford University Press.

Moras, K., Di Nardo, P. A., and Barlow, D. H. (1992). Distinguishing anxiety and depression: Reexamination of the reconstructed Hamilton scales. *Psychological Assessment, 4,* 224-227.

Morey, L. C. (1999). Personality Assessment Inventory. In M. E. Maruish (Ed.), *The use of psychological testing for treatment planning and outcome assessment* (pp. 1083-1122). Hillsdale, NJ: Erlbaum.

Morey, L. C., and Henry, W. (1994). Personality Assessment Inventory. In M. E. Maruish (Ed.), *The use of psychological testing for treatment planning and outcome assessment* (pp.185-216). Hillsdale, NJ: Erlbaum.

Newcomer, P. L., Barenbaum, E. M., and Bryant, B. R. (1994). *Depression and Anxiety in Youth Scale.* Austin, TX: Pro-Ed.

Omer, H. (1985). Fulfillment of therapeutic tasks as a precondition for acceptance in therapy. *American Journal of* Psychotherapy, 56, 496-501.

Peplau, I. A., and Perlman, D. (Eds.).(1982). *Loneliness: A sourcebook of current theory, research, and therapy.* New York:Wiley.

Perry, J. N., and Butcher, J. N. (1999). Butcher Treatment Planning Inventory (BTPI): An objective guide to treatment planning. In M. E. Maruish (Ed.), *The use of psychological testing for treatment planning and outcomes assessment* (pp. 1157-117). Mahvah, N. J. Erlbaum.

Peterson, C., and Seligman, M. E. P. (2004). *Character strengths and virtues: A Sourcebook of classification.* New York: Oxford.

Sanford, K. (2003). Problem-solving conversations in marriage: Does it matter what topics couples discuss? *Personal Relationships, 10*, 97-112.

Schachner, D. A., and Shaver, P. R. (2004). Attachment dimensions and sexual motives. *Personal Relationships, 11*, 179-195.

Schroeder, M., Wormworth, J., and Livesley, W. (1992). Dimensions of personality disorder and their relationships to the Big Five dimensions of personality. *Psychological Assessment, 4*, 47-53.

Shadish, W. R., Matt, G. E., Navarro, A. M., and Phillips, G. (2000). The effects of psychological therapies under clinically representative conditions: A meta-analysis. *Psychological Bulletin, 126*, 512-529.

Smyth, J. M., and L'Abate, L. (2001). A meta-analytic evaluation of Workbook effectiveness in physical and mental health. In L. L'Abate (Ed.), *Distance writing and computer-assisted interventions in psychiatry and mental heath* (pp. 77-90). Westport, CT: Ablex.

Snyder, D. K. (1981). *Manual for Marital Satisfaction Inventory.* Los Angeles, CA: Western Psychological Services.

Snyder, D. K., and Aikman, G. G. (1999). Marital Satisfaction Inventory–Revisited. In M. E. Maruish (Ed.), *The use of psychological testing for treatment planning and outcome assessment* (pp.1173-1219). Hillsdale, NJ: Erlbaum.

Spielberger, C. D., Reheiser, E. R., and Sydeman, S. J. (1995). Measuring the experience, expression, and control of anger. In H. Kassinove (Ed.), *Anger disorders: Definition, diagnosis, and treatment* (pp. 49-67). Philadelphia, PA: Taylor and Francis.

Sternberg, R. J. (1996). *Successful intelligence.* New York: Simon and Schuster.

Sternberg, R. J., and Kaufman, J. C. (1998). Human abilities. *Annual Review of Psychology, 49,* 479-502.

Taylor, S. Koch, K., Koch, W. J., Crockett, D. J., and Passey, G. (1998). The structure of posttraumatic stress concerns. *Journal of Abnormal Psychology, 107*, 154-160.

Weeks, G. R., and L'Abate, L. (1982). *Paradoxical psychotherapy: Theory and practice with individual couples, and families.* New York: Brunner/Mazel.

Weston, H. E., Boxer, P., and Heatherington, L. (1998). Children's attributions about family arguments: Implications for family therapy. *Family Process, 17*, 35-49.

In: Handbook of Technology in Psychology ...
Editor: Luciano L'Abate and David A. Kasier

ISBN: 978-1-62100-004-4

Chapter 19

EFFECTIVENESS OF INTERNET PSYCHOLOGICAL TREATMENTS IN MENTAL HEALTH DISORDERS

M. J. Gallego and P. M. G. Emmelkamp[1*]
[1]Universit of Van Amersterdam, The Netherlands

ABSTRACT

The present chapter describes the state of the art of the efficacy and effectiveness of Internet-based psychological treatments. First, we present a historical background of these treatments and their purpose. Then a systematic review of the controlled studies that use Internet-based psychological treatments to treat anxiety disorders and depression will be presented; we describe the nature of the participants and the results achieved. In addition, we mention recent applications of Internet-based treatments to other areas like substance abuse, eating disorders and health problems. Finally, we discuss not only what has been achieved, but discuss also what needs to be done in the field of Internet psychological treatments, including problems associated with implementation in community settings.

HISTORICAL AND SYSTEMATIC BACKGROUND

Emotional disorders as anxiety disorders and mood disorders are highly prevalent in epidemiologic studies. The lifetime prevalence in the general population is 16.6 % for any anxiety disorder (Kessler et al., 2006) and 6.7% for major depressive disorder and 3.6% for a slightly milder chronic variant: dysthymic disorder (Waraich, Goldner, Somers and Hsu, 2004). The most prevalent eating disorders in the general community are bulimia nervosa and binge eating disorder with estimates for prevalence in females around 1% for each of these disorders (Hoek and Van Hoeken, 2003). Clinical guidelines around the world (e.g. NICE, 2004) recommend behaviour therapy or cognitive behaviour therapy (CBT) as the psychological treatments of choice for anxiety disorders, depression, bulimia and binge eating disorder. This is supported by recent meta-analyses (anxiety disorders: Deacon and

*E-mail: P.M.G.Emmelkamp@uva.nl

Abramowitz, 2004; Norton and Price, 2007; depression: Ekers, Richards and Gilbody, 2008; eating disorders: Wilson, Grilo and Vitousek, 2007). In spite of the existence of effective psychological treatments for anxiety disorders and depression, Andrews and Carter (2001) found that less than half of individuals with these disorders in the Australian population seek help to solve these problems. The same applies to women suffering from eating disorders (Roness, Mykletun and Dahl, 2005). Further, many people that receive treatment do not receive evidence-based behavioral or CBT interventions as recommended by clinical guidelines (Issakidis and Andrews, 2002). This may be related to the difficulties associated with disseminating of CBT in clinical practice (Gaston, Abbott, Rapee and Neary, 2006; Powers and Emmelkamp, 2009). Issues that are related to dissemination include the high costs of face to face CBT and time investments of the patients associated with regularly visiting a therapist. The use of self-help programs and the use of new technologies may be two alternatives to traditional CBT face-to-face treatments that reduces the treatment costs and can reach a lot of people with mental problems. There is some evidence that self-help programs are effective, although less effective than face-to-face therapy. For example, in a meta-analysis conducted by Menchola, Arkowitz and Burke (2007) self-help programs were found to be effective in patients suffering from depression or anxiety. Although generally these self-help programs were more effective than control conditions, these programs were less effective than face to face therapy. Sysko and Walsh (2008) in reviewing the studies evaluating the efficacy of self-help programs in the treatment of bulimia nervosa and binge eating disorder found self-help programs to be more effective than waiting list control groups. Unfortunately, hardly any study compared self-help programs with evidence-based face to face therapy, which prevents conclusions to be drawn. Since the early seventies of the last century, technologies like television, telephone, fax, radio, computer, videoconference, virtual reality and Internet have been used to provide psychiatric and psychological assistance. To give just one example, in 1972 television was used to make referral notes between the Medical Center of the Nebraska University and the State Hospital (Wittson and Benschoter, 1972). A relatively new technology that is a mixture of television and computer is virtual reality, it has been used to treat fear of heights (e.g. Emmelkamp et al., 2002), claustrophobia (e.g. Botella et al., 1998), driving phobia (e.g. Wald and Taylor, 2000), flying phobia (e.g. Krijn et al., 2007), spider phobia (Carlin, Hoffman and Weghorst, 1997), panic disorder (Vincelli, Choi, Molinari, Wiederhold and Riva, 2000), posttraumatic disorder (Rothbaum et al., 1999) and as a component of a treatment package for eating disorder (Emmelkamp, 2005). The effectiveness of virtual reality has now been firmly established for fear of heights and fear of flying as evidenced in systematic reviews (Krijn, Emmelkamp, Olafsson and Biemond, 2004; Meyerbröker and Emmelkamp, 2010) and a meta-analysis (Powers and Emmelkamp, 2008). Computer-assisted treatments are (cognitive)-behavioral interventions that have been computerized to diminish the contact with the therapist. These treatments have been found to be effective in the treatment of phobias and panic (Marks et al., 2004), obsessive compulsive disorder (Greist et al., 2002), general anxiety disorder (Yates, 1996) and mild to moderate depression (Osgood-Hynes et al., 1998; Proudfoot et al., 2003). When treatment is delivered by stand-alone computers, no personalized feedback from therapists is possible. The Internet enhances the therapeutic possibilities of computers by offering the possibility to deliver more tailor-made treatment, including feedback of therapists. Most current Internet applications deliver self-help interventions in addition to face-to-face contact with a clinician. Patients living in remote areas, physically disabled patients with restricted mobility including severe agoraphobic and obsessive-compulsive patients or patients who are reluctant to seek face-to-face therapy due to anxiety or fear of stigmatization may be reached through the World Wide Web (Emmelkamp, 2005). The start of Internet-based psychological programs was around 2000, when different research groups around the world used this type of programs for the prevention / treatment of a variety of mental disorders. Among the pioneers are the research group of Celio et al., who designed a program to improve body satisfaction and reduce weight and shape concerns (Celio et al., 2000; Winzelberg et al., 2000), Lange et al. (2000), who developed an Internet program to treat posttraumatic stress disorder (PTSD) and Botella et al. (2000), who designed an Internet program for fear of public speaking. Since then many more Internet-based programs have been developed to treat disorders like emotional disorders, eating disorders and drug abuse.

Technology with a Purpose/Setting Goals and Objectives

The participants' lack of time, the often long waiting lists and not knowing where to go for psychological help are barriers that hinder the access to face-to-face psychological support. Internet-based treatments were created in order to overcome the limitations of previous psychological treatments. The advantages from Internet-provided assessment and treatment programs are numerous, including cost-effectiveness, ease and rapidity to get advice and reduced stigma. Sometimes the participants withhold information to their face-to-face clinician because of shame or fear of stigmatization (Postel, de Haan and de Jong, 2008). There is evidence, however, that for some patients lack of face to-face contact facilitates the sharing of sensitive issues, including suicide thoughts; they prefer to confide such sensitive information to the computer rather than to a face-to-face therapist. Further, especially in remote areas with little mental health facilities, the use of the Internet removes the need to travel to a clinician (Emmelkamp, 2005; Landau, 2001; Zabinski, Celio, Wilfley and Taylor, 2003). In addition, the demand for evidence-based therapy exceeds the supply of trained therapists and Internet-based assessment and treatment may help shorten waiting lists. Internet-treatment can be used as a first step in stepped care of mental disorders. For example, Green and Iverson (2009) suggested to use the Internet-based assessment and treatment as a first-line step in a Stepped Care Model; in case that Internet-based treatment is contraindicated or turns out to be unsuccessful a traditional face-to-face therapy can be carried out. This type of assistance would lead to a reduction of waiting lists and to saving of time for more complicated cases. Dissemination is still an important target for the Internet-based treatments. In spite of effectiveness of this type of treatments only a minority of clinicians use them. Some of the barriers for the clinicians' resistance are:

- Fear of being replaced by the computer (Emmelkamp, 2005).
- A lack of knowledge about the application of new technologies in the field of clinical psychology (Whitfield and Williams, 2004).
- A lack of training on how to use Internet-based treatment programs.
- Clinician concerns over patient confidentiality on the world wide web (Zabinski et al., 2003).

- The clinician is afraid of taking clinical responsibility of participants that they do not see face-to-face (Whitfield and Williams, 2004).
- Clinician worries about the influence that the Internet-based treatment could have on the development of a productive participant-clinician relationship (Anderson, Jacobs and Rothbaum, 2004; Emmelkamp, 2005).

All the barriers discussed above can be adequately dealt with, although it is necessary to make specific efforts to overcome them. With regard to confidentiality the use of safe servers and the access to the treatment using a login and a password could guarantee the privacy. The clinical responsibility of the therapist is a serious problem, but can be handled by adequate measures such as using well-validated measures, strict exclusion criteria, and –where needed- clinical interview by phone. Further, there is a clear need of training of therapists in the use of these treatments (Anderson et al., 2004), and to inform the clinicians that the goal of the Internet-based treatments is not to replace them. Therefore it is important to create user friendly and attractive Internet-based treatments in order to be well received by the user and the clinician.

Other limitations with respect to the use of Internet as a therapeutic tool are (Zabinski et al., 2003):

- Not everybody has Internet access.
- The therapist cannot observe the facial expression and does not have visual contact through the Internet.
- It is difficult to deal adequately with crisis intervention through the Internet.
- Internet-based cognitive and behavioral treatments for anxiety and depression are characterized by poor adherence (Christensen, Griffiths and Farrer, 2009).

Each limitation is a challenge that researchers have to try to surpass in future studies. With regard to Internet access, this is year 2010 hardly a problem anymore. Daily the number of people that have Internet access is increasing. Some possible solutions to the problem of not being able to view the facial expression are the use of emoticons, a webcam or videoconference. With regard to crisis intervention, Gallego (2007) recommended to use a mobile phone or to receive a message through the PDA (Personal Digital Assistance) when the participant needs help. Finally, the attrition rate of Internet-based treatments could be improved by adding occasional contact with a therapist (Carlbring et al., 2007) or creating short treatments (Christensen et al., 2009).

Nature of Participants Used

By definition, participants in Internet-based treatments have easy access to computer and Internet and are knowledgeable on how to surf the world wide web.

To get an impression of the individuals who make use of Internet-based treatments we looked for characteristics of the samples used in controlled studies that used Internet-based treatments for *mixed anxiety disorders* (see Table 1). Generally, the participants in these studies were young, the mean age was 30.9 (*SD* = 10), around the 70% were women, the mean years of education were 13.08 (*SD* = 1.36) and half of the studies were carried out with university students. Thus, the individuals in these studies are more highly educated than a random sample of the community. Only in three of six controlled studies the sample was clinical: one study recruited participants with panic or phobia (Schneider, Mataix-Cols, Marks and Bachofen, 2005), Titov, Andrews, Robinson et al. (2009) carried out a study with patients with Generalized Anxiety Disorder (GAD), and another study selected phobic participants (Andersson et al., 2009). Two of the six controlled studies focused on subclinical patients who scored higher than 23 on the Anxiety Sensitivity Index (ASI, Peterson and Reiss, 1992; Kenardy, McCafferty and Rosa, 2003, 2006). Finally, one of the studies offered an Internet-based treatment for test anxiety to individuals from the general population (Orbach, Lindsay and Grey, 2007).

Table 1. Scheme of the controlled studies conducted in mixed anxiety

Study	Type of treatment	Subjects per condition	Formal diagnosis	Inclusion criteria	Drop-out rate (%)	Clinical sign. inf.	Follow-up	Efficacy
Kenardy et al. (2003)	Internet-based[wc]	IIG = 43 WL = 40	No	ASI ≥ 24	IIG = 14 WL = 5	No	No	IIG>WL
Kenardy et al. (2006)	Internet-based[wc]	IIG = 36 WL = 38	No	ASI ≥ 24	IIG = 0 WL = 0	No	6 month	IIG>WL
Schneider et al. (2005)	Internet-based[cpe]	IIG = 45 IIG-WE = 23	Yes (phone)	ICD-10 phobia or panic	IIG = 27 IIG-WE= 35	Yes[1]	1 month	IIG = IIG-WE post IIG > IIG-WE $_{FU}$
Andersson et al. (2009)	Internet-based[cfe]	IIG = 13 BLE = 14	Yes (FF)	SCID-I Phobia	-	Yes	1 year	IIG < BLE $_{post}$ IIG = BLE $_{FU}$
Titov, Andrews, Robinson et al. (2009)	Internet-base[ce]	IIG = 25 WL = 23	Yes (phone)	MINI GAD	IIG =20 WL=17	Yes[2]	No	IIG > WL
Orbach et al. (2007)	Internet-based[wc]	IIG = 30 CIP = 28	No	Test anxiety	IIG = 28 CIP = 35	Yes	4 month	IIG = CIP

Note: Clinical sign. inf.: Clinical significance information; [wc]: without contact with the therapist; IIG : Internet-based cognitive-behavioral therapy; WL: Waiting list; ASI: Anxiety Sensitivity Index; A>B: Treatment A superior to treatment B; [cpe]: contact with the therapist by phone or e-mail during the treatment; IIG-WE: Internet cognitive-behavioral therapy without exposure advice; ICD-10: Tenth Revision of the International Classification of Diseases and Related Health Problems; [1]An alternative to Jacobson and Truax (1991); A = B: Treatment A is as effective as treatment B; $_{post}$: posttest; $_{FU}$: Follow-up; [cfe] contact by e-mail when participants failed to send in homework; BLE: Brief live-exposure treatment; FF: face to face assessment; SCID-I: Clinical Interview for DSM-IV Axis I; A<B: Treatment A inferior to treatment B; [ce]: contact with the therapist by e-mail during the treatment; MINI: Mini International Neuropsychiatric Interview; [2] : GAD-7 total scores below the cutoff for a likely diagnosis of GAD; GAD: Generalized Anxiety Disorder; CIP: Control Internet program.

In the controlled studies into Internet treatments for *PTSD* (see Table 2) the mean age of the participants was 33.6 years old (*SD* = 6.7), the percentage of women was around 70% and the level of education in four studies that had this information was high (43.5% had a university degree or were studying for a university degree). Only one of the six controlled studies involved a clinical sample: participants suffered from a terrorist attack and were formally diagnosed with PTSD (Litz, Engel, Bryant and Papa, 2007). The sample in the other studies was subclinical, one study focused on complicated grief (Wagner, Knaevelsrud and Maercker, 2006) and four studies included participants who suffered a traumatic event like the loss of a beloved one, health, house, or job, sexual abuse, incest, rape, physical abuse, robbery, traffic accidents, eye-witnessed traumatic events, divorce or other traumatic events within the family (Hirai and Clum, 2005; Knaevelsrud and Maercker, 2007; Lange, van de Ven, Schrieken and Emmelkamp, 2001; Lange et al., 2003).

Table 2. Scheme of the studies conducted in PTSD

Study	Type of treatment	Subjects per condition	Formal diagnosis	Inclusion criteria	Drop-out rate (%)	Clinical sign. inf.	Follow-up	Efficacy
Lange et al. (2001)	Interapy [c]	IIG = 13 WL = 12	No	traumatic event	IIG = 13.3 WL = 20	Yes	6 weeks	IIG > WL
Lange et al. (2003)	Interapy [c]	IIG = 122 WL = 62	No	traumatic event	IIG = 30.1 WL = 48.4	Yes	6 weeks	IIG > WL
Hirai and Clum (2005)	Internet-based [ce]	IIG = 18 WL = 18	No	PTSD (DSM-IV)*	IIG = 27.8 WL = 22.2	Yes	No	IIG > WL
Wagner et al. (2006)	E-mail	IIG = 29 WL = 26	No	Complicated grief	IIG = 8 WL = 3.9	Yes	3 months	IIG > WL
Knaevelsrud and Maercker (2007)	E-mail	IIG = 49 WL = 47	No	PTSD (DSM-IV)*	IIG = 16.3 WL = 2.1	Yes[1]	3 months	IIG > WL
Litz et al. (2007)	TAIT [cpe]	TAIT = 24 SCI = 21	Yes (FF)	PTSD (DSM-IV)	27.7 for the total sample	Yes[2]	3 and 6 months	TAIT > SCI

Note: Clinical sign. inf.: Clinical significance information; [c]: contact with the therapist during the treatment; IIG: Internet intervention group; WL: Waiting list; Traumatic event: Traumatic event that happened at least 3 months ago; A>B: Treatment A superior to treatment B; PTSD: Posttraumatic stress disorder; DSM-IV: *Diagnostic and statistical manual of mental disorders*; [ce]: contact with the therapist by e-mail during the treatment; * They do not use an structured interview; [1]A cut-off of 35 in the Impact Event Scale; TAIT: Therapist assisted internet treatment in which the 1[rst] session was face-to-face; [cpe]: contact with the therapist by phone and e-mail during the treatment; SCI: Supportative counseling via internet; FF: face to face assessment; [2] Several indicators of clinical significance change.

Regarding to the controlled studies to validate Internet treatments for *panic* (see Table 3) the mean age for the samples was 38.2 (*SD* = 7.2), the percentage of women around 70% and the average years of education 12.6 (*SD* = 0.9). Furthermore, all but one studies used clinical samples, with one exception: Ruwaard, Broeksteeg, Schrieken, Emmelkamp and Lange (2010). In the latter study patients were not assessed with a structured interview. All studies informed if the participants were taking psychoactive medication during the treatment, the average percentage was 44.4%.

Table 3. Scheme of the studies conducted in panic disorder

	Type of treatment	Subjects per condition	Formal diagnosis	Inclusion criteria	Drop-out rate (%)	Clinical sign. inf.	Follow-up	Efficacy
Carlbring et al. (2001)	Internet-based ce	IIG = ± 21 WL = ± 20	Yes (e-mail)	PD (ADIS-IV)	IIG = ± 19.1 WL = ± 5	Yes	-	IIG > WL
Carlbring et al. (2003)	Internet-based ce	IIG = 11 AR ce = 11	Yes (FF)	PD (SCID)	IIG = 27 AR = 18	No	-	IIG = AR
Carlbring et al. (2005)	Internet-based ce	IIG = 25 LIVE = 24	Yes (FF)	PD (SCID)	IIG = 12 LIVE = 12.2	Yes	1 year	IIG = LIVE
Carlbring et al. (2006)	Internet-based cpe	IIG = 30 WL = 30	Yes (phone)	PD (SCID)	IIG = 6.7 WL = 3.3	Yes	9 months	IIG > WL
Klein and Richards (2001)	Internet-based ni	IIG = 11 Self-CC wc=11	Yes (FF)	PD (Prime MD)	4.3 for the total sample	No	-	IIG > Self-CC
Klein et al. (2006)	Internet-based ce	IIG = 19 TAM cp = 18 In-CC cp = 18	Yes (phone)	PD (ADIS-IV)	IIG = 5 TAM = 17 In-CC = 28	Yes[1]	3 months	IIG = TAM > In-CC
Klein et al. (2009)	Internet-based ce	IIG 3ew = 28 IIG 1ew = 29	Yes (phone)	PD (ADIS-IV)	IIG 3ew = 21.4 IIG 1ew = 27.6	Yes[1]	-	IIG 3ew = IIG 1ew
Richards et al. (2006)	Internet-based ce	IIG = 12 IIG-SM ce = 11 In-CC ce = 9	Yes (phone)	PD (ADIS-IV)	IIG = 16.7 IIG-SM ce = 9 In-CC ce = 22	Yes[1]	3 months	IIG = IIG-SM > In-CC
Pier et al. (2008)	IIG-CGP ff	IIG-CGP= 34 IIG-CP ce = 31	Yes (phone)	PD (ADIS-IV)	IIG-CGP= 26.5 IIG-CP = 12.9	Yes[1]	-	IIG-CGP = IIG-CP
Shandley et al. (2008)	IIG-CGP ff	IIG-CGP= 53 IIG-CP ce = 43	Yes (phone)	PD (ADIS-IV)	IIG-CGP= 47.2 IIG-CP = 37.2	Yes[1]	6-months	IIG-CGP = IIG-CP
Kiropoulos et al. (2008)	Internet-based ce	IIG = 46 FFT = 40	Yes (FF)	PD (ADIS-IV)	IIG = 10.9 FFT = 5	Yes[1]	-	IIG = FFT
Ruwaard et al. (2010a)	Internet-based ta	IIG = 27 WL = 31	No	PD (DSM-IV)*	IIG = 11.1 WL = 3.2	Yes	3 years	IIG > WL

Note: Clinical sign. inf.: Clinical significance information; ce: contact with the therapist by e-mail during the treatment; IIG: Internet intervention group; WL: Waiting list; PD: Panic Disorder; ADIS: Anxiety Disorders Interview Schedule; A>B: Treatment A superior to treatment B; AR: applied relaxation; FF: face-to-face assessment; SCID: Structured clinical interview for DSMIV; A = B: Treatment A is as effective as treatment B; LIVE: live therapy; cpe: contact with the therapist by phone and e-mail during the treatment; ni: there is no information about contact during treatment; Self-CC: Self-monitoring control condition; Prime MD: Primary Care Evaluation of Mental Disorder;_TAM: therapist-assisted manual; In-CC: Information-only control condition; [1]high end-state functioning; A= B > C: Treatment A is as effective as B and both treatments are more than C; 3ew: 3 e-mails per week; 1ew: 1 e-mail per week; IIG-SM: Internet-based CBT plus Stress Management; ff: face-to-face contact; IIG-CGP: Internet intervention plus contact with a general practitioner; IIG-CP: Internet intervention plus contact with a psychologist; FFT: face-to-face treatment; ta: therapist-assisted treatment; * They do not use a validated structured interview.

Participants in the controlled studies conducted in *social phobia* (see Table 4) have a mean age of 34.8 (*SD* = 5.2) and the percentage of women is around 60%. The level of education of the participants was high, 66.7% was involved in or had completed university studies. The samples in the various studies can be characterized as clinically relevant: all participants in the ten controlled studies had been diagnosed using a structured interview. Only five studies provided information about the number of participants that took psychoactive medication, the mean of medication use was 25.6% (Titov. Andrews, Choi, Schwencke and Mahoney, 2008; Titov, Andrews, Choi, Schwencke and Johnston, 2009; Titov, Andrews and Schwencke, 2008; Titov, Andrews, Schwencke, Drobny and Einstein, 2008; Titov, Andrews, Schwencke et al., 2009).

Finally, the characteristics of the samples in the Internet-based controlled studies for *depression* were the following: the 71.1% were women, the mean age was 43.1 (SD = 4.7), and 55.3% were higher educated. The studies included in these analyses are summarized in Table 5. It is important to highlight that only two of the nine studies included clinical samples (Spek et al. 2007; Perini, Titov and Andrew, 2009). Three studies provided information about medication use, Andersson et al. (2005), Perini et al. (2009) and Ruwaard et al. (2009), 29%, 51.1% and 19% respectively, were taking psychoactive drugs.

To sum up, the samples used to validate Internet-based treatments are usually young people, mainly females with high level of education. Thus, it is questionable whether results of these studies can be generalized to more diverse samples.

Success and Failures of Approach

In this review we are going to focus on randomized controlled trials (RCTs) focusing on the treatment of a variety of mental disorders. Most controlled studies have been conducted in anxiety disorders and depression. We are going to limit our discussion to studies that included Internet-based therapy and we are not going to review Internet-based programs primarily focusing on prevention.

Anxiety Disorders

Anxiety disorders are common and debilitating psychiatric conditions, they cause considerable suffering and disability for individuals, result in significant health costs and have great impact on family and community. In this section research articles which test the efficacy of Internet-based treatments are discussed by the separate anxiety disorders.

Mixed Anxiety

In this paragraph we will discuss RCTs on the treatment of phobia, panic or GAD. As shown in Table 1 six controlled studies were carried, the control conditions were a waiting list (Kenardy et al., 2003, 2006; Titov, Andrews, Robinson et al., 2009), a CBT Internet treatment without exposure instructions (Schneider et al., 2005), a brief live-exposure treatment (Andersson et al., 2009) or a control treatment via Internet (Orbach et al., 2007) respectively.

Table 4. Scheme of the studies conducted in social phobia

Study	Type of treatment	Subjects per condition	Formal diagnosis	Inclusion criteria	Drop-out rate (%)	Clinical sign. inf.	Follow-up	Efficacy
Andersson et al., 2006	Inter2^{ce}	Inter2 = 32 WL = 32	Yes (FF)	SP (SCID)	Inter2 = 38 WL = 0	Yes	1 year	Inter2 >WL
Carlbring et al. (2007)	Internet-based ^{cp}	IIG = 30 WL = 30	Yes (phone)	SP (SCID)	IIG = 13.3 WL = 6.7	No	1 year 2.5 year[1]	IIG > WL
Tillfors et al. (2008)	Internet-based ^{ce}	IIG = 19 Inter5 ^{ce}= 19	Yes (phone)	SP (SCID)	IIG = 57.9 Inter5= 47.4	Yes	1 year	IIG = Inter5
Titov, Andrews, Schwencke et al. (2008)	Internet-based ^{ce}	IIG = 50 WL = 55	Yes (phone)	SP (CIDI)	IIG = 22 WL = 0	No	6 month[2]	IIG > WL
Titov, Andrews and Schwencke (2008)	Internet-based ^{ce}	IIG = 43 WL = 45	Yes (phone)	SP (CIDI)	IIG = 19.5 WL = 0	No	6 month[2]	IIG > WL
Titov, Andrews, Choi et al. (2008)	Internet-based ^{ce}	ThIIG = 32 SgIIG^{wc}= 31 WL = 35	Yes (phone)	SP (MINI)	ThIIG = 22.6 SgIIG = 66.7 WL = 2.9	No	-	ThIIG > SgIIG = WL
Titov, Andrews, Choi et al. (2009)	Internet-based^{cp}	IIG^{cp} = 84 IIG^{wc} = 84	Yes (phone)	SP (MINI)	IIG^{cp} = 21.4 IIG^{wc} = 32.5	No	-	IIG^{cp} > IIG^{wc}
Titov, Andrews, Schwencke et al. (2009)	Internet-based^{cp}	IIG^{cp} = 43 IIG^{of} = 42	Yes (phone)	SP (MINI)	IIG^{cp} = 20.9 IIG^{of} = 24.4	No	-	IIG^{cp} = IIG^{of}
Berger et al. (2009)	Internet-based ^{ce}	IIG = 31 WL = 21	Yes (FF)	SP (SCID)	IIG = 48.4 WL = 9.5	Yes	-	IIG > WL
Gallego (2007)	Internet-based ^{wc}	IIG = 62 FFT = 36 WL = 29	Yes (FF)	SP (ADIS)	IIG = 51.6 FFT = 38.9 WL = 13.8	Yes*	6 month 1 year[3]	IIG = FFT > WL

Note: Clinical sign. inf.: Clinical significance information; Inter2: Internet-based treatment plus two group exposure sessions; WL: Waiting list; FF: face-to-face assessment; SP: Social Phobia; SCID: Structured clinical interview for DSM-IV; A>B: Treatment A superior to treatment B; ^{cp}: contact with the therapist by phone; IIG: Internet intervention group; [1]: 2.5 years follow-up is given in Carlbring, Nordgren, Furmark and Andersson (2009); ^{ce}: contact with the therapist by e-mail; Inter5: Internet-based treatment plus five group exposure sessions; A = B: Treatment A is as effective as treatment B; CIDI: Composite International Diagnostic Interview; [2]: follow-up was informed in the study of Titov, Andrews, Johnston, Schwencke and Choi (2009); ThIIG: a therapist-assisted Internet intervention; SgIIG: a self-guided Internet intervention; ^{wc}: without contact; MINI: Mini International Neuropsychiatric Interview; A > B = C: Treatment A is more effective than B and treatment B is as effective as treatment C; ^{of}: online discussion forums moderated by a clinician; FFT: face-to-face treatment; ADIS: Anxiety Disorders Interview Schedule; * formal diagnosis after the treatment; [3]: one year follow-up is given in Botella et al. (2010); A = B > C: Treatment A is as effective as treatment B and treatment B is more than treatment C.

Table 5. Scheme of the studies conducted in depression

Study	Type of treatment	Other conditions	Subjects per condition	Formal diagnosis	Inclusion criteria	Drop-out rate (%)	Clinical sign. inf.	Follow-up	Efficacy
Christensen, Griffiths and Jorm (2004)	MG^{cp}	BP^{cp} CGP^{cp}	MG = 182 BP = 166 CGP =178	No	KPDS ≥ 22	MG = 25.3 BP = 15.2 CGP =10.7	No	1 year[1]	MG = BP > CGP
Christensen et al. (2006)	MG^{wc}	$V1^{wc}$, $V2^{wc}$, $V3^{wc}$, $V4^{wc}$, $V5^{wc}$	V1 = 464, V2 = 468, V3 = 465, V4 = 463, V5 = 466, MG = 468	No	GDS of 5.96 (*SD* = 2.09)	*V1 = 75, V2 = 72 , V3 = 73, V4 = 72, V5 = 75, MG = 77	No	-	V1=V2=MG V4=V5>V1 V4>V2=MG V4=V3
Clarke et al. (2002)	$ODIN^{wc}$	UTCG	ODIN= 144 UTCG = 155	No	depressed/ nondepressed	**73.6%	No	7.5 month[2]	ODIN = UTCG [S] ODIN > UTCG [d]
Clarke et al. (2005)	$ODIN^{cpc}$	$ODIN^{cph}$ UTCG	$ODIN^{cpc}$ = 75 $ODIN^{cph}$ = 80 UTCG = 100	No	depressed/ nondepressed	**$ODIN^{cpc}$=27 $ODIN^{cph}$ = 24 UTCG = 7	Yes[a]	3.7 month[2]	$ODIN^{cpc}$ = $ODIN^{cph}$ > UTCG
Andersson et al. (2005)	$IIG+WD^{ce}$	WD^{ce}	IIG+WD = 57 WD = 60	No	CIDI-SF	IIG+WD=36.8 WD = 18.3	No	6 month	IIG+WD > WD
Spek et al. (2007)	$ICBT^{wc}$	GCBT WL	ICBT = 102 GCBT = 99 WL = 100	Yes (FF)	Depression (CIDI)	ICBT = 51.7 GCBT = 5.5 WL = 42	Yes[a]	1 year[3]	ICBT = GCBT > WL
Warmerdam et al. (2008)	$ICBT^{ce}$	$IPST^{ce}$ WL	ICBT = 88 IPST = 88 WL = 87	No	CES-D > 16	ICBT = 61.4 IPST = 62.5 WL = 18.4	Yes	2.8 month[2]	ICBT = IPST > WL
Van Straten et al. (2008)	IIG^{aef}	WL	IIG = 107 WL = 106	No	General Population	IIG = 45 WL = 9.4	Yes	-	IIG > WL
Perini et al. (2009)	IIG^{ce}	WL	IIG = 29 WL = 19	Yes (phone)	Depression (MINI)	IIG = 26 WL = 5.6	Yes[b]	-	IIG > WL
Ruwaard et al. (2009)	TGW	WL	TGW = 36 WL = 18	No	BDI between 10-29	TGW = 9 WL = 11.1	Yes	18 month	TGW > WL

Note: Clinical sign. inf.: Clinical significance information; MG: MoodGym is a CBT website to treat depression; cp: contact with the therapist by phone; BP: BluePages is a website with information about depression; CGP: a control intervention using an attention placebo; KPDS: Kessler Psychological Distress Scale; [1]: one year follow-up in Mackinnon, Griffiths and Christensen (2008); A = B > C: Treatment A is as effective as treatment B and treatment B is more than treatment C; wc: without contact; A = B: Treatment A is as effective as treatment B; V1: Brief CBT; V2: Brief CBT with problem solving; V3: Brief CBT with stress and problem solving; V4: Extended CBT and problem solving; V5: Extended CBT, behavioural strategies and problem solving; GDS: Goldberg Depression Scale;*: The dropout rate in this study is the number of people that have not completed the postest; **: The dropout rate in this study is the number of people that have not completed at least one follow-up; ODIN: Overcoming Depression on the Internet; Depressed: people that received depression medication or psychotherapy in the previous 30 days; Nondepressed: people who did not receive depression medication or psychotherapy; [2]: the length of the follow-up is post-randomization; A>B: Treatment A superior to treatment B; [S]: for the entire sample; [d]: for low depressed participants; [cpc]: contact with the therapist through postcards; [cph]: contact with the therapist by phone; UTCG: usual treatment control group; [a]scores in the CES-D or the Beck Depression Inventory below the cutoff; IIG+WD: internet-based CBT plus participation in a web-based discussion group; [ce]: contact with the therapist by e-mail during the treatment; WD: web-based discussion group only; CIDI-SF: a computerized version of the Composite International Diagnostic Interview-Short Form; ICBT: Internet-based CBT; GCBT: a group CBT treatment; FF: Face-to-face; CIDI: Composite International Diagnostic Interview; [3]: one year follow-up in Spek et al., (2008); CES-D: Center of Epidemilogic Studies Depression Scale; IPST: Internet-based Problem Solving Therapy; [aef]: automated e-mail every week and feedback by e-mail from a psychology student; MINI: Mini International Neuropsychiatric Interview; [b]: Patient Health Questionnaire-9 (PHQ-9) < 10 and a reduction of PHQ-9 scores of at least 50%; TGW: Therapist-guided web-based cognitive behavioural treatment; BDI: Beck Depression Inventory.

Only three of the six controlled studies conducted in mixed anxiety used structured psychiatric interviews based on the DSM-IV (Diagnostic and Statistical Manual of Mental Disorders) or the ICD-10 (World Health Organization, 1992) to assess their participants. Attrition rates in these studies are relatively small and ranged from 0% (Kenardy et al., 2006) to 28% (Orbach et al., 2007) for CBT treatments with exposure instructions; the CBT Internet program without exposure instruction had a higher dropout rate (35%) (Schneider et al., 2005). In these studies the degree of contact with the participants does not seem to be related to the dropout rate (see Table 1). Clinical significance information was given in four of the six controlled studies; two were based on Jacobson and Truax (1991) criteria and the other two used an alternative to these criteria. Two studies reported pre-post data only, while the other four also had a follow-up assessment; the mean length of follow-up was 5.75 (*SD* = 4.66).

The CBT Internet treatment for mixed anxiety has shown to be significantly more effective than the control waiting list in all studies (see Table 1). The CBT treatment with or without exposure instructions have shown to be equally effective at post-test, although at 1 month follow-up the treatment with exposure instructions showed to be significantly more effective (Schneider et al., 2005). In the case of the comparison of a CBT Internet treatment and a brief live-exposure at post-test the in vivo exposure was significantly more effective, but at one year follow-up both treatments were equally effective (Andersson et al., 2009).

Post-Traumatic Stress Disorder (PTSD)

In the area of the Internet-based treatments for PTSD six RCTs have been published (see Table 2). In five of the six studies the control condition was a waiting list control group and the other study used an Internet-based supportive counseling as a comparison condition (Litz et al., 2007). Only in one of these studies (Litz et al., 2007) participants were assessed through a face-to-face structured Interview (PTSD Symptom Scale—Interview Version; Foa and Tolin, 2000), the rest of the studies did not provide a formal diagnosis.

Drop-out rate varies considerably ranging from 30 % Lange et al. (2003) to 8 % (Wagner et al., 2006). All studies provided information with respect to the clinical significance of the improvements achieved; four of the six studies based the clinical significance change on Jacobson and Truax (1991) criteria. As to follow-up, five of the studies reported follow-up data, the mean duration of the follow-up was 3 months (*SD* = 1.84).

Results of the RCTs reveal that Internet-based CBT treatment for PTSD is more effective than no treatment and an Internet-based supportive counseling, but Internet-based CBT treatment has not been compared with a traditional face-to-face CBT. E-mail treatment for PTSD (e.g. Wagner et al., 2006) may result in a lower dropout rate than more structured Internet-based treatments (e.g. Lange et al., 2003), although the reasons for this are unclear and might also be related to differential selection criteria. Finally, even though most of the studies do provide follow-up data, this is still far from the desirable length of one year.

Panic Disorder (PD)

Studies about Internet-based treatments for PD have proliferated in the last decade; there are twelve controlled studies in this area, ten were randomized studies and two non-randomized (Pier et al., 2008; Shandley et al., 2008). The comparison control conditions were in three studies a waiting list control condition (Carlbring et al., 2001, 2006; Rewaard et al., 2010a), two studies used an information-only control condition (Klein, Richards and Austin,

2006; Richards, Klein and Austin, 2006) and another one used a self-monitoring control condition (Klein and Richards, 2001). Internet-based treatments for PD have also been compared with face-to-face treatments in two controlled studies (Carlbring et al., 2005; Kiropoulos et al., 2008). Given the efficacy of Internet-based treatments for PD there are authors that have dedicated their efforts to contrast different Internet treatments (Carlbring, Ekselius and Andersson, 2003; Klein et al., 2009; Richards et al., 2006) or the same treatment with diverse kind of contact (Pier et al., 2008; Shandley et al., 2008). These treatments have also been compared with other type of self-help, Klein et al. (2006) used a therapist-assisted manual. The diagnosis in all but one (Ruwaard et al., 2010a) studies was assessed by structured Interviews based on DSM-IV criteria, either carried out face-to-face, or by phone or via e-mail. Ruwaard et al. (2010a) did not use a validated interview for the diagnosis.

The attrition rate for the Internet-based PD studies has been variable, the highest dropout percentage was 47.2 % for an Internet intervention plus face-to-face contact with a general practitioner (GP) (Shandley et al., 2008), the lowest dropout rate was 5% for an Internet-based treatment plus contact via e-mail with a psychologist (Klein et al., 2006). The mean dropout for all Internet conditions was 19.27%, for the no treatment conditions was 12.31% and for the face-to-face treatments was 8.6%. Furthermore, the frequency of therapist support is independent of the effectiveness of Internet-based therapy (Klein et al., 2009). Information about clinical significance was provided in ten of the controlled studies; four of them based their data on Jacobson and Truax (1991) criteria and the other six studies combined two criteria to define high end-state functioning (panic-free status and clinician severity rating of PD). Follow-up results were reported in six of the twelve studies, the mean length of the follow-up was 11.5 months ($SD = 12.5$).

The Internet-based treatments for PD were found to be more effective than a variety of control conditions (waiting list, self-monitoring and information-only), and to be equally effective as face-to-face treatments and a therapist-assisted self-help manual. Generally, treatment adherence is higher for face-to-face treatments than for Internet-based interventions.

Social Phobia (SP)

Internet-based treatments for SP have been intensively studied in the last five years; ten controlled studies have been carried (see Table 4). In the first controlled study the experimental condition was a combination of an Internet CBT treatment and two group exposure sessions (Andersson et al., 2006); another study compared an Internet-based treatment with an Internet-based treatment combined with five group exposure sessions (Tillfors et al., 2008). Seven of the ten studies used a waiting list as a control condition . Only in one study the comparison condition was a face-to-face treatment (Gallego, 2007). All the studies used validated structured interviews by phone or face-to-face to formally diagnose the social phobia (see Table 4).

The dropout rate for the Internet-based treatment for social phobia ranges from 13.3% (Carlbring et al., 2007) to 66.7% (Titov, Andrews, Choi et al., 2008). In the Carlbring et al. (2007) study short weekly telephone calls were added and in the Titov, Andrews, Choi et al. (2008) participants did not have any contact with the therapist. The mean attrition rate for the Internet conditions without exposure sessions was 33.42% and for the no treatment conditions was 4.7%. Only one study used a face-to-face condition with a dropout rate of 38.9% (Gallego, 2007).

Four of the ten studies provided information about clinical significant change, three based this information on the Jacobson and Truax (1991) criteria. Follow-up assessment was done in six studies and the mean length of this assessment was 13 months (*SD* = 8.8).

Results reveal that Internet-based treatment for social phobia was more effective than waiting list. There was one exception: a self-guided Internet intervention without any contact with the therapist did not better than waiting-list control (Titov, Andrews, Choi et al., 2008). Moreover, added exposure sessions to the treatment did not improve the treatment outcome (Tillfors et al., 2008). Face-to-face treatment for fear of public speaking showed to be as effective and accepted as the same treatment applied over the Internet without any contact with a therapist (Botella et al., 2009; Gallego, 2007). Nevertheless, contact with the therapist during treatment increases treatment compliance and enhances treatment outcome (Titov, Andrews, Choi et al., 2008, 2009).

Depression

Ten controlled studies evaluated the efficacy of Internet interventions for depression symptoms (see Table 5). Eight of those studies used a control condition: in five studies this was a waiting list (Perini et al., 2009; Spek et al. 2007; Ruwaard et al., 2009; Van Straten, Cuijpers and Smits, 2008; Warmerdam, Van Straten, Twisk, Riper and Cuijpers, 2008); in one study attention placebo (Christensen, Griffiths and Jorm, 2004); and in another one (Clarke et al., 2002, 2005) care as usual. Only one study compared the effectiveness of an Internet-based CBT with the effectiveness of group CBT (Spek et al., 2007). Other studies investigated ways to enhance the effects of Internet-based interventions, for example by comparing brief interventions with longer interventions (Christensen, Griffiths, Mackinnon and Brittliffe, 2006) and adding participation in a web-based discussion group (Andersson et al., 2005), or comparing Internet CBT treatment with an Internet problem solving therapy (Warmerdam et al., 2008). Most of the studies did not assess the participants formally; in only two studies participants were assessed with a structured diagnostic interview (Perini et al., 2009; Spek et al., 2007).

The highest attrition rate was in a study of Christensen et al. (2006) in an Internet CBT intervention for depression without any contact with a therapist during the treatment. The mean attrition rate for all Internet-based conditions was 48.4% and for the no treatment conditions was 14.9%. The face-to-face treatment resulted in a dropout rate of 5.5%, but this was one study only (Spek et al., 2007). Data on clinical significance was reported in six of the nine studies (see Table 5). Three studies took into account the Jacobson and Truax (1991) criteria to find out the percentage of participants that improved clinically. However, Clarke et al. (2005) and Spek et al. (2007) considered a clinical significant improvement when scores in the CES-D (Center of Epidemiologic Studies Depression Scale) or the BDI (Beck Depression Inventory) were below the cut-off. Perini et al. (2009) based the clinical improvement on the cut-off of the Patient Health Questionnaire-9 (PHQ-9). Follow-up was done in seven of the nine controlled studies, four of these studies counted the length of follow-up from the post-test, in the other three studies the length of the follow-up was post-randomization (Clarke et al., 2002, 2005; Warmerdam et al., 2008). Based on the studies with follow-up from posttest, the mean was 12 months (*SD* = 4.9). Internet-based treatment for depression was found to be as effective as face-to-face treatment, but this is based on one study only. Internet-based CBT programs were equally effective as Internet-based problem solving therapy and an Internet website with information about depression. A web-based discussion group alone was not found to be effective. As to the length of the Internet-based CBT programs, longer programs were associated with better outcomes. Internet-based treatments have shown to be more effective than other control condition (waiting list, care as usual and attention placebo). One deficiency of the studies into Internet-based treatments for depression is the lack of formal diagnosis in most studies; another deficiency is that the information about the clinical significance of the improvements achieved is given in only 60% of the studies. Further, there is some evidence that face-to-face treatments have lower dropout rates than Internet-based interventions. Briefer treatments and contact with the therapist during the treatment may decrease the attrition rate.

RECENT APPLICATION

Apart from emotional disorders, Internet psychological treatments have been applied to substance abuse, eating disorders and a variety of other health problems. First of all we focus our attention on Internet treatments for substance abuse like alcohol, tobacco and other substances.

Internet-Based Treatments for Substance Abuse

Internet-based treatments for smoking cessation have been used in the last decade. Internet-based treatments were found to be more effective than a waiting list control group at short-term (Swartz, Noell, Schroeder and Ary 2006), but long term follow-up results are disappointing. Muñoz et al. (2009) reported that one year after the self-administration of an Internet treatment for smoking cessation only 20% of the participants were still abstinent. Shahab and McEwen (2008) in a systematic review found that web-based, tailored, interactive smoking cessation interventions were more effective than untailored booklet or e-mail interventions. Two recent controlled studies found an interactive Internet smoking cessation program effective for participants without depressed affect (Seidman et al., 2010; Rabius, Pike, Wiatrek and McAlister, 2008). There is a clear need of studies which compare the effectiveness of Internet-based treatments to give up smoking with other types of therapy (e.g., face-to-face therapy) and studies into predictors of outcome. Further, studies are especially needed in clinically relevant samples such as patients diagnosed with cancer.

As to Internet-based interventions for alcohol abuse, most of the studies were carried out in student populations, which limit the conclusions which can be drawn. Elliott, Carey and Bolles (2008) reviewed research on Internet-based intervention for students and concluded that these interventions were in a number of studies as effective as alternative treatments (e.g., CBT, brief motivational interviews), in some studies more effective than educational interventions, and generally more effective than control conditions. A recent review included Internet-based studies with participants of the general community (Vernon, 2010). Generally, Internet-treatment completers experienced some improvement; however, drop-out rates in these studies were rather high. In a recent meta-analysis (Rooke, Thorsteinsson, Karpin,

Copeland and Allsop, 2010) on computer-delivered interventions for alcohol and tobacco treatment outcome was not related to treatment location, provision of normative feedback, inclusion of a discussion element, number of treatment sessions, inclusion of relapse prevention, amount of therapist participation or length of follow-up.

Internet-Based Treatments for Eating Disorders

Controlled studies in the area of Internet-based interventions for eating disorders have focused primarily on prevention. In most of these studies the Internet treatment was compared with a control condition (waiting list) and generally such prevention programs were more effective than the control condition (e.g. Celio et al., 2000; Heinicke, Paxton, McLean and Wertheim, 2007; Taylor et al., 2006; Zabinski, Wilfley, Calfas, Winzelberg and Taylor, 2004).

Two studies have investigated an Internet-based intervention directed to body dissatisfaction and associated eating disorders. Results were mixed. Gollings and Paxton (2006) found a face-to-face treatment as effective as an Internet intervention, but this was not replicated in a larger study (Paxton, McLean, Gollings, Faulkner andWertheim, 2007).

Internet-based CBT of bulimia nervosa has been studied in six controlled trials (Ljotsson et al., 2007; Nevonen, Mark, Levin, Lindström and Paulson-Karlsson, 2007; Mitchell et al., 2008; Robinson and Serfaty, 2008; Ruwaard et al., 2010b; Fernandez-Aranda et al., 2009). Apart from the study of Robinson and Sefarty (2008) where Internet-based treatment was not effective, effects were large and similar to face-to-face CBT in the studies of Ljotsson et al. (2007) and Mitchell et al.(2008), large and more effective than bibliotherapy (Ruwaard et al., 2010b) and small-to-moderate in Nevonen et al. (2007) and Fernandez-Aranda et al. (2009). Treatment adherence was low across studies, with dropout varying from 26% to 82%.

Internet-Based Treatments for Health Problems

Cognitive-behavioral interventions have shown to have positive effects on health conditions such as pain, sleep problems, headache, cancer and many others. Given that CBT is highly structured it has also been applied through the Internet. Cuijpers, Van Straten and Andersson (2008) conducted a systematic review of randomized controlled trials on Internet-based CBT for health problems. These authors found that Internet-based CBT for pain was more effective than a control group and this was corroborated by a more recent study of Palermo, Wilson, Peters, Lewandowski and Somhegyi (2009). The evidence for the effectiveness of Internet-based CBT for headache is limited, results were small (Ström, Pettersson and Andersson, 2000) or moderate (Devineni and Blanchard, 2005). There are also studies on the effectiveness of Internet-based CBT treatments for chronic diseases (Lorig, Ritter, Laurent and Plant, 2006), breast cancer patients (Owen et al., 2005), tinnitus (Andersson, Strömgren, Ström and Lyttkens, 2002), physical disabilities (Hopps, Pepin and Boisvert, 2003) and pediatric brain injury (Wade, Carey and Wolfe, 2006), again with small to moderate effect size. Finally, Abbott et al. (2009) did not find an Internet CBT program for tinnitus to be more effective than an information program.

In general, Internet CBT programs for health problems may have some promise, but the promise is as yet unfulfilled given that the effect sizes are generally lower than Internet-based CBT for anxiety and depression (Spek et al., 2007).

FUTURE APPLICATIONS: WHAT NEEDS TO BE DONE

Future challenges for Internet-based treatments include many possibilities, some of them predictable and other unforeseen. An important issue is to develop clinically relevant studies with a more rigorous methodology, that is to say, randomized controlled studies, with pre and post-test assessment and short-term and long-term follow-ups up to at least one year post treatment. Such studies should provide also clinically relevant information with respect to the sample included and the clinical significance of the results achieved. The samples should not be restricted to women in their thirties with a high level of education, but be more representative of the community at large. Serious efforts should be undertaken to recruit more men, older and younger people with different levels of education. Furthermore, the sample should be assessed formally using structured clinical interviews to include only clinical participants in the study.

We recommend following the revised CONSORT statement to conduct and report the randomized controlled studies in order to increase the methodological quality (Altman et al., 2001). Apart from carrying out studies to evaluate the efficacy and effectiveness of a full treatment package, it is also needed to assess the efficacy of different modules of a treatment program to find out which component of the program is more effective (Palmqvist, Carlbring and Andersson, 2007). It is also important to study which variables predict treatment outcome. Of note, Andersson, Carlbring and Grimlund (2008) found that participants who scored high on a personality disorder questionnaire (anxious cluster) did worse with Internet-based treatment, but quite well with face-to-face treatment.

Generally, samples in the studies into the effects of Internet-based treatments in anxiety disorders and depression are characterized by highly educated females with an age range from 30 yr to 45 yr. Further, it should be noted that in most currently available programs, patients themselves chose a computer-based treatment rather than face-to-face treatment and are often self-referred. Thus, it is questionable how representative the participants are. Internet-based programs may only be effective for a subset of well-motivated patients who have an interest in this form of communication. There is a clear need of future studies that address characteristics of patients who are suited for tele-therapy.

Improving adherence is another challenge for future Internet-based programs. There is some evidence that contact with a therapist decreases dropout rates (Farvolden, Denisoff, Selby, Bagby and Rudy, 2005) and that shorter programs may have a lower attrition rate. There is a clear need of studies investigating predictors of compliance and treatment adherence.

Another goal for Internet treatment is to tailor them to the participant needs. Comorbidity is very common in clinical setting, for instance the combination of mood disorders and anxiety disorders is not infrequent. An Internet treatment should have modules to deal with co-current problems in order to provide participants an adequate treatment.

Golkaramnay, Bauer, Haug, Wolf and Kordy (2007) have applied group therapy through an Internet Chat as Aftercare. The participants were suffering from different mental disorders like mood disorders, personality or behavior disorders, neurotic, stress or somatoform disorders. This type of treatment had a low dropout rate and the participants improved compared with the control group. Group treatment through chat is a new way to apply treatment that should be investigated further in future studies.

Research in the field of Internet treatments has focused its attention on validating its use as stand alone treatment. Studies are needed in which the combination of Internet-based treatments are combined with other approaches (Andersson et al., 2006; Tillfors et al., 2008; Palmqvist et al., 2007) like pharmacological drugs or life exposure. Internet-based treatments could be applied to other disorders as well like obsessive-compulsive disorders and bipolar disorder. In addition, Internet-based interventions could be developed to support families of individuals with depression, including bipolar disorder, and of anorexics and bulimics.

Although it is generally assumed that Internet-based treatments are more cost-effective than face-to-face therapy, this has hardly been investigated. In studies into cost-effectiveness of Internet-based treatments, the costs of the development and implementation of these programs should be taken into account as well.

Given that Internet-based treatments are efficacious tools to treat some mental health problems, the next step is to introduce these programs in the community settings (Postel et al., 2008). The Stepped Care Model of Green and Iverson (2009) can be useful to establish the use of these treatments in community setting.

Before implementation can be effective, however, therapist needs to be trained in using these treatments. In doing so, we should be prepared to deal with resistance of clinicians towards these technological advances. Many therapists expect that Internet-based treatments may interfere with the natural development of a therapeutic relationship. Among psychotherapists, the value of the therapeutic relationship is felt to be very important. However, as in any therapy, in technology-driven therapies the therapeutic relationship is important (Meyerbröker and Emmelkamp, 2008).

For example, in the Interapy program of Lange et al. (2003) 75% of the patients rated the relationship with their therapist as personal and 88% as pleasant although they never had met the therapist. Of note, in a direct comparison between face-to-face therapy and therapy through the Internet, the quality of the therapeutic relationship was rated as higher in the on-line group (Cook and Doyle, 2002).

CONCLUSION

Internet-based treatments try to overcome some limitations of traditional CBT treatments: they diminish distance between therapist and patient, save therapist (and patient) time, reduce waiting lists, are available for disabled people, allow anonymity and help to face fear of stigmatization (Emmelkamp, 2005; Postel et al., 2008).

Randomized controlled studies that test the efficacy of Internet-based treatments have found that these treatments are as effective as face-to-face therapy for anxiety disorders and depression. Furthermore, Internet-based treatments have shown to be significantly more effective than a waiting list control group in most of the studies into anxiety disorders, depression and eating disorders. It should be noted, however, that participants in these studies were volunteers who applied for psychological treatment through the Internet.

There is some controversy about the need for some face-to-face therapist contact in computer-driven treatment. While there seems to be less need in mild depression, post-traumatic stress and panic disorder, it is questionable whether treatment without any face-to-face support is feasible with severe agoraphobic patients and severely depressed patients. For example, it is much easier not to keep a difficult exposure assignment with an Internet program than with a face-to-face therapist (Emmelkamp, 2005). In addition, contact with the therapist during the Internet-based treatment may prove to be more important than once thought, given that it may increase treatment compliance and improve outcome (Titov, Andrews, Choi et al., 2008, 2009).

Surprisingly, ethical concerns with respect to patients' safety and privacy associated with tele-assessment and therapy are hardly discussed. Clearly, tele-therapy is not suited for all patients (e.g. patients who dissociate, psychotic and suicidal patients) and adequate measures have to be taken that such patients are not enrolled in tele-therapy programs. Further, the system must allow appropriate actions to be taken in case of emergencies. There is a clear and urgent need of guidelines and formal regulation by professional associations of assessment through the Internet and tele-therapy.

REFERENCES

Abbott, J. A. M., Kaldo, V., Klein, B., Austin, D., Hamilton, C., Piterman, L., Williams, B. and Andersson, G. (2009). A cluster randomized trial of an internet-based intervention program for tinnitus distress in an industrial setting. *Cognitive Behaviors Therapy, 38*, 162-173.

Altman, D. G., Schulz, K. F., Moher, D., Egger, M., Davidoff, F., Elbourne, D., Getzsch, P. C. and Lang, T. (2001). The revised CONSORT statement for reporting randomized trials: Explanation and elaboration. *Annals of Internal Medicine, 134*, 663-694.

Anderson, J. J., Jacobs, C. and Rothbaum, B. O. (2004). Computer-Supported Cognitive Behavioral Treatment of Anxiety Disorders. *Journal of Clinical Psychology, 60*, 253-267.

Andersson, G., Bergström, J., Holländare, F., Carlbring, P., Kaldo, V. and Ekselius (2005). L. Internet-based self-help for depression: randomized controlled trial. *British Journal of Psychiatry, 187,* 456–61.

Andersson, G., Carlbring, P. and Grimlund, A. (2008). Predicting treatment outcome in internet versus face to face treatment of panic disorder. *Computers in Human Behavior, 24*, 1790-1801.

Andersson, G., Carlbring, P., Holmström, A., Spartan, L., Furmark, T., Nilsson-Ihrfelt, E., Buhrmen, M. and Ekselius, L. (2006). Internet-based self-help with therapist feedback and in-vivo group exposure for social phobia: a randomised controlled trial. *Journal of Consulting and Clinical Psychology, 74,* 677-686.

Andersson, G., Strömgren, T., Ström, L. and Lyttkens, L. (2002). Randomized controlled trial of internet-based cognitive behavior therapy for distress associated with tinnitus. *Psychosomatic Medicine, 64*, 810-816.

Andersson, G., Waara, J., Jonsson, U., Malmaeus, F., Carlbring, P. and Öst, L. G. (2009). Internet-based self-help versus one-session exposures in the treatment of spider phobia: A randomized controlled trial. *Cognitive Behaviour Therapy, 38*, 114-120.

Andrews, G. and Carter, G. L. (2001). What people say about their general practitioners' treatment of anxiety and depression. *Medical Journal of Australia, 175*, S48-S51.

Berger, T., Hohl, E. and Caspar, F. (2009). Internet-Based treatment for social phobia: A randomized controlled trial. *Journal of Clinical Psychology, 65*, 1021-1035.

Botella, C., Baños, R. M., Guillen, V., Perpiña, C., Alcañiz, M. and Pons, A. (2000). Telepsychology: Public Speaking Fear Treatment on the Internet. *CyberPsychology and Behavior, 3*, 959-968.

Botella, C., Baños, R. M., Perpiña, C., Villa, H., Alcañiz, M. and Rey, B. (1998). Virtual reality treatment of claustrophobia: A case report. *Behaviour Research and Therapy, 36,* 239-246.

Botella, C., Gallego, M. J., Garcia-Palacios, A., Baños, R. M., Quero, S. and Alcañiz, M. (2009). The acceptability of an Internet-based self-help treatment for fear of public speaking. *British Journal of Guidance and Counselling, 37*, 297-311.

Botella, C., Gallego, M. J., Garcia-Palacios, A., Guillen, V., Baños, R. M., Quero, S. and Alcañiz, M. (2010). An internet-based self-help treatment for fear of public speaking: A controlled trial. *Cyberpsychology, Behavior, and Social Networking, 13*, 407-421.

Carlbring, P., Bohman, S., Brunt, S., Buhrman, M.,Westling, B., Ekselius, L. and Andersson, G. (2006). Remote treatment of panic disorder: A randomized trial of Internet-Based Cognitive Behavior Therapy Supplemented with telephone calls. *American Journal of Psychiatry, 163,* 2119-2125.

Carlbring, P., Bohman, S., Brunt, S., Buhrman, M.,Westling, B., Ekselius, L. and Andersson, G. (2007). A randomized trial of Internet-based cognitive behavioural therapy supplemented with telephone calls. *American Journal of Psychiatry, 163*, 2119-2125.

Carlbring, P., Ekselius, L. and Andersson, G. (2003). Treatment of panic disorder via the Internet: a randomized trial of CBT vs. Applied relaxation. *Journal of Behavior Therapy and Experimental Psychiatry, 34,* 129-140.

Carlbring, P., Nilsson-Ihrfelt, E., Waara, J., Ceciclia, K., Buhrman, M., Kaldo, V., Söderberg, M., Ekselius, L. and Andersson, G. (2005). Treatment of panic disorder: live therapy vs. self-help via the Internet. *Behaviour Research and Therapy, 43*, 1321-1333.

Carlbring, P., Nordgren, L. B., Furmark, T. and Andersson, G. (2009). Long-term outcome of internet-delivered cognitive-behavioural therapy for social phobia: A 30 month follow-up. *Behaviour Research and Therapy, 10*, 848-850.

Carlbring, P., Westling, Ljungstrad, P., Ekselius, L., Andersson, G. (2001). Treatment of panic disorder via the Internet: A randomized trial of a self-help program. *Behaviour Therapy, 32*, 751-764.

Carlin, A. S., Hoffman H. G. and Weghorst, S. (1997). Virtual reality and tactile augmentation in the treatment of spider phobia: A case report. *Behaviour Research and Therapy, 35*, 153-158.

Celio, A. A., Winzelberg, A. J., Wilfley, D. E., Springer, E. A., Dev, P. and Taylor, C. B. (2000). Reduciong risk factors for eating disorders: comparison of an internet and a classroom-delivered psychoeducational program. *Journal of Consulting and Clinical Psychology, 68,* 650-657.

Christensen, H., Griffiths, K. M. and Farrer, L. (2009). Adherence in internet interventions for anxiety and depression. *Journal of Medical Internet Research, 11*, e13.

Christensen, H., Griffiths, K. M. and Jorm, A. F. (2004). Delivering interventions for depression by using the Internet: randomised controlled trial. *British Medical Journal, 328*, 265-268.

Christensen, H., Griffiths, K. M., Mackinnon, A. J. and Brittliffe, K. (2006). Online randomized controlled trial of brief and full cognitive behavior therapy for depression. *Psychological Medicine, 36*, 1737-1746.

Clarke, G., Eubanks, D., Reid, E., Kelleher, C., O'Connor, E., DeBar, L. L., Lynch, F., Nunley, S. and Gullion, C. (2005). Overcoming Depression on the Internet (ODIN) (2): A randomized Trial of a Self-Help Depression Skills Program With Reminders. *Journal of Medical Internet Research, 7*, e16.

Clarke, G., Reid, E., Eubanks, D., O'Connor, E., DeBar, L. L., Kelleher, C., Lynch, F. and Nunley, S. (2002). Overcoming Depression on the Internet (ODIN): A Randomized Controlled Trial of an Internet Depression Skills Intervention Program. *Journal of Medical Internet Research, 4*, e14.

Cook, J. E. and Doyle, C. (2002). Working alliance in online therapy as compared to face-to-face therapy: preliminary results. *Cyberpsychology and Behavior, 5*, 95-105,

Cuijpers, P., Van Straten, A. and Andersson, G. (2008). Internet-administered cognitive behavior therapy for health problems: A systematic review. *Journal of Behavioral Medicine, 31*, 169-177.

Deacon, B. J. and Abramowitz, J. S. (2004). Cognitive and Behavioral Treatments for Anxiety Disorders: A Review of Meta-analytic Findings. *Journal of Clinical Psychology, 60*, 429-441.

Devineni, T. and Blanchard, E. B. (2005). A randomized controlled trial of an internet-based treatment for headache. *Behaviour Research and Therapy, 43*, 277-292.

Ekers, D., Richards, D. and Gilbody, S. (2008). A meta-analysis of randomized trials of behavioural treatment of depression. *Psychological Medicine, 38*, 611-623.

Elliott, J. C., Carey, K. B. and Bolles, J. R. (2008). Computer-based interventions for college drinking: A qualitative review. *Addictive Behaviors, 33*, 994-1005.

Emmelkamp, P. M. G. (2005). Technological innovations in clinical assessment and psychotherapy. *Psychotherapy and Psychosomatics, 74*, 336-343.

Emmelkamp, P. M. G., Kjrin, M., Hulsbosch, A. M., de Vries, S., Schuemie, M. J. and van der Mast, C. A. P. G. (2002). Virtual reality treatment versus exposure in vivo: A comparative evaluation in acrophobia. *Behaviour Research and Therapy, 40*, 509-516.

Farvolden, P., Denisoff, E., Selby, P., Bagby, R. M. and Rudy, L. (2005). Usage and longitudinal effectiveness of a web-based self-help cognitive behavioral therapy program for panic disorder. *Journal of Medical Internet Research, 7*, e7.

Fernandez-Aranda, F., Nunez, A., Martinez, C., Krug, I., Cappozzo, M. and Carrard, I. (2009). Internet-based cognitive-behavioral therapy for bulimia nervosa: A controlled study. *CyberPsychology and Behavior*, 12, 37-41

Foa, E. B. and Tolin, D. F. (2000). Comparison of the PTSD Symptoms Scale-Interview Version and the Clinical-Administered PTSD Scale. *Journal of Traumatic Stress, 13*, 181-191.

Gallego, M.J. (2007). *An internet-based self-help treatment for fear of public speaking: A controlled trial.* Doctoral dissertation, Universidad Jaume I of Castellón.

Gaston, J. E., Abbott, M. J., Rapee, R: M. and Neary, S. A. (2006). Do empirically supported treatments generalize to private practice? A benchmark study of a cognitive-behavioural group treatment programme for social phobia? *British Journal of Clinical Psychology, 45*, 33-48.

Golkaramnay, V., Bauer, S., Haug, S., Wolf, M. and Kordy, H. (2007). The exploration of the effectiveness of group therapy through internet chat as aftercare. A controlled naturalistic study. *Psychotherapy and Psychosomatics, 76*, 219-255.

Gollings, E. and Paxton, S. J. (2006). Comparison of internet and face-to-face delivery of a group body image and disordered eating intervention for women: A pilot study. *Eating Disorders, 14*, 1-15.

Green, K. E. and Iverson, K.M. (2009). Computerized cognitive-behavioral therapy in a stepped care model for treatment. *Professional Psychology Research and Practice, 40*, 96-103.

Greist, J. H., Marks, I. M., Baer, L., Kobak, K. A., Wenzel, K. W., Hirsch, M. J., Mantle, J. M. and Clary, C. M. (2002). Behavior therapy for obsessive-compulsive disorder guided by a computer or by a clinician compared with relaxation as a control. *Journal of Clinical Psychology, 63*, 138-145.

Heinicke, B. E., Paxton, S. J., McLean, S. A. and Wertheim, E. H. (2007). Internet-delivered targeted group intervention for body dissatisfaction and disordered eating in adolescents girls. A randomized controlled trial. *Journal of Abnormal Child Psychology, 35*, 379-391.

Hirai, M. and Clum, G. A. (2005). An internet-based self-change program for traumatic event related fear, distress, and maladaptive coping. *Journal of Traumatic Stress, 18*, 631-636.

Hoek, H. W. and Van Hoeken, D. (2003). Review of the prevalence and incidence of eating disorders. *International Journal of Eating Disorders, 43*, 383-396.

Hopps, S. L. Pepin, M. and Boisvert, J. M. (2003). The effectiveness of cognitive-behavioral group therapy for loneliness via interrelay-chat among people with physical disabilities. *Psychotherapy, 40*, 136-147.

Issakidis, C. and Andrews, G. (2002). Service utilization for anxiety in an Australian community sample. *Social Psychiatry and Psychiatric Epidemiology, 37*, 153-163.

Jacobson, N. S. and Truax, P. (1991). Clinical significance: A statistical approach to defining meaningful change in psychotherapy research. *Journal of Consulting and Clinical Psychology, 59*, 12-19.

Kenardy, J., McCafferty, K. and Rosa, V. (2003). Internet-delivered indicated prevention for anxiety disorders: A randomized controlled trial. *Behavioural and Cognitive Psychotherapy, 31*, 279-289.

Kenardy, J., McCafferty, K. and Rosa, V. (2006) Internet-delivered indicated prevention for anxiety disorders: Six-month follow-up. *Clinical Psychologist, 10*, 39-42.

Kessler, R. C., Wai, T.C., Jin, R., Ruscio, A. M., Shear, K. and Walters, E. E: (2006). The epidemiology of panic attacks, panic disorder, and agoraphobia in the National Comorbidity Survey Replication. *Archives of General Psychiatry, 63*, 415-424.

Kiropoulos, L. A., Klein, B., Austin, D.W., Gilson, K., Pier., C., Mitchell, J. and Ciechomski, L. (2008). Is internet-based CBT for panic disorder and agoraphobia as effective as face-to-face CBT? *Journal of Anxiety Disorders, 22*, 1273-1284.

Klein, B., Austin, D., Pier, C., Kiropoulos, L., Shandley, K., Mitchell, J., Gilson, K. and Ciechomski, L. (2009). Internet-based treatment for panic disorder: does frequency of therapist contact make a difference? *Cognitive Behaviour Therapy, 38*, 100-113.

Klein, B. and Richards, J. C. (2001). A brief internet-based treatment for panic disorder. *Behavioural and Cognitive Psychotherapy, 29*, 131-136.

Klein; B., Richards, J. C. and Austin, D. W. (2006). Efficacy of internet therapy for panic disorder. *Journal of Behavior Therapy and Experimental Psychiatry, 37*, 213-238.

Knaevelsrud, C. and Maercker, A. (2007). Internet-based treatment for PTSD reduces distress and facilitates the development of a strong therapeutic alliance: a randomized controlled trial. *BMC Psychiatry, 19*, 7-13.

Krijn, M., Emmelkamp, P. M. G., Olafsson, R. P. and Biemond, R. (2004). Virtual reality exposure therapy of anxiety disorders: A review. *Clinical Psychology Review, 24*, 259-281.

Krijn, M., Emmelkamp, P.M.G., Olafsson, R.P., Bouwman, M., van Gerwen, L., Spinhoven, P., Schuemie, M.J. and van der Mast, C.A.P.G.(2007) Fear of flying treatment methods: Virtual reality exposure vs. cognitive behavioral therapy. *Aviation, Space and Environmental Medicine,* 78:121-128.

Landau, B. M. (2001). Psychotherapy online in 2001: For psychotherapist new to the Internet. *Journal of Mental Imagery, 25*, 62-82.

Lange, A., Rietdijk, D., Hudcovicova, M., van de Ven, J. P., Schrieken, B. and Emmelkamp, P. M. (2003). Interapy: a controlled randomized trial of the standardized treatment of posttraumatic stress through the internet. *Journal of Counsulting And Clinical Psychology, 71*, 901-9.

Lange, A., Schrieken, B., van de Ven, J. P., Bredeweg, B., Emmelkamp, P. M. G., van der Kolk, J., Lydsdottir, L., Massaro, M. and Reuvers, A. (2000). "Interapy": the effects of a short protocolled treatment of posttraumatic stress and pathological grief through the Internet. *Behavioural and Cognitive Psychotherapy, 28*, 175-192.

Lange, A, van de Ven, J. P., Schrieken, B. and Emmelkamp, P. M. G. (2001). Interapy. Treatment of posttraumatic stress through the Internet: a controlled trial. *Journal of Behavior Therapy and Experimental Psychiatry, 32*, 73-90.

Litz, B. T., Engel, C. C., Bryant, R. A. and Papa, A. (2007). A randomized, controlled proof-of-concept trial of an internet-based, therapist-assissted self-management treatment for posttraumatic stress disorder. *American Journal of Psychiatry, 164*, 1676-1683.

Ljotsson, B., Lundin, C., Mitsell, K., Carlbring, P., Ramklint, M. and Ghaderi, A.(2007). Remote treatment of bulimia nervosa and binge eating disorder: A randomized trial of internet-assisted cognitive behavioural therapy. *Behaviour Research and Therapy, 45*, 649-661.

Lorig, K. R., Ritter, P. L., Laurent, D. D. and Plant, K. (2006). Internet-based chronic disease self-management: A randomized trial. *Medical Care, 44*, 964-971.

Mackinnon, A., Griffiths, K. M. and Christensen, H. (2008). Comparative randomized trial of online cognitive-behavioural therapy and an information website for depression: 12-Month outcomes. *British Journal of Psychiatry, 192*, 130-134.

Marks, I. M., Kenwright, M., McDonough, M., Whittaker, M., O'Brien, T. and Mataix-Cols, D. (2004). Saving clinicians' time by delegating routine aspects of therapy to a computer: a randomised controlled trail in panic/ phobia disorder. *Psychological Medicine, 34*, 9-17.

Menchola, M., Arkowitz, H. S. and Burke, B. L. (2007). Efficacy of self-administered treatments for depression and anxiety. *Professional Psychology Research and Practice, 38*, 421-429.

Meyerbröker, K. and Emmelkamp, P.M.G. (2008). Therapeutic processes in virtual reality exposure therapy: The role of cognitions and the therapeutic alliance. *Journal of CyberTherapy and Rehabilitation, 1*, 247-257.

Meyerbröker, K. and Emmelkamp, P.M.G. (2010).Virtual Reality Exposure Therapy in Anxiety Disorders: A Systematic Review of Process- and Outcome Studies. *Depression and Anxiety, 27, 933-944.*

Mitchell, J. E., Crosby, R. D., Wonderlich, S. A., Crow, S., Lancaster, K. and Simonich, H. (2008). A randomized trial comparing the efficacy of cognitive-behavioral therapy for bulimia nervosa delivered via telemedicine versus face-to-face. *Behaviour Research and Therapy, 46,* 581-592.

Muñoz, R. F., Barrera, A. Z., Delucchi, K., Penilla, C., Torres, L.D. and Perez-Stable, E. J. (2009). *Nicotine and Tobacco Research, 11*, 1925-1034.

National Institute for Health and Clinical Excellence (2004). *Anxiety: management of anxiety (panic disorder, with or without agoraphobia, and generalized anxiety disorder) in adults in primary, secondary and community care.* London: National Institute for Health and Clinical Excellence.

Nevonen, L., Mark, M., Levin, B., Lindström, M. and Paulson-Karlsson, G. (2007). Evaluation of a new internet-based self-help guide for patients with bulimic symptoms in Sweden. *Nordic Journal of Psychiatry, 60*, 463-468.

Norton, P. J. and Price, E. C. (2007). A meta-analytic review of adult cognitive-behavioral treatment outcome across the anxiety disorders. *Journal of Nervous and Mental Disease, 195*, 521-531.

Orbach, G., Lindsay, S. and Grey, S. (2007). A randomised placebo-controlled trial of a self-help internet-based intervention for test anxiety. *Psychoanalytic Psychology, 24*, 150-166.

Osgood-Hynes, D. J., Greist, J. H., Marks, I. M., Baer, L., Heneman, S. W., Wenzel, K. W., Manzo, P. A., Parkin, J. R., Spierings, C. J., Dottl, S. L. and Vitse, H. M. (1998). Self-administered psychotherapy for depression using a telephone- accessed computer system plus booklets: An open U.S.-U.K. study. *Journal of Clinical Psychiatry, 59*, 358-365.

Owen, J. E., Klapow, J. C., Roth, D. L., Shuster, J. L., Bellis, J., Meredith, R. and Tucker, D. C. (2005). Randomized pilot of a self-guided internet coping group for women with early-stage breast cancer. *Annals of Behavioral Medicine*, 30, 54-64.

Palermo, T. M., Wilson, A. C., Peters, M., Lewandowski, A. and Somhegyi, H. (2009). Randomized controlled trial of an internet-delivered family cognitive-behavioral therapy intervention for children and adolescents with chronic pain. *Pain, 146*, 205-213.

Palmqvist, B., Carlbring, P. and Andersson, G. (2007). Internet-delivered treatments with or without therapist input: Does the therapist factor have implications for efficacy and cost? *Expert Review of Pharmacoeconomics and Outcomes, 7*, 291-297.

Paxton, S. J., McLean, S. A., Gollings, E. K., Faulkner, C. and Wertheim, E. H.. (2007). Comparison of face-to-face and internet interventions for body image and eating problems in adult women: An RCT. *International Journal of Eating Disorders, 40*, 692-704.

Perini, S., Titov, N. and Andrews, G. (2009). Clinician-assisted internet-based treatment is effective for depression: randomized controlled trial. *Australian and New Zealand Journal of Psychiatry, 43*, 571-578.

Peterson, R. A. and Reiss, S. (1992). *The Anxiety Sensitivity Index* (2nd ed). Worthington, OH: International Diagnostic Systems.

Pier, C., Austin, D. W., Klein, B., Mitchell, J., Schattner, P., Ciechomski, L., Gilson, K. J., Pierce, D., Shandley, K. and Wade, V. (2008). A controlled trial of internet-based cognitive-behavioural therapy for panic disorder with face-to-face support from a general practitioner or e-mail support from a psychologist. *Mental Health in Family Medicine, 5*, 29-39.

Postel, M. G., de Haan, H. A. and de Jong, C. A. J. (2008). E-therapy for mental health problems: A systematic review. *Telemedicine and e-Health, 14*, 707-714.

Powers, M. B. and Emmelkamp, P. M. G. (2008). Virtual reality exposure therapy for anxiety disorders: A meta-analysis. *Journal of Anxiety Disorders, 22*, 561-569.

Powers M,B. and Emmelkamp, P.M.G. (2009). Dissemination of Research Findings. In D.C.S. Richard and S.K.Huprich (Eds.). *Clinical psychology: assessment, treatment and research*. New York: Academic Press (pp 495-524).

Proudfoot, J., Goldberg, D., Mann, A., Everitt, B., Marks, I. and Gray, J. A. (2003). Computerized, interactive, multimedia cognitive-behavioural program for anxiety and depression in general practice. *Psychological Medicine, 33*, 217-227.

Rabius, V., Pike, K. J., Wiatrek, D. and McAlister, A. L. (2008). Comparing internet assistance for smoking cessation: 13-month follow-up of a six-arm randomized controlled trial. *Journal of Medical Internet Research, 10*, e45.

Richards, J. C., Klein, B. and Austin, D. W. (2006). Internet cognitive behavioural therapy for panic disorder: Does the inclusion of stress management information improve end-state functioning? *Clinical Psychologist, 10*, 2-15.

Robinson, P., Serfaty, M. (2008). Getting better byte by byte: A pilot randomised controlled trial of email therapy for bulimia nervosa and binge eating disorder. *European Eating Disorder Review, 16*, 84-93.

Roness, A., Mykletun, A. and Dahl, A. A. (2005). Help-seeking behaviour in patients with anxiety disorder and depression. *Acta Psychiatrica Scandinavica, 111*, 51-58.

Rooke, S., Thorsteinsson, E., Karpin, A., Copeland, J. and Allsop, D. (2010). Computer-delivered interventions for alcohol and tobacco use: A meta-analysis. *Addiction, 105*, 1381-1390.

Rothbaum, B. O., Hodges, L., Alarcon, R., Ready, D., Ready, D., Shahar, F., Graap, K., Pair, J., Hebert, P., Gotz, D., Wills, B. and Baltzell, D. (1999). Virtual reality exposure therapy for PTSD Vietnam veterans: A case study. *Journal of Traumatic Stress, 12*, 263-271.

Ruwaard, J., Broeksteeg, J., Schrieken, B., Emmelkamp, P. M. G., and Lange, A. (2010a). Web-based therapist-assissted cognitive behavioral treatment of panic symptoms: A randomized controlled trial with a three-year follow-up. *Journal of Anxiety Disorders, 24*, 387-396.

Ruwaard, J., Lange, A., Broeksteeg, J., Renteria-Agirre' A.' Schrieken, B., Dolan, C., and Emmelkamp, P. M. G.(2010b). Online cognitive behavioral treatment of bulimia nervosa: A randomized controlled trial with a one-year follow-up. (Submitted).

Ruwaard, J., Schrieken, B., Schrijver, M., Broeksteeg, J., Dekker, J., Vermeulen, H. and Lange, A. (2009). Standardized web-based cognitive behavioural therapy of mild to moderate depression: A randomized controlled trial with a long-term follow-up. *Cognitive Behaviour Therapy, 38*, 206-221.

Schneider, A. J., Mataix-Cols, D., Marks, I.M. and Bachofen, M. (2005). Internet-guided self-help with or without exposure therapy for phobic and panic disorders: A randomized controlled trial. *Psychotherapy and Psychosomatics, 74*, 154-164.

Seidman, D. F., Westmaas, J. L., Goldband, S., Rabius, V., Katkin, E. S., Pike, K. J., Wiatrek, D. and Sloan, R. P. (2010). Randomized controlled trial of an interactive internet smoking cessation program with long-term follow-up. *Annals of Behavioral Medicine, 39*, 1-13.

Shahab, L. and McEwen, A. (2008). Online support for smoking cessation: A systematic review of the literature. *Addiction, 104*, 1792-1804.

Shandley, K., Austin, D. W., Klein, B., Pier, C., Schattner, P., Pierce, D. and Wade, V. (2008). Therapist-assissted, internet-based treatment for panic disorder: Can general practitioners achieve comparable patient outcomes to psychologists? *Journal of Medical Internet Research, 10*, e14.

Spek, V., Cuijpers, P., Nyklíček, I., Smits, N., Riper, H., Keyzer, J. and Pop, V. (2008). One-year follow-up results of a randomized controlled trial on internet-based cognitive behavioural therapy for subthreshold depression in people over 50 years. *Psychological Medicine, 38*, 635-639.

Spek, V., Nyklíček, I., Smits, N., Cuijpers, P., Riper, H., Keyzer, J. and Pop, V. (2007). Internet-based cognitive behavioural therapy for subthreshold depression in people over 50 years old: A randomized controlled clinical trial. *Psychological Medicine, 37*, 1797-1806.

Ström, L., Pettersson, R. and Andersson, G. (2000). A controlled trial of self-help treatment of recurrent headache conducted via the Internet. *Journal of Consulting and Clinical Psychology, 68*, 722-727.

Sysko, R. and Walsh, B.T. (2008). A critical evaluation of the efficacy of self-help interventions for the treatment of bulimia nervosa and binge-eating disorder. *International Journal of Eating Disorders, 41*, 97-112.

Swartz, L. H. G., Noell, J. W., Schroeder, S.W. and Ary, D. V. (2006). An randomised control study of a fully automated internet based smoking cessation programme. *Tobbaco Control, 15*, 7-12.

Taylor, C., Bryson, S., Luce, K., Cunning, D., Celio, A., Abascal, L. B., Rockwell, R., Dev, P., Winzelberg, A. J. and Wilfley, D. E., (2006). Prevention of Eating Disorders in at-risk college age women. *Archives of General Psychiatry, 63*, 881-888.

Tillfors, M., Carlbring, P., Furmark, T., Lewenhaupt, S. Spak, M., Eriksson, A., Westling, B.E. and Andersson, G. (2008). Treating university students with social phobia and public speaking fears: Internet delivered self-help with or without live group exposure sessions. *Depression and Anxiety, 25*, 708-717.

Titov. N., Andrews, G., Choi, I., Schwencke, G. and Mahoney, A. (2008). Shyness 3: Randomized controlled trial of guided versus unguided internet-based CBT for social phobia. *Australian and New Zealand Journal of Psychiatry, 42*, 1030-1040.

Titov, N., Andrews, G., Choi, I., Schwencke, G. and Johnston, L. (2009). Randomized controlled trial of web-based treatment of social phobia without clinician guidance. *Australian and New Zealand Journal of Psychiatry, 43*, 913-919

Titov, N., Andrews, G., Johnston, L., Schwencke, G and Choi, I. (2009). Shyness programme: Longer term benefits, cost-effectiveness, and acceptability. *Australian and New Zealand Journal of Psychiatry, 43*, 36-44.

Titov, N., Andrews, G., Robinson, E., Schwencke, G., Johnston, L., Solley, K. and Choi, I. (2009). Clinician-assisted internet-based treatment is effective for generalized anxiety disorder: randomized controlled trial. *Australian and New Zealand Journal of Psychiatry, 43*, 754-759.

Titov, N., Andrews, G. and Schwencke, G. (2008). Shyness 2: Treating social phobia online: Replication and extension. *Australian and New Zealand Journal of Psychiatry, 42*, 585-594.

Titov, N., Andrews, G., Schwencke, G., Drobny, J. and Einstein, D. (2008). Shyness 1: distance treatment of social phobia over the Internet. *Australian and New Zealand Journal of Psychiatry,42*, 585-594.

Titov, N., Andrews, G., Schwencke, G., Solley, K., Johnston, L. and Robinson, E. (2009). An RCT comparing effect of two types of support on severity of symptoms for people completing internet-based cognitive behaviour therapy for social phobia. *Australian and New Zealand Journal of Psychiatry, 43*, 920-926.

Van Straten, A., Cuijpers, P. and Smits, N. (2008). Effectiveness of a web-based self-help intervention for symptoms of depression, anxiety, and stress: Randomized controlled trial. *Journal of Medical Internet Research, 10*, e7.

Vernon, M. L. (2010). A review of computer-based alcohol problem services designed for the general public. *Journal of Substance Abuse, 38*, 203-211.

Vincelli, J., Choi, Y. H., Molinari, E., Wiederhold, B. K. and Riva, G. (2000). Experimental cognitive therapy for the treatment of panic disorder with agoraphobia: Definition of a clinical protocol. *Cyberpsychology and Behavior, 3*, 375-385.

Wade, S. L., Carey, J. and Wolfe, C. R. (2006). An online family intervention to reduce parental distress following pediatric brain injury. *Journal of Consulting and Clinical Psychology, 74*, 445-454.

Wagner, B., Knaevelsrud, C. and Maercker, A. (2006). Internet-based cognitive-behavioral therapy for complicated grief: A controlled randomized trial. *Death Studies, 30*, 429-453.

Wald, J. and Taylor, S. (2000). Efficacy of virtual reality exposure therapy to treat driving phobia: A case report. *Journal of Behaviour Therapy and Experimental Psychiatry, 31,* 249-257.

Waraich, P., Goldner, E. M., Somers, J. M. and Hsu, L. (2004). Prevalence and incidence studies of mood disorders: A systematic review of the literature. *Canadian Journal of Psychiatry, 49*, 124-138.

Warmerdam, L., Van Straten, A., Twisk, J., Riper, H. and Cuijpers, P. (2008). Internet-based treatment for adults with depressive symptoms: Randomized controlled trial. *Journal of Medical Internet Research, 10*, e44.

Whitfield, G. and Williams, C. (2004). If the evidence is so good –Why doesn't anyone use them? A national survey of the use of computerized cognitive behavior therapy. *Behavioural and Cognitive Psychotherapy, 32*, 57-65.

Wilson, G.T., Grilo, C. M. and Vitousek, K. M. (2007). Psychological treatment of eating disorders. *American Psychologist, 62*, 199-216.

Winzelberg, A. J., Eppstein, D., Eldredge, K. L., Wilfley, D. E., Dasmahapatra, R., Dev, P. and Taylor, C. B. (2000). Effectiveness of an Internet-based program for reducing risk factors for eating disorders. *Journal of Consulting and Clinical Psychology, 68,* 346-350.

Wittson, C. L. and Benschoter, R. (1972). Two-way television: Helping the Medical Center reach out. *American Journal of Psychiatry, 129*, 624-627.

Yates, F. (1996). *Restoring the Balance* (CD-ROM). Available from the Mental Health Foundation, 7th Floor, 83 Victoria Street, London SWIH 0HW.

Zabinski, M. F., Celio, A. A., Wilfley, D. E. and Taylor, C. B. (2003). Prevention of Eating Disorders and Obesity via the Internet. *Cognitive Behaviour Therapy, 32*, 137-150.

Zabinski, M. F., Wilfley, D. E., Calfas, K. J., Winzelberg, A. J. and Taylor, C. B. (2004). An interactive psycheducational intervention for women at risk of developing an eating disorder. *Journal of Consulting and Clinical Psychology, 72*, 914-919.

SECTION V. CONCLUSION

In: Handbook of Technology in Psychology … ISBN: 978-1-62100-004-4
Editor: Luciano L'Abate and David A. Kasier

Chapter 20

THE FUTURE OF TECHNOLOGY IN PSYCHOLOGY, PSYCHIATRY, AND NEUROLOGY: IMPLICATIONS FOR TRAINING AND TREATMENT

Savio W. H. Wong and Antoine Bechara
University of Southern California, Los Angeles, US

After years of research, we now know how our heart maintains blood circulation, how our lung exchanges gases, how our stomach and intestines digest food, how our liver breakdowns chemicals and how our kidney filters out the waste inside our body. However, we are still at the early stage of working out how our brain gives us the ability to think and governs our behaviors. Furthermore, relative to physical disorders like cardiovascular and infectious diseases, the causes of neuropsychiatry diseases are less understood. According to the World Health Organization report in 2004 (World Health Organization, 2008), neuropsychiatric conditions are indeed the most important causes of disability that account for around one third of Year Loss due to Disability (YLD). In particular, the unipolar depressive disorders are the leading causes of YLD among all the diseases. Therefore, there is a compelling need to advance our understanding about the brain functioning and the diseases associated with it.

Despite the complexity of the brain functioning, the rapid development of technology in the recent decades give us hope that, one day, technology may help us to untangle the mysteries behind the brain and the causes of various neuropsychiatry diseases. This handbook collected a series of articles which illustrated how the cutting-edge technology has been adopted in the research and treatment of neuropsychiatry diseases. The goal of this chapter is to summarize how technology can be integrated in the various aspects of Psychology, Psychiatry and Neurology and to predict how technology may change the training and treatment practice in these areas and advance our understanding of human behaviors in both healthy and diseased individuals. We divide our discussion into four sections. First, we discuss how technology drastically changes the research habitat and facilitates the investigation of the mechanisms of various mental processes and the possible causes of neuropsychiatry diseases. The second section discusses, based on the new research findings and the availability of the latest technology, how the assessment and treatment of neuropsychiatry diseases are affected. Thirdly, we discuss the impact of technology on the prevention and education of neuropsychiatry disease. At the end, we conclude the chapter with our predictions about the future direction in integrating technology with the diagnosis and treatment of neuropsychiatry diseases.

THE IMPLICATION OF TECHNOLOGY ON MENTAL HEALTH RESEARCH

The most obvious sign of the rapid technological advancement in our daily life is that, for every several months, we have faster, more powerful and less expensive personal computers and other technology gadgets (e.g., cellphone and tablets) available in the market. Nowadays, computer becomes an essential piece of equipment in every research setting, including the three areas, Psychology, Psychiatry and Neurology that are discussed broadly in this handbook. Researchers are now using computer to design experimental paradigms, implement the paradigms, record the responses from the subjects, analyze the collected data, summarize the findings with figures and video, present the results in various meetings and conferences, and write up the study into manuscript. Computers are indispensible in every stage of modern research. Although faster and more advanced computers may have less impact on the later stage of research (i.e., the reporting stage), they have been drastically changing how experimental research can be done. From the very first step of designing experimental paradigm, researchers can now rely on complex computation to identify the optimal experimental design. One of the typical examples of using computer to design experimental paradigms lies in the area of functional neuroimaging studies. Researchers can now estimate and compare the efficiencies of various experimental designs and identify the most efficient design before collecting any data (Miezin, Maccotta, Ollinger, Petersen, and Buckner, 2000; Smith, Jenkinson, Beckmann, Miller, and Woolrich, 2007). An optimal experimental design offers higher sensitivity to the experimental effects and better signal-to-noise ratio in the collected data (Donaldson and Buckner, 2000). Nevertheless, the process of optimizing experimental design involves considering a large number of factors (e.g., the number and order of conditions, the length of each trial and the temporal resolution and duration of the scanning) and the possible combinations across these factors are uncountable. Therefore, it requires sophisticated mathematical algorithms to model the interactions across various factors and to evaluate the efficiencies of different combinations. This process involves intense computations which is only possible with the assistance of modern technology. Optimizing the efficiency of experimental designs is particularly important in clinical studies. Researchers can identify the optimal experimental design in advance and minimize the number of subjects which are usually difficult to recruit and costly to administer in the clinical settings.

Computer not only allows researchers to automate a large number of the experimental processes, more importantly, it can also improve the accuracy and reliability of experimental research and minimize human errors. With the assistance of computer, stimuli presentation and data recording can be synchronized precisely which is crucial in studying time sensitive processes, such as the pattern of neural activity. In addition, many standardized neuropsychological tests and psychological experiments have been computerized so that experimenters can have more control over the timing and the type of responses obtained from the participants. Another advantage of running experiments with computer is that the parameters of the experiment can be modified more systematically and easily. The more powerful computer nowadays enable researchers to create more complex stimuli, like animation (see Chapter 5) and the virtual reality (VR) environment (see Chapter 15). Using computer, researchers can generate a unique perceptual experience to the subjects in a well-controlled experimental environment which may be either impossible or very costly to implement in real life settings. The expanded computational speed and storage capacity of computer also enable researchers to collect multiple measurements (e.g. fMRI, EEG, EMG, heart rate, blood pressure, skin conductance, respiration, body movement and neural activity) simultaneously (Dale and Halgren, 2001; Laufs, Daunizeau, Carmichael, and Kleinschmidt, 2008; Wong, Xue, and Bechara, 2011). The multimodals data collection approach opens up the opportunity to examine multiple factors at the same time which can generate a more comprehensive picture of human behaviors. Other than collecting data from multiple domains, faster computer and more advanced technology also enable researchers to have higher temporal and spatial resolution in the collected data. Comparing with the single cell recording studies in the early 19 centuries, researchers are now capable to record activity of multiple neurons with multielectrode recording (Berger, Song, Chan, and Marmarelis, 2010; Brown, Kass, and Mitra, 2004; Heuschkel, Fejtl, Raggenbass, Bertrand, and Renaud, 2002). The recent breakthrough in non-invasive neuroimaging techniques even enables researchers to monitor the whole brain activity of healthy individuals (Bandettini, Wong, Hinks, Tikofsky, and Hyde, 1992; Friston, 2009; Kwong et al., 1992; Logothetis, 2008; T. Ogawa et al., 1992).

Among various neuroimaging techniques, the functional magnetic resonance imaging (fMRI) has been the most popular and promising technique due to its superior spatial resolution and non-invasive nature. In an fMRI study, the blood-oxygen-level-dependent (BOLD) signals from hundred thousands of voxels which cover the whole brain are recorded almost simultaneously which enable researchers to estimate the neural activities across the brain in every 2-3s (Bandettini and Cox, 2000). Recently, the advance in fMRI data analysis algorithms and computational speed of computers further empower researchers to perform real-time fMRI studies that on-line brain activity can be extracted for neurofeedback training (DeCharms et al., 2005) and developing brain-computer interface (BCI) with multivariate analysis approaches (LaConte, 2010; Weiskopf et al., 2004). In addition to the multivariate analysis of fMRI data, multimodal approach has been adopted in the recent neuroimaging studies that psychophysiological measurements can be integrated with fMRI studies because of the availability of more powerful and advanced noise filtering techniques (Gray et al., 2009; Wong, et al., 2011). From the early years of James-Lange theory to the recent Somatic Marker Hypothesis proposed by Antonio Damasio (Bechara, Damasio, Tranel, and Damasio, 2005; Damasio, 1994), researchers are aware that there is a close relationship between the body and the mind. The advance in technology makes it feasible to record data simultaneously from multiple domains and enables researchers to examine the body-brain interaction more closely and to derive a more comprehensive model of human behaviors.

With the advanced analysis algorithm and more powerful computer that can handle much more data at the same time, researchers can, on the one hand, acquire a more complete picture about human behaviors and on the other hand investigate the subtle individual differences across participants. In addition to the traditional group level analysis, the continuous multimodal recording (Dale and Halgren, 2001; Wong, et al., 2011) and multivariate analysis (Haxby et al., 2001; Haynes et al., 2007; Kamitani and Tong, 2005) generate a rich set of data about a specific individual which can be used to create a personalized profile of individual behaviors. This is especially important in developing tailor-made individual treatment program which is discussed in more details in the next section.

The advance in technology does not limit to creating faster and more powerful computers, it also changes the way of how information is transferred. The increasing popularity and accessibility of the Internet and wireless communications across the globe bring a new era of experimental research. Experimenters can now collect data through the Internet and conduct experiments online in such a way that participants can provide their responses whenever and wherever they can (Lang, 2011). Ideas and data can be communicated between researchers throughout the world more efficiently and rapidly via Internet, email, instant messages and video-conferencing (see Chapter 18; (Akil, Martone, and Van Essen, 2011)). Using wireless devices, experimenters can monitor the behavior and physiological responses of the participants during their daily life activities (see Chapter 11). The latest technology enables researchers to minimize the size of experimental equipment without affecting its reliability and accuracy. For instance, the traditional bulky psychophysiology equipment can now be compacted into a wearable headband or wristband that can simultaneously monitor electroencephalography (EEG), electromyography (EMG), skin conductance, heart rate, blood pressure, body movement and posture and transmit the collected data in real time through cellphone applications to the workstations of the experimenters (Fletcher et al., 2010; Thatte et al., 2011). This is going to be a major milestone in the field of behavioral research as experiments are no longer restricted to the artificial laboratory settings which have been criticized heavily for their limited implication in the real life situation. Experimenters can now examine the dynamic behaviors in real life situations, such as the psychophysiological responses of normal and autistic children during social interactions (see Chapter 4 and 9). This is extremely important in clinical studies as data collected in real-life situation provide a more realistic understanding about the symptoms and development of neuropsychiatry diseases which are invaluable information in guiding treatment plan.

In addition to advancing the existing research techniques, the latest technology also provides a window for researchers to visualize the process that have not been seen beforehand. The discovery and development of the non-invasive functional magnetic resonance imaging (fMRI) technique in the early 1990s enables researchers to visualize the brain activity of healthy individuals without posing risk to the participants (Bandettini, et al., 1992; Kwong, et al., 1992; S. Ogawa et al., 1992). The emergent of fMRI technique did not only create a new research area in neuroscience (i.e., neuroimaging) but also lays the foundations for many new research areas such as cognitive and social neurosciences (Friston, 2009). In Chapter 13, Moser and Moran reviewed the event-related potential (ERP) technique which has lower spatial resolution but much better temporal resolution relative to fMRI. Recent studies have integrated ERP with fMRI which will provide exceptional spatial and temporal resolution (Dale and Halgren, 2001; Laufs, et al., 2008). In Chapter 14, Muller and colleagues discussed the diffusion tensor imaging (DTI) technique which has been used to visualize in vivo the nerve fibers in the human brain and to examine the anatomical connectivity across brain regions. Before the availability of the latest neuroimaging technique, these sorts of anatomical studies had only been done in animal or postmortem brains. Also, technology enables researchers to study the biochemical process in the nervous system at the molecular level. The action of various neurotransmitters, neuropeptides, hormones and drugs at the different part of the nervous system can be examined in vivo (Brabant, Cain, Jackson, and Kreitschmann-Andermahr, 2011; Dreosti and Lagnado, 2011). Nevertheless, many of these imaging techniques are still in their early stage of development and need to overcome a lot of technical challenges. However, with the exponential growth of technology advancement, these challenges will soon be solved and there will be newer and better imaging techniques with even higher temporal and spatial resolution. On the other hand, these new technologies will be able to accelerate the growth of research if they become more accessible, in terms of their availability and cost, to the researchers. We foresee that the imaging technology will play a leading role in directing the future of neuroscience research and be recognized as one of the cornerstones in the study of Psychology, Psychiatry and Neurology.

In sum, technology has drastically improved the way of conducting neuroscience and mental health research. The enhanced computational power enables researchers to collect more data from different sources with higher accuracy and better temporal and spatial resolution. The emergent of new techniques enable researchers to visualize the process that we have never seen before. The improved data analysis algorithm allows researchers to consider multiple factors simultaneously and to generate more comprehensive models of human behaviors. In the future, we expect that technology will enable researchers to examine individualized behaviors during real-life situation and to integrate contextual information in generating individual-specific behavior profiles. The new information generated from these cutting-edge of research will not only deepen our understanding about mental processes and behaviors, they will guide the development of diagnosis and treatment of neuropsychiatry diseases which is discussed in more details in the next section

Technology in Assessment and Treatment of Neuropsychiatry Diseases

Neuropsychiatry diseases are no exception than other diseases that the earlier the assessment is done, the better the prognosis. One of the key directions in mental health research is to identify the biomarkers of neuropsychiatry diseases (Singh and Rose, 2009). Relative to the self-report given by the patients, biomarkers provide more objective and precise accounts of the status and severity of neuropsychiatry diseases. New technologies with superior sensitivity in detecting the subtle changes of biomarkers may help clinicians to identify sub-threshold symptoms even before patients become aware of them. In this section, we discuss the impact of new technology on identifying various biomarkers and the new opportunities that biotechnology and information technology bring to the diagnosis and treatment of neuropsychiatry diseases.

The latest imaging technique does not only revolutionize the research in neuroscience as we have discussed in the last section, it is also a powerful tool for identifying abnormalities in the brain. Recent brain imaging research has revealed the link between neuropsychiatry diseases and the functional and/or structural abnormalities in the brain (Drevets et al., 1997; Giedd and Rapoport, 2010; Pezawas et al., 2005). The latest imaging technique hence becomes an important diagnostic tool for neuropsychiatry diseases. The non-invasive nature of some of these imaging techniques allows physicians to monitor the progress of the diseases regularly. Physicians can base on the functional and structural changes in the brain to evaluate the effectiveness of the treatment plan and to make adjustments to the treatment strategy whenever it is necessary.

In addition to serving as a diagnostic tool, the latest imaging technique has a direct role in the treatment of neuropsychiatry diseases. As mentioned before, neuroimaging techniques have been combined with the real-time analysis algorithm for neurofeedback training. Neurofeedback is a variant of biofeedback and demonstrated to be effective in alleviating symptoms of attention deficit hyperactivity disorder (ADHD) (Heinrich, Gevensleben, and Strehl, 2007). Real-time fMRI neurofeedback training has also been used in pain management (DeCharms, et al., 2005). The capability to decode neural activity in the brain in real-time also advance the development of brain-computer interface (BCI). With the latest BCI technology, patients can control prosthesis and computer through generating certain brain activity patterns which are decoded in real-time with sophisticated analysis algorithm (LaConte, 2010; Weiskopf, et al., 2004). This promising technique is going to help not only neuropsychiatry patients but also amputee and paralysis patients to regain movement and the ability to interact with others. Furthermore, researchers have demonstrated with fMRI that a selective group of vegetative patients are indeed conscious and can communicate with others through changing their brain activity patterns according to the instructions of the researchers (Owen and Coleman, 2008). This radical observation challenges the long-held belief that consciousness is compromised in vegetative patients and sheds light on the possibility that some of the vegetative patients may regain the ability to interact with the world through the assistance of new technology.

Recent research in genetics has identified the relationship between certain genotypes and the development of neuropsychiatry diseases (Homberg and Lesch, 2010; Pezawas, et al., 2005). These findings are in line with the results of earlier twin studies that most of the neuropsychiatry diseases have their genetic roots (Lohoff, 2010). Although genetic may only be one of the many factors contributing to the development of neuropsychiatry diseases, the ability to identify this vulnerability in individuals as early as they are in the womb of the mother would definitely improve the assessment and intervention of the diseases. Gene therapy has been proposed to alter the disease related genes and to prevent the progression of the diseases (Alexander et al., 2010). Nevertheless, research in gene therapy is still in its infancy stage. We foresee that it would be one of the most promising areas in the prevention and intervention of neuropsychiatry diseases in the near future.

Another major issue in the treatment of neuropsychiatry disease is the advancement of neuropharmacology. The latest pharmacological research reveals the biochemical mechanisms underlying human behaviors, including the relationship between neuropsychiatry diseases and the malfunction of neurotransmitters action (Snyder and Ferris, 2000). These research findings have been translated into the development of new drug which can alleviate the symptoms associated with neuropsychiatry diseases. Relative to the traditional behavioral therapy, medication is regarded as more cost efficient but with the drawback of potential side-effect from the drugs. Nevertheless, the newer generation of neuropsychiatry drugs may cause less side-effect (Leucht et al., 2009). The advancement in neuropharmacology treatment is again another proof of how the latest biomedical technology can revolutionize the treatment of neuropsychiatry diseases.

Neurosurgical procedures have been used as the last remedy in treating neuropsychiatry disease due to its irreversibility (Lipsman, Neimat, and Lozano, 2007; Mayberg et al., 2005). One of the future research directions is to develop non-invasive procedures that can yield the similar treatment efficacies as the invasive procedures. Transcranial magnetic stimulation (TMS) and transcranial direct current stimulation (tDCS) have been used to treat certain neuropsychiatry diseases and showed some initial success (Avery et al., 2006; George and Aston-Jones, 2010; O'Reardon et al., 2007). The major challenge for these transcranial techniques is that they have limited penetrating ability for targeting the more medial structures, such as the limbic system and the midbrain structures which are more closely linked with neuropsychiatry diseases than other neocortical structures. Future techniques which can non-invasively alter the malfunction neural network in the brain will provide an alternative to behavioral therapy and pharmacology treatment.

Similar to the treatment of other diseases, stem cell therapy is regarded as a promising area of treatment development of neurodegenerative diseases. One of the major challenges in treating neurodegenerative diseases and brain trauma is that most neurons within the nervous system are not regenerative. Once a neuron is gone, it is gone forever. Therefore, the regenerative capability of stem cell brings promising hope to mental health research that stem cell may be used to repair and replace the damaged neurons within the nervous system. In particular, there is hope that manipulation of endogenous neural stem cells and stem cell transplantation may be applied to treat Alzheimer's disease and aged-related cognitive decline (Lazarov, Mattson, Peterson, Pimplikar, and van Praag, 2010). Nevertheless, stem cell therapy is still in its laboratory stage and its development has already raised a lot of controversies. We expect that the advancement of stem cell therapy will encounter not only technical challenges but also ethical one.

Other than the advancement in biomedical technology, the rapid growth of information technology also plays a significant role in improving the assessment and treatment of neuropsychiatry diseases. The traditional face-to-face consultation with the therapists is no longer the only way for patients to receive diagnosis and advices. Patients can now communicate with the therapists via various channels, such as email, short message service (SMS), instant messages and video conferencing (see Chapter 18). In Chapter 20, Gallego and Emmelkamp have discussed extensively the pros and cons of integrating web-based communications in the treatment of neuropsychiatry diseases. In brief, information technology helps to reduce the logistic challenges (i.e., scheduling and travelling between home and the clinic) and the embarrassment that the patients may have in seeking help due to the stigmatization of neuropsychiatry diseases. Besides, patients and care-givers may share their experiences and exchange their concerns and worries with others who have gone through similar process through online social networks, forums, blogs and wikis. Information technology can therefore be used to promote more effective communications not only between patients and therapists but also between patients and patients that can foster mutual support and gain better understanding of the diseases. Also there are numerous computer-assisted therapy and self-practice packages available for patients. For instance, virtual reality (VR) has been used effectively to re-create the traumatic situation in a well-controlled environment for exposure treatment in post-traumatic stress disorder (PTSD) patients (see Chapter 15). Research showed that computer-assisted psychotherapy can deliver similar efficacies as face-to-face cognitive behavioral therapy (CBT) in the interventions for anxiety, mood, substance use and eating disorder (see Chapter 2, also see (Newman, Szkodny, Llera, and Przeworski, 2011 and Przeworski 2011)). In the long run, information technology will make the treatment of neuropsychiatry disease more accessible and affordable to more people.

Therapists and patients are expecting that new technologies can bring more precise diagnoses and more effective treatments for neuropsychiatry diseases. However, comorbidity is very common in neuropsychiatry patients that each patient may display a unique combination of symptoms. The heterogeneity of symptoms across patients poses significant challenges in making generalized diagnosis. For instance, there is a popular saying in the autism community that "if you have met one person with autism, you have met one person with autism" (see Chapter 9). Therefore, an important future direction in optimizing diagnoses is to obtain multiple measurements across different domains from the patients and create an individualized profile for each patient. Researchers have been collaborating with engineers on developing compact and wearable biosensors that can wirelessly monitor multiple biomarkers (e.g., EEG, autonomic nerve activity, blood chemistry) and record the behaviors of the patients over an extended period of time (Fletcher, et al., 2010; Thatte, et al., 2011). The data recorded from these wearable biosensors can be transferred with wireless devices (e.g., smartphone) and analyzed in real-time on the server with multivariate algorithms to generate a comprehensive picture of the characteristics and severity of the symptoms.

In summary, the advancement of biomedical technology and information technology revolutionize the diagnosis and treatment of neuropsychiatry diseases. New biomedical technologies bring more sensitive and objective diagnosis of neuropsychiatry diseases while information technology enables clinicians to provide more comprehensive and individualized assessments and interventions. Technology also helps to expand the accessibility of treatment and lower the cost of treatment which not only benefit the patients but also the society as a whole.

Technology in the Prevention and Education of Neuropsychiatry Diseases

As the causes of neuropsychiatry diseases were largely unknown in the past, it was impossible to discuss the prevention of the diseases. Some of the misunderstandings of neuropsychiatry diseases had mystified the diseases and led to prejudice and discrimination against the patients. The training for clinicians was restricted due to the insufficient information about the diseases. As discussed above, the rapid development of technology helped researchers to identify the biological root of neuropsychiatry diseases and to generate a more comprehensive and accurate picture about the causes, the symptoms and the prognosis of these diseases for the general public. In addition, the latest information technology provides new opportunities in integrating multimedia and virtual reality in computer-assisted and web-based trainings for the clinicians and therapists. These new training approaches can reduce the cost and training time and increase the flexibility and opportunities for the trainees to explore different treatment possibilities before practicing on patients (Williams, Aubin, Harkin, and Cottrell, 2001). For instance, virtual reality (VR) is adopted in the training for psychotherapists (Beutler and Harwood, 2004) in which trainees can practice extensively under various simulated situations. The utilization of animation and multimedia materials enriched the learning experience for the trainees and better prepared the trainees in making diagnosis and treating neuropsychiatry patients which are often characterized by the heterogeneity of their symptoms. On the other hand, information technology also improves the availability of the information about neuropsychiatry diseases to the general public. As a result, the awareness of the diseases is heightened in the community and persons who may suffer from neuropsychiatry diseases may be able to identify the symptoms and seek clinical diagnosis earlier. In the other words, technology helps to change the impression of the public about the diseases, to create a more accepting and understanding community, to provide more support to the patients and their family and to improve the prevention and intervention of neuropsychiatry diseases through more informative education to the general public.

Conclusion

It is an exciting era for researchers, therapists, and patients in experiencing the enormous benefits that technology brings to the field of mental health care. In the past decades, there is tremendous development in the biomedical and information technology that enables researchers to study the mystical organ in our body, the human brain, more thoroughly. This chapter summarizes how technology revolutionizes the research, diagnosis, treatment, prevention and education of neuropsychiatry diseases. To conclude this chapter, we make two predictions on the future of technology in Psychology, Psychiatry, and Neurology.

First, we expect that new technology will continue to assist researchers and clinicians to identify the biomarkers associated with various neuropsychiatry diseases. New biomarker-based diagnostic tools will provide more objective and accurate diagnosis even at the early stage of the development of the diseases which would improve the intervention and prognosis of the diseases significantly. Although new technologies tend to be relatively more expensive and less available when they first emerge, we learned from the history that technology will eventually lower the overall cost of mental health care and their availability will be increased over time.

Second, we predict that individualized diagnosis and treatment which are the key directions of mental health research will be benefited enormously from the advancement of technology. 24/7 continuous monitoring of the progression of neuropsychiatry disease will be made possible with more compact but also more powerful biosensors and data collection system. More sophisticated and automated analysis techniques will be used to analyze on-the-fly the multimodal measurements that are recorded from the patients. These individualized data which will be used to facilitate the diagnosis and treatment can be transmitted with portable wireless devices or the smartphone of the patients to the computer of the therapists. With the assistance of technology, clinicians and therapists will be able to develop individual-specific treatment plan and to make immediate response to the patients. These individualized diagnosis and treatment plan which account for the individual differences of the heterogeneity of neuropsychiatry diseases and provide a more comprehensive view of the disease will definitely improve the prognosis of the patients.

REFERENCES

Akil, H., Martone, M. E., and Van Essen, D. C. (2011). Challenges and Opportunities in Mining Neuroscience Data. *Science, 331*, 708-712.

Alexander, B., Warner-Schmidt, J., Eriksson, T., Tamminga, C., Arango-Llievano, M., Ghose, S., . . . Kaplitt, M. G. (2010). Reversal of Depressed Behaviors in Mice by p11 Gene Therapy in the Nucleus Accumbens. *Science Translational Medicine, 2*, 54ra76.

Avery, D. H., Holtzheimer, P. E., Fawaz, W., Russo, J., Neumaier, J., Dunner, D. L., . . . Roy-Byrne, P. (2006). A controlled study of repetitive transcranial magnetic stimulation in medication-resistant major depression. *Biological psychiatry, 59*, 187-194.

Bandettini, P. A., and Cox, R. W. (2000). Event-related fMRI contrast when using constant interstimulus interval: theory and experiment. *Magnetic resonance in medicine, 43*, 540-548.

Bandettini, P. A., Wong, E. C., Hinks, R. S., Tikofsky, R. S., and Hyde, J. S. (1992). Time course EPI of human brain function during task activation. *Magnetic Resonance in Medicine, 25*, 390-397.

Bechara, A., Damasio, H., Tranel, D., and Damasio, A. R. (2005). The Iowa Gambling Task and the somatic marker hypothesis: some questions and answers. *Trends in Cognitive Sciences, 9*, 159-162.

Berger, T. W., Song, D., Chan, R. H. M., and Marmarelis, V. Z. (2010). The Neurobiological Basis of Cognition: Identification by Multi-Input, Multioutput Nonlinear Dynamic Modeling: A method is proposed for measuring and modeling human long-term memory formation by mathematical analysis and computer simulation of nerve-cell dynamics. *Proceedings of the IEEE. Institute of Electrical and Electronics Engineers, 98*, 356-374.

Beutler, L. E., and Harwood, T. M. (2004). Virtual reality in psychotherapy training. *Journal of clinical psychology, 60*, 317-330.

Brabant, G., Cain, J., Jackson, A., and Kreitschmann-Andermahr, I. (2011). Visualizing hormone actions in the brain. *Trends in endocrinology and metabolism, 22*, 153-163.

Brown, E. N., Kass, R. E., and Mitra, P. P. (2004). Multiple neural spike train data analysis: state-of-the-art and future challenges. *Nature neuroscience, 7*, 456-461.

Dale, A. M., and Halgren, E. (2001). Spatiotemporal mapping of brain activity by integration of multiple imaging modalities. *Current opinion in neurobiology, 11*, 202-208.

Damasio, A. R. (1994). *Descartes' error : emotion, reason, and the human brain*. New York: G.P. Putnam.

Decharms, R. C., Maeda, F., Glover, G. H., Ludlow, D., Pauly, J. M., Soneji, D., . . . Mackey, S. C. (2005). Control over brain activation and pain learned by using real-time functional MRI. *Proceedings of the National Academy of Sciences of the United States of America, 102*, 18626-18631.

Donaldson, D. I., and Buckner, R. L. (2000). Effective Paradigm Design. In P. Matthews}, P. Jezzard and A. Evans (Eds.), *Functional Magnetic Resonance Imaging of the Brain: Methods for Neuroscience.*: Oxford: Oxford University Press.

Dreosti, E., and Lagnado, L. (2011). Optical reporters of synaptic activity in neural circuits. *Experimental physiology, 96*, 4-12.

Drevets, W. C., Price, J. L., Simpson, J. R., Todd, R. D., Reich, T., Vannier, M., and Raichle, M. E. (1997). Subgenual prefrontal cortex abnormalities in mood disorders. *Nature, 386*, 824-827.

Fletcher, R. R., Dobson, K., Goodwin, M. S., Eydgahi, H., Wilder-Smith, O., Fernholz, D., . . . Picard, R. W. (2010). iCalm: wearable sensor and network architecture for wirelessly communicating and logging autonomic activity. *IEEE transactions on information technology in biomedicine, 14*, 215-223.

Friston, K. J. (2009). Modalities, modes, and models in functional neuroimaging. *Science, 326*, 399-403.

George, M. S., and Aston-Jones, G. (2010). Noninvasive techniques for probing neurocircuitry and treating illness: vagus nerve stimulation (VNS), transcranial magnetic stimulation (TMS) and transcranial direct current stimulation (tDCS). *Neuropsychopharmacology, 35*, 301-316.

Giedd, J. N., and Rapoport, J. L. (2010). Structural MRI of Pediatric Brain Development: What Have We Learned and Where Are We Going? *Neuron, 67*, 728-734.

World Health Organization (2008). *The Global Burden of Disease: 2004 update*. Retrieved from http://www.who.int/healthinfo/global_burden_disease/2004_report_update/en/index.html.

Gray, M. A., Minati, L., Harrison, N. A., Gianaros, P. J., Napadow, V., and Critchley, H. D. (2009). Physiological recordings: basic concepts and implementation during functional magnetic resonance imaging. *NeuroImage, 47*, 1105-1115.

Haxby, J. V., Gobbini, M. I., Furey, M. L., Ishai, A., Schouten, J. L., and Pietrini, P. (2001). Distributed and overlapping representations of faces and objects in ventral temporal cortex. *Science, 293*, 2425-2430.

Haynes, J.-D., Sakai, K., Rees, G., Gilbert, S., Frith, C., and Passingham, R. E. (2007). Reading hidden intentions in the human brain. *Current Biology, 17*, 323-328.

Heinrich, H., Gevensleben, H., and Strehl, U. (2007). Annotation: neurofeedback - train your brain to train behaviour. *Journal of Child Psychology and Psychiatry, 48*, 3-16.

Heuschkel, M. O., Fejtl, M., Raggenbass, M., Bertrand, D., and Renaud, P. (2002). A three-dimensional multi-electrode array for multi-site stimulation and recording in acute brain slices. *Journal of Neuroscience Methods, 114*, 135-148.

Homberg, J. R., and Lesch, K.-P. (2010). Looking on the Bright Side of Serotonin Transporter Gene Variation. *Biological psychiatry, 69*, 513-519.

Kamitani, Y., and Tong, F. (2005). Decoding the visual and subjective contents of the human brain. *Nature Neuroscience, 8*, 679-685.

Kwong, K. K., Belliveau, J. W., Chesler, D. A., Goldberg, I. E., Weisskoff, R. M., Poncelet, B. P., . . . Turner, R. (1992). Dynamic magnetic resonance imaging of human brain activity during primary sensory stimulation. *Proceedings of the National Academy of Sciences of the United States of America, 89*, 5675-5679.

Laconte, S. M. (2010). Decoding fMRI brain states in real-time. *NeuroImage, 56*, 440-454.

Lang, T. (2011). Advancing Global Health Research Through Digital Technology and Sharing Data. *Science, 331*, 714-717.

Laufs, H., Daunizeau, J., Carmichael, D. W., and Kleinschmidt, A. (2008). Recent advances in recording electrophysiological data simultaneously with magnetic resonance imaging. *NeuroImage, 40*, 515-528.

Lazarov, O., Mattson, M. P., Peterson, D. A., Pimplikar, S. W., and Van Praag, H. (2010). When neurogenesis encounters aging and disease. *Trends in Neurosciences, 33*, 569-579.

Leucht, S., Corves, C., Arbter, D., Engel, R., Li, C., and Davis, J. (2009). Second-generation versus first-generation antipsychotic drugs for schizophrenia: a meta-analysis. *The Lancet, 373*, 31-41.

Lipsman, N., Neimat, J. S., and Lozano, A. M. (2007). Deep brain stimulation for treatment-refractory obsessive-compulsive disorder: the search for a valid target. *Neurosurgery, 61*, 1-11.

Logothetis, N. K. (2008). What we can do and what we cannot do with fMRI. *Nature, 453*, 869-878.

Lohoff, F. W. (2010). Overview of the genetics of major depressive disorder. *Current Psychiatry Reports, 12*, 539-546.

Mayberg, H. S., Lozano, A. M., Voon, V., Mcneely, H. E., Seminowicz, D., Hamani, C., . . . Kennedy, S. H. (2005). Deep brain stimulation for treatment-resistant depression. *Neuron, 45*, 651-660.

Miezin, F. M., Maccotta, L., Ollinger, J. M., Petersen, S. E., and Buckner, R. L. (2000). Characterizing the hemodynamic response: effects of presentation rate, sampling procedure, and the possibility of ordering brain activity based on relative timing. *NeuroImage, 11*, 735-759.

Newman, M. G., Szkodny, L. E., Llera, S. J., and Przeworski, A. (2011). A review of technology-assisted self-help and minimal contact therapies for anxiety and depression: Is human contact necessary for therapeutic efficacy? *Clinical Psychology Review, 31*, 89-103.

O'reardon, J. P., Solvason, H. B., Janicak, P. G., Sampson, S., Isenberg, K. E., Nahas, Z., ... Sackeim, H. A. (2007). Efficacy and safety of transcranial magnetic stimulation in the acute treatment of major depression: a multisite randomized controlled trial. *Biological Psychiatry, 62*, 1208-1216.

Ogawa, S., Tank, D. W., Menon, R., Ellermann, J. M., Kim, S. G., Merkle, H., and Ugurbil, K. (1992). Intrinsic signal changes accompanying sensory stimulation: functional brain mapping with magnetic resonance imaging. *Proceedings of the National Academy of Sciences of the United States of America, 89*, 5951-5955.

Ogawa, T., Spina, R. J., Martin, W. H., Kohrt, W. M., Schechtman, K. B., Holloszy, J. O., and Ehsani, A. A. (1992). Effects of aging, sex, and physical training on cardiovascular responses to exercise. *Circulation, 86*, 494-503.

Owen, A. M., and Coleman, M. R. (2008). Functional neuroimaging of the vegetative state. *Nature Review Neuroscience, 9*, 235-243.

Pezawas, L., Meyer-Lindenberg, A., Drabant, E. M., Verchinski, B. A., Munoz, K. E., Kolachana, B. S., . . . Weinberger, D. R. (2005). 5-HTTLPR polymorphism impacts human cingulate-amygdala interactions: a genetic susceptibility mechanism for depression. *Nature Neuroscience, 8*, 828-834.

Singh, I., and Rose, N. (2009). Biomarkers in psychiatry. *Nature, 460*, 202-207.

Smith, S., Jenkinson, M., Beckmann, C., Miller, K., and Woolrich, M. (2007). Meaningful design and contrast estimability in FMRI. *NeuroImage, 34*, 127-136.

Snyder, S. H., and Ferris, C. D. (2000). Novel neurotransmitters and their neuropsychiatric relevance. *American Journal of Psychiatry, 157*, 1738-1751.

Thatte, G., Li, M., Lee, S., Emken, A., Mitra, U., Spruijt-Metz, D., and Annavaram, M. (2011). KNOWME : An Energy-Efficient , Multimodal Body Area Network for Physical Activity Monitoring. *ACM Transactions on Embedded Computing Systems.*

Weiskopf, N., Mathiak, K., Bock, S. W., Scharnowski, F., Veit, R., Grodd, W., . . . Birbaumer, N. (2004). Principles of a brain-computer interface (BCI) based on real-time functional magnetic resonance imaging (fMRI). *IEEE transactions on bio-medical engineering, 51*, 966-970.

Williams, C., Aubin, S., Harkin, P., and Cottrell, D. (2001). A randomized, controlled, single-blind trial of teaching provided by a computer-based multimedia package versus lecture. *Medical Education, 35*, 847-854.

Wong, S. W. H., Xue, G., and Bechara, A. (2011). Integrating fMRI with Psychophysiological Measurements in the Study of Decision Making. *Journal of Neuroscience, Psychology, and Economics, 4*, 85-94.

INDEX

B

C

E

F

G

H

I

J

K

L

M

N

O

P

Q

R

S

T

U

V

W

X

Y